GEOGRAPHY

The World and Its People

VOLUME 1

☐ NATIONAL GEOGRAPHIC

SENIOR AUTHOR
Richard G. Boehm, Ph.D.

David G. Armstrong, Ph.D.

Francis P. Hunkins, Ph.D.

 Glencoe McGraw-Hill

New York, New York
Columbus, Ohio
Woodland Hills, California
Peoria, Illinois

ABOUT THE AUTHORS

NATIONAL GEOGRAPHIC

The National Geographic Society, founded in 1888 for the increase and diffusion of geographic knowledge, is the world's largest nonprofit scientific and educational organization. Since its earliest days, the Society has used sophisticated communication technologies, from color photography to holography, to convey geographic knowledge to a worldwide membership. The School Publishing Division supports the Society's mission by developing innovative educational programs—ranging from traditional print materials to multimedia programs including CD-ROMS, videos, and software.

SENIOR AUTHOR
Richard G. Boehm

Richard G. Boehm, Ph.D., was one of seven authors of *Geography for Life,* national standards in geography, prepared under Goals 2000: Educate America Act. He was also one of the authors of the *Guidelines for Geographic Education,* in which the five themes of geography were first articulated. In 1990 Dr. Boehm was designated "Distinguished Geography Educator" by the National Geographic Society. In 1991 he received the George J. Miller award from the National Council for Geographic Education (NCGE) for distinguished service to geographic education. He was President of the NCGE and has twice won the *Journal of Geography* award for best article. He has received the NCGE's "Distinguished Teaching Achievement" award and presently holds the Jesse H. Jones Distinguished Chair in Geographic Education at Southwest Texas State University in San Marcos, Texas.

David G. Armstrong

David G. Armstrong, Ph.D., is Dean of the School of Education at the University of North Carolina at Greensboro. A social studies education specialist with additional advanced training in geography, Dr. Armstrong was educated at Stanford University, University of Montana, and University of Washington. He taught at the secondary level in the state of Washington before beginning a career in higher education. Dr. Armstrong has written books for students at the secondary and university levels, as well as for teachers and university professors. He maintains an active interest in travel, teaching, and social studies education.

Francis P. Hunkins

Francis P. Hunkins, Ph.D., is Professor of Education at the University of Washington. He began his professional career as a teacher in Massachusetts. He received his masters degree in education from Boston University and his doctorate from Kent State University with a major in general curriculum and a minor in geography. Dr. Hunkins has written numerous books and articles dealing with general curriculum, social studies, and questioning and thinking for students and educators at elementary, middle school, high school, and university levels.

Contributing Writer

Bob Haddad, ethnomusicologist, developed the Music features. He also chose the music selections for Glencoe's *World Music: A Cultural Legacy* program and authored the accompanying teacher guide. Mr. Haddad is founder of Music of the World and president of Owl's Head Music.

Glencoe/McGraw-Hill

A Division of The McGraw-Hill Companies

Printed in the United States of America.
Send all inquiries to:
Glencoe/McGraw-Hill
8787 Orion Place
Columbus, Ohio 43240-4027

ISBN 0-07-824940-6 (Student Edition)   ISBN 0-07-824694-6 (Teacher Wraparound Edition)
5 6 7 8 9   027/043   05 04 03

CONSULTANTS

TEACHER REVIEWERS

Volume One

Contents

Volume Two
Contents

Features

Volume 1

Volume 2

Volume 1

Volume 2

Geography Skills

Volume 1

Volume 2

Technology Skills

Volume 1

Volume 2

Critical Thinking Skills

Volume 1

Volume 2

Study and Writing Skills

Volume 1

Volume 2

Making Connections

Teen from Senegal ▶

Features

Who, What, Where in the World?

Cultural Close-Up

GeoLAB Activity

Komodo dragon ▶

Maps

Maps

Unit 4   Europe

Unit 5   Russia

Volume 2

Unit 6   North Africa, Southwest Asia, and Central Asia

Maps

Unit 7 Africa South of the Sahara

Unit 8 Asia

Maps

Charts and Graphs

Charts and Graphs

Be an Active Reader!

How Should I Read My Textbook? Reading your social studies book is different than other reading you might do. Your textbook has a great amount of information in it. It is an example of nonfiction writing—it describes real-life events, people, ideas, and places.

Here are some reading strategies that will help you become an active textbook reader. Choose the strategies that work best for you. If you have trouble as you read your textbook, look back at these strategies for help.

✔ Before You Read

Set a Purpose
- Why are you reading the textbook?
- How might you be able to use what you learn in your own life?

Preview
- Read the chapter title to find out what the topic will be.
- Read the subtitles to see what you will learn about the topic.
- Skim the photos, charts, graphs, or maps.
- Look for vocabulary words that are boldfaced. How are they defined?

Draw From Your Own Background
- What do you already know about the topic?
- How is the new information different from what you already know?

If You Don't Know What A Word Means...
- think about the setting, or *context*, in which the word is used.
- check if prefixes such as *un, non,* or *pre* can help you break down the word.
- look up the word's definition in a dictionary or glossary.

✓ As You Read

Question
- What is the main idea?
- How well do the details support the main idea?
- How do the photos, charts, graphs, and maps support the main idea?

Connect
- Think about people, places, and events in your own life. Are there any similarities with those in your textbook?

Predict
- Predict events or outcomes by using clues and information that you already know.
- Change your predictions as you read and gather new information.

Visualize
- Use your imagination to picture the settings, actions, and people that are described.
- Create graphic organizers to help you see relationships found in the information.

Reading Do's

Do . . .
- ✔ establish a purpose for reading.
- ✔ think about how your own experiences relate to the topic.
- ✔ try different reading strategies.

Reading Don'ts

Don't . . .
- ⊘ ignore how the textbook is organized.
- ⊘ allow yourself to be easily distracted.
- ⊘ hurry to finish the material.

✓ After You Read

Summarize
- Describe the main idea and how the details support it.
- Use your own words to explain what you have read.

Assess
- What was the main idea?
- Did the text clearly support the main idea?
- Did you learn anything new from the material?
- Can you use this new information in other school subjects or at home?

REFERENCE ATLAS

NATIONAL
GEOGRAPHIC
SOCIETY

ATLAS KEY

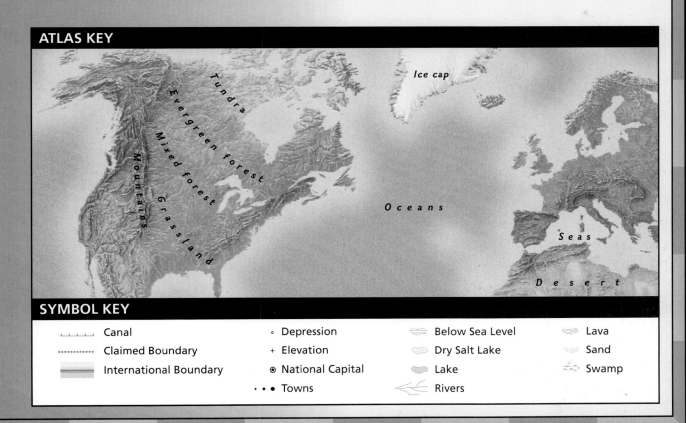

SYMBOL KEY

⊥⊥⊥⊥ Canal	○ Depression	⬱ Below Sea Level	▨ Lava
·········· Claimed Boundary	+ Elevation	⬱ Dry Salt Lake	▨ Sand
▨▨▨ International Boundary	⊛ National Capital	⬱ Lake	▨ Swamp
	·•· Towns	⟨ Rivers	

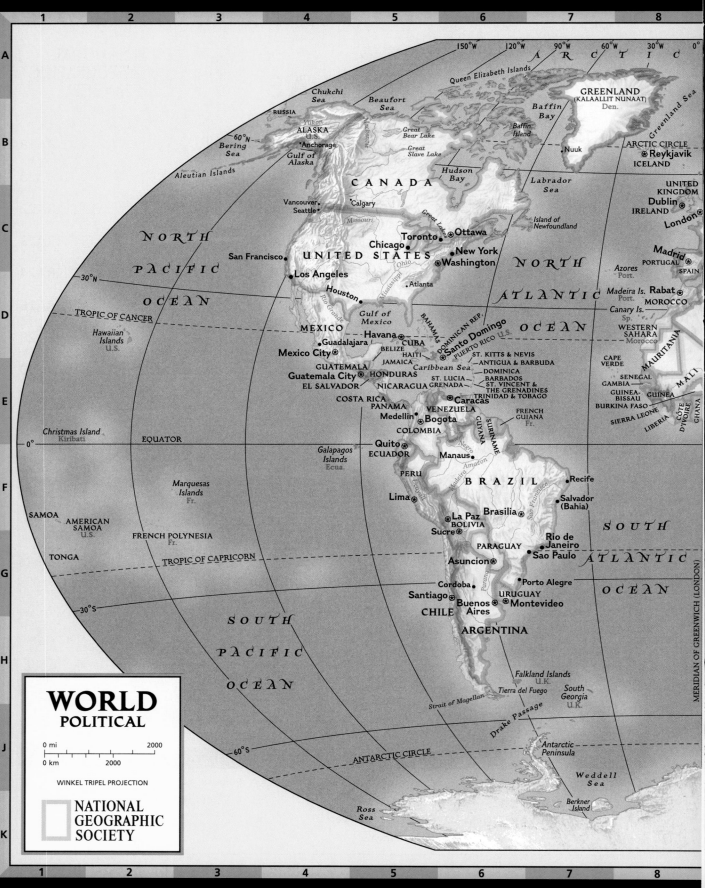

A R C T I C

150°W   120°W   90°W   60°W   30°W   0°

Queen Elizabeth Islands

GREENLAND
(KALAALLIT NUNAAT)
Den.

Chukchi
Sea
RUSSIA
ALASKA
U.S.
•Anchorage
Gulf of
Alaska

Beaufort
Sea

Baffin
Bay

Baffin
Island

Greenland Sea

Bering
Sea

60°N

Aleutian Islands

Great
Bear Lake

Great
Slave Lake

CANADA

Hudson
Bay

Nuuk

ARCTIC CIRCLE
⊛Reykjavik
ICELAND

Labrador
Sea

Island of
Newfoundland

UNITED
KINGDOM
Dublin⊛
IRELAND
London⊛

NORTH

PACIFIC

OCEAN

Vancouver•
Seattle•
San Francisco•

Calgary•

Missouri

UNITED STATES
Chicago•
Ohio

Toronto•
•Ottawa
New York•
⊛Washington

NORTH

ATLANTIC

Madrid⊛
PORTUGAL SPAIN
Azores
Port.

30°N

•Los Angeles
Houston•
Rio Grande

•Atlanta
Mississippi

OCEAN

Madeira Is. Rabat⊛
Port. MOROCCO
Canary Is.
Sp. ─Morocco

TROPIC OF CANCER

Hawaiian
Islands
U.S.

MEXICO
•Guadalajara
Mexico City⊛

•Havana
CUBA
BELIZE
HAITI
JAMAICA

Gulf of
Mexico

BAHAMAS

DOMINICAN REP.
Santo Domingo⊛
PUERTO RICO U.S.
ST. KITTS & NEVIS
—ANTIGUA & BARBUDA
DOMINICA
BARBADOS
ST. LUCIA
ST. VINCENT &
THE GRENADINES
TRINIDAD & TOBAGO

WESTERN
SAHARA
Morocco

CAPE
VERDE

MAURITANIA

SENEGAL
GAMBIA
GUINEA-
BISSAU GUINEA
BURKINA FASO
SIERRA LEONE
LIBERIA

MALI

CÔTE
D'IVOIRE
GHANA

GUATEMALA
Guatemala City⊛
EL SALVADOR

HONDURAS
NICARAGUA
COSTA RICA
PANAMA

Caribbean Sea

GRENADA•
⊛Caracas
VENEZUELA

FRENCH
GUIANA
Fr.

Medellin• ⊛Bogota
COLOMBIA

GUYANA
SURINAME

Christmas Island
Kiribati

0°

EQUATOR

Galapagos
Islands
Ecua.

Quito⊛
ECUADOR
PERU

Negro

Manaus•

Amazon

Madeira

BRAZIL

•Recife

Marquesas
Islands
Fr.

Lima⊛

Purus

São Francisco

•Salvador
(Bahia)

SAMOA
AMERICAN
SAMOA
U.S.

FRENCH POLYNESIA
Fr.

⊛La Paz
BOLIVIA
Sucre⊛

Brasilia•

Rio de
Janeiro•
•Sao Paulo

SOUTH

ATLANTIC

TONGA

TROPIC OF CAPRICORN

PARAGUAY
Asuncion⊛

Paraná

•Porto Alegre
URUGUAY
Montevideo⊛

OCEAN

30°S

SOUTH

Cordoba•
Santiago⊛
CHILE

Buenos
Aires⊛

ARGENTINA

PACIFIC

OCEAN

Falkland Islands
U.K.
Tierra del Fuego
Strait of Magellan

South
Georgia
U.K.

Drake Passage

MERIDIAN OF GREENWICH (LONDON)

60°S

ANTARCTIC CIRCLE

Antarctic
Peninsula

Ross
Sea

Weddell
Sea

Berkner
Island

WORLD
POLITICAL

0 mi   2000
0 km   2000

WINKEL TRIPEL PROJECTION

NATIONAL
GEOGRAPHIC
SOCIETY

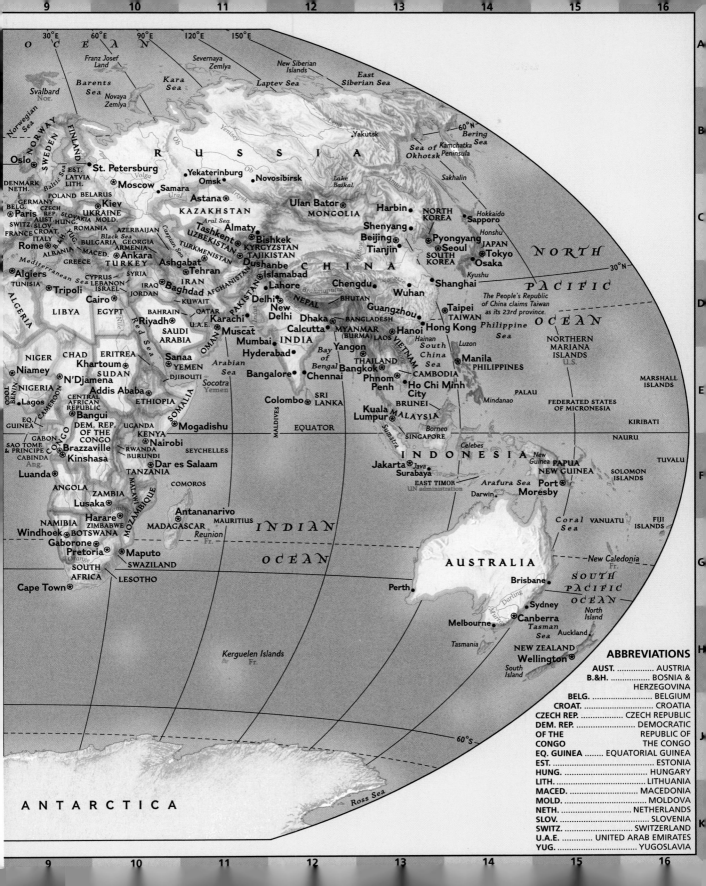

ABBREVIATIONS

AUST.	AUSTRIA
B.&H.	BOSNIA & HERZEGOVINA
BELG.	BELGIUM
CROAT.	CROATIA
CZECH REP.	CZECH REPUBLIC
DEM. REP. OF THE CONGO	DEMOCRATIC REPUBLIC OF THE CONGO
EQ. GUINEA	EQUATORIAL GUINEA
EST.	ESTONIA
HUNG.	HUNGARY
LITH.	LITHUANIA
MACED.	MACEDONIA
MOLD.	MOLDOVA
NETH.	NETHERLANDS
SLOV.	SLOVENIA
SWITZ.	SWITZERLAND
U.A.E.	UNITED ARAB EMIRATES
YUG.	YUGOSLAVIA

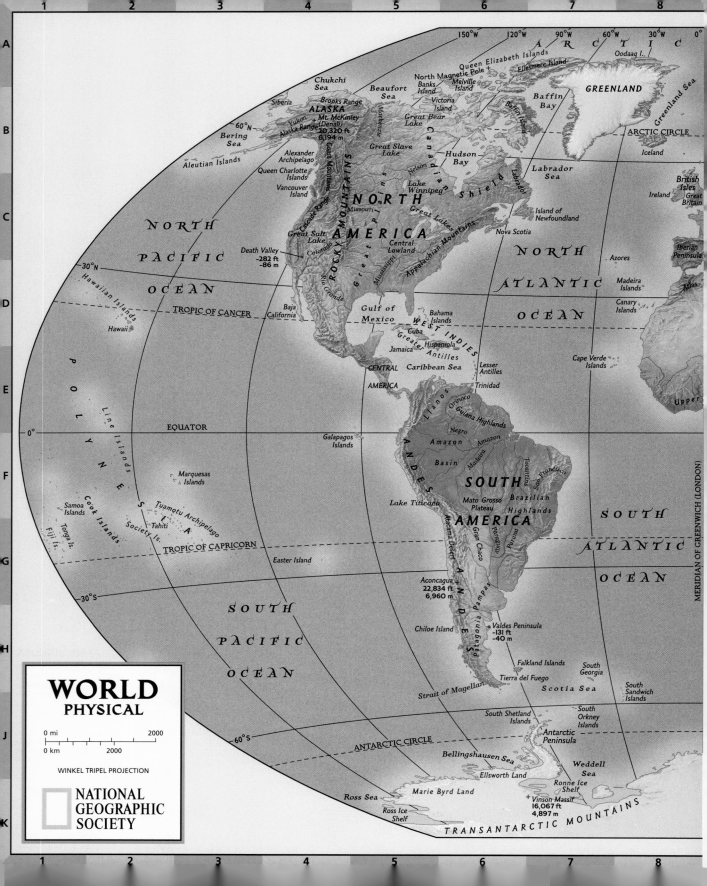

WORLD
PHYSICAL

0 mi   2000

0 km   2000

WINKEL TRIPEL PROJECTION

NATIONAL GEOGRAPHIC SOCIETY

150°W   120°W   90°W   60°W   30°W   0°

A R C T I C

Queen Elizabeth Islands

North Magnetic Pole +

Ellesmere Island

Oodaaq I.

GREENLAND

Chukchi Sea

Beaufort Sea

North Magnetic Pole +

Banks Island

Melville Island

Victoria Island

Baffin Island

Baffin Bay

ARCTIC CIRCLE

Greenland Sea

Siberia

Brooks Range

ALASKA

+ Mt. McKinley (Denali) 20,320 ft 6,194 m

Alaska Range

Yukon

60°N

Bering Sea

Aleutian Islands

Mackenzie

Great Bear Lake

Great Slave Lake

Canadian Shield

Hudson Bay

Labrador Sea

Iceland

Alexander Archipelago

Queen Charlotte Islands

Vancouver Island

Coast Mountains

Cascade Range

ROCKY MOUNTAINS

Great Plains

Lake Winnipeg

Nelson

Labrador

NORTH AMERICA

Missouri

Great Lakes

Island of Newfoundland

Nova Scotia

British Isles

Ireland

Great Britain

NORTH

Great Salt Lake

Colorado

Death Valley -282 ft -86 m

Central Lowland

Mississippi

Appalachian Mountains

NORTH

Azores

PACIFIC

Rio Grande

ATLANTIC

Madeira Islands

Iberian Peninsula

Atlas

30°N

Hawaiian Islands

OCEAN

TROPIC OF CANCER

Baja California

Gulf of Mexico

WEST INDIES

Bahama Islands

OCEAN

Canary Islands

Hawaii

Cuba

Greater Antilles

Jamaica

Hispaniola

Cape Verde Islands

CENTRAL AMERICA

Caribbean Sea

Lesser Antilles

Trinidad

Upper

Llanos

Orinoco

Guiana Highlands

EQUATOR

Galapagos Islands

ANDES

Negro

Amazon

Amazon

0°

P O L Y N E S I A

Line Islands

Marquesas Islands

Basin

Madeira

Tocantins

São Francisco

SOUTH

Mato Grosso Plateau

Brazilian Highlands

SOUTH

Samoa Islands

Cook Islands

Tahiti

Tuamotu Archipelago

Society Is.

Lake Titicaca

AMERICA

Paraguay

Paraná

ATLANTIC

Tonga Is.

Fiji Is.

TROPIC OF CAPRICORN

Atacama Desert

Gran Chaco

Easter Island

Aconcagua + 22,834 ft 6,960 m

Paraná

Pampas

OCEAN

30°S

SOUTH

ANDES

Chiloe Island

Valdes Peninsula -131 ft -40 m

PACIFIC

Patagonia

Strait of Magellan

Falkland Islands

South Georgia

OCEAN

Tierra del Fuego

Scotia Sea

South Sandwich Islands

South Shetland Islands

South Orkney Islands

Antarctic Peninsula

60°S

ANTARCTIC CIRCLE

Bellingshausen Sea

Weddell Sea

Ellsworth Land

Ronne Ice Shelf

Ross Sea

Marie Byrd Land

+ Vinson Massif 16,067 ft 4,897 m

Ross Ice Shelf

TRANSANTARCTIC MOUNTAINS

MERIDIAN OF GREENWICH (LONDON)

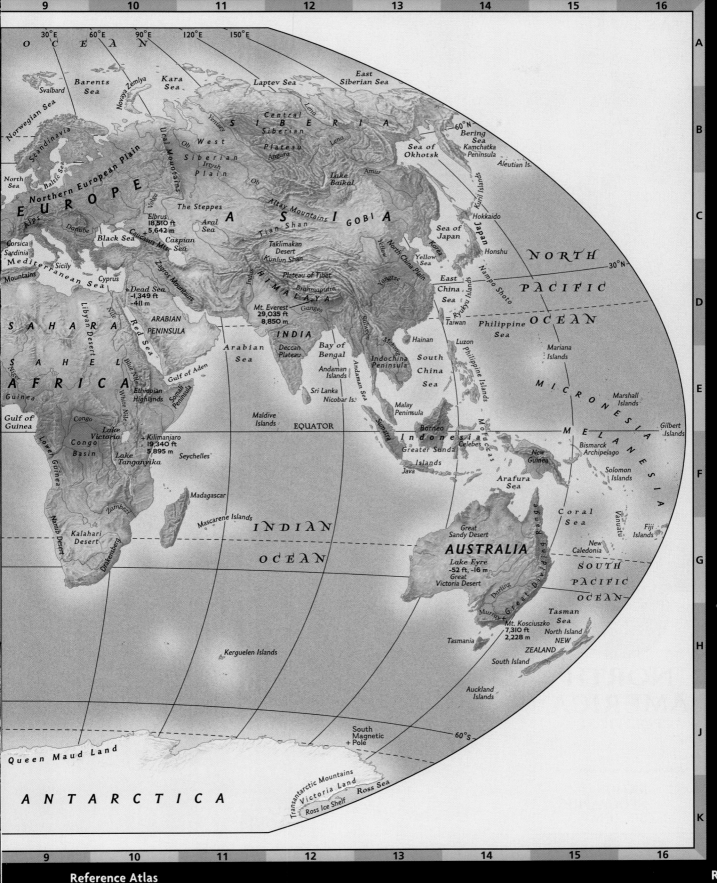

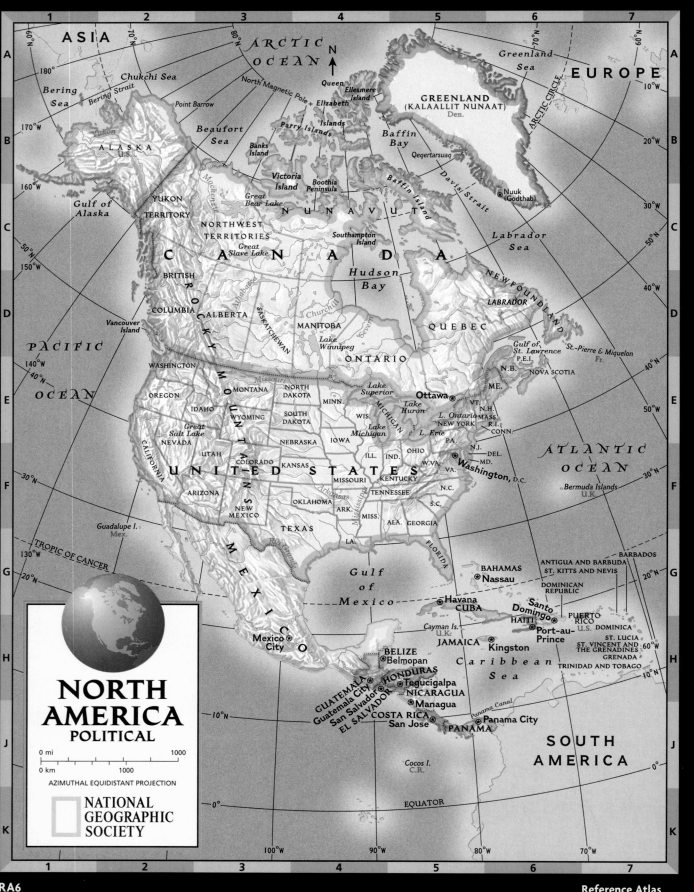

NORTH AMERICA
POLITICAL

0 mi   1000
0 km   1000

AZIMUTHAL EQUIDISTANT PROJECTION

NATIONAL GEOGRAPHIC SOCIETY

ASIA

EUROPE

ARCTIC OCEAN

N

Greenland Sea

ARCTIC CIRCLE

180°

170°W

160°W

150°W

140°W

130°W

TROPIC OF CANCER

10°W

20°W

30°W

40°W

50°N

60°N

Bering Sea

Chukchi Sea

Bering Strait

Point Barrow

North Magnetic Pole

Queen Elizabeth Islands

Ellesmere Island

GREENLAND (KALAALLIT NUNAAT) Den.

Beaufort Sea

Parry Islands

Baffin Bay

Banks Island

Qeqertarsuaq

Nuuk (Godthab)

ALASKA U.S.

Yukon

Victoria Island

Boothia Peninsula

Baffin Island

Davis Strait

YUKON TERRITORY

Great Bear Lake

NUNAVUT

Labrador Sea

Gulf of Alaska

NORTHWEST TERRITORIES

Mackenzie

Southampton Island

C A N A D A

BRITISH COLUMBIA

Great Slave Lake

Hudson Bay

NEWFOUNDLAND

PACIFIC OCEAN

Vancouver Island

ALBERTA

SASKATCHEWAN

Athabasca

Churchill

MANITOBA

Severn

LABRADOR

QUEBEC

Gulf of St. Lawrence

St.-Pierre & Miquelon Fr.

WASHINGTON

R O C K Y   M O U N T A I N S

Missouri

ONTARIO

Lake Winnipeg

P.E.I.

N.B.

NOVA SCOTIA

OREGON

MONTANA

NORTH DAKOTA

MINN.

Lake Superior

Lake Huron

Ottawa

ME.

N.H.

VT.

MASS.

IDAHO

WYOMING

SOUTH DAKOTA

WIS.

MICHIGAN

Lake Michigan

L. Ontario

NEW YORK

R.I.

CONN.

Great Salt Lake

NEBRASKA

IOWA

L. Erie

PA.

N.J.

ATLANTIC OCEAN

CALIFORNIA

NEVADA

UTAH

COLORADO

KANSAS

U N I T E D   S T A T E S

MISSOURI

ILL.

IND.

OHIO

W.VA.

VA.

DEL.

MD.

Washington, D.C.

ARIZONA

NEW MEXICO

OKLAHOMA

Arkansas

KENTUCKY

TENNESSEE

N.C.

S.C.

Bermuda Islands U.K.

TEXAS

ARK.

MISS.

Mississippi

ALA.

GEORGIA

LA.

FLORIDA

Guadalupe I. Mex.

M E X I C O

Gulf of Mexico

BAHAMAS

Nassau

ANTIGUA AND BARBUDA

ST. KITTS AND NEVIS

BARBADOS

Havana

CUBA

Cayman Is. U.K.

DOMINICAN REPUBLIC

Santo Domingo

PUERTO RICO U.S.

DOMINICA

Mexico City

JAMAICA

Kingston

HAITI

Port-au-Prince

ST. LUCIA

ST. VINCENT AND THE GRENADINES

GRENADA

BELIZE

Belmopan

Caribbean Sea

TRINIDAD AND TOBAGO

GUATEMALA

Guatemala City

HONDURAS

Tegucigalpa

NICARAGUA

Managua

Panama Canal

San Salvador

EL SALVADOR

COSTA RICA

San Jose

PANAMA

Panama City

SOUTH AMERICA

Cocos I. C.R.

EQUATOR

60°N

70°N

80°N

60°N

50°N

40°N

30°N

20°N

10°N

0°

100°W

90°W

80°W

70°W

60°W

30°N

20°N

10°N

0°

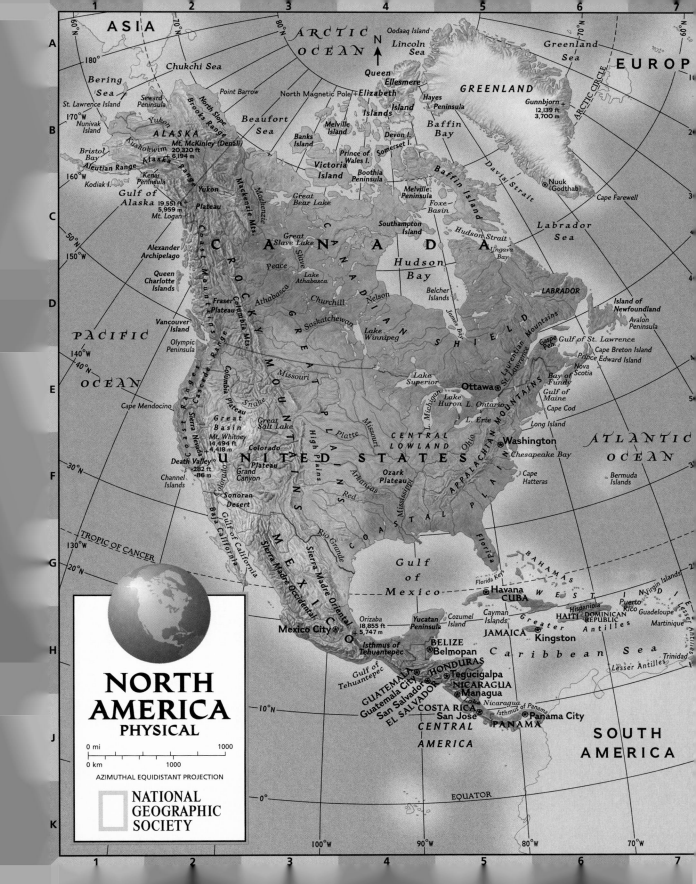

Grid coordinates (top and bottom): 1 2 3 4 5 6 7

Grid coordinates (left): A B C D E F G H J K

ASIA

ARCTIC OCEAN

N

Oodaaq Island

Lincoln Sea

Greenland Sea

EUROPE

Chukchi Sea

80°N

Queen Ellesmere

GREENLAND

Bering Sea

Point Barrow

North Magnetic Pole

Island

Hayes Peninsula

Gunnbjorn 12,139 ft 3,700 m

ARCTIC CIRCLE

70°N

Seward Peninsula

North Slope

Beaufort Sea

Queen Elizabeth Islands

Baffin Bay

St. Lawrence Island

170°W

ALASKA

Brooks Range

Banks Island

Melville Island

Devon I.

Nuuk (Godthab)

Nunivak Island

Yukon

Prince of Wales I.

Somerset I.

Davis Strait

Cape Farewell

160°W

Bristol Bay

Kuskokwim

Mt. McKinley (Denali) 20,320 ft 6,194 m

Alaska Range

Victoria Island

Boothia Peninsula

Melville Peninsula

Baffin Island

Labrador Sea

Aleutian Range

Kenai Peninsula

Kodiak I.

Gulf of Alaska

19,551 ft 5,959 m Mt. Logan

Yukon Plateau

Mackenzie Mts.

Great Bear Lake

Foxe Basin

Hudson Strait

Ungava Bay

50°N

150°W

Alexander Archipelago

C A N A D A

Great Slave Lake

Slave

Southampton Island

Hudson Bay

LABRADOR

Queen Charlotte Islands

Coast Mountains

Peace

Athabasca

Lake Athabasca

C A N A D I A N

Belcher Islands

Island of Newfoundland

140°W

PACIFIC

Vancouver Island

Olympic Peninsula

Fraser Plateau

Columbia Mts.

Saskatchewan

Churchill

Nelson

S H I E L D

Gaspe Pen.

Gulf of St. Lawrence

Cape Breton Island

Prince Edward Island

Avalon Peninsula

40°N

OCEAN

Cascade Range

Columbia Plateau

R O C K Y

G R E A T

Lake Winnipeg

James Bay

Laurentian Mountains

St. Lawrence

Nova Scotia

Cape Mendocino

Snake

Missouri

Lake Superior

Lake Michigan

Ottawa

Bay of Fundy

Gulf of Maine

Coast Ranges

Sierra Nevada

Great Basin

Great Salt Lake

M O U N T A I N S

P L A I N S

Lake Huron

L. Ontario

L. Erie

Cape Cod

Long Island

ATLANTIC

30°N

Mt. Whitney 14,494 ft 4,418 m

U N I T E D   S T A T E S

Missouri

Platte

C E N T R A L

Ohio

APPALACHIAN MOUNTAINS

OCEAN

Death Valley −282 ft −86 m

Colorado Plateau

High Plains

L O W L A N D

Washington

Chesapeake Bay

Channel Islands

Grand Canyon

Colorado

Arkansas

Ozark Plateau

Mississippi

Cape Hatteras

Bermuda Islands

Sonoran Desert

Red

COASTAL PLAIN

130°W

TROPIC OF CANCER

Baja California

Gulf of California

Rio Grande

Florida

BAHAMAS

20°N

M E X I C O

Sierra Madre Occidental

Sierra Madre Oriental

Gulf of Mexico

Florida Keys

Havana

CUBA

W E S T

N. Virgin Islands

Puerto Rico

Guadeloupe

Lesser Antilles

Mexico City

Orizaba 18,855 ft 5,747 m

Yucatan Peninsula

Cozumel Island

Cayman Islands

Greater Antilles

Hispaniola

HAITI

DOMINICAN REPUBLIC

Martinique

JAMAICA

Kingston

Isthmus of Tehuantepec

BELIZE

Belmopan

Caribbean Sea

Trinidad

Gulf of Tehuantepec

GUATEMALA

Guatemala City

HONDURAS

Tegucigalpa

NICARAGUA

Managua

Lesser Antilles

San Salvador

EL SALVADOR

COSTA RICA

San Jose

Lake Nicaragua

Isthmus of Panama

PANAMA

Panama City

SOUTH AMERICA

10°N

CENTRAL AMERICA

NORTH AMERICA PHYSICAL

0 mi   1000
0 km   1000

AZIMUTHAL EQUIDISTANT PROJECTION

NATIONAL GEOGRAPHIC SOCIETY

0°

EQUATOR

100°W   90°W   80°W   70°W

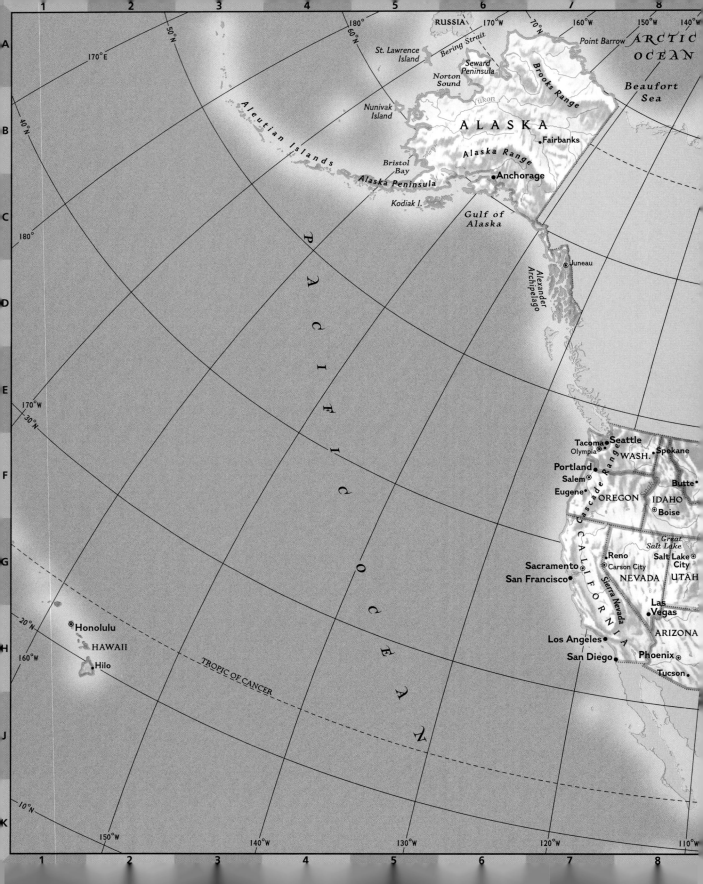

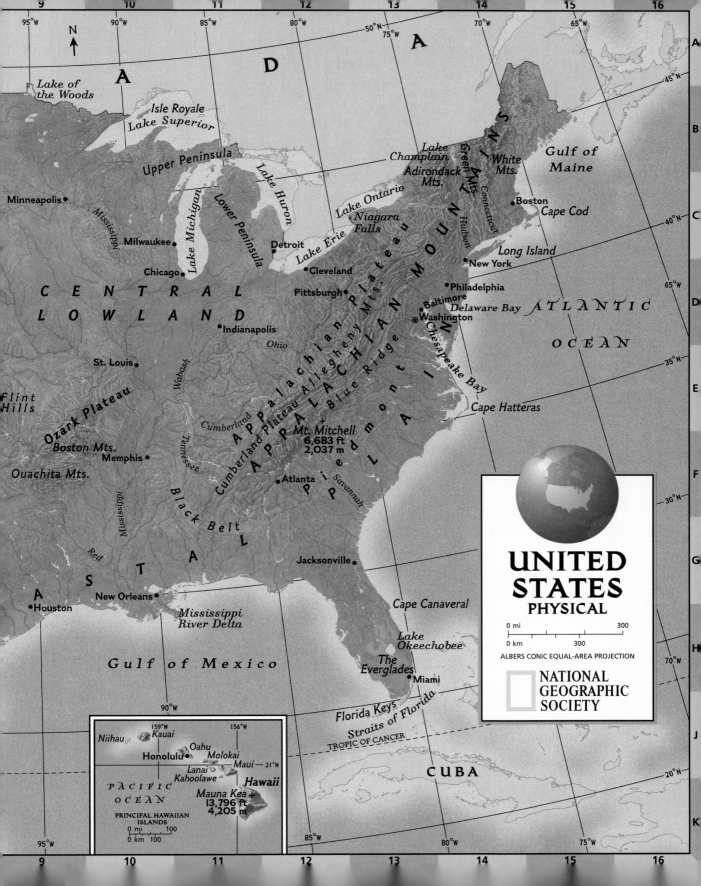

Lake of
the Woods

N

Isle Royale
Lake Superior

Upper Peninsula

A

D

A

50°N

Lake
Champlain

Green
Mts.

White
Mts.

Gulf of
Maine

Minneapolis

Mississippi

Milwaukee

Lake Michigan

Lower Peninsula

Lake Huron

Lake Ontario

Niagara
Falls

Adirondack
Mts.

Connecticut

Boston

Cape Cod

Chicago

Detroit

Lake Erie

Cleveland

Hudson

New York

Long Island

CENTRAL

Pittsburgh

Appalachian Plateau

Allegheny Mts.

Philadelphia

Baltimore

Delaware Bay

ATLANTIC

LOWLAND

Indianapolis

Ohio

Washington

Chesapeake Bay

OCEAN

St. Louis

Wabash

Cumberland Plateau

Blue Ridge

Piedmont

Cape Hatteras

Flint
Hills

Ozark Plateau

Cumberland

Tennessee

APPALACHIAN MOUNTAINS

Boston Mts.

Memphis

Mt. Mitchell
6,683 ft
2,037 m

Ouachita Mts.

Mississippi

Black Belt

Atlanta

Savannah

COASTAL

Red

Jacksonville

PLAIN

New Orleans

Houston

Mississippi
River Delta

Cape Canaveral

Lake
Okeechobee

Gulf of Mexico

The
Everglades

Miami

Florida Keys

Straits of Florida

TROPIC OF CANCER

CUBA

UNITED STATES PHYSICAL

0 mi 300
0 km 300

ALBERS CONIC EQUAL-AREA PROJECTION

NATIONAL GEOGRAPHIC SOCIETY

Niihau

Kauai

Oahu

Honolulu

Molokai

Lanai

Maui — 21°N

Kahoolawe

Hawaii

PACIFIC
OCEAN

Mauna Kea
13,796 ft
4,205 m

PRINCIPAL HAWAIIAN
ISLANDS

0 mi 100
0 km 100

CANADA
PHYSICAL/POLITICAL

0 mi — 400
0 km — 400

AZIMUTHAL EQUIDISTANT PROJECTION

NATIONAL
GEOGRAPHIC
SOCIETY

ICELAND

GREENLAND
(KALAALLIT NUNAAT)
Den.

Ellesmere
Island

Devon Island

Baffin
Bay

Davis Strait

Melville
Peninsula

Foxe
Basin

Baffin Island

NUNAVUT

Iqaluit

Southampton
Island

Hudson Strait

Ungava
Bay

Labrador
Sea

Hudson
Bay

Belcher
Islands

NEWFOUNDLAND

LABRADOR

Cartwright

Scheffervile

Happy Valley-
Goose Bay

Smallwood
Reservoir

"Churchill Falls

Island of
Newfoundland

QUEBEC

James Bay

SHIELD

Manicouagan
Reservoir
Sept-Iles

Anticosti I.

Gulf of
St. Lawrence

St. John's
Avalon
Peninsula

St.-Pierre & Miquelon
Fr.

ONTARIO

Lake
Nipigon

Thunder
Bay

Lake
Superior

Chicoutimi

Rouyn-Noranda

Quebec
City

St. Lawrence

Gaspe
Pen.

PRINCE
EDWARD
ISLAND

Charlottetown

NEW
BRUNSWICK

Fredericton

Saint John

Cape Breton I.

NOVA
SCOTIA

Halifax

ATLANTIC

Sudbury

Ottawa

Montreal

Bay of Fundy

OCEAN

Lake
Huron

Toronto

L. Ontario

Lake Michigan

Niagara Falls

London

L. Erie

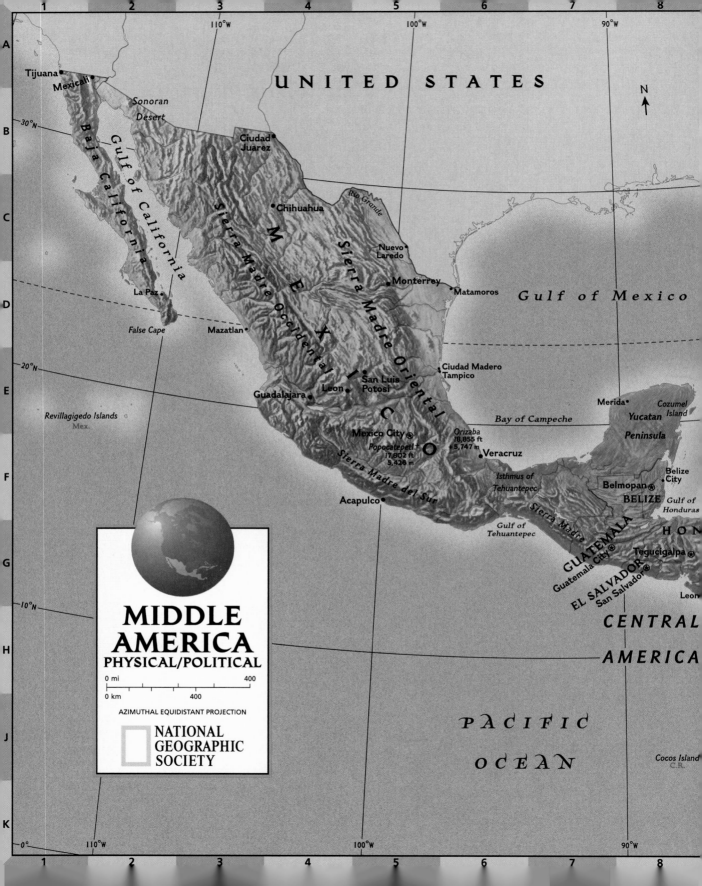

1   2   3   4   5   6   7   8

A

B

C

D

E

F

G

H

J

K

UNITED STATES

N

Tijuana
Mexicali

*Sonoran
Desert*

Ciudad
Juarez

30°N

Gulf of California

Baja California

Sierra Madre Occidental

Chihuahua

Rio Grande

Nuevo
Laredo

Monterrey

Matamoros

Gulf of Mexico

La Paz

M E X I C O

Sierra Madre Oriental

False Cape

Mazatlan

Ciudad Madero
Tampico

20°N

Guadalajara

Leon

San Luis
Potosi

Revillagigedo Islands
Mex.

Bay of Campeche

Merida

*Cozumel
Island*

*Yucatan
Peninsula*

Mexico City
Popocatepetl
17,802 ft
5,426 m

Orizaba
18,855 ft
5,747 m

Veracruz

Sierra Madre del Sur

Acapulco

*Isthmus of
Tehuantepec*

Belmopan

Belize
City

BELIZE

*Gulf of
Honduras*

Sierra Madre

GUATEMALA

Tegucigalpa

HON

*Gulf of
Tehuantepec*

Guatemala City

EL SALVADOR

San Salvador

Leon

10°N

MIDDLE AMERICA
PHYSICAL/POLITICAL

0 mi   400

0 km   400

AZIMUTHAL EQUIDISTANT PROJECTION

NATIONAL GEOGRAPHIC SOCIETY

CENTRAL

AMERICA

PACIFIC

OCEAN

*Cocos Island
C.R.*

0°

110°W

100°W

90°W

1   2   3   4   5   6   7   8

9   10   11   12   13   14   15   16

80°W   70°W   60°W   30°N

A

B

ATLANTIC

OCEAN

C

TROPIC OF CANCER   20°N

•Freeport

⊕ Nassau

B A H A M A S

W E S T   I N D I

Turks & Caicos Islands U.K.

D

Straits of Florida

Andros Island

ST. KITTS & NEVIS

⊕ Havana

CUBA

•Camaguey

•Holguin

•Santiago de Cuba

Santiago•

Hispaniola

Santo Domingo•

San Juan ⊕

ANTIGUA & BARBUDA

Virgin Islands U.S. U.K.

Isle of Youth

HAITI

DOMINICAN REPUBLIC

Puerto Rico U.S.

Guadeloupe Fr.

Lesser A

E

Cayman Islands U.K.

G r e a t e r

Port-au-⊕ Prince

DOMINICA

Bird I. Venez.

Martinique Fr.

Montego Bay•

⊗ Kingston

JAMAICA

A n t i l l e s

ST. LUCIA

BARBADOS

n t i l l e s

ST. VINCENT & THE GRENADINES

F

C a r i b b e a n   S e a

GRENADA

TRINIDAD & TOBAGO

Tobago

10°N

Port of Spain ⊕

Trinidad

60°W

Neth.

Curacao Bonaire

Aruba Neth.

Lesser Antilles

G

DURAS

Coco

Mosquito Coast

NICARAGUA

⊛ Managua

Lake Nicaragua

COSTA

Puerto Limon

Gulf of Mosquitos

Panama

•Panama

⊗ Panama City

H

San Jose ⊗

RICA

P A N A M A

David•

Isthmus of

Gulf of Panama

SOUTH

J

AMERICA

0°

80°W

K

70°W

EQUATOR

9   10   11   12   13   14   15   16

SOUTH AMERICA POLITICAL

Map labels (grid coordinates 1–7 across the top and bottom, A–K down the sides):

A
Caribbean Sea
Santa Marta
Barranquilla
Cartagena
Maracaibo
Barquisimeto
Caracas
Valencia
Lake Maracaibo
80°W
10°N
N

B
Bucaramanga
San Cristobal
Medellin
Bogota
Cali
VENEZUELA
Ciudad Guayana
GUYANA
Georgetown
SURINAME
Paramaribo
Cayenne
FRENCH GUIANA
Fr.
ATLANTIC OCEAN
Malpelo I. Col.
COLOMBIA
Orinoco
Boa Vista
Boundary claimed by Suriname
40°W
60°W
50°W

C
Esmeraldas
Quito
ECUADOR
Guayaquil
Iquitos
Manaus
Santarem
Marajo Island
Belem
EQUATOR
Fortaleza
Teresina
Natal
Negro
Amazon
Amazon (Solimões)
Madeira
Tapajos
Xingu
AMAZON BASIN
0°
Marañon

D
Rio Branco
Porto Velho
BRAZIL
Campina Grande
Recife
Salvador (Bahia)
Purus
Teles Pires
Tocantins
Sao Francisco
10°S

E
Callao
Lima
Ayacucho
Cuzco
Machu Picchu
Arequipa
Trinidad
BOLIVIA
La Paz
Oruro
Sucre
Santa Cruz
Goiania
Brasilia
Uberandia
Uberaba
Belo Horizonte
Campo Grande
Lake Titicaca
PERU

F
Arica
Iquique
Tarija
PARAGUAY
Asuncion
Campinas
Londrina
Sao Paulo
Santos
Rio de Janeiro
Nova Iguacu
Curitiba
Antofagasta
Salta
TROPIC OF CAPRICORN
San Felix I. San Ambrosio I. Chil.
Paraguay
20°S

G
La Serena
Coquimbo
Cordoba
Valparaiso
Santiago
Mendoza
Rosario
Buenos Aires
La Plata
URUGUAY
Montevideo
Uruguaiana
Santa Maria
Porto Alegre
Rio de la Plata
Juan Fernandez Is. Chil.
San Miguel de Tucuman
Parana
ANDES
ARGENTINA
30°S

H
Concepcion
Puerto Montt
Comodoro Rivadavia
Mar del Plata
Bahia Blanca
Negro
40°S

J
PACIFIC OCEAN
Falkland Islands (Islas Malvinas)
Stanley
Administered by United Kingdom
(Claimed by Arg.)

K
Rio Gallegos
Punta Arenas
Strait of Magellan
TIERRA DEL FUEGO
Ushuaia
Cape Horn
South Georgia I. U.K.
50°S

0 mi 800
0 km 800
AZIMUTHAL EQUIDISTANT PROJECTION

NATIONAL GEOGRAPHIC SOCIETY

100°W 90°W 80°W 70°W 60°W 50°W 40°W 30°W 20°W

Caribbean Sea

N

A T L A N T I C
O C E A N

80°W   70°W   60°W   50°W   40°W

10°N

Lake
Maracaibo
Caracas

V E N E Z U E L A
Orinoco

GUYANA
Georgetown
SURINAME
Paramaribo
Cayenne
FRENCH
GUIANA

Bogota
Angel Falls
Total drops
3,212 ft 979 m

COLOMBIA

GUYANA HIGHLANDS

Boundary claimed
by Suriname

Malpelo I.

EQUATOR

0°

Quito
ECUADOR

Negro

Amazon
Marajo
Island

A M A Z O N

Selvas

B A S I N

Amazon

Madeira

Tapajos

Xingu

Tocantins

São Francisco

Purus

Teles
Pires

B R A Z I L

10°S

Lima

Machu Picchu

Lake Titicaca

BOLIVIA

MATO GROSSO

PLATEAU

B R A Z I L I A N

Brasilia

La Paz

H I G H L A N D S

Altiplano

Sucre

20°S

Salar
de Uyuni

PARAGUAY

Iguazu
Falls

Asuncion

GRAN CHACO

TROPIC OF CAPRICORN

San Felix I.   San Ambrosio I.

Paraguay

Parana

Uruguay

30°S

Aconcagua 22,834 ft
6,960 m

Juan Fernandez Is.

Santiago

Buenos
Aires

URUGUAY
Montevideo

Rio de la Plata

PAMPAS

A N D E S

C O R D I L L E R A

A R G E N T I N A

Negro

40°S

Chiloe Island

-131 ft
-40 m

Valdes Peninsula

PATAGONIA

Gulf of
San Jorge

P A C I F I C

O C E A N

Taitao
Peninsula

Wellington I.

Falkland Islands
(Islas Malvinas)

Stanley

50°S

Strait of Magellan
Tierra del Fuego

Cape Horn

South Georgia I.

SOUTH
AMERICA
PHYSICAL

| 0 mi | | 800 |
| 0 km | | 800 |

AZIMUTHAL EQUIDISTANT PROJECTION

NATIONAL
GEOGRAPHIC
SOCIETY

100°W   90°W   80°W   70°W   60°W   50°W   40°W   30°W   20°W

EUROPE
POLITICAL

0 mi   400

0 km   400

AZIMUTHAL EQUIDISTANT PROJECTION

NATIONAL
GEOGRAPHIC
SOCIETY

Akureyri

Reykjavik

ICELAND

ARCTIC CIRCLE

Tromso

Norwegian Sea

Trondheim

Are

Alesund

Sundsvall

Faroe Islands
Den.

Torshavn

Bergen

Gulf of

Rockall
U.K.

Shetland
Islands

Lerwick

Orkney Islands

Oslo

Uppsala

Stavanger

Stockholm

Isle of Lewis

Skagerrak

Goteborg

Inverness

Gotland

UNITED

Aberdeen

Baltic

SCOTLAND

Glasgow

Edinburgh

*North
Sea*

DENMARK

Arhus

Kiel

Malmo

Gdansk

NORTHERN
IRELAND

Belfast

IRELAND

*Irish
Sea*

Liverpool

Copenhagen

Dublin

Manchester

NETH.

Hamburg

Bydgoszcz

Cork

KINGDOM

WALES

Birmingham

Amsterdam

Berlin

POLAND

Cardiff

ENGLAND

The
Hague

GERMANY

*Celtic
Sea*

London

Brussels

Lodz

Land's End

Southampton

BELGIUM

Bonn

Frankfurt

Wroclaw

English Channel

Le Havre

LUX.

Prague

Brest

Paris

CZECH REP.

Rennes

Bratislava

SLOVAKIA

*ATLANTIC
OCEAN*

Nantes

Strasbourg

Munich

Vienna

La Rochelle

*Bay of
Biscay*

Limoges

FRANCE

Zurich

LIECH.

AUSTRIA

Budapest

Bordeaux

Bern

A L P S

SWITZERLAND

SLOVENIA

HUNGARY

A Coruna

Geneva

Lyon

Milan

Ljubljana

Zagreb

Vigo

Donostia-
San Sebastian

Turin

Venice

CROATIA

Porto

Bilbao

Toulouse

MONACO

Genoa

SAN.
MARINO

**BOSNIA &
HERZEGOVINA**

Coimbra

Pyrenees

Nice

Sarajevo

ANDORRA

Marseille

MONTENEGRO

Lisbon

Madrid

Zaragoza

*Corsica
Fr.*

ITALY

SPAIN

Barcelona

VATICAN
CITY

Rome

PORTUGAL

Valencia

Palma

*Sardinia
It.*

Tirana

ALBANIA

Cape
St. Vincent

Cordoba

Murcia

*Balearic
Islands
Sp.*

Naples

Seville

Cartagena

Cadiz

Cagliari

*Tyrrhenian
Sea*

*Ionian
Sea*

GIBRALTAR
U.K.

Malaga

Strait of Gibraltar

Mediterranean

Palermo

Messina

Sicily

Catania

Valletta

AFRICA

MALTA

60°N

40°W

30°W

20°W

10°W

70°N

0°

10°E

30°W

50°N

20°W

40°N

30°N

10°W

0°

10°E

MERIDIAN OF GREENWICH (LONDON)

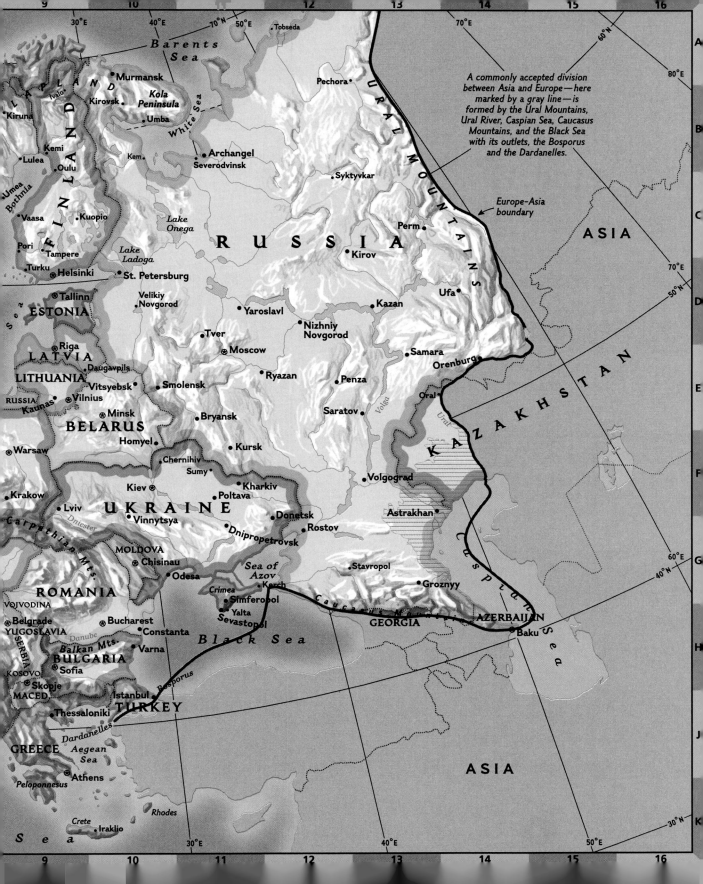

EUROPE
PHYSICAL

0 mi 400
0 km 400

AZIMUTHAL EQUIDISTANT PROJECTION

NATIONAL
GEOGRAPHIC
SOCIETY

ICELAND
Reykjavik

ARCTIC CIRCLE

MERIDIAN OF GREENWICH (LONDON)

Norwegian Sea

N

S C A N D I N A V I A

Gulf of Bothnia

SWEDEN

Oslo

Stockholm

Baltic Sea

Faroe Islands

Shetland Islands

Orkney Islands

Outer Hebrides

Highlands

British Isles

Edinburgh

Belfast

UNITED

IRELAND
Dublin

Irish Sea

Great Britain

KINGDOM

Cardiff

London

North Sea

Jutland

DENMARK
Copenhagen

Zealand

ATLANTIC OCEAN

English Channel

Brittany

Seine
Paris

Loire

FRANCE

Amsterdam
NETH.

BELGIUM
Brussels

LUX.

Rhine

N O R T H

GERMANY

Berlin

Elbe

Prague
CZECH REP.

POLAND

Oder

Danube

Bratislava
Vienna
SLOVAKIA

LIECH.

AUSTRIA
Budapest

HUNGARY

Bay of Biscay

Mont Blanc
15,771 ft
4,807 m

Bern
SWITZ.

A L P S

Massif Central

Rhône

MONACO

Riviera

Po

SLOVENIA
Ljubljana
Zagreb

Drava

CROATIA

Sava

Danube

Cantabrian Mountains

Pyrenees

ANDORRA

A p e n n i n e s

SAN MARINO

BOSNIA &
Sarajevo
HERZEGOVINA

Adriatic Sea

Douro

IBERIAN

Ebro

Corsica

VATICAN
CITY

ITALY

Rome

PORTUGAL

Madrid

Tagus

SPAIN

Lisbon

PENINSULA

Baetic Mountains

GIBRALTAR
Strait of Gibraltar

Balearic Islands

Sardinia

Tyrrhenian Sea

Tirana
ALBANIA

Ionian Sea

M e d i t e r r a n e a n

Sicily
Etna
10,902 ft
3,323 m

Valletta
MALTA

AFRICA

60°N

40°W

30°W

20°W

10°W
70°N

0°

10°E

30°N
50°N

20°W
40°N

10°W

0°

10°E

30°N

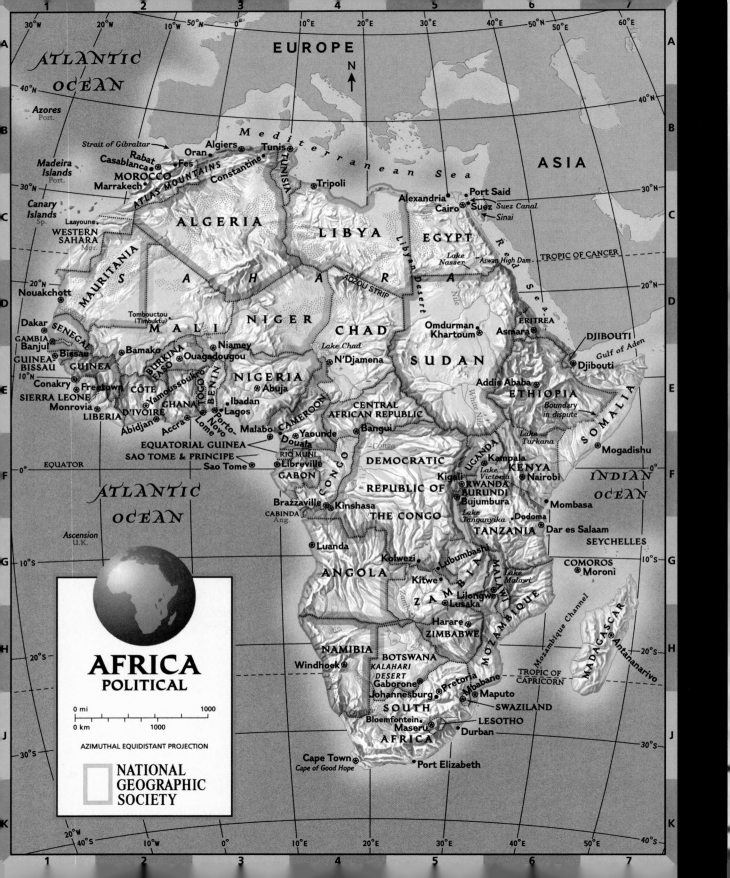

AFRICA
POLITICAL

0 mi   1000
0 km   1000

AZIMUTHAL EQUIDISTANT PROJECTION

NATIONAL GEOGRAPHIC SOCIETY

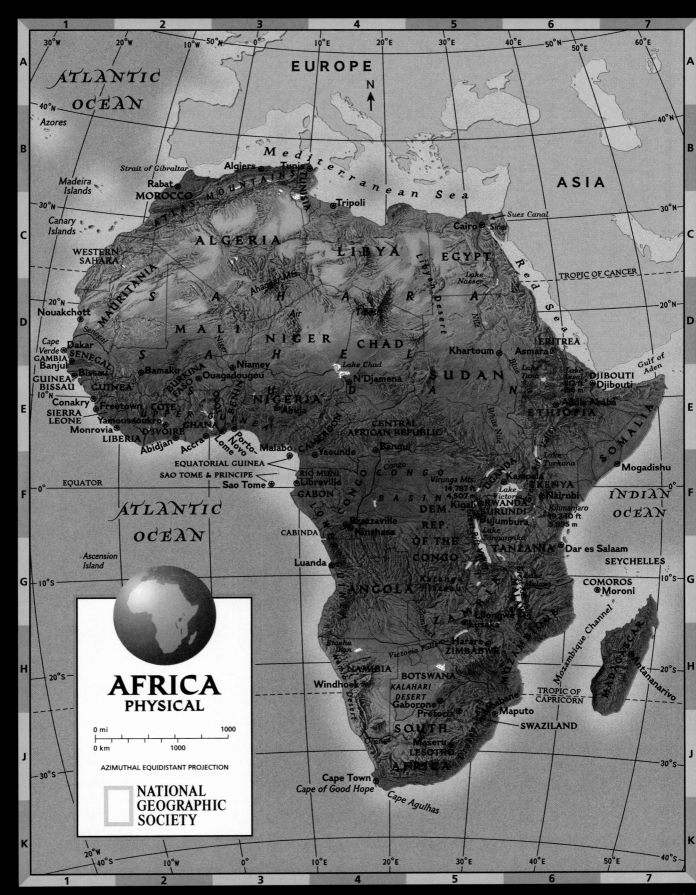

AFRICA PHYSICAL

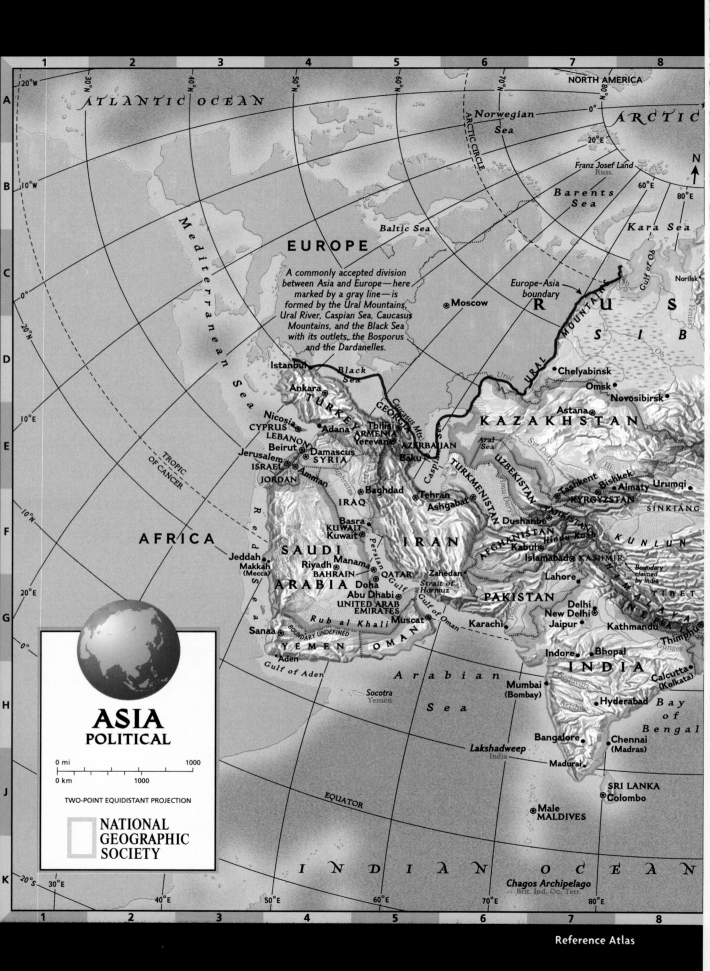

9 **10** **11** **12** **13** **14** **15** **16**

North Pole

NORTH AMERICA

Bering Strait

OCEAN

Chukchi Sea

Wrangel I.

East Siberian Sea

Gulf of Anadyr

Anadyr

Bering Sea

North Land

New Siberian Islands

120°E

100°E

Laptev Sea

Cherskiy Range

Koryma Range

Commander Is.

Kamchatka Peninsula

Magadan

Verkhoyansk Range

S I A

E R I A

Yakutsk

Sea of Okhotsk

Sakhalin

Kuril Islands

Lake Baikal

Irkutsk

Hokkaido

Sapporo

Vladivostok

Sea of Japan

JAPAN

Tokyo

Honshu

Kyoto

Osaka

Hiroshima

Kyushu

Marcus I.
Jap.

TROPIC OF CANCER

MONGOLIA

Ulan Bator

ALTAY MTS.

GOBI

Changchun

Shenyang

NORTH KOREA

Pyongyang

Seoul

SOUTH KOREA

Beijing

Qingdao

Shijiazhuang

Yellow Sea

Xuzhou

*Bonin Is.
Jap.*

*Volcano Is.
Jap.*

SHAN

Lanzhou

Xian

C H I N A

Nanjing

Shanghai

East China Sea

Ryukyu Islands

Okinawa

Parece Vela
Jap.

P A C I F I C   O C E A N

Nanchang

Chengdu

Fuzhou

Changsha

Taipei

TAIWAN

Guiyang

Kunming

Guangzhou

Hong Kong

Macau

The People's Republic of China claims Taiwan as its 23rd province.

Philippine

BHUTAN

BANGLADESH

Dhaka

Hanoi

Haiphong

South China Sea

Hainan

Luzon

Quezon City

Manila

Samar

Sea

MYANMAR (BURMA)

Yangon (Rangoon)

Vientiane

LAOS

THAILAND

Da Nang

Mindoro

PHILIPPINES

Leyte

Bangkok

CAMBODIA

VIETNAM

Phnom Penh

Ho Chi Minh City

Palawan

Panay

Negros

Mindanao

Andaman Islands
India

Andaman Sea

Gulf of Thailand

Bandar Seri Begawan

SABAH

BRUNEI

Biak

Jayapura

New Guinea

Nicobar Islands
India

Kuala Lumpur

Medan

MALAYSIA

SARAWAK

Halmahera

Morotai

Kepi

Merauke

Dolak

Aru
Is.

Mentawai Islands

SINGAPORE

I N D O N E S I A

Borneo

Celebes

Moluccas

Buru

Ceram

Following East Timor's vote for independence from Indonesia, the United Nations, in October 1999, established a UN transitional administration to help East Timor's 870,000 people toward independence.

Sumatra

Jambi

GREATER SUNDA ISLANDS

Tanimbar Is.

Dili

EAST TIMOR

Timor

UN admin.

Timor Sea

Jakarta

Java Sea

Java

Kupang

AUSTRALIA

100°E

110°E

120°E

130°E

140°E

150°E

9 **10** **11** **12** **13** **14** **15** **16**

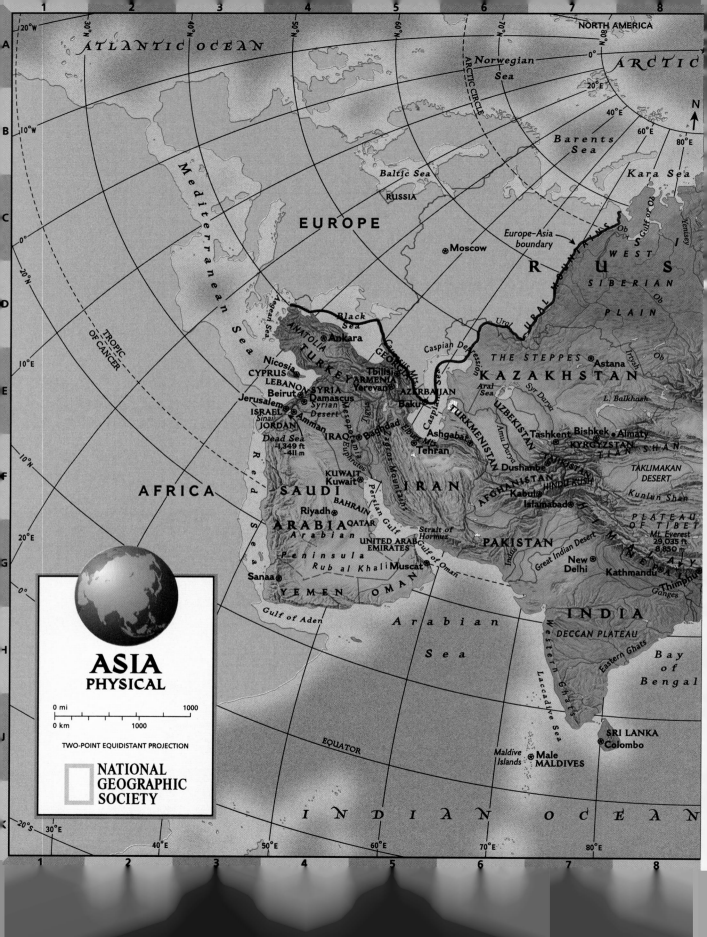

ASIA
PHYSICAL

0 mi   1000
0 km   1000

TWO-POINT EQUIDISTANT PROJECTION

NATIONAL GEOGRAPHIC SOCIETY

ATLANTIC OCEAN

Mediterranean Sea

EUROPE

RUSSIA

Baltic Sea

Norwegian Sea

ARCTIC CIRCLE

NORTH AMERICA

ARCTIC

Barents Sea

Kara Sea

⊛ Moscow

Europe-Asia boundary

Gulf of Ob

Ob

Yenisey

WEST SIBERIA N

SIBERIAN PLAIN

Ob

Irtysh

Ob

TROPIC OF CANCER

AFRICA

Black Sea

ANATOLIA

TURKEY

⊛ Ankara

Aegean Sea

Nicosia
CYPRUS ⊛
LEBANON
Beirut ⊛
Jerusalem ⊛
ISRAEL
Sinai
JORDAN
Amman ⊛
Damascus ⊛
SYRIA
Syrian Desert

Caucasus Mts.

GEORGIA
Tbilisi ⊛
ARMENIA
Yerevan ⊛
AZERBAIJAN
Baku ⊛

Caspian Depression

Urai

URAL MOUNTAINS

Caspian Sea

THE STEPPES

KAZAKHSTAN

Astana ⊛

Aral Sea

Syr Darya

L. Balkhash

UZBEKISTAN

Amu Darya

Tashkent ⊛
Bishkek ⊛ Almaty ⊛
KYRGYZSTAN
TIAN SHAN

Dead Sea
-1,349 ft
-411 m

Mesopotamia

IRAQ ⊛

Baghdad ⊛

Euphrates

Zagros Mountains

Elburz Mts.

Tehran ⊛

Ashgabat ⊛
TURKMENISTAN

Dushanbe ⊛
TAJIKISTAN
AFGHANISTAN
HINDU KUSH
Kabul ⊛
Islamabad ⊛

TAKLIMAKAN DESERT

Kunlun Shan

KUWAIT
Kuwait ⊛

SAUDI

Riyadh ⊛
ARABIA

Arabian

Peninsula

BAHRAIN
QATAR

Persian Gulf

IRAN

UNITED ARAB EMIRATES

Strait of Hormuz

Gulf of Oman

PAKISTAN

Indus

Great Indian Desert

New ⊛ Delhi

PLATEAU OF TIBET

Mt. Everest
29,035 ft
8,850 m

HIMALAYA

NEPAL

Kathmandu ⊛

BHUTAN

Thimphu ⊛

Ganges

Sanaa ⊛

Rub al Khali

YEMEN

OMAN

Muscat ⊛

Red Sea

Gulf of Aden

Arabian

Sea

INDIA

DECCAN PLATEAU

Western Ghats

Eastern Ghats

Bay of Bengal

EQUATOR

Laccadive Sea

SRI LANKA
Colombo ⊛

Maldive Islands

Male ⊛
MALDIVES

INDIAN OCEAN

N

20°W
10°W
0°
20°N
10°E
0°
10°N
20°E
0°
20°S
30°E
40°E
50°E
60°E
70°E
80°E

30°N
40°N
50°N
60°N
70°N
80°N

0°
20°E
40°E
60°E
80°E

A B C D E F G H J K

1 2 3 4 5 6 7 8

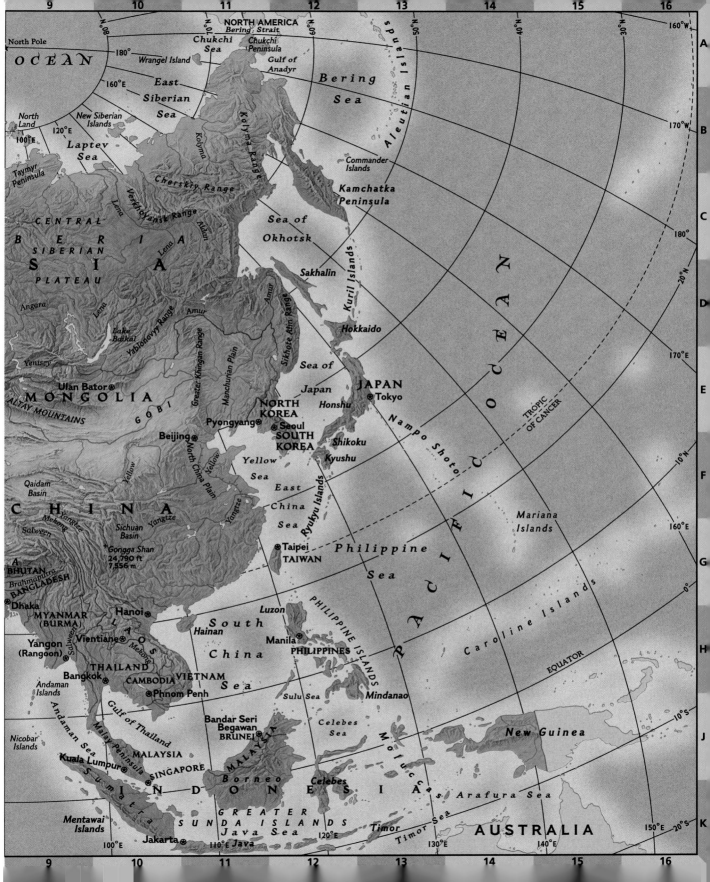

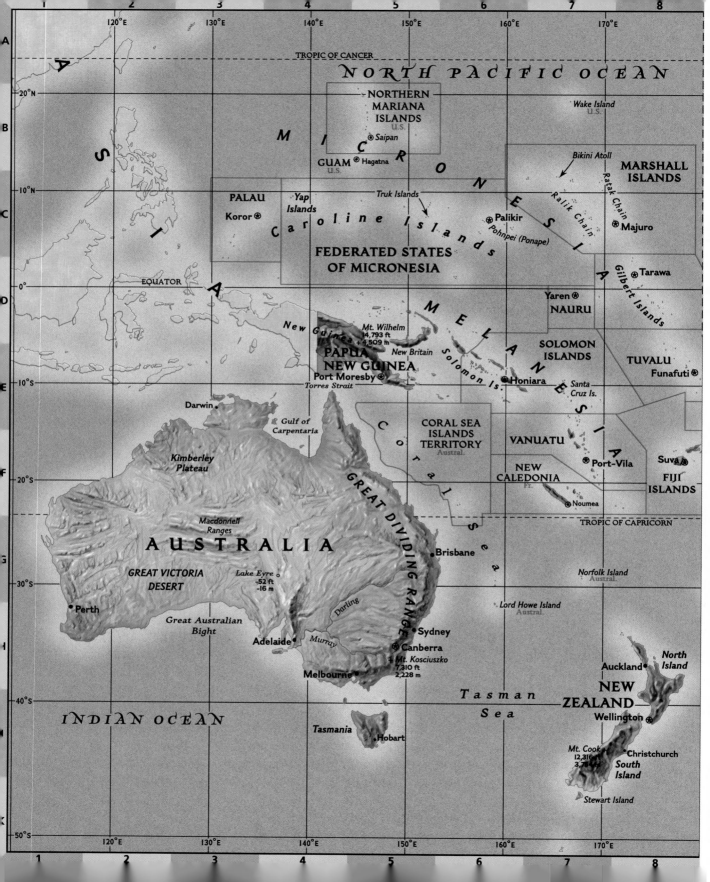

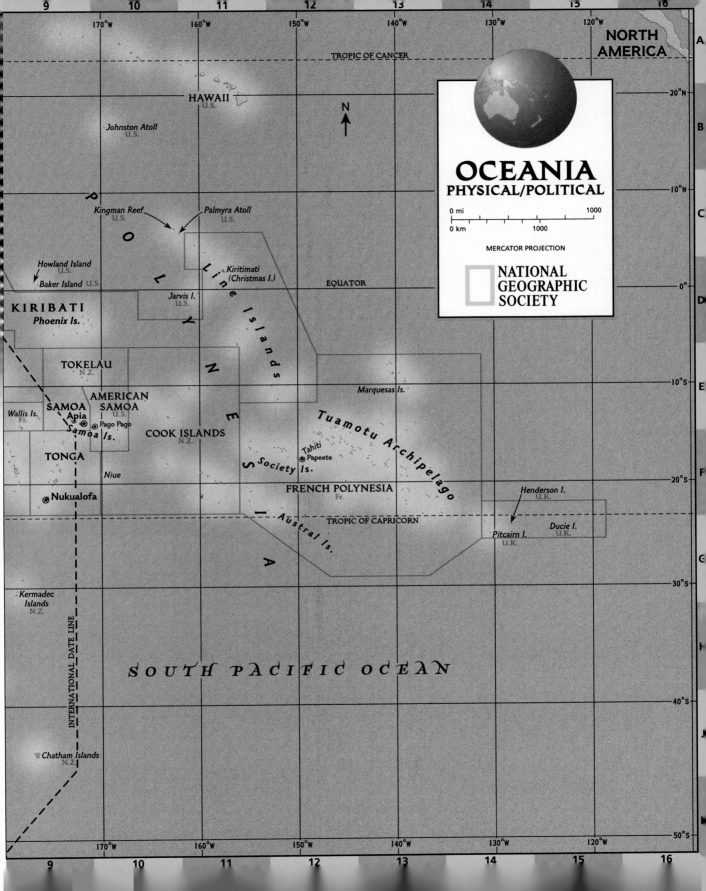

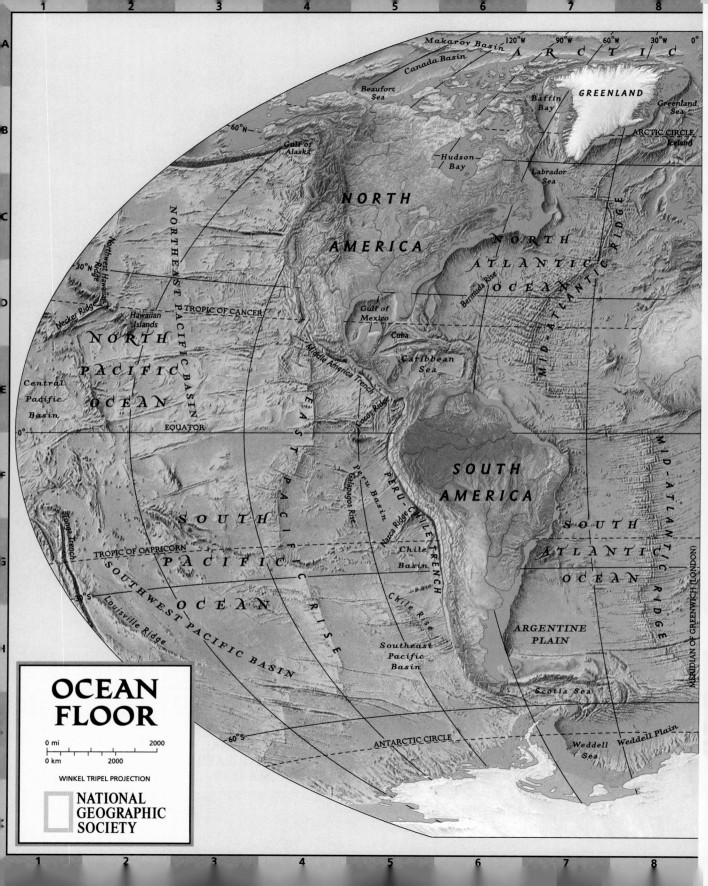

OCEAN FLOOR

0 mi   2000
0 km   2000

WINKEL TRIPEL PROJECTION

NATIONAL GEOGRAPHIC SOCIETY

Makarov Basin
Canada Basin
A R C T I C
Beaufort Sea
GREENLAND
Baffin Bay
Greenland Sea
60°N
ARCTIC CIRCLE
Iceland
Gulf of Alaska
Hudson Bay
Labrador Sea
NORTH AMERICA
NORTH ATLANTIC OCEAN
Bermuda Rise
MID-ATLANTIC RIDGE
NORTHWEST HAWAIIAN RIDGE
NORTHEAST PACIFIC BASIN
30°N
Necker Ridge
TROPIC OF CANCER
Hawaiian Islands
NORTH PACIFIC OCEAN
Gulf of Mexico
Cuba
Caribbean Sea
Central Pacific Basin
Middle America Trench
Cocos Ridge
EQUATOR
0°
EAST PACIFIC
Galapagos Rise
Peru Basin
PERU-CHILE TRENCH
SOUTH AMERICA
MID-ATLANTIC RIDGE
SOUTH
PACIFIC
OCEAN
Nazca Ridge
Chile Basin
SOUTH ATLANTIC OCEAN
TROPIC OF CAPRICORN
Chile Rise
RISE
Tonga Trench
Louisville Ridge
SOUTHWEST PACIFIC BASIN
30°S
Southeast Pacific Basin
ARGENTINE PLAIN
MERIDIAN OF GREENWICH (LONDON)
60°S
ANTARCTIC CIRCLE
Scotia Sea
Weddell Sea
Weddell Plain

120°W   90°W   60°W   30°W   0°

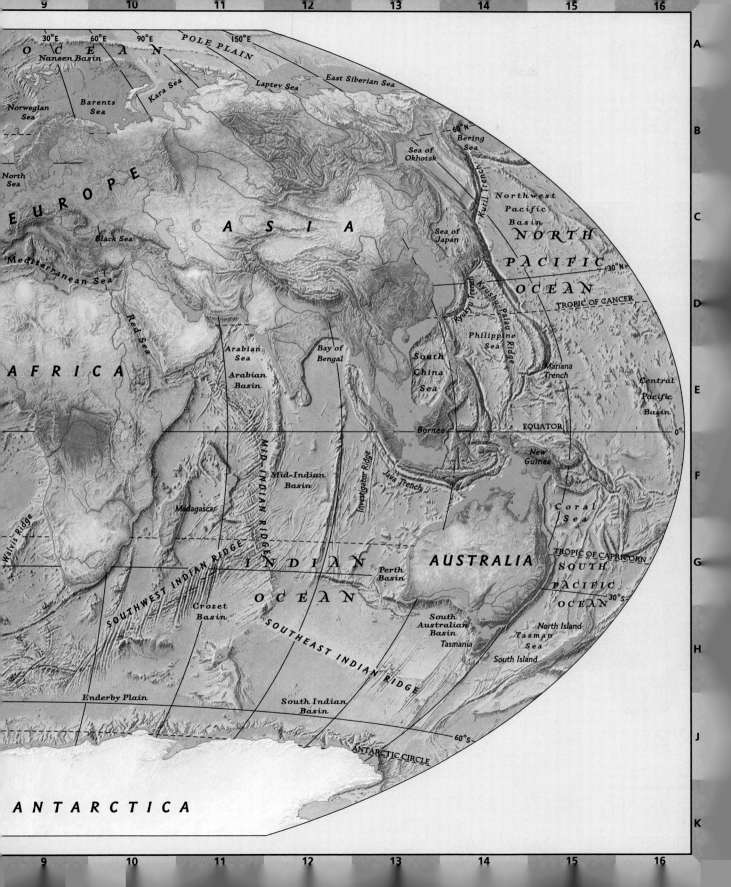

OCEAN

30°E   60°E   90°E   POLE PLAIN   150°E

Nansen Basin

Norwegian Sea

Barents Sea

Kara Sea

Laptev Sea

East Siberian Sea

60°N

Bering Sea

North Sea

EUROPE

ASIA

Sea of Okhotsk

Kuril Trench

Northwest Pacific Basin

NORTH

Black Sea

Mediterranean Sea

Sea of Japan

PACIFIC

30°N

Red Sea

AFRICA

Arabian Sea

Arabian Basin

Bay of Bengal

South China Sea

OCEAN

TROPIC OF CANCER

Ryukyu Trench

Kyushu-Palau Ridge

Philippine Sea

Mariana Trench

Central Pacific Basin

Borneo

EQUATOR   0°

Madagascar

MID-INDIAN RIDGE

Mid-Indian Basin

Investigator Ridge

Java Trench

New Guinea

Coral Sea

Walvis Ridge

SOUTHWEST INDIAN RIDGE

INDIAN

OCEAN

Perth Basin

AUSTRALIA

TROPIC OF CAPRICORN

SOUTH

PACIFIC

OCEAN

30°S

Crozet Basin

SOUTHEAST INDIAN RIDGE

South Australian Basin

Tasmania

North Island

Tasman Sea

South Island

Enderby Plain

South Indian Basin

60°S

ANTARCTIC CIRCLE

ANTARCTICA

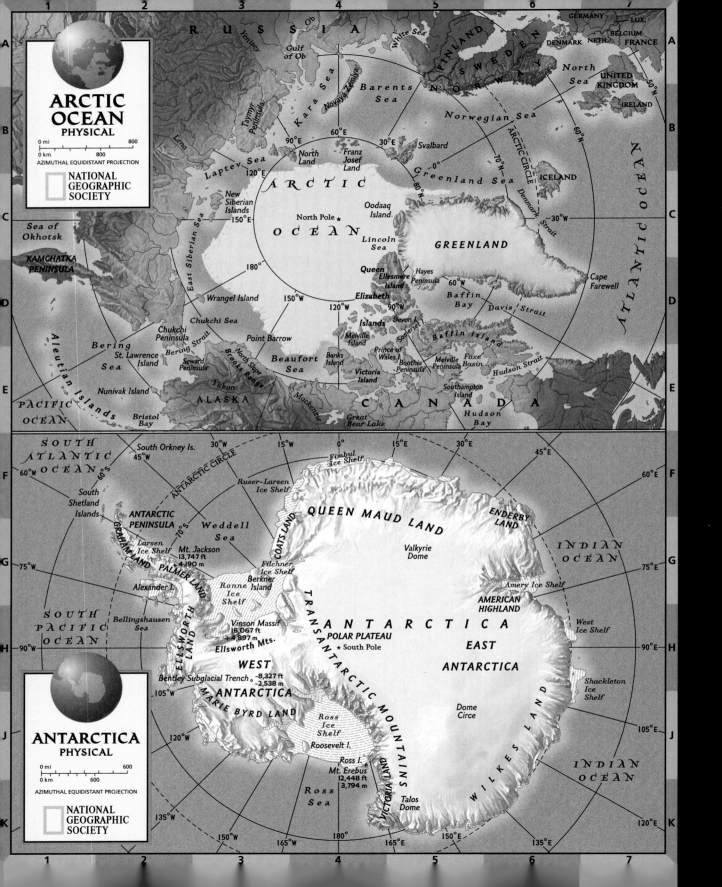

NATIONAL
GEOGRAPHIC

GEOGRAPHY HANDBOOK

A geographer is a person who studies the earth and its people. Have you ever wondered if you could be a geographer? One way to learn how is to use this **Geography Handbook**. It will show you that geography is more than studying facts and figures. It also means doing some of your own exploring of the earth. By learning how to use geographic tools, such as globes, maps, and graphs, you will get to know and appreciate the wonders of our planet.

Geologists studying a volcano ▲

Table of Contents

Polar ice cap ▼

◄ Lava flow

How Do I Study

Everything you see, touch, use, and even hear is related to geography—the study of the world's people, places, and environments. How can you possibly study such a huge amount of information in your geography class? Where do you start?

Geographers—people who study geography—ask themselves this question, too. To understand how our world is connected, some geographers have broken down the study of geography into five themes. The Five Themes of Geography are (1) location, (2) place, (3) human/environment interaction, (4) movement, and (5) regions. These themes are highlighted in blue throughout this textbook.

Most recently, as suggested in the *Geography Standards for Life,* geographers have begun to look at geography a different way. They break down the study of geography into Six Essential Elements, which are explained for you below. Being aware of these elements will help you sort out what you are learning about geography.

ELEMENT 2 Places and Regions

Geographers also look at places and regions. Place includes those features and characteristics that give an area its own identity or personality. These can be physical characteristics—such as landforms, climate, plants, and animals—or human characteristics—such as language, religion, architecture, music, politics, and way of life.

To make sense of all the complex things in the world, geographers often group places or areas into regions. Regions are united by one or more common characteristics.

Des Moines, Iowa, is a place ▲ characterized by farms. It is also part of a region known as the Corn Belt.

ELEMENT 1 The World in Spatial Terms

Geographers first take a look at where a place is located. Location serves as a starting point by asking "Where is it?" Knowing the location of places helps you to orient yourself in space and to develop an awareness of the world around you.

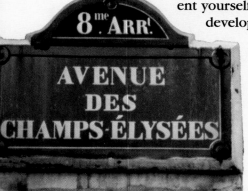

8ᵐᵉ. ARRᵗ
AVENUE DES CHAMPS-ÉLYSÉES

◀ This street sign is located in Paris, France.

ELEMENT 3 Physical Systems

Why do some places have mountains and other places have flat deserts? When studying places and regions, geographers analyze how physical systems—such as volcanoes, glaciers, and hurricanes—interact and shape the earth's surface. They also look at ecosystems, or communities of plants and animals that are dependent upon one another and their particular surroundings for survival.

◀ A glacier carved this deep valley in New Zealand.

Geography?

Human Systems

Geographers also examine human systems, or how people have shaped our world. They look at how boundary lines are determined and analyze why people settle in certain places and not in others. An ongoing theme in geography is the continual movement of people, ideas, and goods.

People, vehicles, and goods move quickly through Asmara, the capital of Eritrea.

Environment and Society

The study of geography includes looking at human/environment interaction, or how and why people change their surroundings. Throughout history, people have cut forests and dammed rivers to build farms and cities. Some activities have led to air and water pollution. The physical environment affects human activities as well. The type of soil and amount of water in a place determines if crops can be grown. Earthquakes and floods also affect human life.

Romanian farmers work in a field near a nuclear power plant.

The Uses of Geography

Understanding geography, and knowing how to use the tools and technology available to study it, prepares you for life in our modern society. Individuals, businesses, and governments use geography and maps of all kinds on a daily basis. For example, geographic information systems (GIS) are special software programs that help geographers gather and use information about a place.

A cartographer uses GIS to make a map.

Using Globes and Maps

Guide To Reading

Main Idea

Globes and maps provide different ways of showing features of the earth.

Terms to Know

- globe
- hemisphere
- latitude
- longitude
- grid system
- absolute location
- great circle route
- projection

What Is a Globe?

A globe is a model of the earth that shows the earth's shape, lands, distances, and directions as they truly relate to one another. A world globe can help you find your way around the earth. By using one, you can locate places and determine distances.

Hemispheres

To locate places on the earth, geographers use a system of imaginary lines that crisscross the globe. One of these lines, the Equator, circles the middle of the earth like a belt. It

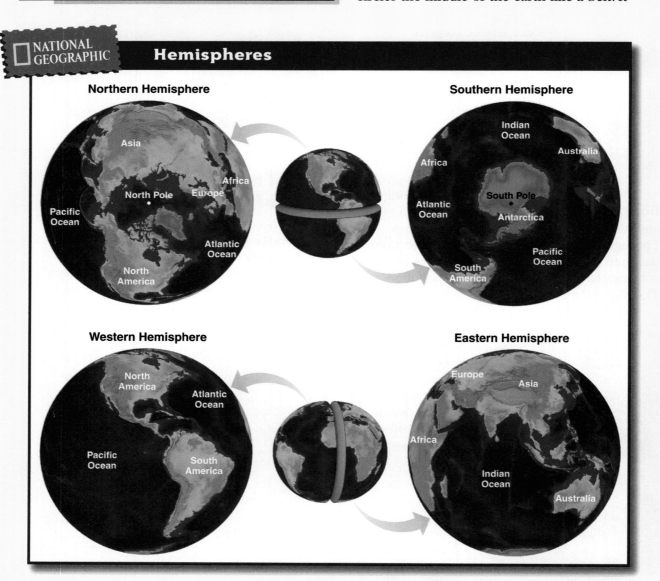

NATIONAL GEOGRAPHIC

Hemispheres

Northern Hemisphere

Asia
Africa
Europe
North Pole
Pacific Ocean
Atlantic Ocean
North America

Southern Hemisphere

Indian Ocean
Australia
Africa
South Pole
Atlantic Ocean
Antarctica
Pacific Ocean
South America

Western Hemisphere

North America
Atlantic Ocean
Pacific Ocean
South America

Eastern Hemisphere

Europe
Asia
Africa
Indian Ocean
Australia

divides the earth into "half spheres," or **hemispheres.** Everything north of the Equator is in the Northern Hemisphere. Everything south of the Equator is in the Southern Hemisphere.

Another imaginary line runs from north to south and helps divide the earth into half spheres in the other direction. Find this line—called the Prime Meridian or the Meridian of Greenwich—on a globe. Everything east of the Prime Meridian for 180 degrees is in the Eastern Hemisphere. Everything west of the Prime Meridian for 180 degrees is in the Western Hemisphere. In which hemispheres is North America located? It is found in both the Northern Hemisphere and the Western Hemisphere.

Latitude and Longitude

The Equator and the Prime Meridian are the starting points for two sets of lines used to find any location. *Parallels* circle the earth like stacked rings and show latitude, or distance measured in degrees north and south of the Equator. The letter *N* or *S* following the degree symbol tells you if the location is north or south of the Equator. The North Pole, for example, is at 90°N (North) latitude, and the South Pole is at 90°S (South) latitude.

Two important parallels in between the poles are the Tropic of Cancer at 23½°N latitude and the Tropic of Capricorn at 23½°S latitude. You can also find the Arctic Circle at

NATIONAL GEOGRAPHIC

Latitude and Longitude

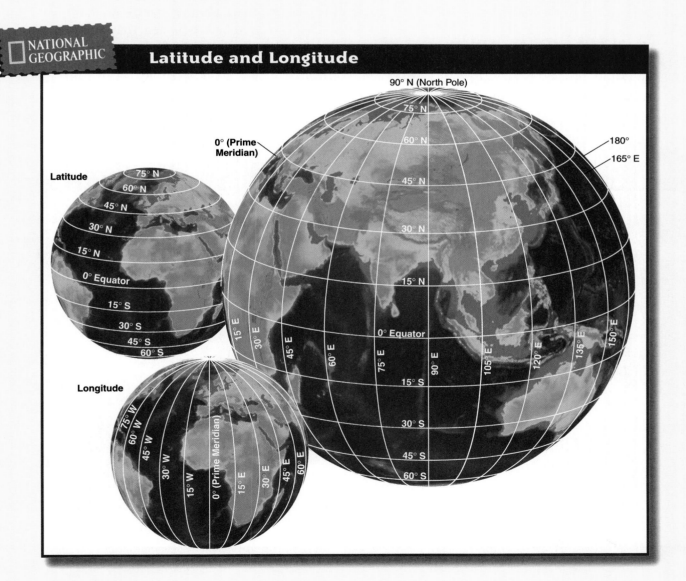

66½°N latitude and the Antarctic Circle at 66½°S latitude.

Meridians run from pole to pole and crisscross parallels. Meridians signify longitude, or distance measured in degrees east *(E)* or west *(W)* of the Prime Meridian. The Prime Meridian, or 0° longitude, runs through Greenwich, England. On the opposite side of the earth is the 180° meridian, also called the International Date Line.

Lines of latitude and longitude cross each other in the form of a grid system. You can find a place's absolute location by naming the latitude and longitude lines that cross exactly at that place. For example, the city of Tokyo, Japan, is located at 36°N latitude and 140°E longitude.

🌐 Great Circle Route

A straight line of true direction—one that moves directly from west to east, for example—is not always the shortest distance between two points on the earth. To find the shortest distance between any two places, take a piece of string and stretch it around a globe from one point to another. The string will form part of a *great circle,* or an imaginary line that follows the curve of the earth. A line drawn along the Equator is an example of a great circle. Traveling along a great circle is called following a great circle route. Airplane pilots use great circle routes to reduce travel time and to save fuel.

The idea of a great circle shows one important difference between using a globe and using a map. Because a globe is round, it accurately shows great circles. However, on a flat map the great circle route between two points may not appear to be the shortest distance. On map A below, the great circle distance (dotted line) between Tokyo and Los Angeles appears to be far longer than the true direction distance (solid line). In fact, the great circle distance is 345 miles (555 km) shorter, which is evident on map B.

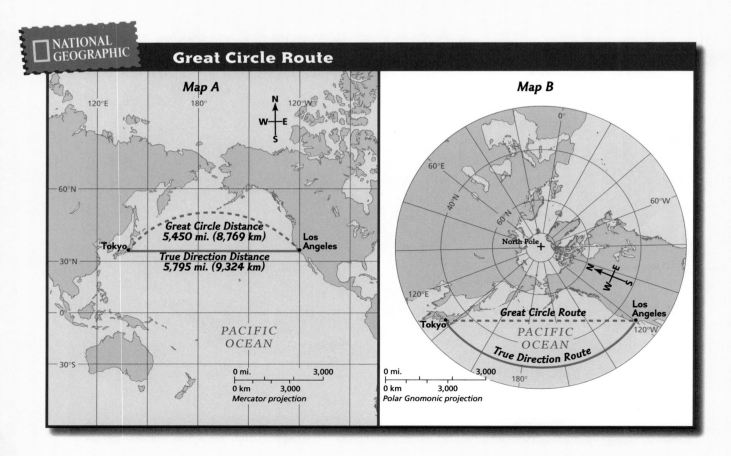

Great Circle Route

Map A
Great Circle Distance 5,450 mi. (8,769 km)
True Direction Distance 5,795 mi. (9,324 km)
PACIFIC OCEAN
0 mi. 3,000
0 km 3,000
Mercator projection

Map B
North Pole
Great Circle Route
True Direction Route
PACIFIC OCEAN
0 mi. 3,000
0 km 3,000
Polar Gnomonic projection

🌐 Why Use Maps?

Globes are the best, most accurate way to show the round earth. Using a globe has its difficulties, though. First, a globe is too big and awkward to carry around. Also, a globe cannot show you the whole world at one time. For these reasons, geographers use maps instead. A map is made by taking data from a round globe and placing it on a flat surface. It is important to remember that the earth's features, which are shown accurately on a globe, become distorted when the curves of a globe become straight lines on a flat map.

🌐 Map Projections

Imagine taking the whole peel from an orange and trying to flatten it on a table. You would either have to cut it or stretch parts of it. Mapmakers face a similar problem in showing the surface of the round earth on a flat map. When the earth's surface is flattened, big gaps open up. To fill in the gaps, mapmakers stretch parts of the earth. They choose to show either the correct shapes of places or their correct sizes. It is impossible to show both. As a result, mapmakers have developed different projections, or ways of showing the earth on a flat piece of paper. Each projection has its strengths and weaknesses. None is a completely accurate representation of the earth, but all prove useful in one way or another.

Mercator Projection   The *Mercator projection* shows land shapes fairly accurately, but not size or distance. Areas that are located far from the Equator are quite distorted on this projection. Alaska, for example, appears much larger on a Mercator map than it does on a globe. The Mercator projection does show true directions, however. This makes it very useful for sea travel.

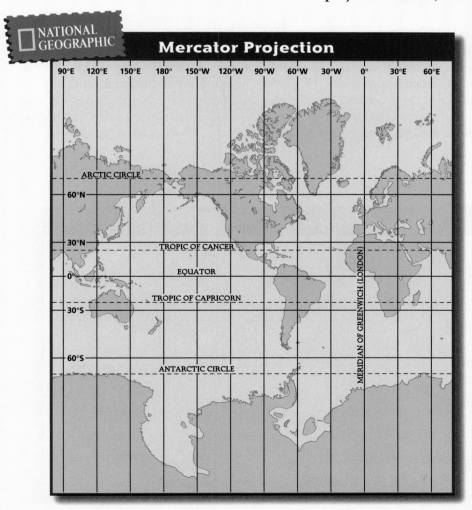

NATIONAL GEOGRAPHIC

Mercator Projection

90°E 120°E 150°E 180° 150°W 120°W 90°W 60°W 30°W 0° 30°E 60°E

ARCTIC CIRCLE
60°N
30°N
TROPIC OF CANCER
0°
EQUATOR
TROPIC OF CAPRICORN
30°S
60°S
ANTARCTIC CIRCLE

MERIDIAN OF GREENWICH (LONDON)

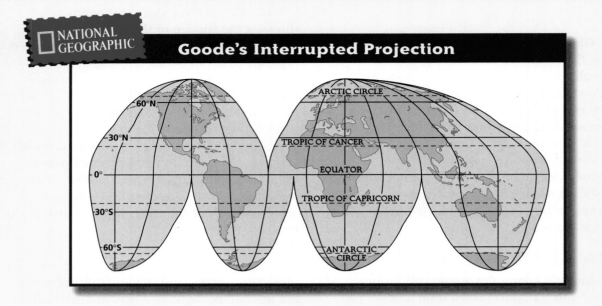

NATIONAL GEOGRAPHIC

Goode's Interrupted Projection

Goode's Interrupted Projection Take a second look at your peeled, flattened orange. You might have something that looks like a map based on *Goode's interrupted projection*. A map with this projection shows continents close to their true shapes and sizes. Distances—especially in the oceans—are less accurate. The Goode's projection would be helpful if you wanted to compare land area data about the continents.

Robinson Projection A map using the *Robinson projection* shows size and shape with less distortion than does a Mercator map. Land on the western and eastern sides of the Robinson map appear much as they do on a globe. The areas most distorted on this projection are near the North and South poles. You may notice that many atlases use Robinson projections.

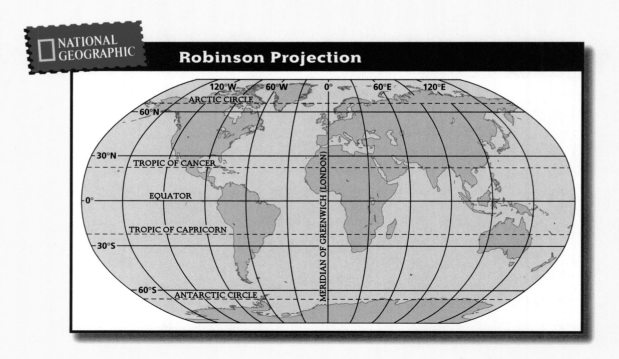

NATIONAL GEOGRAPHIC

Robinson Projection

Winkel Tripel Projection

Mapmakers are always looking for a more accurate way to show the round earth on flat paper. The *Winkel Tripel projection* gives a good overall view of the continents' shapes and sizes. You may notice a close similarity between the Winkel Tripel and Robinson projections. Land areas in a Winkel Tripel projection are not as distorted near the poles as they are in the Robinson projection. In 1998 the National Geographic Society began using the Winkel Tripel projection for its reference maps of the world.

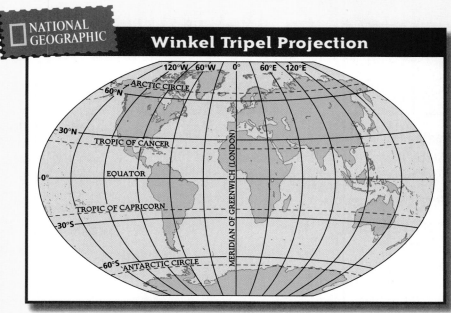

NATIONAL GEOGRAPHIC

Winkel Tripel Projection

Section 1 Assessment

Defining Terms

1. Define globe, hemisphere, latitude, longitude, grid system, absolute location, great circle route, projection.

Recalling Facts

2. What imaginary line divides the earth into the Northern and the Southern Hemispheres?

3. What imaginary line divides the earth into the Eastern and the Western Hemispheres?

4. What is the best way to find the shortest distance between two places?

Critical Thinking

5. Drawing Conclusions Why do map projections distort some parts of the earth?

6. Making Comparisons What map projection has fairly accurate shapes in the center but increasing distortion toward the edges?

Graphic Organizer

7. Organizing Information Create a chart like this one. On the left, list three words or phrases that describe globes. On the right, do the same for maps.

Globes	Maps

Applying Geography Skills

8. Analyzing Globes Look at the globe showing latitude on page 5. Where on the globe do parallels, or lines of latitude, become shorter?

Section 2 Learning Map Basics

Guide To Reading

Main Idea
Maps have several basic features that help you understand what they are showing.

Terms to Know
- geographic information systems (GIS)
- map key
- cardinal directions
- compass rose
- intermediate directions
- scale bar
- scale
- relief
- elevation
- contour line

🌐 How Maps Are Made

For more than 4,000 years, people have made maps to organize their knowledge of the world. The reason for producing maps has not changed over the centuries, but the tools of mapmaking have. Today satellites located thousands of miles in space gather data about the earth below. The data are then sent back to the earth, where computers change the data into images of the earth's surface. Mapmakers analyze and use these images to produce maps.

For modern mapmakers, computers have replaced pen and paper. Most mapmakers use computers with software programs called geographic information systems (GIS). With GIS,

each kind of information on a map is kept as a separate electronic "layer" in the map's computer files. Because of this modern technology, mapmakers are able to make maps—and change them—more quickly and easily than before.

🌐 How to Read a Map

Maps can direct you down the street, across the country, or around the world. An ordinary map holds all kinds of information. Learn the map's code, and you can read it like a book.

Map Key The map key explains the lines, symbols, and colors used on a map. Look at the map of Spain below. Its key shows that dots mark major cities. A circled star indicates the national capital—in Spain's case, the city of Madrid. Some keys tell which lines stand for national boundaries, roads, or railroads. Other

NATIONAL GEOGRAPHIC

Spain: Political

- — National boundary
- ⊛ National capital
- • Major city

Lambert Azimuthal Equal-Area projection

map symbols may represent human-made or natural features, such as canals, forests, or natural gas deposits.

Compass Rose

An important step in reading any map is to find the direction marker. A map has a symbol that tells you where the cardinal directions—north, south, east, and west—are positioned. Sometimes all of these directions are shown with a compass rose. An intermediate direction, such as southeast, may also be on the compass rose. Intermediate directions fall between the cardinal directions.

Latitude and Longitude Lines

Like globes, maps have lines of latitude and longitude that form a grid. Every place on the earth has a unique position or "address" on this grid. Knowing this address makes it easier for you to locate cities and other places on a map. For example, what is the grid address of

Madrid, Spain? The map on page 10 shows you that the address is 41°N latitude, 4°W longitude.

Scale

A measuring line, often called a scale bar, helps you determine distance on a map. The map's scale tells you what distance on the earth is represented by the measurement on the scale bar. For example, 200 miles on the earth may be represented by 1 inch on the map. Knowing the scale allows you to see how large an area is. Map scale is usually given in both miles and kilometers.

Each map has its own scale. What scale a mapmaker uses depends on the size of the area shown on the map. If you were drawing a map of your backyard, you might use a scale of 1 inch equals 5 feet. In contrast, the scale bar on the map above of Austin, Texas, shows that about ⅝ inch represents 8 miles. Scale is important when you are trying to compare the size of one area to another.

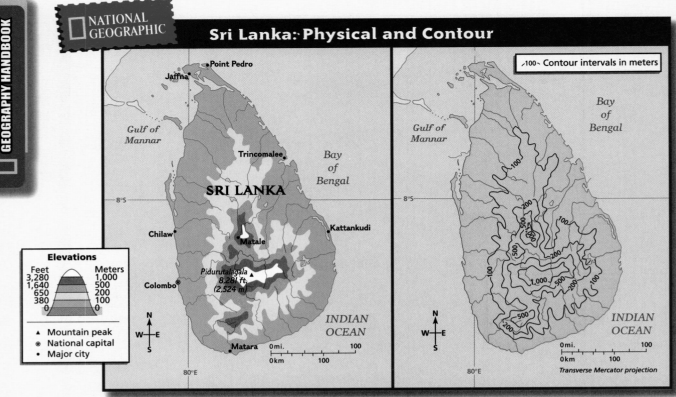

NATIONAL GEOGRAPHIC

Sri Lanka: Physical and Contour

100 Contour intervals in meters

Point Pedro
Jaffna
Gulf of Mannar
Trincomalee
SRI LANKA
Bay of Bengal
8°S
Chilaw
Matale
Kattankudi
Pidurutalagala
8,281 ft.
(2,524 m)
Colombo
Matara
INDIAN OCEAN
80°E

Elevations

Feet	Meters
3,280	1,000
1,640	500
650	200
380	100
0	0

▲ Mountain peak
⊛ National capital
• Major city

0 mi. 100
0 km 100

N W E S

Gulf of Mannar
Bay of Bengal
8°S
INDIAN OCEAN
80°E
Transverse Mercator projection

0 mi. 100
0 km 100

N W E S

🌐 General Purpose Maps

Maps are amazingly useful tools. You can use them to preserve information, to display data, and to make connections between seemingly unrelated things. Geographers use many different types of maps. Maps that show a wide range of general information about an area are called *general purpose maps.* Two of the most common general purpose maps are political and physical maps.

Political Maps   *Political maps* show the names and boundaries of countries and often identify major physical features. The political map of Spain on page 10, for example, shows the boundaries between Spain and other countries. It also shows cities and rivers within Spain and bodies of water surrounding Spain.

Physical Maps   *Physical maps* call out landforms and water features. The physical map of Sri Lanka above shows rivers and mountains. The colors used on physical maps include brown or green for land, and blue for water.

These colors and shadings may show relief—or how flat or rugged the land surface is. In addition, physical maps may use colors to show elevation—the height of an area above sea level. A key explains what each color and symbol stands for.

Contour Maps   One kind of physical map, called a *contour map,* also shows elevation. A contour map has contour lines—one line for each major level of elevation. All the land at the same elevation is connected by a line. These lines usually form circles or ovals—one inside the other. If contour lines come very close together, the surface is steep. If the lines are spread apart, the land is flat or rises very gradually. Compare the contour map of Sri Lanka above to its physical map.

🌐 Special Purpose Maps

Some maps are made to present specific kinds of information. These are called *special purpose maps.* They usually show patterns, often emphasizing one subject or theme. Special purpose maps may present climate, natural resources, or population density. They may also display historical information, such as battles or territorial changes. The map's title tells what kind of special information it shows. Colors and symbols in the map key are especially important on these types of maps.

One type of special purpose map uses colors to show population density, or the average number of people living in a square mile or square kilometer. As with other maps, it is important to first read the title and the key. The population density map of Egypt to the right gives a striking picture of differences in population density. The Nile River valley and delta are very densely populated. In contrast, the desert areas east and west of the river are home to few people.

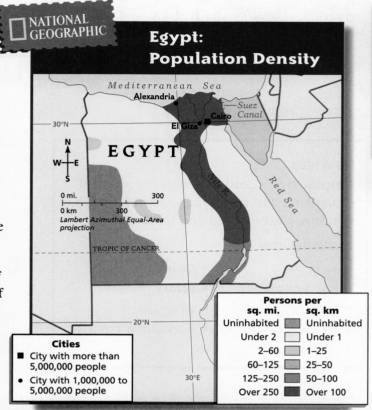

NATIONAL GEOGRAPHIC

Egypt: Population Density

Mediterranean Sea
Alexandria
Suez Canal
Cairo
30°N   El Giza
EGYPT
Nile R.
Red Sea

0 mi.   300
0 km   300
Lambert Azimuthal Equal-Area projection

TROPIC OF CANCER

20°N

30°E

Cities
■ City with more than 5,000,000 people
● City with 1,000,000 to 5,000,000 people

Persons per	
sq. mi.	sq. km
Uninhabited	Uninhabited
Under 2	Under 1
2–60	1–25
60–125	25–50
125–250	50–100
Over 250	Over 100

Section 2 Assessment

Defining Terms

1. Define geographic information systems (GIS), map key, cardinal directions, compass rose, intermediate directions, scale bar, scale, relief, elevation, contour line.

Recalling Facts

2. Why do people make maps?

3. What are the four cardinal directions?

4. What are two of the most common types of general purpose maps?

Critical Thinking

5. Comparing and Contrasting Describe the similarities and differences between physical maps and contour maps.

6. Synthesizing Information Draw a compass rose that shows the cardinal and intermediate directions.

Graphic Organizer

7. Organizing Information Create a diagram like the one below. In each of the outer ovals, write an example of a feature that you would find on a typical physical map.

Map Features

Applying Geography Skills

8. Analyzing Maps Look at the map of Egypt above. At what latitude and longitude is Alexandria located? Use the key to describe the population density of Alexandria and its surrounding area.

Using Graphs, Charts, and Diagrams

Guide To Reading

Main Idea
Graphs, charts, and diagrams are ways of organizing and displaying information so that it is easier to see and understand.

Terms to Know

- axis
- bar graph
- line graph
- circle graph
- pictograph
- climograph
- chart
- diagram
- elevation profile

What Is a Graph?

A graph is a way of summarizing and presenting information visually. Each part of a graph gives useful information. First read the graph's title to find out its subject. Then read the labels along the graph's axes—the vertical line along the left side of the graph and the horizontal line along the bottom of the graph. One axis will tell you what is being measured. The other axis tells what units of measurement are being used.

Bar, Line, and Circle Graphs

Bar Graphs   Graphs that use bars or wide lines to compare data visually are called bar graphs. Look carefully at the bar graph below, which compares world languages. The vertical axis lists the languages. The horizontal axis gives speakers of the language, in millions. By comparing the lengths of the bars, you can quickly tell which language is spoken by the most people. Bar graphs are especially useful for comparing quantities, and they may show the bars rising up from the bottom of the graph or extending out from the vertical axis.

Line Graphs   A line graph is a useful tool for showing changes over a period of time. The amounts being measured are plotted on the grid above each year, and then are connected by a line. Line graphs sometimes have two or more lines plotted on them. The line graph on page 15 shows that the number of farms in the United States has decreased since 1940. The vertical axis lists the number of farms in millions. The horizontal axis shows the passage of time in ten-year periods from 1940 to 1998.

NATIONAL GEOGRAPHIC

Comparing World Languages

Language	Numbers of Speakers (in millions)
Chinese (Mandarin)	885
English	322
Spanish	266
Bengali	189
Hindi	182
Portuguese	170
Russian	170
Japanese	125
German	98
Chinese (Wu)	77

Numbers of Speakers (in millions)

Source: *National Geographic Atlas of the World*, 1999.

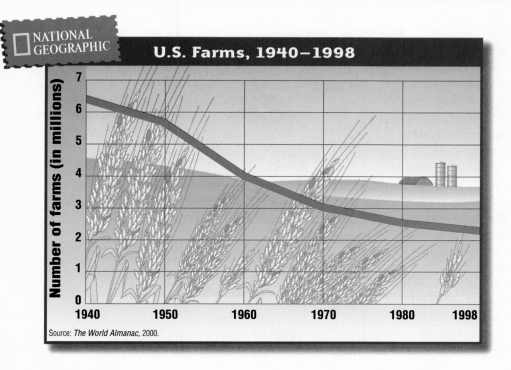

U.S. Farms, 1940–1998

Number of farms (in millions)

7
6
5
4
3
2
1
0

1940   1950   1960   1970   1980   1998

Source: *The World Almanac*, 2000.

Circle Graphs   You can use **circle graphs** when you want to show how the *whole* of something is divided into its *parts*. Because of their shape, circle graphs are often called pie graphs. Each "slice" represents a part or percentage of the whole "pie." On the circle graph below, the whole circle represents the world's population in 1999. The slices show how this population is divided among the world's five largest continents.

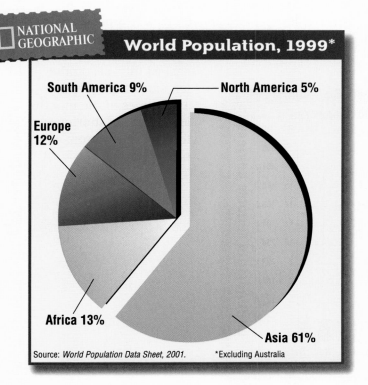

World Population, 1999*

South America 9%

North America 5%

Europe 12%

Africa 13%

Asia 61%

Source: *World Population Data Sheet, 2001.*   *Excluding Australia

Charts

Charts present related facts and numbers in an organized way. They arrange data, especially numbers, in rows and columns for easy reference. Look at the chart on page 85. To interpret the chart, first read the title. It tells you what information the chart contains. Next, read the labels at the top of each column and on the left side of the chart. They explain what the numbers or data on the chart are measuring. One kind of chart, a *flowchart*, joins certain elements of a chart and a diagram. It can show the order of how things happen or how they are related to each other. The flowchart on page 132 presents the branches of the United States government. Notice how the chart shows the relationship among the tasks and the offices or bodies of each branch.

🌐 Pictographs

Like bar and circle graphs, pictographs are good for making comparisons. Pictographs use rows of small pictures or symbols, with each picture or symbol representing an amount. The pictograph on the right shows the number of automobiles produced in the world's five major automobile-producing countries. The key tells you that one car symbol stands for 1 million automobiles. Pictographs are read like a bar graph. The total number of car symbols in a row adds up to the auto production in each selected country.

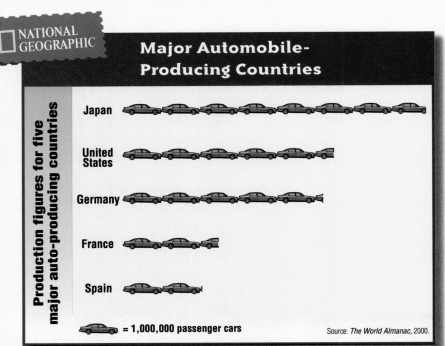

NATIONAL GEOGRAPHIC

Major Automobile-Producing Countries

Production figures for five major auto-producing countries

Japan
United States
Germany
France
Spain

🚗 = 1,000,000 passenger cars

Source: *The World Almanac*, 2000.

🌐 Climographs

A climograph, or climate graph, combines a line graph and a bar graph. It gives an overall picture of the climate—the long-term weather patterns—in a specific place. Because climographs include several kinds of information, you need to read them carefully.

Note that the vertical bars on the climograph below represent average amounts of precipitation (rain, snow, or sleet) in each month of the year. These bars are measured against the axis on the right side of the graph. The line plotted above the bars represents changes in the average monthly temperature. You measure this line against the axis on the left side of the graph. The names of the months are shown in shortened form on the bottom axis of the graph.

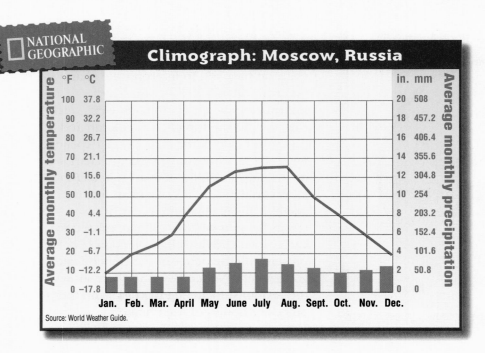

NATIONAL GEOGRAPHIC

Climograph: Moscow, Russia

Average monthly temperature

°F	°C
100	37.8
90	32.2
80	26.7
70	21.1
60	15.6
50	10.0
40	4.4
30	−1.1
20	−6.7
10	−12.2
0	−17.8

Average monthly precipitation

in.	mm
20	508
18	457.2
16	406.4
14	355.6
12	304.8
10	254
8	203.2
6	152.4
4	101.6
2	50.8
0	0

Jan. Feb. Mar. April May June July Aug. Sept. Oct. Nov. Dec.

Source: World Weather Guide.

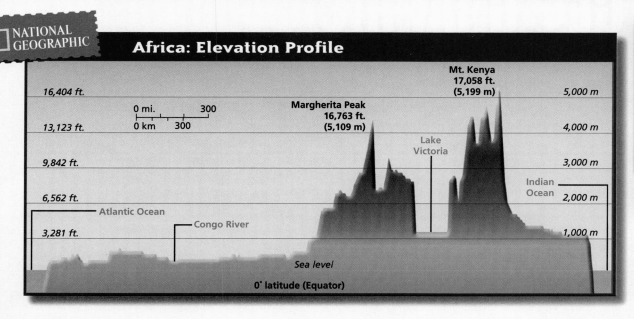

Africa: Elevation Profile

Mt. Kenya
17,058 ft.
(5,199 m)

16,404 ft.

0 mi.   300
0 km   300

13,123 ft.

Margherita Peak
16,763 ft.
(5,109 m)

5,000 m

4,000 m

9,842 ft.

Lake
Victoria

3,000 m

Indian
Ocean

6,562 ft.

2,000 m

Atlantic Ocean

3,281 ft.

Congo River

1,000 m

Sea level

0° latitude (Equator)

🌐 Diagrams

Diagrams are drawings that show steps in a process, point out the parts of an object, or explain how something works. You can use a diagram to assemble a stereo. The diagram on page 212 shows how locks enable ships to move through a canal. An **elevation profile** is a type of diagram that can be helpful when comparing the elevations of an area. It shows a profile, or side view, of the land as if it were sliced and you were viewing it from the side. (For reasons of scale, the profile is exaggerated.) The elevation profile of Africa above clearly shows low areas and mountains. The line of latitude at the bottom tells you where this profile was "sliced."

Section 3 Assessment

Defining Terms

1. **Define** axis, bar graph, line graph, circle graph, pictograph, climograph, chart, diagram, elevation profile.

Recalling Facts

2. How does a bar graph differ from a line graph?
3. What percentage does the whole circle in a circle graph always represent?
4. What two features does a climograph show?

Critical Thinking

5. **Synthesizing Information** Draw and label a flowchart showing the steps in some simple process—for example, making a sandwich or doing laundry.

Graphic Organizer

6. **Organizing Information** Create a chart like the one below. In the left column, list the types of graphs that are discussed in this section. In the right column, list what each type of graph is most useful for showing.

Types of Graphs	Useful for showing . . .

Applying Geography Skills

7. **Analyzing Graphs** Look at the bar graph on page 14. Which language is the most widely spoken? About how many people speak it?

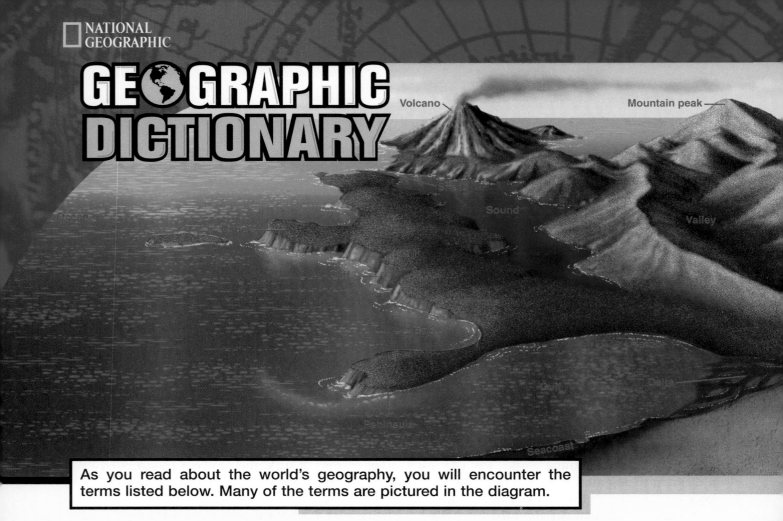

NATIONAL GEOGRAPHIC

GEOGRAPHIC DICTIONARY

Volcano · Mountain peak ·

Ocean

Sound · Valley

Cliff

Delta

Peninsula

Seacoast

As you read about the world's geography, you will encounter the terms listed below. Many of the terms are pictured in the diagram.

absolute location exact location of a place on the earth described by global coordinates

basin area of land drained by a given river and its branches; area of land surrounded by lands of higher elevations

bay part of a large body of water that extends into a shoreline, generally smaller than a gulf

canyon deep and narrow valley with steep walls

cape point of land that extends into a river, lake, or ocean

channel wide strait or waterway between two landmasses that lie close to each other; deep part of a river or other waterway

cliff steep, high wall of rock, earth, or ice

continent one of the seven large landmasses on the earth

delta flat, low-lying land built up from soil carried downstream by a river and deposited at its mouth

divide stretch of high land that separates river systems

downstream direction in which a river or stream flows from its source to its mouth

elevation height of land above sea level

Equator imaginary line that runs around the earth halfway between the North and South Poles; used as the starting point to measure degrees of north and south latitude

glacier large, thick body of slowly moving ice

gulf part of a large body of water that extends into a shoreline, generally larger and more deeply indented than a bay

harbor a sheltered place along a shoreline where ships can anchor safely

highland elevated land area such as a hill, mountain, or plateau

hill elevated land with sloping sides and rounded summit; generally smaller than a mountain

island land area, smaller than a continent, completely surrounded by water

isthmus narrow stretch of land connecting two larger land areas

lake a sizable inland body of water

latitude distance north or south of the Equator, measured in degrees

longitude distance east or west of the Prime Meridian, measured in degrees

lowland land, usually level, at a low elevation

map drawing of the earth shown on a flat surface

meridian one of many lines on the global grid running from the North Pole to the South Pole; used to measure degrees of longitude

18   **Geography Handbook**

Image labels: Mountain range, Glacier, Source of river, Channel, Highland, Lake, Plateau, Hills, Mouth of river, Canyon, Desert, Upstream, Downstream, Plain, Lowland, Basin, Tributary

mesa broad, flat-topped landform with steep sides; smaller than a plateau

mountain land with steep sides that rises sharply (1,000 feet or more) from surrounding land; generally larger and more rugged than a hill

mountain peak pointed top of a mountain

mountain range a series of connected mountains

mouth (of a river) place where a stream or river flows into a larger body of water

ocean one of the four major bodies of salt water that surround the continents

ocean current stream of either cold or warm water that moves in a definite direction through an ocean

parallel one of many lines on the global grid that circles the earth north or south of the Equator; used to measure degrees of latitude

peninsula body of land jutting into a lake or ocean, surrounded on three sides by water

physical feature characteristic of a place occurring naturally, such as a landform, body of water, climate pattern, or resource

plain area of level land, usually at low elevation and often covered with grasses

plateau area of flat or rolling land at a high elevation, about 300–3,000 feet high

Prime Meridian line of the global grid running from the North Pole to the South Pole through Greenwich, England; starting point for measuring degrees of east and west longitude

relief changes in elevation over a given area of land

river large natural stream of water that runs through the land

sea large body of water completely or partly surrounded by land

seacoast land lying next to a sea or an ocean

sound broad inland body of water, often between a coastline and one or more islands off the coast

source (of a river) place where a river or stream begins, often in highlands

strait narrow stretch of water joining two larger bodies of water

tributary small river or stream that flows into a large river or stream; a branch of the river

upstream direction opposite the flow of a river; toward the source of a river or stream

valley area of low land between hills or mountains

volcano mountain created as liquid rock and ash erupt from inside the earth

Unit

Athlete at Olympic Games, Sydney, Australia

City of Hong Kong, China

NATIONAL GEOGRAPHIC SOCIETY

The World

Y ou are about to journey to dense rain forests, bleak deserts, bustling cities and marketplaces, and remote villages. You are entering the world of geography—the study of the earth and all its people. Imagine that you could visit any place in the world. Where would you want to go? What would you want to see?

Hot air balloon floating over cultivated fields, Egypt

NGS ONLINE
www.nationalgeographic.com/education

21

Looking at the Earth

The World and Its People — NATIONAL GEOGRAPHIC

To learn more about Earth's structure and landforms, view **The World and Its People Chapter 1** video.

Geography Online

Chapter Overview Visit the **Geography: The World and Its People** Web site at gwip.glencoe.com and click on **Chapter 1— Chapter Overviews** to preview information about Earth.

Thinking Like a Geographer

Guide to Reading

Main Idea

Geographers use various tools to understand the world.

Terms to Know

- geography
- landform
- environment
- region
- Global Positioning System (GPS)
- geographic information systems (GIS)

Places to Locate

- China
- Yangtze River
- United States
- Andes
- Antarctica
- California

Reading Strategy

Create a chart like this one and write three details or examples for each heading.

How Geographers View the World 1. 2. 3.
Tools of Geography 1. 2. 3.
Uses of Geography 1. 2. 3.

NATIONAL GEOGRAPHIC Exploring Our World

How would *you* go about making an accurate map of the world? Scientists decided the best way to map the earth was to see it from space. In February 2000, the space shuttle *Endeavour* used a radar camera to take pictures of the land below. By using radar, the camera was not hampered by clouds or darkness.

Why do geographers want to know exactly what the earth looks like? Think about the following: In **China,** the spring flooding of the **Yangtze** (YANG•SEE) **River** threatens people and crops every year. In 1998 the floods killed more than 4,000 people. After studying the land and the climate, officials in China's government sought solutions. They built dams to hold some of the floodwaters back. Recently, the number of people who died from the floods fell to 700.

This is just one example of how people around the world use geographic knowledge collected from various sources. **Geography** is the study of the earth in all its variety. When you study geography, you learn about the earth's land, water, plants, and animals. This is physical geography. You also study people—where they live, how they live, how they change and are influenced by their environment, and how different groups compare to one another. This is human geography.

◀ Skydiving over Key West, Florida

A Geographer's View of Place

Geographers look at major issues—like the flooding of the Yangtze, which affects millions of people. They also look at local issues—like where the best place is for a company to build a new store in town. Whether major or local, geographers try to understand both the physical and human characteristics, or features, of an issue.

Physical Characteristics   Geographers study places. They look at *where* something is located on the earth. They also try to understand what the place is *like.* They ask: What features make a place similar to or different from other places?

To answer this question, geographers identify the landforms of a place. Landforms are individual features of the land, like mountains and valleys. Geographers also look at water. Is the place near the ocean or on a river? Does it have plentiful or very little freshwater? They consider whether the soil will produce crops. They see how much rain the place usually receives and how hot or cold the area is. They find out whether the place has minerals, metals, trees, or other resources.

Human Characteristics   Geographers also look at the human characteristics of the people living in the place. Do many or only a few people live there? Do they live close together or far apart? Why? What kind of government do they have? What religion do they follow? What kinds of work do they do? What languages do they speak? From where did the people's ancestors come?

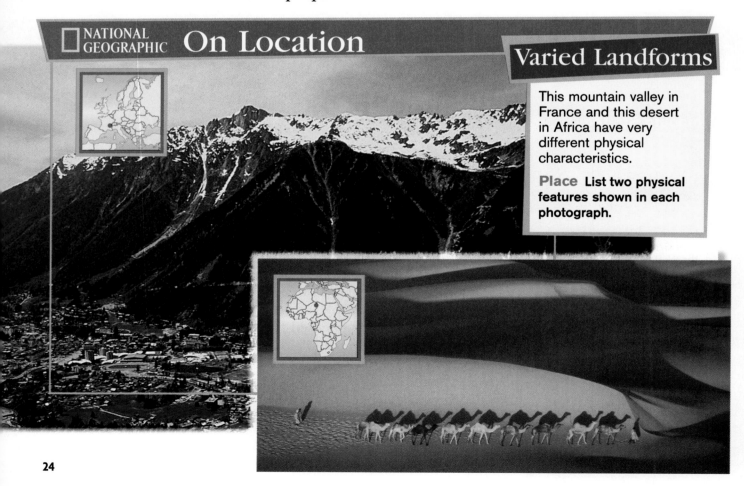

NATIONAL GEOGRAPHIC On Location

Varied Landforms

This mountain valley in France and this desert in Africa have very different physical characteristics.

Place List two physical features shown in each photograph.

People and Places Geographers are especially interested in how people interact with their environment, or natural surroundings. People can have a major impact on the environment. Remember how the Chinese built dams along the Yangtze? When they did so, they changed the way the river behaved in flood season.

Where people live often has a strong influence on *how* they live. People near the sea might catch fish and build ships for trade. Those living inland might farm or take up ranching. Of course, as the people of the world use more computers and other technology in their work, the surrounding environment may have less of an influence on what types of work they do.

Regions Geographers carefully study individual cities, rivers, and other landforms. They also look at the big picture, or how individual places relate to other places. In other words, geographers look at a region, or an area that shares common characteristics. Regions can be relatively small—like your state or town. They can also be huge—like all of the western **United States.** Some regions may even include several countries because these countries have similar environments, or because their people follow similar ways of life and speak the same language. The countries along the western side of South America are often discussed as a region—the Andean countries—because the **Andes,** a series of mountain ranges, run through all of them.

✓Reading Check **What do geographers study to determine the human characteristics of a place?**

The Tools of Geography

Geographers need tools to study people and places. Maps and globes are the main tools they use. As you read in the Geography Handbook on page 11, geographers use many different types of maps. Each type gives geographers a particular kind of information about a place.

Collecting Data for Mapping Earth How do geographers gather information so they can make accurate maps? One way is to take photographs from high above the earth. These are called LANDSAT photos and show details such as the shape of the land, what plants cover an area, and how land is being used. Special cameras can even reveal hidden information. Photos of **Antarctica** taken from radar cameras show rivers of ice 500 miles (805 km) long—all hidden by snow.

How do geographers accurately label the exact locations of places on a map? Believe it or not, the best way to find a location is from outer space. A group of satellites traveling around the earth make up the Global Positioning System (GPS). A GPS receiver is a special device that receives signals from these satellites. When the receiver is put at a location, the GPS satellite can tell the exact latitude and longitude of that place. As a result, a mapmaker can know where exactly on the earth the particular area is located. GPS devices are even installed in vehicles to help drivers find their way.

Looking at the Earth

Mt. Everest

GPS satellites locate and *measure* places on the earth. A GPS receiver placed on top of Mt. Everest, the tallest mountain in the world, showed that it is 7 feet (2.1 m) higher than people had previously thought.

Location Why is it important for geographers to know exactly where places are located on the earth?

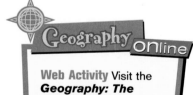

Web Activity Visit the *Geography: The World and Its People* Web site at gwip.glencoe.com and click on **Chapter 1— Student Web Activities** to learn more about geographic information systems.

Geographic Information Systems Today geographers use another powerful tool in their work—computers. Special computer software called geographic information systems (GIS) helps geographers gather many different kinds of information about the same place. After typing in all the data they collect, geographers use the software to combine and overlap the information on special maps.

As a result, different kinds of information about the same place can be compared. People then use that information to make decisions. For example, in the 1990s, GIS technology was used to help settle an argument over how to use resources in northern **California.** A logging company wanted to cut down parts of a forest. Environmental groups said that doing so would destroy the nesting areas of some rare birds. Geographers created maps that showed both the forest and the nesting sites—and then overlapped these maps. People could then see which areas had to be protected and which could be cut.

✓Reading Check What is the difference between GPS and GIS?

Uses of Geography

Have you ever gone on a long-distance trip in a car or taken a subway ride? If you used a road map or subway map to figure out where you were going, you were using geography. This is just one of the many uses of geographic information.

Geographic information is used in planning. Government leaders use geographic information to plan new services in their communities. They might plan how to handle disasters or how much new housing to allow in an area. Companies can see where people are moving in a region to make their plans for expanding.

In addition, the availability of geographic information helps people make sound decisions. Perhaps a question arises over whether a new building should be constructed. City leaders look at street use to see if the area can handle additional traffic. They make sure the area has the power, water, and sewage systems the building will need.

Businesses use geographic information for making decisions, too. Many businesses study population trends to see what areas need new products or services. Some businesses offer geographic information to their customers. Suppose you were looking for an apartment to rent. Some realtors have a computer program that can identify all the apartments of a certain size and price in an area. The program can create a map so you can see exactly where each apartment is located.

Finally, geographic information helps people manage resources. Many natural resources, such as oil or coal, are available only in limited supply. Geographic information can help people find more of those resources. Other resources, such as trees or water, can be replaced or renewed. People can use geographic information to show them how to manage these resources so they are not all used up.

✓ Reading Check Why do people have to manage resources carefully?

Assessment

Defining Terms
1. **Define** geography, landform, environment, region, Global Positioning System (GPS), geographic information systems (GIS).

Recalling Facts
2. **Place** What two kinds of characteristics of a place do geographers study?
3. **Technology** What are the main tools of geography?
4. **Human/Environment Interaction** What are three uses for geography?

Critical Thinking
5. **Understanding Cause and Effect** How have the physical characteristics of your region affected the way people live there?
6. **Categorizing Information** Give five examples of regions. Begin with an area near you that shares common characteristics, then look for larger and larger regions.

Graphic Organizer
7. **Organizing Information** Draw a diagram like this one. In the center, write the name of a place you would like to visit. In the outer ovals, identify the types of geographic information you would like to learn about this place.

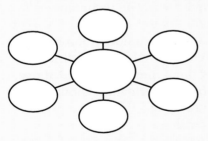

Applying Geography Skills

8. **Analyzing Maps** Find Egypt on the map on page RA23 of the **Reference Atlas.** Along what physical feature do you think most Egyptians live? Why? Turn to the population map of Egypt on page 13 of the **Geography Handbook** to see if you are correct.

Geographic Information Systems

What if a farmer could save money by applying fertilizer only to the crops that needed it? Today, thanks to computer technology called geographic information systems (GIS), farmers can do just that.

The Technology

Geographic information systems (GIS) use computer software to combine and display a wide range of information about an area. GIS programs start with a map showing a specific location on the earth. This map is then linked with other information about that same place, such as satellite photos, amounts of rainfall, or where houses are located.

Think of geographic information systems as a stack of overhead transparencies. Each transparency shows the same general background but highlights different information. The first transparency may show a base map of an area. Only the borders may appear. The second transparency may show only rivers and highways. The third may highlight mountains and other physical features, buildings, or cities.

In a similar way, GIS technology places layers of information onto a base map. It can then switch each layer of information on or off, allowing data to be viewed in many different ways. In the case of the farmer mentioned above, GIS software combines information about soil type, plant needs, and last year's crop to pinpoint exact areas that need fertilizer.

How It Is Used

GIS technology allows users to quickly pull together data from many different sources and construct maps tailored to specific needs. This helps people analyze past events, predict future scenarios, and make sound decisions.

A person who is deciding where to build a new store can use GIS technology to help select the best location. The process might begin with a list of possible sites. The store owner gathers information about the areas surrounding each place. This could include shoppers' ages, incomes, and educations; where shoppers live; traffic patterns; and other stores in the area. The GIS software then builds a computerized map composed of these layers of information. The store owner can use the information to decide on a new store location.

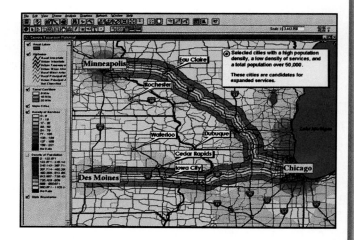

Graphic image created using ArcView® GIS software, and provided courtesy of Environmental Systems Research Institute, Inc.

➤ Making the Connection

1. What is GIS technology?
2. How do GIS programs analyze data in a variety of ways?
3. **Drawing Conclusions** How could a school district use GIS technology to locate the best place to add a new school?

The Earth in Space

Guide to Reading

Main Idea

The earth has life because of the sun. The earth has different seasons because of the way it tilts and revolves around the sun.

Terms to Know

- solar system
- orbit
- atmosphere
- axis
- revolution
- leap year
- summer solstice
- winter solstice
- equinox

Places to Locate

- Earth
- sun
- moon

Reading Strategy

Draw a diagram like this one and list three facts about the sun in the first column. In the second, write how these facts contribute to life on the earth.

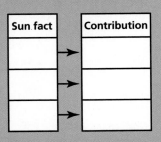

Sun fact	Contribution
	→
	→
	→

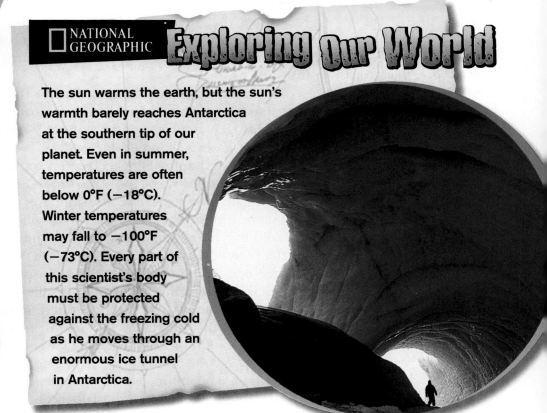

NATIONAL GEOGRAPHIC

Exploring Our World

The sun warms the earth, but the sun's warmth barely reaches Antarctica at the southern tip of our planet. Even in summer, temperatures are often below 0°F (−18°C). Winter temperatures may fall to −100°F (−73°C). Every part of this scientist's body must be protected against the freezing cold as he moves through an enormous ice tunnel in Antarctica.

The sun's heat provides life on our planet. **Earth,** eight other planets, and thousands of smaller bodies all revolve around the **sun.** Together with the sun, these bodies form the solar system. Look at the diagram of the solar system on page 30. As you can see, Earth is the third planet from the sun.

The Solar System

Each planet travels along its own path, or orbit, around the sun. The paths they travel are ellipses, which are like stretched-out circles. Each planet takes a different amount of time to complete one full trip around the sun. Earth makes one trip in 365¼ days. Mercury orbits the sun in just 88 days. Far-off Pluto takes almost 250 years!

Planets can be classified into two types—those that are like Earth and those that are like Jupiter. Earthlike planets are Mercury, Venus, Mars, and Pluto. These planets are solid and small. They have few or no moons. They also rotate, or spin, fairly slowly.

The other four planets—Jupiter, Saturn, Neptune, and Uranus—are huge. Uranus, the smallest of the four, is 15 times the size of Earth.

Solar Eclipse

One of the most spectacular sights in the sky is a solar eclipse. This event takes place when the moon passes between Earth and the sun and covers some or all of the sun. The photograph here shows a total eclipse, when the moon completely blocks the sun. When the moon blocks the sun's light, a large shadow is cast on part of Earth.

These planets are more like balls of gas than rockier Earthlike planets. They spin rapidly and have many moons. Surrounding each one is a series of rings made of bits of rock and dust.

Sun, Earth, and Moon   The sun—about 93 million miles (150 million km) from Earth—is made mostly of intensely hot gases. Reactions that occur inside the sun make it as hot as 27 million degrees Fahrenheit (about 15 million degrees Celsius). As a result, the sun gives off light and warmth. Life on Earth could not exist without the sun.

The layer of air surrounding Earth—the **atmosphere**—also supports life. This cushion of gases measures about 1,000 miles (1,609 km) thick. Nitrogen and oxygen form about 99 percent of the atmosphere, with other gases making up the rest.

Humans and animals need oxygen to breathe. The atmosphere is important in other ways, too. This protective layer holds in enough of the sun's heat to make life possible, just as a greenhouse keeps in enough heat to protect plants. Without this protection, Earth would be too cold for most living things. At the same time, the atmosphere also reflects some heat back into space with the result that Earth does not become too warm. Finally, the atmosphere shields living things. It screens out some rays from the sun that are dangerous. You will learn more about the atmosphere in Chapter 2.

Earth's nearest neighbor in the solar system is its **moon.** The moon orbits Earth, taking about 30 days to complete each trip. A cold, rocky sphere, the moon has no water and no atmosphere. The moon also gives off no light of its own. When you see the moon shining, it is actually reflecting light from the sun.

NATIONAL GEOGRAPHIC

The Solar System

Analyzing the Diagram

Earth and eight other planets in our solar system travel around the sun.

Movement Between what two planets' orbits is Earth's orbit?

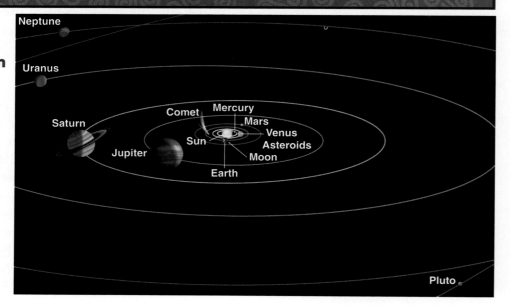

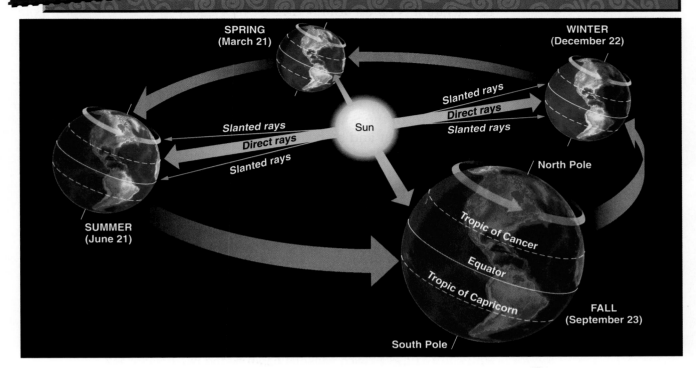

SPRING (March 21)

WINTER (December 22)

Slanted rays

Direct rays

Slanted rays

Sun

Slanted rays

Direct rays

Slanted rays

North Pole

SUMMER (June 21)

Tropic of Cancer

Equator

Tropic of Capricorn

FALL (September 23)

South Pole

Earth's Movement Like all the planets, Earth rotates, or spins, on its axis. The **axis** is an imaginary line that runs through Earth's center between the North and South Poles. Earth takes 24 hours to finish one complete spin on its axis. As a result, one day is 24 hours. As Earth turns, different parts of the planet are in sunlight or in darkness. The part facing the sun has day, and the part facing away has night.

Earth has another motion, too. The planet makes one **revolution**, or complete orbit around the sun, in 365¼ days. This period is what we define as one year. Every four years, the extra one-fourths of a day are combined and added to the calendar as February 29. A year that contains one of these extra days is called a **leap year.**

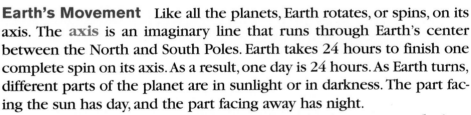 Reading Check How does Earth's orbit affect you?

The Sun and the Seasons

Earth is tilted 23½ degrees on its axis. As a result, seasons change as Earth makes its year-long orbit around the sun. To see why this happens, look at the four globes in the diagram above. Notice how sunlight falls directly on the northern or southern halves of Earth at different times of the year. Direct rays from the sun bring more warmth than the slanted rays. When the people in a hemisphere receive those direct rays from the sun, they enjoy the warmth of summer. When they receive only indirect rays, they experience winter, which is colder.

 Analyzing the Diagram

Because Earth is tilted, different areas receive direct rays from the sun at different times of the year.

Movement How does this fact cause changes in seasons?

Looking at the Earth

31

Solstices and Equinoxes Four days in the year have special names because of the position of the sun in relation to Earth. These days mark the beginnings of the four seasons. On or about June 21, the North Pole is tilted toward the sun. On noon of this day, the sun appears directly overhead at the line of latitude called the Tropic of Cancer (23½°N latitude). This day is the **summer solstice,** the day in the Northern Hemisphere with the most hours of sunlight and the fewest hours of darkness. It is the beginning of summer—but only in the Northern Hemisphere. In the Southern Hemisphere, it is the day with the fewest hours of sunlight and marks the beginning of winter.

Six months later—on or about December 22—the North Pole is tilted away from the sun. At noon, the sun's direct rays strike the line of latitude known as the Tropic of Capricorn (23½°S latitude). In the Northern Hemisphere, this day is the **winter solstice**—the day with the fewest hours of sunlight. This same day, though, marks the beginning of summer in the Southern Hemisphere.

Spring and autumn begin midway between the two solstices. These are the **equinoxes,** when day and night are of equal length in both hemispheres. On or about March 21, the *vernal equinox* (spring) occurs. On or about September 23, the *autumnal equinox* occurs. On both of these days, the noon sun shines directly over the Equator.

Reading Check Which seasons begin on the two equinoxes?

Assessment

Defining Terms
1. **Define** solar system, orbit, atmosphere, axis, revolution, leap year, summer solstice, winter solstice, equinox.

Recalling Facts
2. **Region** What bodies make up the solar system?
3. **Science** List two gases that are found in the atmosphere.
4. **Movement** What two motions does Earth make in space?

Critical Thinking
5. **Analyzing Information** How does the position of Earth determine whether a day is one of the solstice or equinox days?
6. **Summarizing Information** In a paragraph, describe why you experience seasonal changes.

Graphic Organizer
7. **Organizing Information** Draw two diagrams like those below. In the first, list the effects of Earth's rotation on human, plant, and animal life. In the second, list effects if Earth were to stop rotating.

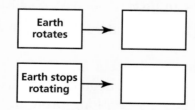

Applying Geography Skills

8. **Analyzing Diagrams** Look at the diagram on page 31. When the sun's direct rays hit the Tropic of Capricorn, what season is it in the Northern Hemisphere?

Geography Skill

Using a Map Key

To understand what a map is showing, you must read the **map key,** or legend. The map key explains the meaning of special colors, symbols, and lines on the map.

Learning the Skill

Colors in the map key may represent different elevations or heights of land, climate areas, or languages. Lines may stand for rivers, streets, or boundaries.

Maps also have a **compass rose** showing directions. The **cardinal directions** are north, south, east, and west. North and south are the directions of the North and South Poles. If you stand facing north, east is the direction to your right. West is the direction on your left. The compass rose might also show **intermediate directions,** or those that fall between the cardinal directions. For example, the intermediate direction *northeast* falls between north and east. To use a map key, follow these steps:

- Read the map title.
- Read the map key to find out what special information it gives.

- Find examples of each map key color, line, or symbol on the map.
- Use the compass rose to identify the four cardinal directions.

Practicing the Skill

Look at the map of Washington, D.C., below to answer the following questions.

1. What does the red square represent?
2. What does the blue square represent?
3. Does the Washington Monument lie east or west of the Lincoln Memorial?
4. From the White House, in what direction would you go to get to the Capitol?

Applying the Skill

Find a map in a newspaper or magazine. Use the map key to explain three things the map is showing.

GO TO

Practice key skills with **Glencoe Skillbuilder Interactive Workbook, Level 1.**

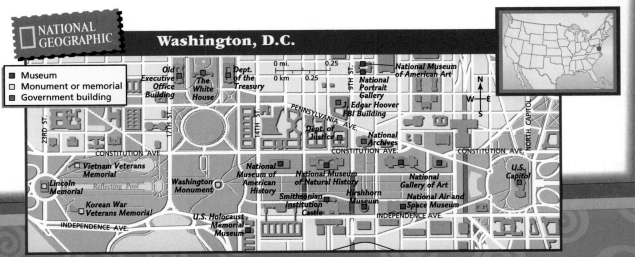

NATIONAL GEOGRAPHIC

Washington, D.C.

■ Museum
□ Monument or memorial
■ Government building

The Earth's Structure

Guide to Reading

Main Idea

Forces both inside the earth and on its surface affect the shape of the land.

Terms to Know

- core
- mantle
- magma
- crust
- continent
- plate tectonics
- earthquake
- tsunami
- fault
- weathering
- erosion
- glacier

Places to Locate

- South America
- Africa
- Himalaya
- Asia

Reading Strategy

Create a chart like this one. Write five forces that change the shape of the land. Then write five effects these forces can have on the earth.

Forces	Effects
→	
→	
→	
→	
→	

NATIONAL GEOGRAPHIC Exploring Our World

Forces beneath the earth's surface shape the land and the lives of the people who live on it. Here in the Azores Islands, a volcano makes cooking easy. People wrap pots of meat and vegetables in a cloth and bury the bundle in a hole where heat from deep inside the earth rises to the surface. The temperature reaches 200°F (93°C), which is hot enough to steam the food.

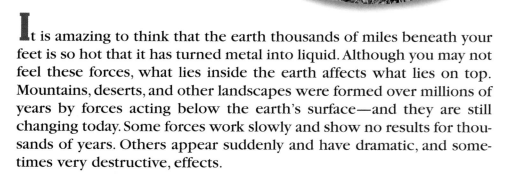

It is amazing to think that the earth thousands of miles beneath your feet is so hot that it has turned metal into liquid. Although you may not feel these forces, what lies inside the earth affects what lies on top. Mountains, deserts, and other landscapes were formed over millions of years by forces acting below the earth's surface—and they are still changing today. Some forces work slowly and show no results for thousands of years. Others appear suddenly and have dramatic, and sometimes very destructive, effects.

Inside the Earth

Scientists have been able to study only the top layer of the earth firsthand. Still, they have developed a picture of what lies inside the earth. They have found that the inside of the earth has three layers—the core, the mantle, and the crust. Have you ever seen a cantaloupe cut in half? The earth's core is like the center of the cantaloupe, where you find the seeds. The mantle is like the flesh of the fruit, sandwiched between the core and the skin. The earth's crust, or topmost layer, is like the melon's skin. Let us take a closer look at these three layers.

In the center of the earth is a dense **core** of hot iron mixed with other metals. The very center is solid, but the outer core is so hot that the metal has melted into liquid. Surrounding the core is the **mantle,** a layer of rock about 1,800 miles (2,897 km) thick. Like the core, the mantle also has two parts. The section nearest the core remains solid, but the rock in the outer mantle sometimes melts. If you have seen photographs of an active volcano, then you have seen this melted rock, called **magma,** when it flowed to the surface in a volcanic eruption.

The uppermost layer of the earth, the **crust,** is relatively thin. It reaches only 31 to 62 miles (50 to 100 km) deep. The crust includes the ocean floors. It also includes seven massive land areas known as **continents.** The crust is thinnest on the ocean floor. It is thicker below the continents. Turn to the map on page 41 to see where the earth's seven continents are located.

Reading Check What layer of the earth is thinnest?

NATIONAL GEOGRAPHIC **Earth's Layers**

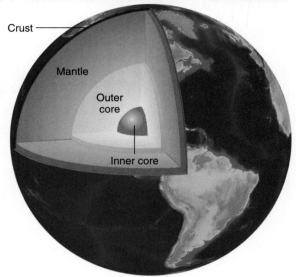

Crust

Mantle

Outer core

Inner core

Analyzing the Diagram

Hot rock and metal—some of it liquid—fill the center of the earth.

Region What is the innermost layer inside the earth called? In what layer do you find the continents?

Forces Beneath the Earth's Crust

Most of you have probably watched science shows about earthquakes and volcanoes. You have probably also seen news on television discussing the destruction caused by earthquakes. These events result from forces at work inside the earth.

Plate Movements Scientists have developed a theory about the earth's structure called **plate tectonics.** This theory states that the crust is not an unbroken shell but consists of plates, or huge slabs of rock, that move. The plates float on top of the liquid rock in the upper part of the mantle. They move—but often in different directions. Oceans and continents sit on these gigantic plates, as the diagram on page 36 shows.

Have you ever noticed that the eastern part of **South America** seems to fit into the western side of **Africa?** That is because these two continents were once joined together in a landmass that scientists call Pangaea. Millions of years ago, however, the continents moved apart. Tectonic activity caused them to move. The plates are still moving today, but they move so slowly that you do not feel it. The plate under the Pacific Ocean moves to the west at the rate of about 4 inches (10 cm) a year. That is about the same rate that a man's beard grows. The plate along the western edge of South America moves east at the rate of about 1.8 inches (5 cm) a year. That is a little faster than your fingernails grow. Turn to page 45 to see what Pangaea looked like before and after it experienced this movement, known as *continental drift.*

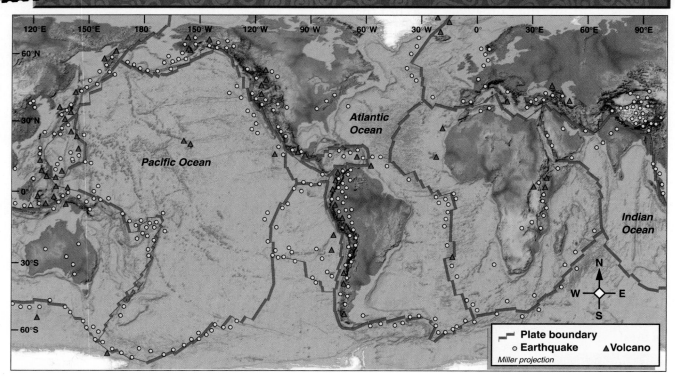

120°E 150°E 180° 150°W 120°W 90°W 60°W 30°W 0° 30°E 60°E 90°E

60°N

30°N

0°

30°S

60°S

Pacific Ocean

Atlantic Ocean

Indian Ocean

N
W   E
S

Plate boundary
○ **Earthquake** ▲ **Volcano**
Miller projection

Analyzing the Diagram

Most of North America sits on one plate.

Region What pattern do you see among plate boundaries, earthquakes, and volcanoes?

When Plates Meet The movements of the earth's plates have actually shaped the surface of the earth. Sometimes the plates spread, or pull away from each other. That type of tectonic action separated South America and Africa millions of years ago. Sometimes, though, the plates push against each other. When this happens, one of three events occurs, depending on what kinds of plates are involved.

If two continental plates smash into each other, the collision produces high mountain ranges. This kind of collision produced the **Himalaya** in South **Asia.**

If a continental plate and an ocean plate move against each other, the thicker continental plate slides over the thinner ocean plate. The edge of the lower plate melts into the mantle, causing molten rock to build up and perhaps erupt in a volcano. Another result may occur from the pressure that builds up between the two sliding plates. This pressure may cause one plate to move suddenly. The result is called an **earthquake,** or a violent and sudden movement of the earth's crust.

Earthquakes can be very damaging to both physical structures and human lives. They can collapse buildings, destroy bridges, and break apart underground water or gas pipes. Undersea earthquakes can cause huge waves called **tsunamis** (tsu•NAH•mees). These waves may reach as high as 98 feet (30 m). Such waves can cause severe flooding of coastal towns.

Sometimes two plates do not meet head-on but move alongside each other. To picture this, put your hands together and then move them in opposite directions. When this action occurs in the earth, the two plates slide against each other. This movement creates **faults,** or cracks in the earth's crust. Violent earthquakes can happen near these faults. In 1988, for example, an earthquake struck the country of Armenia. About 25,000 people were killed, and another 500,000 lost their homes. One of the most famous faults in the United States is the San Andreas Fault in California. The earth's movement along this fault caused a severe earthquake in San Francisco in 1906 and another less serious quake in 1989.

Reading Check What happens when two continental plates collide?

Forces Shaping Landforms

The forces under the earth's crust that move tectonic plates cause volcanoes and earthquakes to change the earth's landforms. Once formed, however, these landforms will continue to change because of forces that work on the earth's surface.

Weathering Weathering is the process of breaking surface rock into boulders, gravel, sand, and soil. Water and frost, chemicals, and even plants cause weathering. Water seeps into cracks of rocks and then freezes. As it freezes, the ice expands and splits the rock. Sometimes entire sides of cliffs fall off because frost has wedged the rock apart. Chemicals, too, cause weathering when acids in air pollution mix with rain and fall back to the earth. The chemicals eat away the surfaces of stone structures and natural rocks. Even tiny seeds that fall into cracks can spread out roots, causing huge boulders to eventually break apart.

Architecture

In earthquake-prone parts of the world, engineers design new buildings to stand up to tremors, or shaking of the earth. Flexible structures allow buildings to sway rather than break apart. Placing a building on pads or rollers cushions the structure from the motion of the ground. Some so-called intelligent buildings automatically respond to tremors, shifting their weight or tightening and loosening joints.

Looking Closer How can studying earthquake-damaged buildings help designers improve future construction?

San Francisco, California, 1989 ▶

Erosion Erosion is the process of wearing away or moving weathered material. Water, wind, and ice are the greatest factors that erode, or wear away, surface material. Moving water in oceans, rivers, streams, and rain can erode even the hardest stone over time. Rainwater working its way to streams and rivers picks up and moves soil and sand. These particles make the river water similar to a giant scrub brush that grinds away at riverbanks and any other surface in the water's path.

Wind is also a major cause of erosion as it lifts weathered soil and sand. The areas that lose soil often become unable to grow crops and support life. The areas that receive the windblown soil often benefit from the additional nutrients to the land. When wind carries sand, however, it acts as sandpaper. Rock and other structures are carved into smooth shapes.

The third cause of erosion is ice. Giant, slow-moving sheets of ice are called **glaciers.** Forming high in mountains, glaciers change the land as they inch over it. Similar to windstorms, glaciers act like sandpaper as they pick up and carry rocks down the mountainside, grinding smooth everything beneath them. Some glaciers are thousands of feet thick. The weight and pressure of thousands of feet of ice also cut deep valleys at the mountain's base.

✓ Reading Check **List three things that can cause weathering.**

Assessment
Section 3

Defining Terms
1. Define core, mantle, magma, crust, continent, plate tectonics, earthquake, tsunami, fault, weathering, erosion, glacier.

Recalling Facts
2. Region What are the three layers inside the earth?

3. Movement In what three ways can tectonic plates move?

4. Science What are the three greatest factors that cause erosion?

Critical Thinking
5. Making Comparisons How does water play a role in both processes of weathering and erosion?

6. Understanding Cause and Effect How does erosion hurt some areas yet benefit others?

Graphic Organizer
7. Organizing Information Draw a diagram like this one, then label the inner arrows with inside forces that shape landforms. Label the outer arrows with surface forces that change the earth's landforms.

Applying Geography Skills

8. Analyzing Diagrams Look at the diagram of tectonic plate boundaries on page 36. Why might it be a problem that most of the world's population lives along the western edge of the Pacific Ocean?

Landforms

Guide to Reading

Main Idea

Landforms in all their variety affect how people live.

Terms to Know

- elevation
- plain
- plateau
- isthmus
- peninsula
- island
- continental shelf
- trench
- strait
- channel
- delta

Places to Locate

- Mt. Everest
- North European Plain
- Plateau of Tibet
- Grand Canyon
- Mariana Trench
- Strait of Magellan

Reading Strategy

Draw a diagram like this one. In each of the surrounding circles, write the name of a landform and a fact about it.

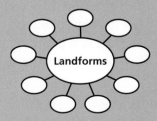

Landforms

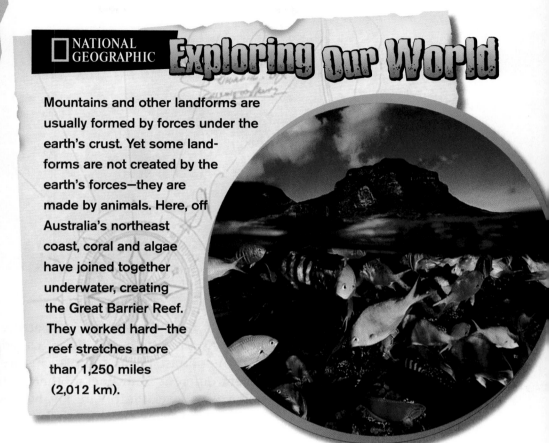

NATIONAL GEOGRAPHIC Exploring Our World

Mountains and other landforms are usually formed by forces under the earth's crust. Yet some landforms are not created by the earth's forces—they are made by animals. Here, off Australia's northeast coast, coral and algae have joined together underwater, creating the Great Barrier Reef. They worked hard—the reef stretches more than 1,250 miles (2,012 km).

The earth's land surface consists of seven continents. These are North America, South America, Europe, Africa, Asia, Australia, and Antarctica. All have a variety of landforms, or individual features of the land—even icy Antarctica.

Types of Landforms

Look at the illustration on page 18 of the **Geography Handbook.** Notice the many different forms that the land may take. Which ones are familiar to you? Which ones are new to you?

On Land Mountains are huge towers of rock formed by the collision of the earth's tectonic plates or by volcanoes. Some mountains may be a few thousand feet high. Others can soar higher than 20,000 feet (6,096 m). The world's tallest mountain is **Mt. Everest,** located in South Asia's Himalaya mountain ranges. It towers at 29,035 feet (8,850 m)—nearly 5.5 miles (8.9 km) high.

Mountains often have high peaks and steep, rugged slopes. Hills are lower and more rounded, though they are still higher than the land

Valleys vs. Canyons

The Great Rift Valley in Africa is surrounded by mountains (above). Canyons, like the Grand Canyon in Arizona (right), are carved from plateaus.

Place How are valleys and canyons similar?

around them. Some hills form at the foot, or base, of mountains. As a result, these hills are called foothills.

In contrast, plains and plateaus are mostly flat. What makes them different from one another is their elevation, or height above sea level. Plains are low-lying stretches of flat or gently rolling land. Many plains reach from the middle of a continent to the coast. The **North European Plain** is an example. Plateaus are also flat but have higher elevation. With some plateaus, a steep cliff forms on one side where the plateau rises above nearby lowlands. With others, such as the **Plateau of Tibet** in Asia, the plateau is surrounded by mountains.

Between mountains and hills lie valleys. A valley is a long stretch of land lower than the land on either side. You often find rivers at the bottom of valleys. Canyons are steep-sided lowlands that rivers have cut through a plateau. One of the most famous canyons is the **Grand Canyon** in Arizona. For millions of years, the Colorado River flowed over a plateau and carved through rock, forming the Grand Canyon.

Geographers describe some landforms by their relationship to larger land areas or to bodies of water. An isthmus is a narrow piece of land that connects two larger pieces of land. A peninsula is a piece of land with water on three sides. A body of land smaller than a continent and completely surrounded by water is an island.

Under the Oceans If you were to explore the oceans, you would see landforms under the water that are similar to those on land. The map on page RA32 of the **Reference Atlas** shows you what the ocean floors look like. Off each coast of a continent lies a plateau called a continental shelf that stretches for several miles underwater. At the edge of the shelf, steep cliffs drop down to the ocean floor.

Tall mountains and very deep valleys line the ocean floor. Valleys here are called trenches, and they are the lowest spots in the earth's crust. The deepest one, in the western Pacific Ocean, is called the **Mariana Trench.** This trench plunges 35,840 feet (10,924 m) below sea level. How deep is that? If Mt. Everest were placed into this trench, the mountain would have to grow 1.3 miles (2 km) higher just to reach the ocean's surface.

Landforms and People   Humans have settled on all types of landforms. Some people live at high elevations in the Andes mountain ranges of South America. The people of Bangladesh live on a low coastal plain. Farmers in Ethiopia work the land on a plateau called the Ethiopian Highlands.

Why do people decide to live in a particular area? Climate—the average temperature and rainfall of a region—is one reason. You will read more about climate in the next chapter. The availability of resources is another reason. People settle where they can get freshwater and where they can grow food, catch fish, or raise animals. They might settle in an area because it has good supplies of useful items such as trees for building, iron for manufacturing, or petroleum for making energy. You will read more about resources in Chapter 3.

✔ **Reading Check**   How are plains and plateaus similar? How are they different?

Applying Map Skills

1. What are the names of the seven large landmasses on the earth?

2. What are the earth's four major oceans?

Find NGS online map resources @ www.nationalgeographic.com/maps

NATIONAL GEOGRAPHIC

World Continents and Oceans

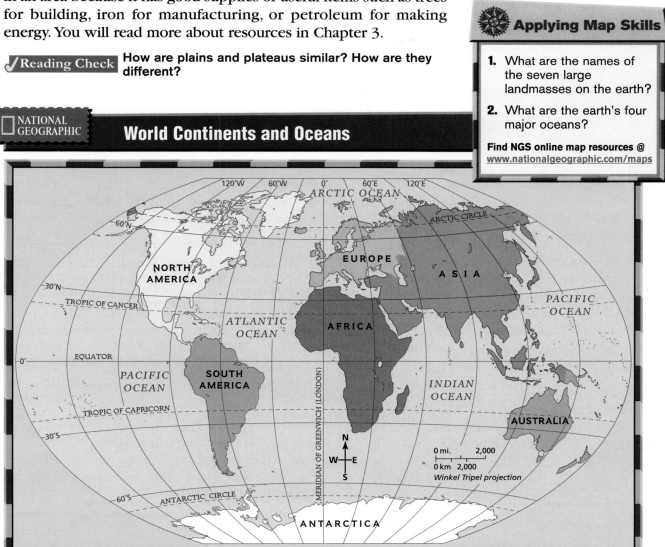

Bodies of Water

About 70 percent of the earth's surface is water. Most of that water is salt water, which people and most animals cannot drink. Only a small percentage is freshwater, which is drinkable.

Oceans, consisting of salt water, are the earth's largest bodies of water. Smaller bodies of salt water are connected to oceans but are at least partly enclosed by land. These bodies include seas, gulfs, and bays.

Two other kinds of water form passages that connect two larger bodies of water. A **strait** is a narrow body of water between two pieces of land. The **Strait of Magellan** flows between the southern tip of South America and an island called Tierra del Fuego (tee•EHR•uh DEHL fyu•AY•goh). This strait connects the Atlantic and the Pacific Oceans. A wider passage is called a **channel.** The Mozambique Channel separates southeastern Africa from the island of Madagascar.

Bodies of freshwater appear on the world's continents and islands. They include larger bodies like lakes and rivers as well as smaller ones such as ponds and streams. The point at which a river originates—usually high in the mountains—is called its source. The mouth of a river is where it empties into another body of water. As you learned in Section 3, rivers carry soil and sand. They eventually deposit this soil at the mouth, which builds up over time to form a **delta.**

✓ Reading Check **What is the difference between the source and the mouth of a river?**

Assessment

Defining Terms
1. **Define** elevation, plain, plateau, isthmus, peninsula, island, continental shelf, trench, strait, channel, delta.

Recalling Facts
2. **Place** What is the difference between mountains and hills?
3. **Place** How are straits and channels similar? How are they different?
4. **Culture** What are two reasons people decide to settle in a particular area?

Critical Thinking
5. **Analyzing Information** What two landforms are created by rivers?
6. **Making Inferences** Why do people often settle on the edges of rivers?

Graphic Organizer
7. **Organizing Information** Make a chart like this and give three examples for each item.

Landforms			
Landforms Under the Ocean			
Types of Bodies of Water			

Applying Geography Skills

8. **Analyzing Maps** Look at the world map on page RA4 of the **Reference Atlas.** Find at least two of the following: plain, plateau, isthmus, peninsula, island, strait, and channel. List their specific names.

Reading Review

Section 1 — Thinking Like a Geographer

Terms to Know
geography
landform
environment
region
Global Positioning System (GPS)
geographic information systems (GIS)

Main Idea
Geographers use various tools to understand the world.
✓ **Place** Geographers study the physical and human characteristics of places.
✓ **Culture** Geographers are especially interested in how people interact with their environment.
✓ **Technology** To study the earth, geographers use maps, globes, photographs, the Global Positioning System, and geographic information systems.
✓ **Economics** People can use information from geography to plan, make decisions, and manage resources.

Section 2 — The Earth in Space

Terms to Know
solar system   leap year
orbit   summer solstice
atmosphere   winter solstice
axis   equinox
revolution

Main Idea
The earth has life because of the sun. The earth has different seasons because of the way it tilts and revolves around the sun.
✓ **Science** The sun's light and warmth allow life to exist on Earth.
✓ **Science** The atmosphere is a cushion of gases that protects Earth and provides air to breathe.
✓ **Movement** Earth spins on its axis to make day and night.
✓ **Movement** The tilt of Earth and its revolution around the sun cause changes in seasons.

Section 3 — The Earth's Structure

Terms to Know
core   plate tectonics
mantle   tsunami
magma   fault
crust   weathering
continent   erosion
earthquake   glacier

Main Idea
Forces both inside the earth and on its surface affect the shape of the land.
✓ **Region** Earth has an inner and outer core, a mantle, and a crust.
✓ **Movement** The continents are on large plates of rock that move.
✓ **Movement** Earthquakes and volcanoes can reshape the land.
✓ **Movement** Wind, water, and ice can change the look of the land.

Section 4 — Landforms

Terms to Know
elevation   continental shelf
plain   trench
plateau   strait
isthmus   channel
peninsula   delta
island

Main Idea
Landforms in all their variety affect how people live.
✓ **Location** Mountains, plateaus, valleys, and other landforms are found on land and under the oceans.
✓ **Science** About 70 percent of the earth's surface is water.
✓ **Culture** People have adapted in order to live on various landforms.

Looking at the Earth

Chapter 1 Assessment and Activities

Using Key Terms

Match the terms in Part A with their definitions in Part B.

A.

1. elevation
2. landform
3. summer solstice
4. plate tectonics
5. geographic information systems
6. Global Positioning System
7. erosion
8. equinox
9. fault
10. weathering

B.

a. height above sea level
b. wearing away of the earth's surface
c. theory that the earth's crust consists of huge slabs of rock that move
d. a group of satellites around the earth
e. special software that helps geographers gather and use information
f. when day and night are of equal length
g. a process that breaks surface rocks into gravel, sand, or soil
h. a crack in the earth's crust
i. the day with the most hours of sunlight
j. particular features of the land

Reviewing the Main Ideas

Section 1 Thinking Like a Geographer

11. **Place** Give three examples of the physical characteristics of a place.
12. **Region** How is a region different from a place?
13. **Human/Environment Interaction** Give an example of how people use geographic knowledge.

Section 2 The Earth in Space

14. **Region** How many planets are in the solar system?
15. **Movement** What movement of Earth causes day and night?
16. **Movement** How does Earth's revolution around the sun relate to the seasons?

Section 3 The Earth's Structure

17. **Movement** How do the plates in the earth's crust move?
18. **Movement** Give an example of erosion.

Section 4 Landforms

19. **Place** Which has a higher elevation—plains or plateaus?
20. **Movement** What two reasons lead people to settle in a particular region?

NATIONAL GEOGRAPHIC The World

Place Location Activity

On a separate sheet of paper, match the letters on the map with the numbered places listed below.

1. North America
2. Pacific Ocean
3. Africa
4. South America
5. Antarctica
6. Australia
7. Atlantic Ocean
8. Asia

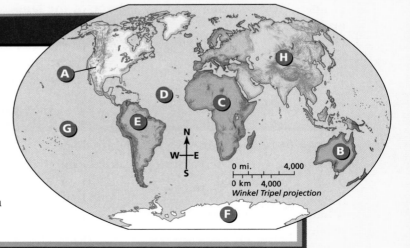

Self-Check Quiz Visit the *Geography: The World and Its People* Web site at gwip.glencoe.com and click on **Chapter 1— Self-Check Quizzes** to prepare for the Chapter Test.

Critical Thinking

21. **Making Comparisons** Why do people in Australia snow-ski during the Northern Hemisphere's summer months?

22. **Understanding Cause and Effect** Create a diagram like this one. In the left box, write "plate movements." In the right box, describe the effect that this force has on the earth. Draw five more pairs of boxes and do the same for the other forces that shape the earth: earthquakes, volcanoes, weathering, and erosion.

GeoJournal Activity

23. **Writing a Paragraph** Write a description of how the earth looks in your area. Then sketch a picture of your area showing all the landforms and bodies of water you described. Label each landform and body of water that you show.

Mental Mapping Activity

24. **Focusing on the Region** Draw a simple outline map of the earth, then label the following:

- core
- mantle
- crust
- atmosphere

Technology Skills Activity

25. **Building a Database** Use a word processing program to make a database like the following. In each row of the left column, list one of the following landforms: mountain, hill, plain, plateau, valley, canyon, isthmus, peninsula, and island. In the other columns, write a fact about each landform and give an example.

Landform	Facts	Example

Standardized Test Practice

Directions: Study the maps below, then answer the question that follows.

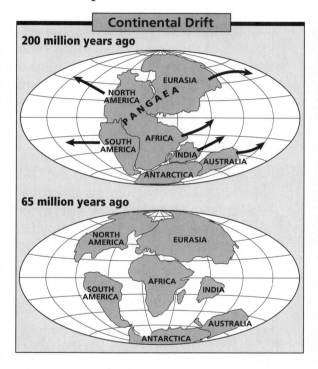

1. What "supercontinent" do many scientists believe existed 200 million years ago?

   A Eurasia
   B Pangaea
   C Gondwana
   D Antarctica

Test-Taking Tip: Use information *on the maps* to answer this question. Read the title above the maps and then the two subtitles. If you reread the question, you see it is asking about a certain time period. Make sure you use the correct map above to answer the question.

45

GeoLAB ACTIVITY

Tornadoes: Swirling Fury

1 Background

It is a stormy, humid day. You see a funnel of twisting, black clouds coming down from the sky and stirring up dust on the ground below. Tornado warning: Take cover! The United States has more tornadoes each year than anywhere else on the planet. Over the central part of our country—nicknamed Tornado Alley—warm, moist air from the Gulf of Mexico meets cold, dry air sweeping down from Canada. These are perfect conditions for tornadoes to form. Learn more about these storms by making and experimenting with a model of a tornado.

2 Materials

- 2 empty 2-liter clear soda bottles
- black marker
- 1 rubber washer that is the same size as the bottle openings
- duct tape
- water
- paper towels
- glitter

A Tornado Hits Oakfield, Wisconsin

Believe It or Not!

The United States has the most violent weather in the world. Our country experiences 1,000 tornadoes, 5,000 floods, and 10,000 strong thunderstorms in an average year.

3 What to Do

1. Use the marker to label one bottle "A" and the other "B."
2. Fill Bottle A three-fourths full of water. Dry the top and outside of the bottle completely. Add a small amount of glitter to the water.
3. Tape the rubber washer to the opening of Bottle A. Do not cover the washer hole with tape.
4. Put Bottle B upside down on Bottle A. Tape the mouths of both bottles together. Make sure the seal is tight and use several layers of tape.
5. Turn the bottles over so that Bottle A is on top. Then quickly swirl the bottles a couple of times, as you would stir something in a bowl.
6. Set the bottles on a table with Bottle A on top and observe the "tornado." The spinning funnel you see is called a *vortex*.
7. Repeat the experiment several times. Compare how water near the center and edges of the vortex moves each time. Observe what happens when you spin the bottles faster or slower.

4 LAB ACTIVITY REPORT

1. Describe what you saw when you swirled the bottles and set them down.
2. Does water near the center of the vortex move any differently than water at the edges of the vortex? Explain.
3. What did you observe when you swirled the bottles faster or slower?
4. **Drawing Conclusions** Did the vortex always spin in the same direction? What should you do to make it spin in the opposite direction?

▲ Glitter added to the water allows you to see the vortex more clearly.

5 Extending the Lab

Activity

Find out what you should do if a tornado warning is announced in your area. Discover what to do if you are outside, in a car, or inside a building. Use the information you gather to create a tornado safety poster. Present your poster to the class.

Water, Climate, and Vegetation

The World and Its People NATIONAL GEOGRAPHIC

To learn more about water, climate, and vegetation, view **The World and Its People Chapter 2** video.

Geography ONLine

Chapter Overview Visit the **Geography: The World and Its People** Web site at gwip.glencoe.com and click on **Chapter 2— Chapter Overviews** to preview information about water, climate, and vegetation.

The Water Planet

Guide to Reading

Main Idea

Water is one of the earth's most precious resources.

Terms to Know

- water vapor
- water cycle
- evaporation
- condensation
- precipitation
- collection
- glacier
- groundwater
- aquifer

Places to Locate

- Pacific Ocean
- Atlantic Ocean
- Indian Ocean
- Arctic Ocean

Reading Strategy

Make a diagram like this one. Starting at the top, write the steps of the water cycle—each in a separate square—in the correct sequence.

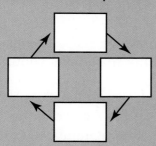

NATIONAL GEOGRAPHIC **Exploring Our World**

Humans, trees, other plants, and animals all need water. We cannot survive without it. Here, a swimmer enjoys a remarkable sight—a permanent pool of water in the middle of Mexico's Chihuahuan Desert. The water bubbles to the surface from an underground spring. The clear water does more than attract swimmers, though. It supports a variety of animal and plant life.

Some people call Earth "the water planet." Why? Water covers about 70 percent of the earth's surface. Water exists all around you in many different forms. Streams, rivers, lakes, seas, and oceans contain water in liquid form. The atmosphere holds **water vapor,** or water in the form of gas. Glaciers and ice sheets are masses of water that have been frozen solid. As a matter of fact, the human body itself is about 60 percent water.

The Water Cycle

The total amount of water on the earth does not change, but it does not stay in one place, either. Instead, the water moves constantly. In a process called the **water cycle,** the water goes from the oceans to the air to the ground and finally back to the oceans.

Look at the diagram on page 50 to see how the water cycle works. The sun drives the cycle by evaporating water mostly from the surface of oceans, but also from lakes and streams. In **evaporation,** the sun's heat turns liquid water into water vapor—also called *humidity.* The amount of water vapor that the air holds depends on the air

The Water Cycle

Analyzing the Diagram

The water cycle involves evaporation, condensation, precipitation, and the collection of water above and below the ground.

Movement How does water get from the ground to the oceans?

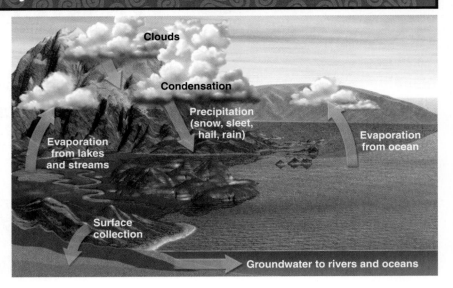

Clouds

Condensation

Precipitation (snow, sleet, hail, rain)

Evaporation from lakes and streams

Evaporation from ocean

Surface collection

Groundwater to rivers and oceans

temperature. Warm air can hold more humidity than cool air, which you have probably felt on warm, muggy summer days.

In addition, warm air tends to rise. As warm air rises higher in the atmosphere, it cools, thus losing its ability to hold as much humidity. As a result, the water vapor changes back into a liquid in a process called **condensation.** Tiny droplets of water come together to form clouds. Eventually, the water falls back to the earth as some form of **precipitation**—rain, snow, sleet, or hail—depending on the temperature of the surrounding air.

When this precipitation reaches the earth's surface, it soaks into the ground and collects in streams and lakes. During **collection,** streams and rivers both above and below the ground carry the water back to the oceans, and the cycle begins again.

Reading Check What kind of air—warm or cold—holds the most water vapor?

Water Resources

It is a hot day, and you rush home for a glass of water. Like all other people—and all plants and animals—you need water to survive. Think about the many ways you use water in just a single day—to bathe, to brush your teeth, to cook your food, and to quench your thirst. People and most animals need freshwater to live. Many other creatures, however, make their homes in the earth's much larger kind of liquid water: salt water.

Freshwater Only about 2 percent of the water on the earth is freshwater. Eighty percent of that freshwater is frozen in **glaciers,** or giant sheets of ice. Only a tiny fraction of the world's freshwater—not even four-hundredths of a percent—is found in lakes and rivers.

When you think of freshwater, you probably think of mighty rivers and huge lakes. People can get freshwater from another source, though. **Groundwater** is water that fills tiny cracks and holes in the rock layers below the surface of the earth. This is a vital source of water because there is 10 times more groundwater than there is water in rivers and lakes. Groundwater can be tapped by wells. Some areas have **aquifers,** or underground rock layers that water flows through. In regions with little rainfall, both farmers and city dwellers sometimes have to depend on aquifers and other groundwater for most of their water supply.

Salt Water   All the oceans on the earth are part of a huge, continuous body of salt water—almost 98 percent of the planet's water. Look at the map on page 58. You will see that the four major oceans are the **Pacific Ocean,** the **Atlantic Ocean,** the **Indian Ocean,** and the **Arctic Ocean.**

The Pacific Ocean is the largest and deepest of these four oceans. It covers almost 64 million square miles (166 million sq. km)—more than all the land areas of the earth combined. As you learned in Chapter 1, bodies of salt water smaller than the oceans are called seas, gulfs, bays, or straits. Look back at the diagram on page 18 of the **Geography Handbook** to see these features again.

✓ Reading Check **What is the difference between groundwater and aquifers?**

Assessment

Defining Terms
1. **Define** water vapor, water cycle, evaporation, condensation, precipitation, collection, glacier, groundwater, aquifer.

Recalling Facts
2. **Region** What percentage of the earth is covered by water?
3. **Movement** In which part of the water cycle does water return to the earth?
4. **Region** What are the world's four oceans?

Critical Thinking
5. **Understanding Cause and Effect** How does the temperature of the air affect the amount of humidity that you feel? How does the air's temperature also influence the form of precipitation that falls?

6. **Drawing Conclusions** Why do you think it is important to keep groundwater free of dangerous chemicals?

Graphic Organizer
7. **Organizing Information** Draw a diagram like this one. List at least four sources of freshwater and salt water on the lines under each heading.

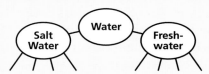

Applying Geography Skills

8. **Analyzing Diagrams** Look at the diagram of the water cycle on page 50. From where does water evaporate?

Water, Climate, and Vegetation

Making Connections

ART   SCIENCE   LITERATURE   TECHNOLOGY

Exploring Earth's Water

More than two-thirds of the earth's surface is covered with water, yet scientists know more about the surface of the moon than they do about the ocean floor. Using an AUV, or autonomous underwater vehicle, called *Autosub,* researchers hope to gain new understanding about the earth's watery surface.

What It Does

It looks like a giant torpedo, but *Autosub* is really a battery-powered robotic submarine that is 23 feet (7 m) long. Its mission is to explore parts of the ocean that are beyond the reach of other research vessels or are too dangerous for humans. Although it is still being tested, *Autosub* has already conducted hundreds of underwater missions.

Exploring Ice Shelves

One of the most promising areas of research for *Autosub* lies in seawater under the ice shelves near Greenland in the Arctic and near Antarctica at the southern extreme of the globe. Traditional submarines are unable to explore these places safely. Satellite photographs show that the area of the ice shelves is changing. Scientists want to use *Autosub*'s technology to measure changes in the thickness of sea ice. They believe that this information may give important clues about the possible rise in the earth's temperature.

Sea ice plays an important role in keeping the earth's climate stable. It acts as insulation—a kind of protection—between the ocean and the atmosphere. Sea ice reflects light, so it limits the amount of heat absorbed into the water and keeps the ocean from getting too warm. In winter, sea ice helps prevent heat from escaping the warmer oceans into the atmosphere.

What the Future Holds

So far, *Autosub*'s missions have been fairly short. Scientists hope to someday program *Autosub* to make long voyages, sampling seawater and collecting data from ocean floors. The information that *Autosub* provides will help scientists make better predictions about the earth's climate.

▲ *Autosub* can be launched from shore, towed out to sea by a small boat, or lowered by a crane into the water.

► Making the Connection

1. What is *Autosub*?
2. Why do scientists want to use *Autosub* to explore under the ice shelves?
3. **Understanding Cause and Effect** How could a loss of sea ice affect the earth's climate?

Climate

Guide to Reading

Main Idea

Wind and water carry rainfall and the sun's warmth around the world to create different climates.

Terms to Know

- weather
- climate
- tropics
- monsoon
- tornado
- hurricane
- typhoon
- drought
- El Niño
- La Niña
- current
- local wind
- rain shadow
- greenhouse effect
- rain forest

Places to Locate

- Equator
- Tropic of Cancer
- Tropic of Capricorn

Reading Strategy

Make a chart like this one. Write at least two details that explain how each force contributes to climate.

Sun	Wind	Water

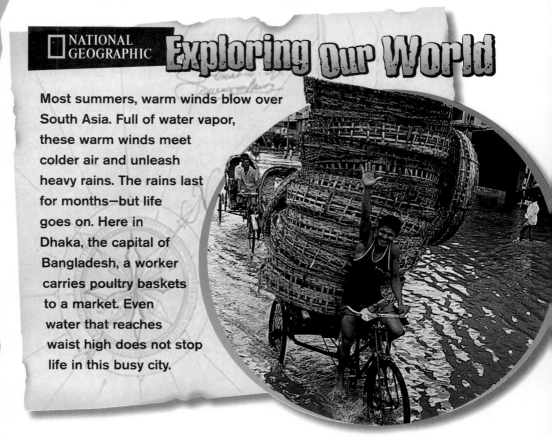

NATIONAL GEOGRAPHIC

Exploring Our World

Most summers, warm winds blow over South Asia. Full of water vapor, these warm winds meet colder air and unleash heavy rains. The rains last for months—but life goes on. Here in Dhaka, the capital of Bangladesh, a worker carries poultry baskets to a market. Even water that reaches waist high does not stop life in this busy city.

Why are some areas of the world full of lush forests, while others are covered by bone-dry deserts? Why do some people struggle through chilling winters, while others enjoy a day at the beach? To understand these mysteries, you need to unlock the secrets of climate.

Weather and Climate

As you learned in Chapter 1, the earth is surrounded by the atmosphere, which holds a combination of gases we call air. The atmosphere's many layers protect life on the earth from harmful rays of the sun. The layer of atmosphere immediately surrounding the earth is also where you will find weather patterns. Suppose a friend calls you and asks what it is like outside. You might say, "It's a beautiful day—warm and sunny!" You are describing the weather. Weather refers to the unpredictable changes in air that take place over a short period of time.

Suppose that a cousin from another part of the country asks what summers and winters are like in your area. You might say, "Summers are usually hot and rainy, and winters are cool but dry." Your answer describes not the weather but your area's climate. Climate is the usual, predictable pattern of weather in an area over a long period of time. It is affected by the sun, the wind, the oceans and other bodies of water, landforms, and even people.

You might also say to your cousin, "Usually we get a little rain in the summer. Last year, though, we went two months without any rain!" You recognize that the weather does not always follow its *usual* climate pattern. Some years it rains more than others; some years the temperature is lower than others.

To understand climate, scientists have to look at the extremes of *temperature* and *precipitation* that an area can have. Consider a country that has an average precipitation of 38 inches (97 cm) a year. In any year, it may be as low as 13 inches (33 cm) or as high as 63 inches (160 cm). Imagine if you were a farmer trying to grow food there. How would your crops be affected if you received only one-third of the rain you expected?

✓ **Reading Check** What is the difference between weather and climate?

The Sun and Climate

What causes climate? The original source of climate is the sun. It gives off energy and light that all plants and animals need to survive. The sun's rays warm the air, water, and land on our planet. Warm gases and liquids are lighter than cool gases and liquids. Because they are lighter, the warmer gases and liquids rise. Then wind and water carry this warmth around the globe, distributing the sun's heat.

NATIONAL GEOGRAPHIC On Location

Washington, D.C.

A cross-country skier braves a blizzard in our nation's capital.

Place When scientists study climate, what two factors do they analyze?

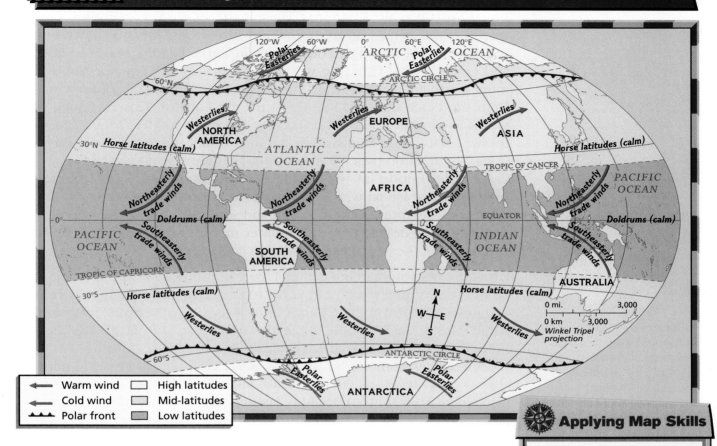

Legend:
- ← Warm wind
- ← Cold wind
- ▲▲▲ Polar front
- ☐ High latitudes
- ☐ Mid-latitudes
- ▨ Low latitudes

Latitude and Climate   Climate is also affected by the angle at which the sun's rays hit the earth. As you learned in Chapter 1, because of the earth's tilt and revolution around the sun, the sun's rays hit various places at different angles at different times of the year. The sun's rays hit places in low latitudes—regions near the **Equator**— more directly than places at higher latitudes. The areas near the Equator, known as the tropics, lie between the **Tropic of Cancer** (23½°N latitude) and the **Tropic of Capricorn** (23½°S latitude). If you lived in the tropics, you would almost always experience a hot climate, unless you lived high in the mountains where temperatures are cooler. Find the low-latitude tropics on the map above. (To learn how to use latitude and longitude, turn to page 62.)

Outside the tropics, the sun is never directly overhead. The midlatitudes extend from the tropics to about 60° both north and south of the Equator. When the North Pole is tilted toward the sun, the sun's rays fall more directly on the Northern Hemisphere. This affects our climate by giving us warm summer days. Six months later, the South Pole is tilted toward the sun, and the seasons are reversed. At the high latitudes near the North and South Poles, the sun's rays hit *very* indirectly. Climates in these regions are always cool or cold.

✓**Reading Check** How does the tilt of the earth affect climate?

The Wind's Effect on Climate

Movements of air are called winds. From year to year, winds follow *prevailing*, or typical, patterns. These patterns are very complex. One reason is that winds do more than move east and west or north and south. They also go up and down. As you learned earlier, warm air rises and cold air falls. Thus, the warmer winds near the Equator rise and move north and south toward the Poles of the earth. The colder winds from the Poles sink and move toward the Equator. This exchange of winds is complicated by the fact that the earth rotates, which causes the winds to curve. Winds, then, are in constant motion in many directions. The map on page 55 shows you prevailing wind patterns.

Another important wind pattern is the monsoon. Monsoons are tremendous seasonal winds that blow over continents for months at a time. They are found mainly in Asia and some areas in Africa. Although they often are destructive, the summer monsoons in South Asia bring much-needed heavy rains.

Storms   As you read in Section 1, part of the water cycle is rain and other types of precipitation that fall to the earth. A little rain may ruin a picnic or spoil a ball game, but it is not a serious problem. Sometimes, though, people suffer through fierce storms. Why is that? What causes these destructive events?

When warm, moist air systems meet cold air systems, thunderstorms may develop. These storms include thunder, lightning, and heavy rain. They tend to be short, lasting only about 30 minutes. Some areas are more likely to see thunderstorms than others. In central Florida, as many as 90 days a year may experience thunderstorms.

A thunderstorm may produce another danger—a tornado. Tornadoes are funnel-shaped windstorms that sometimes form during severe thunderstorms. They occur all over the world, but the United States has more tornadoes than any other area. Winds in tornadoes often reach 250 miles (402 km) per hour.

Hurricanes, or violent tropical storm systems, form over the warm Atlantic Ocean in late summer and fall. Hurricanes bring high winds that can reach more than 150 miles (241 km) per hour. They also produce rough seas and carry drenching rain. Hurricanes strike North America and the islands in the Caribbean Sea. They also rip through Asia, although in that region they are called typhoons. These storms can do tremendous damage. Their strong winds destroy buildings and snap power lines. Heavy rains can flood low-lying areas.

El Niño and La Niña   In 1998 the world experienced unusual weather. Heavy rains brought floods to Peru, washing away whole villages. Europe, eastern Africa, and most of the southern United States also had severe flooding. In the western Pacific, normally heavy rains never came. Indonesia suffered a drought, a long period of extreme dryness. The land there became so dry that forest fires burned thousands of acres of trees. Thick smoke from the fires forced drivers to put their headlights on at noon!

Where in the World?

Mt. Pinatubo

Mt. Pinatubo (PEE•nah•TOO•boh) is a volcanic mountain in the Philippine Islands. Its eruption in the early 1990s had a tremendous impact on the world's climate. The powerful explosion shot massive amounts of ash and sulfur dioxide into the earth's atmosphere. The ash and chemicals blocked some of the sun's rays from reaching the earth. As a result, the world's climate was cooler for two years after the volcano's blast.

Why did these disasters take place? They resulted from a combination of temperature, wind, and water effects in the Pacific Ocean called **El Niño** (ehl NEE•nyoh). The name "El Niño" was coined by early Spanish explorers in the Pacific. They use the phrase—which refers to the Christ child—because the effect hits South America around Christmas.

El Niños form when cold winds from the east are weak. Without these cold winds, the central Pacific Ocean grows warmer than usual. More water evaporates, and more clouds form. The thick band of clouds changes wind and rain patterns. Some areas receive heavier than normal rains and others have less than normal rainfall.

Does El Niño come every year? Scientists have found that El Niño occurs about every three years. They also found that in some years, the opposite kind of unusual weather takes place. This event is called **La Niña** (lah NEE•nyah), Spanish for "girl," because the effects are the opposite of those in El Niño. Winds from the east become very strong, cooling more of the Pacific. When this happens, heavy clouds form in the western Pacific.

✓ Reading Check Why do El Niños occur?

Ocean Currents

Winds carry large masses of warm and cool air around the earth. At the same time, moving streams of water called **currents** carry warm or cool water through the world's oceans. Look at the map on page 58. As you can see, these currents follow certain patterns. Notice how the warm currents tend to move along the Equator or from the Equator to the Poles. The cold currents carry cold polar water toward the Equator.

These currents affect the climate of land areas. Look at the warm current called the Gulf Stream. It flows from the Gulf of Mexico along the east coast of North America. Then it crosses the Atlantic Ocean toward Europe, where it is called the North Atlantic Current. Winds that blow over these warm waters bring warm air to western Europe. Because these winds blow from west to east, areas in Europe enjoy warmer weather than areas lying west of the Gulf Stream in Canada.

✓ Reading Check What areas of the world would be affected by a change in the Gulf Stream?

El Niño

NORMAL CONDITIONS

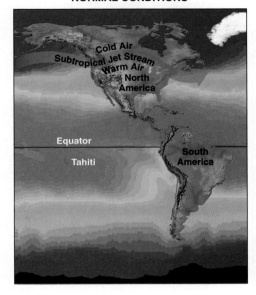

Cold Air
Subtropical Jet Stream
Warm Air
North America
Equator
Tahiti
South America

EL NIÑO CONDITIONS

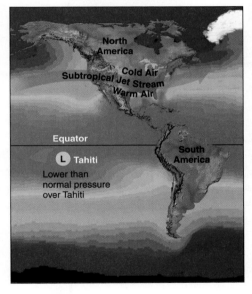

North America
Subtropical Jet Stream
Cold Air
Warm Air
Equator
L Tahiti
Lower than normal pressure over Tahiti
South America

Analyzing the Diagram

The temperature of the oceans varies from warm (dark red) to very cold (dark purple).

Movement What happens to the Jet Stream during El Niño conditions?

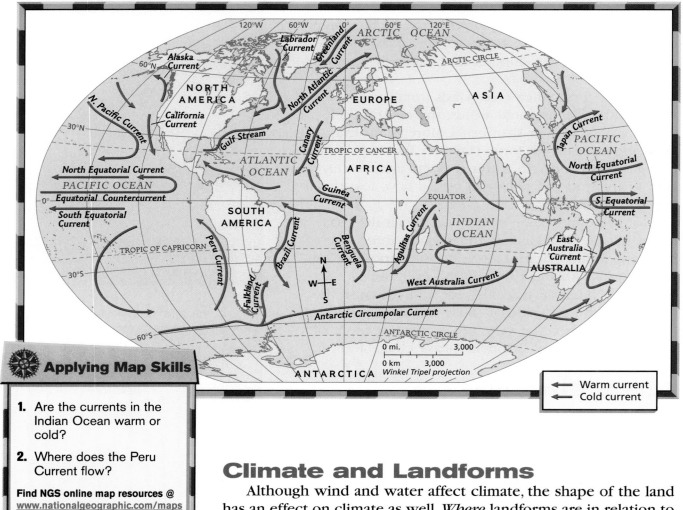

CHAPTER 2

Applying Map Skills

1. Are the currents in the Indian Ocean warm or cold?

2. Where does the Peru Current flow?

Find NGS online map resources @ www.nationalgeographic.com/maps

Climate and Landforms

Although wind and water affect climate, the shape of the land has an effect on climate as well. *Where* landforms are in relation to one other and to water influences climate, too.

Landforms and Local Winds Geographers study major wind patterns that blow over the earth. When they study the climate of a given area, however, geographers also look at what they call local winds. Local winds are patterns of wind caused by landforms in a particular area.

Some local winds occur because land warms and cools more quickly than water does. As a result, cool sea breezes keep coastal areas cool during the day. After the sun sets, the opposite occurs. The air over the land cools more quickly than the air over water. At night, then, a land breeze blows from the land out to the sea.

A similar effect occurs near mountains. Air warmed by the sun rises up mountain slopes during the day. At night, cooler air moves down the mountain into the valley below. Have you ever seen fog lying on a valley floor on a cool morning? That fog was caused by the cool air that came down the mountain during the night.

Mountains, Temperature, and Rainfall   The higher the elevation of a particular place, the lower the temperature that place will have. In high mountains, the air becomes thinner and cannot hold as much heat from the sun. The temperature drops. Even near the Equator, where the sun's rays strike the earth directly, snow covers the peaks of high mountains.

The cooling effect of elevation is not as strong on high plateaus, however. Why is this so? As you learned in Chapter 1, plateaus are flat surfaces, so there are no rising air currents as there are along mountainsides. Some plateaus—as in the southwestern United States—can reach as much as a mile above sea level. Even these high plateaus can hold the sun's warmth.

Mountains also have an effect on rainfall. When warm moist winds blow inland from the ocean toward a coastal mountain range, the winds are forced upward over the mountains. As these warm winds rise, the air cools and loses its moisture. Rain or snow falls on the mountains. The climate on this *windward*—or wind-facing—side of mountain ranges is moist and often foggy. Trees and vegetation are thick and green.

By the time the air moves over the mountain peaks, it is cool and dry. This creates a **rain shadow,** a dry area on the side of the mountains facing away from the wind. Geographers call this side the *leeward* side. The dry air of a rain shadow warms up again as it moves down the leeward side, giving the region a dry or desert climate.

A rain shadow occurs along the western coast of the United States. Winds moving east from the Pacific Ocean lose their moisture as they move upward on the windward slopes of the coastal mountains. Great deserts and dry basins are located on the leeward side of these ranges.

✓Reading Check **Why are areas of higher elevation often cooler?**

 NATIONAL GEOGRAPHIC   **Rain Shadow**

 Analyzing the Diagram

Rain shadows usually occur on the inland sides of mountain ranges near oceans.

Location What is the term for the side of a mountain where precipitation falls?

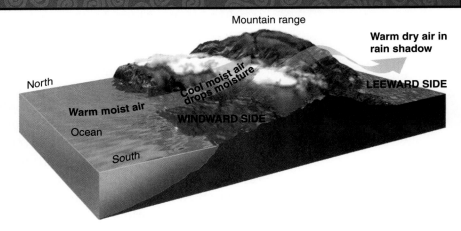

Mountain range

Warm dry air in rain shadow

North

Cool moist air drops moisture

LEEWARD SIDE

Warm moist air

WINDWARD SIDE

Ocean

South

Water, Climate, and Vegetation

Shanghai, China

Shanghai, China (right), like other modern cities, is generally warmer than the rural areas around it (above).

Human/Environment Interaction Why are cities warmer than surrounding areas?

The Impact of People on Climate

People's actions can affect climate. You may have noticed that temperatures in large cities are generally higher than those in nearby rural areas. Why is that? The city's streets and buildings absorb more of the sun's rays than do the plants and trees of rural areas.

Cities are warmer even in winter. People burn fuels to warm houses, power industry, and move cars and buses along the streets. This burning raises the temperature in the city. The burning also releases a cloud of chemicals into the air. These chemicals blanket the city and hold in more of the sun's heat, creating a so-called *heat island*.

The Greenhouse Effect The burning of fuels may be creating a worldwide problem. In the past two hundred years, people have burned coal, oil, and natural gas as sources of energy. Burning these fuels releases certain gases into the air.

Some scientists warn that the buildup of these gases presents dangers. It creates a greenhouse effect—like a greenhouse, the gases prevent the warm air from rising and escaping into the atmosphere. As a result, the overall temperature of the earth will increase. Some scientists predict disastrous results from this global warming. The ice at the North and South Poles will melt, they say. Ocean levels will rise and flood coastal cities. In addition, global warming will make some areas that are now fertile unable to grow crops. It may become much harder to feed all the world's people.

Not all scientists agree about the greenhouse effect. Some argue that the world is not warming. Others say that even if it is, the predictions of disaster are extreme. Many scientists are studying world temperature trends closely. They hope to be able to discover whether the greenhouse effect is a real threat.

Clearing the Rain Forests   Along the Equator, you can see dense forests called rain forests that receive high amounts of rain each year. In some countries, people are clearing large areas of these forests. They want to sell the lumber from the trees. They also want to use the land to grow crops or as pasture for cattle. Clearing the rain forests, though, can hurt the world's climate.

One reason is related to the greenhouse effect. People often clear the forests by burning down the trees. This burning releases more gases into the air, just like burning oil or natural gas does. Another danger of clearing the rain forests is related to rainfall. Remember the water cycle discussed in Section 1? Water on the earth's surface evaporates into the air and then falls as rain. In the rain forests, much of this water evaporates from the leaves of trees. If the trees are cut, less water will evaporate. As a result, less rain will fall. Scientists worry that, over time, the area that now holds rain forests will actually become dry and unable to grow anything.

✔ Reading Check   **What is the greenhouse effect?**

Web Activity Visit the *Geography: The World and Its People* Web site at gwip.glencoe.com and click on **Chapter 2—Student Web Activities** to learn more about the destruction of the rain forests.

Assessment

Defining Terms
1. **Define** weather, climate, tropics, monsoon, tornado, hurricane, typhoon, drought, El Niño, La Niña, current, local wind, rain shadow, greenhouse effect, rain forest.

Recalling Facts
2. **Movement** What five elements affect climate?
3. **Location** Between what two lines of latitude are the tropics?
4. **Place** Provide an example of how landforms influence climate.

Critical Thinking
5. **Making Comparisons** How does the amount of rainfall on the windward side of a mountain differ from that on the leeward side?

6. **Summarizing Information** What general patterns do wind and currents follow?

Graphic Organizer
7. **Organizing Information** Draw a diagram like this one. In the first box, list three human actions that lead to the greenhouse effect. In the third box, list four possible results of the greenhouse effect.

Human Actions → Greenhouse Effect → Results of Greenhouse Effect

Applying Geography Skills

8. **Analyzing Maps** Look at the world ocean currents map on page 58. Which two continents lie completely outside the tropics?

Geography Skill

Using Latitude and Longitude

Learning the Skill

To find an exact location, geographers use a set of imaginary lines. One set of lines—**latitude** lines—circles the earth's surface east and west. The starting point for numbering latitude lines is the Equator, which is 0° latitude. Every other line of latitude is numbered from 1° to 90° and is followed by an N or S to show whether it is north or south of the Equator. Latitude lines are also called parallels.

A second set of lines—**longitude** lines—runs vertically from the North Pole to the South Pole. Each of these lines is also called a meridian. The starting point—0° longitude—is called the Prime Meridian (or Meridian of Greenwich). Longitude lines are numbered from 1° to 180° followed by an E or W—to show whether they are east or west of the Prime Meridian.

To find latitude and longitude, choose a place on a map. Identify the nearest parallel, or line of latitude. Is it located north or south of the Equator? Now identify the nearest meridian, or line of longitude. Is it located east or west of the Prime Meridian?

Practicing the Skill

1. On the map below, what is the exact location of Washington, D.C.?
2. What cities on the map lie south of 0° latitude?
3. What city is located near 30°N, 30°E?

Applying the Skill

Turn to page RA2 of the **Reference Atlas.** List the latitude and longitude for one city. Ask a classmate to use the information to find and name the city.

NATIONAL GEOGRAPHIC

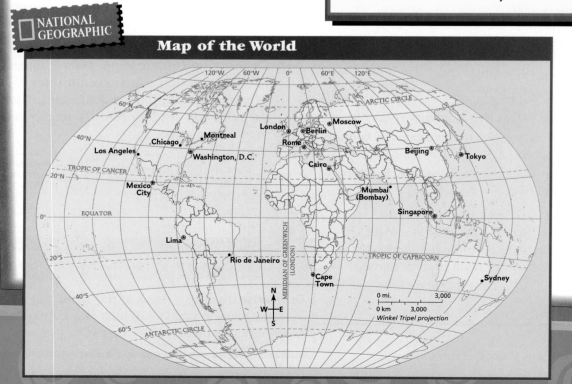

Map of the World

Climate Zones and Vegetation

Guide to Reading

Main Idea

Geographers divide the world into different climate zones, each of which has special characteristics.

Terms to Know

- canopy
- savanna
- marine west coast climate
- Mediterranean climate
- humid continental climate
- humid subtropical climate
- subarctic
- tundra
- permafrost
- steppe
- timberline

Reading Strategy

Complete a chart like the one below by listing the categories of each type of climate next to the correct headings.

Climate Type	Categories
Tropical	
Mid-Latitude	
High Latitude	
Dry	
Highland	

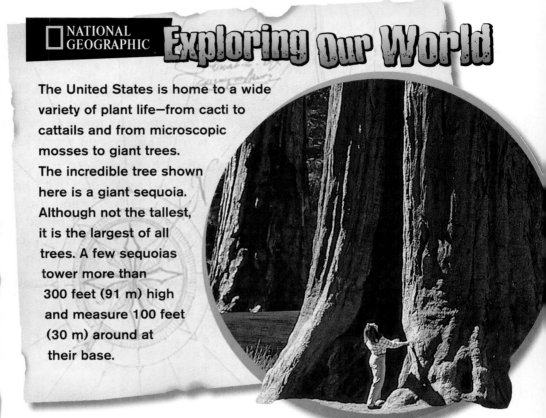

NATIONAL GEOGRAPHIC

Exploring Our World

The United States is home to a wide variety of plant life—from cacti to cattails and from microscopic mosses to giant trees. The incredible tree shown here is a giant sequoia. Although not the tallest, it is the largest of all trees. A few sequoias tower more than 300 feet (91 m) high and measure 100 feet (30 m) around at their base.

Why do you think a photograph of a giant tree is in a chapter on climate? The reason is that climate and vegetation go together. Consider this: The state of Washington sits next to the state of Idaho. The plant life in western Washington, however, is much more similar to that of the United Kingdom, which is thousands of miles away, than it is to the plant life in eastern Washington and Idaho, which touch each other. Why? The patterns of temperature, wind, and precipitation in western Washington and the United Kingdom are similar.

Scientists use these patterns to group climates into many different types. They have put the world's climates into five major groups: tropical, mid-latitude, high latitude, dry, and highland. Three of these groups—tropical, mid-latitude, and high latitude—are based on an area's latitude, or distance from the Equator. Some of these major groups have subcategories of climate zones within them. In addition, each climate zone has particular kinds of plants that grow in it.

Tropical Climates

The tropical climate gets its name from the tropics—the areas along the Equator reaching from 23½°N to 23½°S. If you like warm weather all during the year, you would love a tropical climate. The tropical climate region can be separated into two types—tropical rain forest and tropical savanna. The tropical rain forest climate receives up to 100 inches (254 cm) of rain a year. As a result, the rain forest climate is wet in most months. The tropical savanna climate has two distinct seasons—one wet and one dry.

Tropical Rain Forest Climate   Year-round rains in some parts of the tropics produce lush vegetation and thick rain forests. These forests are home to millions of kinds of plant and animal life. Tall hardwood trees such as mahogany, teak, and ebony form the canopy, or top layer of the forest. The vegetation at the canopy layer is so thick that little sunlight reaches the forest floor. The Amazon Basin in South America is the world's largest rain forest area.

Tropical Savanna Climate   In other parts of the tropics, such as southern India and eastern Africa, most of the year's rain falls in just a few months of the year. This is called the wet season. The rest of the year is hot and dry. Savannas, or broad grasslands with few trees, occur in this climate region. Find the tropical savanna climate areas on the map on page 65.

√ Reading Check   **Where are the tropical climate zones found?**

NATIONAL GEOGRAPHIC **On Location**

Tropical Vegetation

The **tropical rain forest** climate remains wet most of the year (far left), whereas the **tropical savanna** climate has a distinct wet and dry season (below).

Region What is a savanna?

World Climate Regions

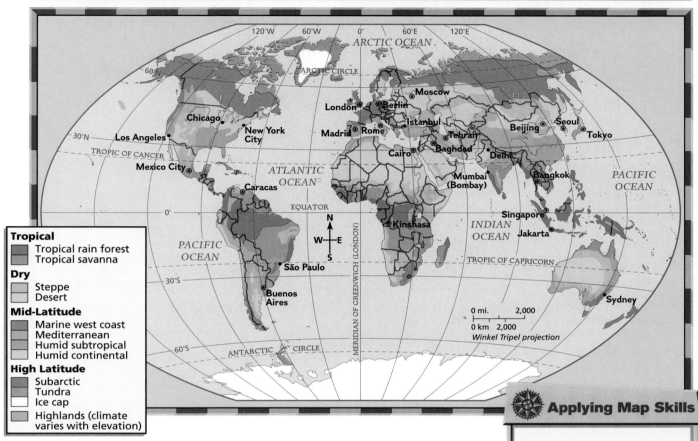

Tropical
- Tropical rain forest
- Tropical savanna

Dry
- Steppe
- Desert

Mid-Latitude
- Marine west coast
- Mediterranean
- Humid subtropical
- Humid continental

High Latitude
- Subarctic
- Tundra
- Ice cap
- Highlands (climate varies with elevation)

0 mi. 2,000
0 km 2,000
Winkel Tripel projection

Applying Map Skills

1. Which climate covers most of the southeastern United States?

2. Which climate is most common in countries directly on the Equator?

Find NGS online map resources @ www.nationalgeographic.com/maps

Mid-Latitude Climates

Mid-latitude, or moderate, climates are found in the middle latitudes of the Northern and Southern Hemispheres. They extend from about 23½° to 60° both north and south of the Equator. Most of the world's people—probably including you—live within these two bands around the earth. They are called mid-latitude because they are in the middle of both the Northern Hemisphere and the Southern Hemisphere. The mid-latitude climates are neither as close to the Equator as the tropics nor as close to the Poles as the high latitude climates.

The mid-latitude region includes more and different climate zones than other regions. This variety results from a mix of air masses. As you remember from Section 2, warm air comes from the tropics, and cool air comes from the polar regions. In most mid-latitude climates, the temperature changes with the seasons. Sometimes the climate zones in this region are called *temperate climates.*

Marine West Coast Climate   Coastal areas that receive winds from the ocean usually have a mild marine west coast climate. If you lived in one of these areas, your winters would be rainy and mild, and your summers would be cool. These areas usually have strong growth

Water, Climate, and Vegetation

World Natural Vegetation Regions

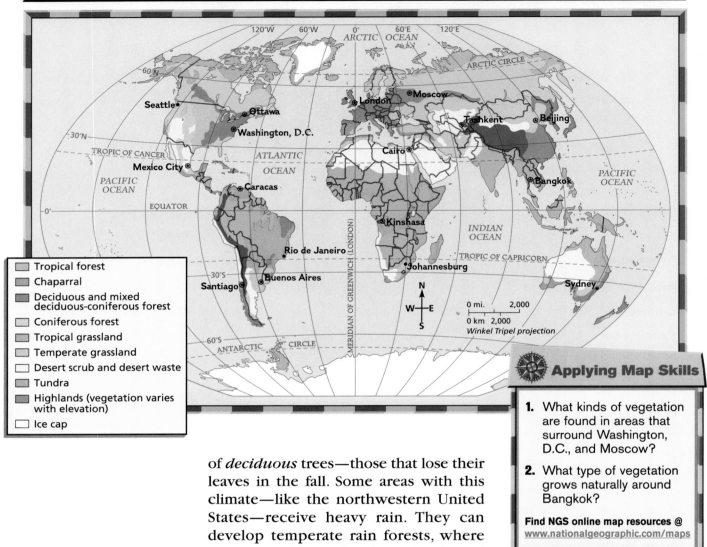

Legend:
- Tropical forest
- Chaparral
- Deciduous and mixed deciduous-coniferous forest
- Coniferous forest
- Tropical grassland
- Temperate grassland
- Desert scrub and desert waste
- Tundra
- Highlands (vegetation varies with elevation)
- Ice cap

Applying Map Skills

1. What kinds of vegetation are found in areas that surround Washington, D.C., and Moscow?

2. What type of vegetation grows naturally around Bangkok?

Find **NGS** online map resources @ www.nationalgeographic.com/maps

of *deciduous* trees—those that lose their leaves in the fall. Some areas with this climate—like the northwestern United States—receive heavy rain. They can develop temperate rain forests, where evergreens like cedar, fir, and redwood grow.

Mediterranean Climate Another mid-latitude coastal climate is called a Mediterranean climate because it is similar to the climate found around the Mediterranean Sea. This climate has mild, rainy winters like the marine west coast climate. Instead of cool summers, however, people living in a Mediterranean climate experience hot, dry summers. The vegetation that grows in this climate includes shrubs and short trees. Some are evergreens, but others lose their leaves in the dry season.

Humid Continental Climate If you live in inland areas of North America, Europe, or Asia, you usually face a harsher humid continental climate. In these areas, winters can be long, cold, and snowy. Summers are short but may be very hot. Deciduous trees grow in forests, and vast grasslands flourish in some areas of this zone.

Humid Subtropical Climate   Mid-latitude regions close to the tropics have a humid subtropical climate. Rain falls throughout the year but is heaviest during the hot and humid summer months. Humid subtropical winters are generally short and mild. Trees like oak, magnolia, and palms grow in this zone.

✓ Reading Check   **What causes the mid-latitude region to have more and different climate zones than other regions?**

High Latitude Climates

High latitude climate regions lie mostly in the high latitudes of each hemisphere, from 60°N to the North Pole and 60°S to the South Pole. These climates are generally cold, but some are more severely cold than others.

Subarctic Climate   In the high latitudes nearest the mid-latitude zones, you will find the subarctic climate. The few people living here face severely cold and bitter winters, but temperatures do rise above freezing during summer

Mid-Latitude Vegetation

Fir trees (bottom left) thrive in a **marine west coast** climate. Shrubs and olive trees (top right) grow in a **Mediterranean** climate. Deciduous trees (bottom right) flourish in a **humid continental** climate. Palm trees are common in **humid subtropical** zones.

Place Which type of vegetation is most common in your area?

NATIONAL GEOGRAPHIC   **On Location**

Water, Climate, and Vegetation

months. Huge evergreen forests called *taiga* (TY•guh) grow in the subarctic region, especially in northern Russia.

Tundra Climate   Closer to the Poles than the subarctic zone lie the **tundra** areas. The tundra is a vast rolling plain without trees. The climate in this zone is harsh and dry. In parts of the tundra and subarctic regions, the lower layers of soil are called **permafrost** because they stay permanently frozen. Only the top few inches of the ground thaw during summer months. Because the surface is flat and the soil below is frozen, water tends to stay on the land when it melts. This provides the moisture that plants need to grow. You will find sturdy grasses and low-growing berry bushes in the tundra.

Ice Cap Climate   On the polar ice caps and the great ice sheets of Antarctica and Greenland, the climate is bitterly cold. Monthly temperatures average below freezing. Temperatures in Antarctica have been measured at −128°F (−89°C)! Although no other vegetation grows here, *lichens*—or funguslike plants and mosses—can live on rocks.

✓ Reading Check   What are the three types of high latitude climates?

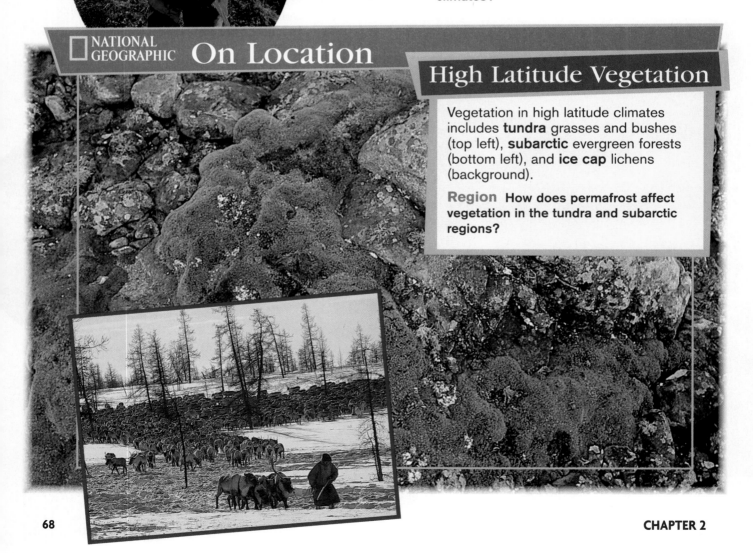

On Location

High Latitude Vegetation

Vegetation in high latitude climates includes **tundra** grasses and bushes (top left), **subarctic** evergreen forests (bottom left), and **ice cap** lichens (background).

Region How does permafrost affect vegetation in the tundra and subarctic regions?

Dry Climate Vegetation

The vegetation that survives **desert** and **steppe** climates includes cacti (above) and short grasses (left).

Region Are very dry climates always hot? Explain.

Dry Climates

Dry climate refers to dry or partially dry areas that receive little or no rainfall. Temperatures can be extremely hot during the day and very cold at night. Dry climates can also have severely cold winters. You can find dry climate regions at any latitude.

Desert Climate The driest climates have less than 10 inches (25 cm) of rainfall a year. Regions with such climates are called deserts. Only scattered plants such as cacti can survive a desert climate. With roots close to the surface, cacti can collect any rain that falls. Most cacti are found only in North America. In other countries, however, small areas of thick plant life dot the deserts. These arise along rivers or where underground springs reach the surface.

Steppe Climate Many deserts are surrounded by partly dry grass-lands and prairies known as **steppes.** The word *steppe* comes from a Russian word meaning "treeless plain." The steppes get more rain than deserts, averaging 10 to 20 inches (25 to 51 cm) a year. Bushes and short grasses cover the steppe landscape. The Great Plains of the United States has a steppe climate.

✓Reading Check Where are steppe climate zones often located?

Water, Climate, and Vegetation

Highland Vegetation

The wildflowers and shrubs that grow in meadows above the timberline are often called *alpine* vegetation. This name refers to the Alps mountain ranges in Europe.

Location How does elevation affect climate?

Highland Climate

As you read in Section 2, the elevation of a place changes its climate dramatically. Mountains tend to have cool climates—and the highest mountains have very cold climates. This is true even for mountains that are on the Equator. A highland, or mountain, climate has cool or cold temperatures year-round.

If you climb a mountain, you will reach an area called the timberline. The timberline is the elevation above which no trees grow. Once you reach the timberline, you will find only small shrubs and wildflowers growing in meadows.

Reading Check What is the timberline?

Section 3 Assessment

Defining Terms

1. **Define** canopy, savanna, marine west coast climate, Mediterranean climate, humid continental climate, humid subtropical climate, subarctic, tundra, permafrost, steppe, timberline.

Recalling Facts

2. **Region** What are the five types of climate regions?

3. **Region** How do the climate zones in the mid-latitude region differ?

4. **Region** What kind of vegetation grows in the tundra climate zone?

Critical Thinking

5. **Making Comparisons** What do the tropical savanna and humid continental climates have in common?

6. **Drawing Conclusions** How can snow exist in the tropics along the Equator?

Graphic Organizer

7. **Organizing Information** Draw a globe like this one. Label the three climate regions that are based on latitude, then identify the lines of latitude that separate the climate regions.

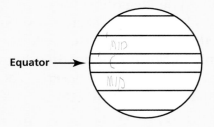

Equator →

Applying Geography Skills

8. **Analyzing Maps** Look at the world natural vegetation regions map on page 66. What type of natural vegetation thrives around Cairo?

Section 1 · The Water Planet

Terms to Know

water vapor	glacier
water cycle	groundwater
evaporation	aquifer
condensation	
precipitation	
collection	

Main Idea

Water is one of the earth's most precious resources.

✓ Region Water covers about 70 percent of the earth's surface.

✓ Movement Water follows a cycle of evaporation, condensation, precipitation, and collection on and beneath the ground.

✓ Science Humans and most animals need freshwater to live. Only a small fraction of the world's water is found in rivers and lakes.

Section 2 · Climate

Terms to Know

weather	La Niña
climate	current
tropics	local wind
monsoon	rain shadow
tornado	greenhouse
hurricane	effect
typhoon	rain forest
drought	
El Niño	

Main Idea

Wind and water carry rainfall and the sun's warmth around the world to create different climates.

✓ Region Climate is the usual pattern of weather over a long period of time. It includes extremes in temperature and rainfall.

✓ Region The tropics, near the Equator, receive more of the sun's warmth than other regions.

✓ Location Landforms and position near water affect climate in a local area.

✓ Culture Human actions like building cities, burning fuels, and clearing the rain forests can affect climate.

Section 3 · Climate Zones and Vegetation

Terms to Know

canopy
savanna
marine west coast climate
Mediterranean climate
humid continental climate
humid subtropical climate
subarctic
tundra
permafrost
steppe
timberline

Main Idea

Geographers divide the world into different climate zones, each of which has special characteristics.

✓ Region The world has five main climate regions that are based on latitude, amount of moisture, and/or elevation. These regions are tropical, mid-latitude, high latitude, dry, and highland.

✓ Region Each climate zone has particular kinds of vegetation.

Fishing on the Pacific Ocean ▶

Using Key Terms

Match the terms in Part A with their definitions in Part B.

A.

1. evaporation
2. savanna
3. monsoon
4. tundra
5. condensation

6. greenhouse effect
7. rain forest
8. El Niño
9. precipitation
10. current

B.

a. moving streams of water in the oceans

b. treeless plain in which only the top few inches of ground thaw in summer

c. weather pattern in the Pacific Ocean

d. seasonal wind that blows over a continent

e. buildup of certain gases in the atmosphere that hold the sun's warmth

f. water that falls back to the earth

g. dense forest that receives much rain

h. water vapor changes back into a liquid

i. sun's heat turns water into water vapor

j. broad grassland in the tropics

Reviewing the Main Ideas

Section 1 The Water Planet

11. **Movement** What are the four steps in the water cycle?
12. **Region** What percentage of the world's water is freshwater?
13. **Region** What has more freshwater—lakes and rivers or groundwater?

Section 2 Climate

14. **Movement** How do wind and water affect climate?
15. **Location** How do mountains affect rainfall?
16. **Human/Environment Interaction** Why are cities warmer than nearby rural areas?

Section 3 Climate Zones and Vegetation

17. **Region** Which climate region has the most climate zones? Why?
18. **Place** What kind of vegetation grows in Mediterranean climates?
19. **Place** In what climate zone would you find large grasslands?

NATIONAL GEOGRAPHIC   **World Oceans and Currents**

Place Location Activity

On a separate sheet of paper, match the letters on the map with the numbered places listed below.

1. Arctic Ocean
2. Atlantic Ocean
3. California Current

4. Japan Current
5. Indian Ocean
6. Gulf Stream

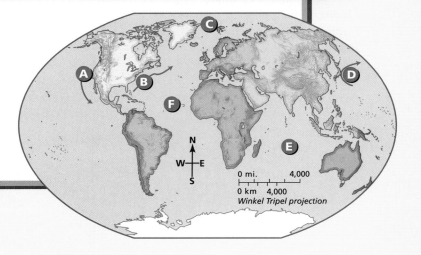

Self-Check Quiz Visit the *Geography: The World and Its People* Web site at gwip.glencoe.com and click on **Chapter 2— Self-Check Quizzes** to prepare for the Chapter Test.

Critical Thinking

20. **Analyzing Information** From where does the freshwater in your community come? How can you find out?

21. **Categorizing Information** Create five webs like the one shown here. In each large oval, write the name of a climate region. In the medium-sized ovals, write the name of each climate zone in that region. For each zone, fill in the three small ovals with the usual weather in summer, the usual weather in winter, and the kind of vegetation.

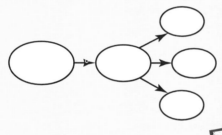

GeoJournal Activity

22. **Writing a Poem** Make a list of all the types of precipitation that you can think of. Write a poem about precipitation and climate using the words in your list.

Mental Mapping Activity

23. **Focusing on the Region** Draw a freehand map of the world's oceans and continents. Label the following items:

- Equator
- Pacific Ocean
- high latitude climate regions
- tropical climate regions
- North America
- Africa

Technology Activity

24. **Using the Internet** Research a recent hurricane or tornado. Find out when and where it occurred, how much force the storm had, and what damage it caused.

Standardized Test Practice

Directions: Study the graph, then answer the following question.

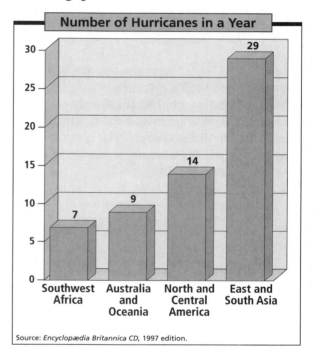

Number of Hurricanes in a Year

Southwest Africa: 7
Australia and Oceania: 9
North and Central America: 14
East and South Asia: 29

Source: *Encyclopædia Britannica CD*, 1997 edition.

1. **How many more hurricanes does East and South Asia experience in a year than North and Central America?**

   F 29

   G 14

   H 9

   J 15

Test-Taking Tip: Make sure you read the question carefully. It is not asking for the total number of hurricanes in East and South Asia. Instead, the question asks how many *more* hurricanes one region has than another.

THE OCEANS
Deep Trouble

Endangered Seas Next time you are in the grocery store, walk past the fresh seafood section. You will see fish, shrimp, and other seafood on ice. The abundance and variety of seafood available makes it seem as if the oceans can provide us with food forever. Unfortunately, that is not true. Oceans worldwide are in deep trouble because of human activities. There are several trouble spots around the world.

- Grand Banks — Overfishing in the northwest Atlantic Ocean has seriously harmed once-rich populations of cod, halibut, and other fish.

- Gulf of Mexico — Pollution from the United States's rivers causes a "dead zone" in the Gulf each summer. Fish die due to a lack of oxygen in the water.

- Yellow Sea — China's industries have dumped poisons into these Pacific Ocean waters. This sea may have the world's highest levels of heavy metals.

- Kara Sea — The former Soviet Union dumped huge quantities of radioactive wastes into these Arctic waters.

Saving Oceans It is not too late to preserve the oceans. Some steps have been taken.

- More than 1,200 protected zones exist along coasts worldwide.

- Countries are assigning individual catch limits to fishing boats.

- Science organizations are collecting data on oceans and urging government leaders to pass laws to protect marine environments.

Sea stars

Workers clean a beach after an oil spill.

Bottlenose dolphin

Making a Difference

Underwater Explorer Imagine exploring where no one has ever been before. Dr. Sylvia Earle has! Earle explores deep in the oceans, discovering new kinds of animals, plants, and systems of life. Earle is a marine biologist and former chief scientist of the National Oceanic and Atmospheric Administration (NOAA). Now she is leading a research project called the Sustainable Seas Expeditions. Using a special, one-person submersible called *DeepWorker,* Earle and her team are studying deepwater habitats in the United States's National Marine Sanctuary system. This system includes 12 underwater parks that lie in coastal waters around the continental United States and in Hawaii and American Samoa.

DeepWorker has a clear acrylic dome that covers the pilot's head and shoulders. Mechanical arms allow the pilot to work underwater. The expedition team takes photographs and collects data on marine organisms and their habitats in the sanctuaries. Sylvia Earle and her team hope that their research will help people worldwide realize how important the oceans are and why we need to protect them.

Marine biologist
Sylvia Earle

What Can You Do?

🐚 Get Involved
How can you help the oceans? Recycle and do not pollute! About 77 percent of all pollutants in the oceans originate on land. Plastic and other trash in ocean waters kill fish, sea birds, turtles, seals, and whales.

🐚 Find Out More
Learn more about the 12 national marine sanctuaries. Find out where the sanctuaries are located and the types of marine organisms found in each. Visit the Web site www.sanctuaries.nos.noaa.gov. Locate the sanctuaries on a world map.

🐚 Use the Internet
Follow the progress of the Sustainable Seas Expeditions. Check the project's "Sanctuary Log" at http://sustainableseas.noaa.gov/missions/missions.html. You can also see what it is like to pilot *DeepWorker,* by visiting www.nationalgeographic.com/monterey

Earle studying
animals in Hawaiian waters

Chapter 3

The World's People

The World and Its People NATIONAL GEOGRAPHIC

To learn more about the world's culture regions, view *The World and Its People* Chapter 3 video.

Geography online

Chapter Overview Visit the *Geography: The World and Its People* Web site at gwip.glencoe.com and click on **Chapter 3– Chapter Overviews** to preview information about the world's people.

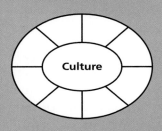

Culture

NATIONAL GEOGRAPHIC

Exploring Our World

Three thousand years ago, the Olmec people lived in Mexico. They sometimes wore skins of jaguars, cats that were sacred to them. This young boy lives in an area where the jaguar is still honored. He is preparing for a jaguar dance. An object from modern culture—a soft drink bottle—is used to make the "jaguar" spots of ash on the boy's clay-covered skin.

If you wake up to rock music, put on denim jeans, speak English, and celebrate the Fourth of July, those things are part of your culture. If you eat a tortilla, or flat bread, for breakfast, speak Spanish, and take part in celebrations honoring the jaguar, those things are part of your culture.

What Is Culture?

As used in geography, culture is the way of life of a group of people who share similar beliefs and customs. In studying a society's culture, geographers look at eight elements called *traits*. They study what groups the society is divided into, what language the people speak, and what religion they follow. They examine people's daily lives. They consider what history the people share and what artworks they have created. They also look at how the society is governed and how the people make a living.

Social Groups One way of studying cultures is by looking at the different groups of people in the society. For instance, geographers study how many people are rich, poor, and in the middle class. They

◀ Thousands of Indians meet for an annual Hindu festival.

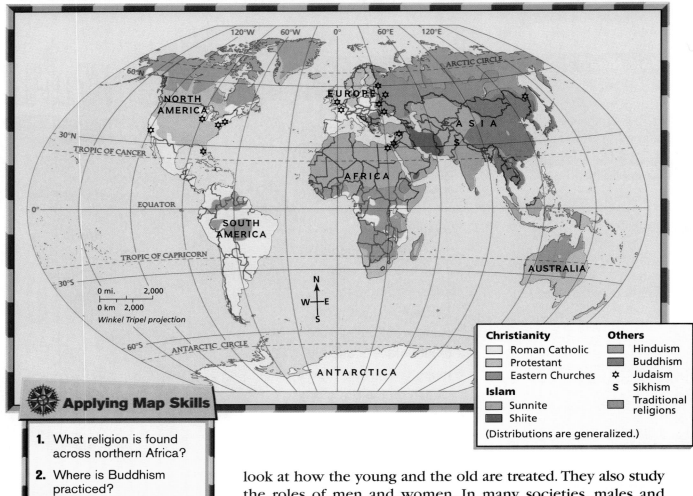

Christianity
- Roman Catholic
- Protestant
- Eastern Churches

Islam
- Sunnite
- Shiite

Others
- Hinduism
- Buddhism
- ✿ Judaism
- S Sikhism
- Traditional religions

(Distributions are generalized.)

Applying Map Skills

1. What religion is found across northern Africa?

2. Where is Buddhism practiced?

Find NGS online map resources @ www.nationalgeographic.com/maps

look at how the young and the old are treated. They also study the roles of men and women. In many societies, males and females have different rights and responsibilities. Understanding those differences is a vital part of understanding a particular culture.

Geographers also examine how people in the culture treat others who are different from themselves. Most countries include people who belong to different ethnic groups. An **ethnic group** is a group of people who share a common culture, language, or history. Many societies have people who have moved there from another place. Many have people who practice different religions.

Language People use language to share information. Sharing a language is one of the strongest unifying forces for a culture. Even within a culture, though, geographers find language differences. Some people may speak a **dialect,** or a local form of a language that differs from the language in other areas. The differences may include pronunciation and the meaning of words. For example, people in the northeastern United States say "soda," whereas people in the Midwest say "pop." Both groups are referring to soft drinks, however.

Religion Another important part of culture is religion. In many cultures, religion helps people answer basic questions about life's meaning. Religious beliefs vary significantly around the world. Struggles over religious differences are a problem in many countries. Some of the major world religions are Buddhism, Christianity, Hinduism, Islam, and Judaism. The map on page 78 shows you the main areas where these religions are practiced.

Daily Life Do you enjoy eating pizza, tacos, yogurt, and eggrolls? All of these foods came from different cultures. What people eat and *how* they eat it—whether with their fingers, silverware, chopsticks, and so on—reflect their culture. What people wear also reflects cultural differences. The same is true of how people build traditional homes in their societies.

History A culture group has a shared history, and that history shapes how they view the world. People remember the successes of the past and often celebrate holidays to honor the heroes and heroines who brought about those successes. Stories about these heroes reveal the personal characteristics that the people think are important. A group also remembers the dark periods of history, when they met with disaster or defeat. These experiences, too, influence how a group of people sees itself.

Arts People express their culture through the arts. Art is not just paintings and sculptures, but also architecture, dance, music, theater, and literature. By viewing the arts of a culture, you can gain insight into what the people of that culture think is beautiful and important.

Government The kind of government, or political system, a society has reflects its culture. Some countries are led by individuals. In a monarchy, kings and queens inherit the right to rule. In others, dictators take control of the government and rule the country as they wish.

In many countries today, power rests with the people of the nation. Citizens choose their leaders by voting for them. When the people of a country hold the power of government, we call that government a democracy. Some places have mixed forms of government. For example, the United Kingdom is both a monarchy and a democracy. The queen is the symbolic head of the country, but the power to rule is in the hands of elected leaders.

NATIONAL GEOGRAPHIC **On Location**

Celebrations

In most cultures, people often bring out their most beautiful clothes for events like weddings. This wedding guest is from Morocco (upper left), and this bride is from Mauritius.

Place What aspects of daily life besides clothing reflect culture?

The World's People

The Economy People must make a living, whether in farming or in industry or by providing services such as designing a Web page or preparing food. Geographers look at how people in a culture earn a living. They also look at the culture's economic system.

An economic system sets rules for how people decide what goods and services to produce and how they are exchanged. In a *traditional economy,* things are done "the way they have always been done." Economic decisions are based on customs and beliefs—often religious—handed down from generation to generation. For example, if your grandparents and parents fished for a living, you will fish for a living.

In a *market economy,* individuals make decisions about what to produce. People who own businesses make what they think customers want. Customers have the freedom to choose what products they will buy. A market economy is based on *free enterprise.* This is the idea that you have the right to own property or businesses and make a profit without the government interfering. People are free to choose what jobs they will do and for whom they will work.

In a *command economy,* however, the government owns businesses and controls decisions about what goods and services will be produced and who will receive them. A command economy is often called *socialism* or *communism,* depending on how much the government is involved. In some command economies, the government even decides which people receive training for particular jobs.

✓ Reading Check **What is culture?**

Sri Lanka

Some people in Sri Lanka harvest tea (left). Others work in factories making dolls to be sold around the world (below).

Place **On what is a market economy based?**

NATIONAL GEOGRAPHIC **On Location**

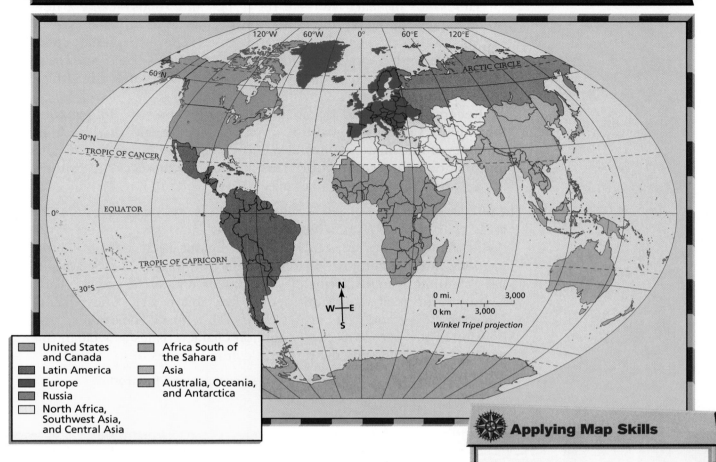

United States and Canada
Latin America
Europe
Russia
North Africa, Southwest Asia, and Central Asia
Africa South of the Sahara
Asia
Australia, Oceania, and Antarctica

Applying Map Skills

1. Which culture region includes most nations of Africa?

2. What culture region is on the continents of both Africa and Asia?

Find NGS online map resources @
www.nationalgeographic.com/maps

Cultural Change

Cultures do not remain the same. Humans constantly invent new ideas and technologies and create new solutions to problems. Trade, the movement of people, and war can spread these changes to other cultures. The process of spreading new knowledge and skills to other cultures is called cultural diffusion. Today the Internet is making cultural diffusion take place more rapidly than ever before.

Culture Over Time Historians have traced the tremendous changes that humans have made in their cultures. In the first human societies, people lived by hunting animals and gathering fruits and vegetables. They were *nomadic,* or lived in small groups that moved from place to place to follow sources of food.

Starting about 10,000 years ago, people learned to grow food by planting seeds. This change brought about the Agricultural Revolution. Groups stayed in one place and built settlements. Their societies became more complex. As a result, four civilizations, or highly developed cultures, arose in river valleys in present-day **Iraq, Egypt, India,** and **China.** These civilizations included cities, complex governments and religions, and systems of writing. The map on page 83 shows you where these civilizations were located.

Thousands of years later—in the 1700s and 1800s—came a new set of changes in the world. Some countries began to *industrialize,* or use machines and factories to make goods. These machines could work harder, faster, and longer than people or animals could. As a result of the Industrial Revolution, people began to live longer, healthier, more comfortable lives.

Recently, the world began a new revolution—the Information Revolution. Computers make it possible to store and process huge amounts of information. They also allow people to instantly send this information all over the world. This revolution connects the cultures of the world more closely than ever before.

Culture Regions   As you recall, geographers use the term *regions* for areas that share common characteristics. Today geographers often divide the world into areas called culture regions. Each **culture region** includes different countries that have traits in common. They share similar economic systems, forms of government, and social groups. Their languages are related, and the people may follow the same religion. Their history and art are similar. The food, costumes, and housing of the people all have common characteristics as well. In this textbook, you will study the different culture regions of the world.

✓ Reading Check **What three revolutions have changed the world?**

Section 1 Assessment

Defining Terms
1. **Define** culture, ethnic group, dialect, monarchy, dictator, democracy, economic system, cultural diffusion, civilization, culture region.

Recalling Facts
2. **Culture** What kinds of social groups do geographers study?
3. **Government** What are the different forms of government a society may have?
4. **Culture** In what ways does cultural diffusion occur?

Critical Thinking
5. **Understanding Cause and Effect** How does history shape a people's culture?
6. **Making Comparisons** Describe two kinds of economic systems.

Graphic Organizer
7. **Organizing Information** Create a diagram like this one that describes features of your culture. On the lines write the types of food, clothing, language, music, and so on.

Your Culture

Applying Geography Skills

8. **Analyzing Maps** Look at the map on page 81. In which culture region do you live? In which culture region(s) did your ancestors live?

Geography Skill

Reading a Special Purpose Map

Special purpose maps concentrate on a single theme. This theme may be to show the battles of a particular war or endangered species, for example.

Learning the Skill

To read a special purpose map, follow these steps:

- Read the map title. It tells what kind of special information the map shows.
- Find the map's scale to determine the general size of the area.
- Read the key. Colors and symbols in the map key are especially important on this type of map.
- Analyze the areas on the map that are highlighted in the key. Look for patterns.

Practicing the Skill

Look at the map below to answer the following questions.

1. What is the title of the map?
2. Read the key. What four civilizations are shown on this map?
3. Which civilization was farthest west? East?
4. What do the locations of each of these civilizations have in common?

Applying the Skill

Find a special purpose map in a newspaper or magazine. Write three questions about the map's purpose, then have a classmate answer the questions.

GO TO

Practice key skills with **Glencoe Skillbuilder Interactive Workbook, Level 1.**

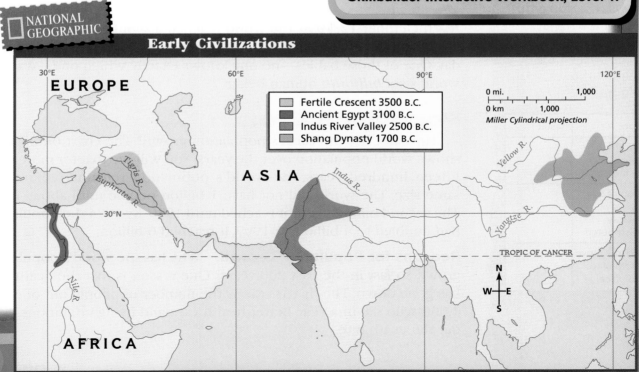

NATIONAL GEOGRAPHIC

Early Civilizations

Fertile Crescent 3500 B.C.
Ancient Egypt 3100 B.C.
Indus River Valley 2500 B.C.
Shang Dynasty 1700 B.C.

0 mi. 1,000
0 km 1,000
Miller Cylindrical projection

EUROPE

ASIA

AFRICA

Tigris R.
Euphrates R.
Nile R.
Indus R.
Yellow R.
Yangtze R.

30°E 60°E 90°E 120°E

30°N

TROPIC OF CANCER

N W E S

Guide to Reading

Main Idea

The world's population is growing rapidly, and how and where people live are changing, too.

Terms to Know

- death rate
- birthrate
- famine
- population density
- urbanization
- emigrate
- refugee

Places to Locate

- Afghanistan
- Nepal
- Mexico City
- Buenos Aires

Reading Strategy

Draw a chart like this one. In the right column, write a result of the fact listed in the left column.

Fact	Result
World population is increasing.	
Population is not evenly distributed.	
People move from place to place.	

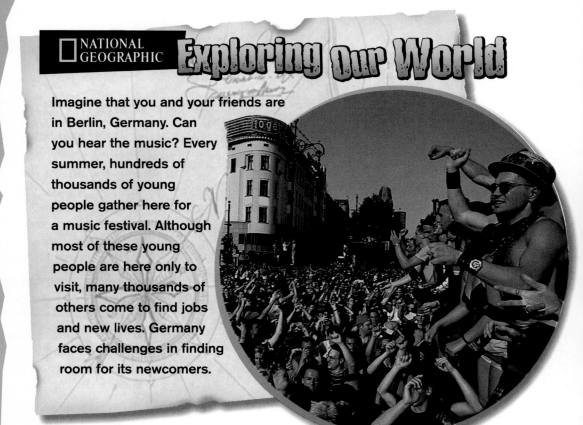

NATIONAL GEOGRAPHIC Exploring Our World

Imagine that you and your friends are in Berlin, Germany. Can you hear the music? Every summer, hundreds of thousands of young people gather here for a music festival. Although most of these young people are here only to visit, many thousands of others come to find jobs and new lives. Germany faces challenges in finding room for its newcomers.

On October 12, 1999, the world reached a significant turning point in its history. About 370,000 babies were born around the world that day. One of those babies—no one knows exactly which one—was the world's six *billionth* human being.

Population Growth

How fast has the earth's population grown? The chart on page 85 shows world population over the years. You will see that for more than fifteen hundred years, the world's population remained about the same size. The world did not have 1 billion people until about 1800. The second billion was not reached until 1930. By 1974 the population had doubled to 4 billion. In 1999, it reached 6 billion.

Reasons for Population Growth Why has the world's population grown so fast in the past 200 years? One reason is that the death rate has gone down. The **death rate** is the number of people out of every 1,000 who die in a year. Better health care and living conditions have cut the death rate.

POPULATION GROWTH

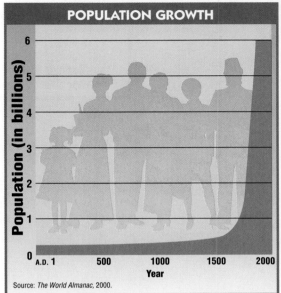

Population (in billions)

6
5
4
3
2
1
0

A.D. 1   500   1000   1500   2000

Year

Source: *The World Almanac*, 2000.

MOST POPULOUS COUNTRIES

Country	Millions of People
China	1,254.1
India	1,000.8
United States	281.0
Indonesia	211.8
Brazil	168.0
Pakistan	146.5
Russia	146.5

Sources: *National Geographic Atlas of the World; The World Almanac*, 2000.

Analyzing the Graph and Chart

The world's population is expected to reach about 9 billion by 2050.

Place Which country has the second-largest number of people?

Visit gwip.glencoe.com and click on **Chapter 3—Textbook Updates.**

Another reason for the fast growth in the world's population is that in some regions of the world the birthrate is high. The **birthrate** is the number of children born each year for every 1,000 people. In Asia, Africa, and Latin America, families traditionally are large because children help with farming. High numbers of births have combined with low death rates to increase population growth in these areas. As a result, population in these continents has doubled every 25 years or so.

Challenges From Population Growth   Rapid population growth presents many challenges. An increase in the number of people means that more food is needed. Fortunately, since 1950 world food production has increased faster than population on all continents except Africa. Because so many people there need food, bad weather or war can ruin crops and bring disaster. Millions may suffer from **famine,** or lack of food.

Also, populations that grow rapidly may use resources more quickly than populations that do not grow as fast. Some countries face shortages of water and housing. Population growth also puts a strain on economies. More people means a country must create more jobs. Some experts claim that rapid population growth could badly affect the planet. Others are optimistic. They predict that as the number of humans rises, the levels of technology and creativity will also rise.

✓Reading Check   **How do the definitions of death rate and birthrate differ?**

Where People Live

Where do all the people live? Actually, the world's people live on a surprisingly small part of the earth. As you learned in Chapter 2, land covers only about 30 percent of the earth's surface. Half of this land is

The World's People

World Population Density

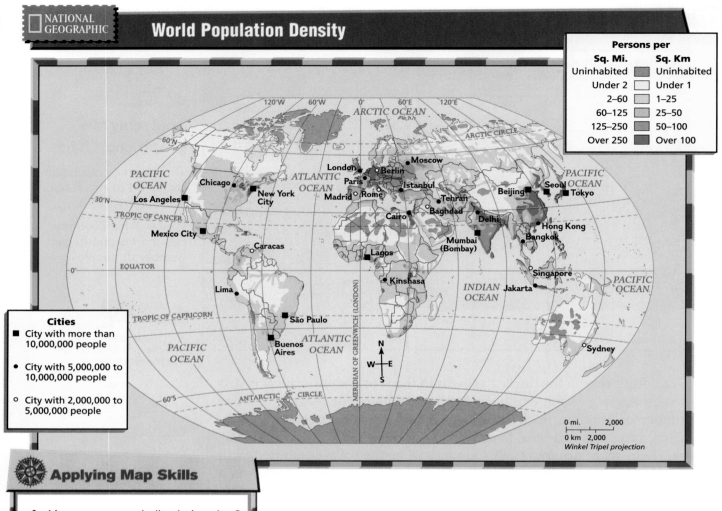

Persons per

Sq. Mi.		Sq. Km
Uninhabited		Uninhabited
Under 2		Under 1
2–60		1–25
60–125		25–50
125–250		50–100
Over 250		Over 100

Cities

■ City with more than 10,000,000 people

● City with 5,000,000 to 10,000,000 people

○ City with 2,000,000 to 5,000,000 people

Applying Map Skills

1. How many people live in London?

2. What cities have more than 10 million people?

Find NGS online map resources @
www.nationalgeographic.com/maps

not usable by humans, however. Large numbers of people cannot survive on land covered with ice, deserts, or high mountains. The world's people, then, live on a small fraction of the earth's surface.

Population Distribution Even on the usable land, population is not distributed, or spread, evenly. People naturally prefer to live in places that have plentiful water, good land, and a favorable climate. During the industrial age, people moved to places that had important resources such as coal or iron ore to run or make machines. People gather in other areas because these places hold religious significance or because they are government and transportation centers. The table on page 85 shows you the most populous countries in the world. Four of these countries are located on the Asian continent.

Population Density Geographers have a way of determining how crowded a country or region is. They measure population density—the average number of people living in a square mile or square kilometer. To arrive at this figure, the total population is divided by the total land area. For example, the countries of **Afghanistan** and **Nepal**

have about the same number of people. They are very different in terms of population density, though. With a smaller land area, Nepal has 447 people per square mile (173 people per sq. km). Afghanistan has an average of only 103 people per square mile (40 people per sq. km). Nepal, then, is more crowded than Afghanistan.

Remember that population density is an *average.* It assumes that people are distributed evenly throughout a country. Of course, this seldom happens. A country may have several large cities where most of the people actually live. In Egypt, for example, overall population density is 173 people per square mile (67 people per sq. km). In reality, about 99 percent of Egypt's people live within 20 miles (32 km) of the Nile River. The rest of Egypt is desert. Thus, some geographers prefer to figure a country's population density in terms of farmable or usable land rather than total land area. When Egypt's population density is measured this way, it equals about *5,550 people* per square mile. The map on page 13 of the **Geography Handbook** shows how population density can vary within a country. The areas with high density in Egypt follow the path of the Nile River.

✔ Reading Check What is population density?

Population Movement

Throughout the world, people are moving in great numbers from place to place. Some people move from city to city, or suburb to suburb. More and more people are leaving villages and farms and moving to cities. This movement to cities is called **urbanization.**

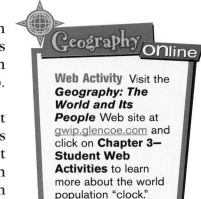

▢ NATIONAL GEOGRAPHIC On Location

Kosovo, Yugoslavia

In 1999 a civil war exploded in Kosovo, a province of Yugoslavia. Thousands of people were forced from their homes.

Movement What causes people to become refugees?

People move to cities for many reasons. The biggest one is to find jobs. Rural populations have grown, but the amount of land that can be farmed has not increased to meet the growing number of people who need to work and to eat. As a result, many people find city jobs in manufacturing or in services like tourism.

Nearly half the world's people live in cities—a far higher percentage than ever before. Between 1960 and 2000, the population of **Mexico City** more than tripled. Other cities in Latin America, as well as cities in Asia and Africa, have seen similar growth. Some of these cities hold a large part of a country's entire population. About one-third of Argentina's people, for instance, live in the city of **Buenos Aires.**

Some population movement is between countries. Some people emigrate, or leave the country where they were born and move to another. They are called *emigrants* in their homeland and referred to as *immigrants* in their new country. In the past 40 years, millions have left Africa, Asia, and Latin America to find jobs in the richer nations of Europe and North America. Some people were forced to flee their country because of wars, political unrest, food shortages, or other problems. They are refugees, or people who flee to another country to escape persecution or disaster.

✓Reading Check **Why do so many people move from rural areas to cities?**

Assessment

Defining Terms
1. **Define** death rate, birthrate, famine, population density, urbanization, emigrate, refugee.

Recalling Facts
2. **Culture** What are three problems caused by overpopulation?

3. **Human/Environment Interaction** Why do people live on only a small fraction of the earth?

4. **Economics** What is the main reason for growing urbanization?

Critical Thinking
5. **Making Comparisons** What is the difference between an emigrant and an immigrant?

6. **Understanding Cause and Effect** Why have populations in areas of Asia, Africa, and Latin America doubled every 25 years or so?

Graphic Organizer
7. **Organizing Information** Draw a diagram like this one, and list three causes of population growth.

```
Causes
_____
_____   ──→   Population Growth
_____
```

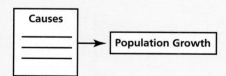
Applying Geography Skills

8. **Analyzing Maps** Look at the population density map on page 86. How would you describe the population density around Tokyo?

Counting Heads

How do we know there are more than 280 million people in the United States? Who counts the people? Every 10 years since 1790, the United States Census Bureau has counted heads in this country. Why and how do they do this?

The First Census

After the American colonies fought the Revolutionary War and won their independence, the new government ordered a census. By knowing how many people were in each state, the government could divide the war expenses fairly. The census would also determine the number of people that each state could send to Congress.

This census began in August 1790, about a year after George Washington became president. The law defined who would be counted and required that every household be visited by census takers. These workers walked or rode on horseback to gather their data. By the time it was completed, the census counted 3.9 million people.

The first census asked for little more than one's name and address. Over time, the census added questions to gather more than just population data. By 1820, there were questions about a person's job. Soon after, questions about crime, education, and wages appeared.

Changing Technology

As the country's population grew and the quantity of data increased, new technology helped census workers. In 1890 clerks began to use a keypunch device, invented by a Census Bureau worker, to add the numbers. The Tabulating Machine, as it was called, used an electric current to sense holes in punched cards and to keep a running total of the data. In 1950 the census used its first computer to process data. Now census data are released over the Internet.

Remarkably, one technology slow to change has been the way the government takes the census. Not until 1960 did the U.S. Postal Service become the major means of conducting the census. Even today, census takers go door-to-door to gather information from those who do not return their census forms in the mail.

▲ The Electric Tabulating Machine processed the 1890 census in 2½ years, a job that would have taken nearly 10 years to complete by hand.

➤ Making the Connection

1. In what two ways were population data from the first census used?

2. How has technology changed the way census data are collected and processed?

3. **Drawing Conclusions** Why do you think the national and state governments want information about people's education and jobs?

Resources and World Trade

Guide to Reading

Main Idea

Because many resources are limited and distributed unevenly, countries must trade for goods.

Terms to Know

- natural resource
- renewable resource
- nonrenewable resource
- export
- import
- tariff
- quota
- free trade
- developed country
- developing country

Places to Locate

- Brazil
- China

Reading Strategy

Draw a chart like the one below, then write in the names of different resources and how they are used.

Resource	Use

NATIONAL GEOGRAPHIC Exploring Our World

About 7,000 windmills stand guard on an 80-square-mile patch of hilly land near San Francisco. They turn in the strong winds that blow through a nearby pass in California's mountains. Why were they put there? These wind vanes generate electricity. In fact, they churn out enough electricity every year to meet the needs of all the homes in San Francisco.

Natural Resources

As you learned in Section 2, people settle in some areas to gain access to resources. Natural resources are products of the earth that people use to meet their needs. Wind, water, and oil are resources that provide energy to power machines. Good soil and fish are resources that people use to produce food. Stones like granite and ores like iron ore are resources people can use for making products.

The value of resources changes as people discover new technology. Trees, for example, have been a valuable resource throughout history. People used wood to stay warm, cook food, and build homes. Oil, on the other hand, was a gooey nuisance until the Industrial Revolution. People soon began to use oil products to run cars and heat homes. Today many countries in Southwest Asia have great pools of underground oil. As a result, they are among the richest countries on the earth.

Renewable Resources   People can use some natural resources as much as they want. These **renewable resources** cannot be used up or can be replaced naturally or grown again. Wind and sun cannot be used up—the wind will continue to blow even if we use a windmill to catch some of its power. Forests, grasslands, plants and animals, and soil can be replaced—*if* people manage them carefully. A lumber company concerned about future growth can plant as many new trees as it cuts. Fishing fleets can limit the number of fish they catch to make sure that enough fish remain to reproduce.

Today many countries are trying to find efficient ways of using renewable energy sources. Some produce *hydroelectric power,* the energy generated by falling water. Even if a river does not flow quickly, it is possible to turn it into hydroelectric power. Engineers can build a dam and then release the water in a powerful stream.

Do you have a solar-powered calculator? If so, you know that the sun can provide energy to run people's machines. *Solar energy* is power produced by the heat of the sun. Making use of this energy on a large scale requires huge pieces of equipment. As a result, this energy source is not yet economical to use.

Nonrenewable Resources   Humans use a great variety of minerals. They make steel from iron and aluminum from bauxite. Can you think of other examples? Metals and other minerals found in the earth's crust are also resources. They are **nonrenewable resources** because the earth provides limited supplies of them and they cannot be replaced. These resources were formed over millions of years by forces within the earth. Thus, it simply takes too long to generate new supplies.

One major nonrenewable source of energy is *fossil fuels*—coal, oil, and natural gas. People burn oil and gas to heat homes or run cars. They burn fossil fuels to generate electricity. Oil and coal are also used as raw materials to make plastics and medicines.

Another nonrenewable energy source is nuclear energy. *Nuclear energy* is power made by creating a controlled atomic reaction. Nuclear energy can be used to produce electricity, but some people fear its use. Nuclear reactions produce dangerous waste products that are difficult to dispose of. They need thousands of years to become safe. Some people worry that there is no safe way to transport and store this waste. Still, some countries rely on nuclear energy to generate power. France, Japan, South Korea, and Taiwan are examples.

✓ Reading Check   What are three fossil fuels?

World Trade

Resources, like people, are not distributed evenly around the world. Some areas have large amounts of one resource. Others have none of that resource but are rich in another one. These differences affect the economies of the world's countries.

What in the World?

Saffron—A Valuable Resource

A resource does not have to produce energy to be valued. The people in the Indian region of Kashmir are picking a resource that is precious to cooks—crocus flowers. Inside each crocus are three tiny orange stalks. When dried, the stalks become a spice called saffron. Cooks use it to add a delicate orange color and flavor to food. Saffron—the world's most expensive spice—is in short supply, though. Producers need nearly 4,700 flowers to produce just 1 ounce (28 g) of saffron.

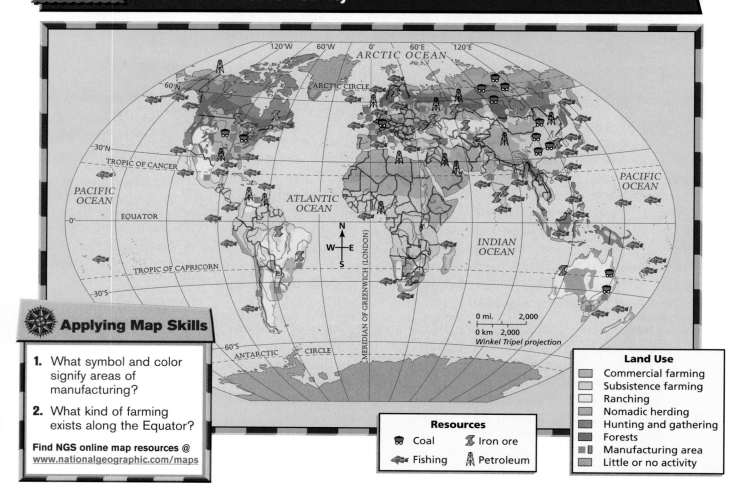

ARCTIC OCEAN

ARCTIC CIRCLE

60°N

30°N

TROPIC OF CANCER

PACIFIC OCEAN

EQUATOR

ATLANTIC OCEAN

PACIFIC OCEAN

TROPIC OF CAPRICORN

INDIAN OCEAN

30°S

MERIDIAN OF GREENWICH (LONDON)

0 mi. 2,000
0 km 2,000
Winkel Tripel projection

60°S ANTARCTIC CIRCLE

120°W 60°W 0° 60°E 120°E

N W E S

Applying Map Skills

1. What symbol and color signify areas of manufacturing?

2. What kind of farming exists along the Equator?

Find NGS online map resources @
www.nationalgeographic.com/maps

Resources

🚃 Coal   🪨 Iron ore
🐟 Fishing   🛢 Petroleum

Land Use
- Commercial farming
- Subsistence farming
- Ranching
- Nomadic herding
- Hunting and gathering
- Forests
- Manufacturing area
- Little or no activity

Look at the map above. Do you see the centers of manufacturing in the northern and eastern United States? There are large supplies of coal in the region and deposits of iron ore nearby. These areas became industrial centers because the people here took advantage of the resources they had.

In the western United States, you see another picture. People use much of the land for ranching. The soil and climate are well suited to raising livestock. *Commercial farming*—or growing food for sale in markets—occurs throughout much of the United States.

These examples show how the world's people respond to the unequal distribution of resources. They *specialize,* or focus on the economic activities best suited to their resources. Parts of **Brazil** have the perfect soil and climate for growing coffee. As a result, Brazil produces more coffee beans than any other country. Cotton grows well in parts of **China,** the world's top producer of that product.

Countries often cannot use all that they produce. What do they do with the extra? They export what they do not need, trading it to other countries. When they cannot produce as much as they need of a good, they import it, or buy it from another country. The world's countries, then, are connected with one another in a complex web of trade.

Barriers to Trade   Governments try to manage their country's trade to benefit their country's economy. Some charge a tariff, or a tax added to the price of goods that are imported. If there is a tariff on cars, for instance, all people who buy an imported car pay more than just the purchase price. Governments often create tariffs hoping to persuade their own people to buy products made in their own country instead of buying imported goods.

Governments sometimes create other barriers to trade. They might put a strict quota, or number limit, on how many items of a particular product can be imported from a particular country. A government may even stop trading with another country altogether as a way to punish it.

Free Trade   In recent years, governments around the world have moved toward free trade. Free trade means taking down trade barriers so that goods flow freely among countries. Several countries have joined together to create free trade agreements in certain parts of the world. For instance, the United States, Mexico, and Canada have agreed to eliminate all trade barriers to one another's goods. These three countries set up the North American Free Trade Agreement (NAFTA). The largest free trade agreement—the European Union (EU)—includes most of the countries of Western Europe.

√ Reading Check   **What are three kinds of barriers to trade?**

NATIONAL GEOGRAPHIC **On Location**

Earning a Living

Factory workers in China sew clothes (left). Most people still farm in Kenya (above). In the United States, more jobs are in service or technology industries.

Region **What regions are considered developed?**

Differences in Development

Look again at the economic activity map on page 92. You see that countries differ in how much manufacturing they have. Countries that have a great deal of manufacturing are called **developed countries.** The countries of Europe and North America are developed countries. So are Australia and Japan. Other countries have only a few—or no—manufacturing centers. Many people in these countries practice *subsistence farming,* or grow only enough food for their own families. These countries, which are working toward industrialization, are called **developing countries.** Developing countries are found throughout Africa, Asia, and Latin America.

Why do countries want manufacturing? Workers in industry typically earn more than farmers. Industrial companies can generally charge more for their products than food companies can. As a result, industrial countries are wealthier than agricultural countries.

In recent decades, the economies of many countries have changed dramatically. Developing countries have invited large companies from developed countries to build factories in their lands. The companies find that developing countries have a valuable resource—people. The spread of industry has created booming economies in places like Hong Kong, Singapore, South Korea, Taiwan, and China.

✓Reading Check **Why do developing countries want more industry?**

Assessment

Defining Terms
1. **Define** natural resource, renewable resource, nonrenewable resource, export, import, tariff, quota, free trade, developed country, developing country.

Recalling Facts
2. **History** How can the value of resources change over time?
3. **Economics** What is the difference between commercial farming and subsistence farming?
4. **Economics** Why do countries specialize in producing certain goods?
5. **Economics** What is the difference between a developing country and a developed country?

Critical Thinking
6. **Drawing Conclusions** Why are tariffs and quotas called "barriers" to trade?

7. **Analyzing Information** How has the world's economy changed in recent decades?

Graphic Organizer
8. **Organizing Information** Draw a chart like this one, listing three examples for each type of resource.

Renewable resources	Nonrenewable resources

Applying Geography Skills

9. **Analyzing Maps** Look at the economic activity map on page 92. What two types of farming are shown on the map?

People and the Environment

Guide to Reading

Main Idea

The actions that people take have a huge effect on the environment.

Terms to Know

- desalinization
- conservation
- pesticide
- ecosystem
- crop rotation
- irrigation
- erosion
- deforestation
- acid rain

Places to Locate

- Saudi Arabia

Reading Strategy

Draw a diagram like this one. On the lines write at least two problems that arise with human use of water, land, and air.

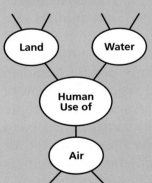

■ NATIONAL GEOGRAPHIC **Exploring Our World**

What happens when we harm the environment? Some plants and animals might be gone forever. This scientist hopes to prevent that. She works at a seed bank. Behind her, stored at −4°F (−20°C), are jars of plant seeds from around the world. Should any of these 4,000 types of plants become extinct, these seeds can start growing them again.

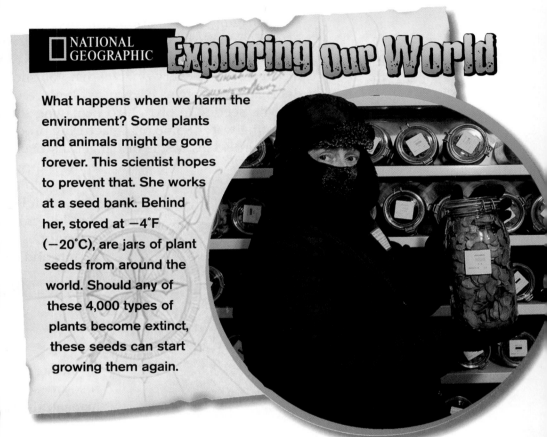

The rapidly growing number of people threatens the delicate balance of life in the world. More and more people use more water. They need more land to live on and to grow more food. Spreading industry fouls the air. Humans must act carefully to be sure not to destroy the earth that gives us life.

Water Use

As you recall from Chapter 2, people, plants, and animals need freshwater to live. People need clean water to drink. They also need water for their crops and their animals. In fact, as much as 70 percent of the water used is for farming. Only a small fraction of the world's water is freshwater, though. Some countries, such as **Saudi Arabia,** obtain drinkable water through desalinization—or removing salt from seawater. This is a process in which seawater is boiled in huge chambers, and the resulting steam is condensed as freshwater. For now, though, this process is expensive and cannot be widely used. Because the earth's supply of water is limited, people must manage freshwater carefully.

Water Management   Some regions receive heavy rainfall in some months of the year and little, or none, in other months. They can manage their water supply by building storage areas to hold the heavy rains for later use. People have to build these storage areas carefully to prevent too much of the water from evaporating.

Managing water supplies involves two main steps. The first step is conservation, or the careful use of resources so they are not wasted. Did you know that 6 or 7 gallons (23 to 27 liters) of water go down the drain every *minute* that you shower? Taking shorter showers is an easy way to prevent wasting water.

The second approach to managing the water supply is to avoid polluting water. Some manufacturing processes use water. Sometimes those processes result in dangerous chemicals or dirt entering the water supply. Many farmers use fertilizers to help their crops grow. Many also use pesticides, or powerful chemicals that kill crop-destroying insects. These substances can seep into the water supply and cause harm.

✓ Reading Check   How can manufacturing and farming harm the water supply?

Land Use

In addition to managing water resources, people must carefully manage the land they use. As humans expand their communities, they invade ecosystems. These are places where the plants and animals are dependent upon one another—and their particular surroundings—for survival. Ecosystems can be found in every climate and vegetation region of the world. For example, some people may want to drain a wet, marshy area to get rid of disease-carrying mosquitoes and make the soil useful for farming. When the area is drained, however, the ecosystem is destroyed. The delicate balance among the insects, reptiles, birds, and water plants is torn apart. In managing land, humans must recognize that some soils are not suited for growing certain crops—and some land is not useful for growing any crops at all.

Soil   To grow food, soil needs to have certain minerals. Farmers add fertilizers to the soil to supply some of these minerals. Some also

NATIONAL GEOGRAPHIC On Location

Disappearing Rain Forests

Workers harvest trees in a rain forest in Ecuador.

Human/Environment Interaction Why is cutting down the rain forests a problem?

practice **crop rotation,** or changing what they plant in a field to avoid using up all the minerals in the soil. Some crops—like beans—actually restore valuable minerals to the soil. Many farmers now plant bean crops every three years to build the soil back up.

In many dry areas, farmers use **irrigation** to deliver water to their crops. Over time, the small amounts of salt in this freshwater can build up on the land. Once that happens, the land is no longer fertile.

If people do not carefully manage the soil, it can erode away. In **erosion,** wind or water carries soil away, leaving the land less fertile than before. Have you ever seen a group of trees alongside a farmer's field? The farmer may have planted those trees to block the wind and prevent erosion. In the tropics, erosion by water presents a problem—especially if farmers plant their crops on sloping land. When heavy rains come, the soil may simply wash down the hillside.

Forests Some areas of the world have thick forests of tall trees. Growing populations in these countries often turn to these lush forests as a source of land to grow food. Yet **deforestation,** or cutting down forests, is a problem—especially in the tropics. Rains in these areas are extremely heavy. When the tree roots are no longer there to hold the soil, the water can wash it away. Often, the cleared land is fertile for just a few years. Then farmers have to move to a new area and cut the trees in that section of forest. As a result, more and more of the forest is lost over time.

Reading Check **What problem can result from irrigation?**

NATIONAL GEOGRAPHIC **On Location**

Taiwan

Burning fossil fuels adds harmful chemicals to the air.

Human/Environment Interaction What are some effects of air pollution?

Air Pollution

Industries and vehicles that burn fossil fuels are the main sources of air pollution. Throughout the world, fumes from cars and other vehicles pollute the air. The chemicals in air pollution can seriously damage people's health.

These chemicals also combine with precipitation. They then fall as **acid rain,** or rain containing high amounts of chemical pollutants. Acid rain kills fish and eats away at the surfaces of buildings. It can even destroy entire forests.

The World's People

Some scientists believe that increasing amounts of pollutants in the atmosphere will cause the earth to warm. You learned about this greenhouse effect in Chapter 2. While not all experts agree, some scientists say that the increase in temperature can have disastrous effects. Glaciers and ice caps may melt, raising the level of the world's seas. Higher seas could flood coastal cities. Warmer temperatures can also make some land no longer able to produce food.

☑ Reading Check **What causes acid rain?**

Balancing People and Resources

Water, land, and air are among people's most precious resources. We need water and air to live. We need land to grow food. Only by caring for these resources can we be sure that we will still have them to use in the future.

Sometimes, though, protecting the environment for the future seems to clash with feeding people in the present. For example, farmers destroy the rain forests not because they want to but because they need to feed their families. They dislike being told by people in other countries that they should save the rain forests. Before they stop cutting down the rain forests, these farmers will need to find new ways to meet their needs.

☑ Reading Check **How does saving the rain forests clash with current human needs?**

Assessment

Defining Terms
1. **Define** desalinization, conservation, pesticide, ecosystem, crop rotation, irrigation, erosion, deforestation, acid rain.

Recalling Facts
2. **Human/Environment Interaction** How do human activities affect ecosystems?
3. **Human/Environment Interaction** What are three ways of managing water?
4. **Economics** Why do farmers practice crop rotation?

Critical Thinking
5. **Making Comparisons** Which resource—water, soil, or air—do you think is most precious to people? Why?

6. **Analyzing Information** What ecosystems were affected by the growth of your community?

Graphic Organizer
7. **Organizing Information** Draw a diagram like this one and list three results of global warming.

Global warming

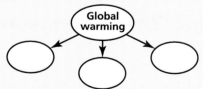

Applying Geography Skills

8. **Analyzing Maps** Look at the vegetation map in Chapter 2 on page 66. In what parts of the world do you see tropical rain forests?

Reading Review

Section 1 Culture

Terms to Know

culture
ethnic group
dialect
monarchy
dictator
democracy
economic
 system
cultural
 diffusion
civilization
culture
 region

Main Idea

People usually live with others who follow similar beliefs learned from the past.

✓Culture *Culture* is the way of life of a group of people who share similar beliefs and customs.

✓Culture Culture includes eight elements or traits: social groups, language, religion, daily life, history, arts, a government system, and an economic system.

✓Culture Cultures change over time and influence other regions.

Section 2 Population

Terms to Know

death rate
birthrate
famine
population
 density
urbanization
emigrate
refugee

Main Idea

The world's population is growing rapidly, and how and where people live are changing, too.

✓History In the past 200 years, the world's population has grown at a very rapid rate.

✓Movement Some areas are more densely populated than others.

✓Culture About 50 percent of the world's people live in cities.

Section 3 Resources and World Trade

Terms to Know

natural
 resource
renewable
 resource
nonrenewable
 resource
export
import
tariff
quota
free trade
developed
 country
developing
 country

Main Idea

Because many resources are limited and distributed unevenly, countries must trade for goods.

✓Human/Environment Interaction Renewable resources cannot be used up or can be replaced fairly quickly.

✓Human/Environment Interaction Some resources—such as fossil fuels and minerals—are nonrenewable.

✓Economics Countries specialize by producing what they can produce best with the resources they have.

✓Economics Countries export their specialized products and import what they need.

Section 4 People and the Environment

Terms to Know

desalinization
conservation
pesticide
ecosystem
crop rotation
irrigation
erosion
deforestation
acid rain

Main Idea

The actions that people take have a profound effect on the environment.

✓Human/Environment Interaction People need to manage water resources because freshwater is not available everywhere.

✓Human/Environment Interaction Air pollution has damaging effects on the land and on people's health.

Chapter 3 Assessment and Activities

Using Key Terms

Match the terms in Part A with their definitions in Part B.

A.

1. culture
2. developed country
3. irrigation
4. crop rotation
5. population density
6. culture region
7. tariff
8. quota
9. developing country
10. cultural diffusion

B.

a. collecting water and bringing it to crops
b. spreading knowledge to other cultures
c. countries working toward industrialization
d. many different countries with cultural traits in common
e. a number limit on imports from a country
f. the average number of people living in a square mile
g. country where much manufacturing is carried out
h. the way of life of a group of people who share similar beliefs and customs
i. a tax added to the price of imported goods
j. alternating what is planted in a field

Reviewing the Main Ideas

Section 1 Culture
11. **Culture** What are the major religions?
12. **Economics** What is the difference between a market economy and socialism?
13. **Movement** Give an example of cultural diffusion.

Section 2 Population
14. **Culture** What has created rapid population growth?
15. **Culture** How do you calculate population density?
16. **Movement** Why have many people moved to cities?

Section 3 Resources and World Trade
17. **Human/Environment Interaction** What are three kinds of renewable energy sources?
18. **Economics** How do countries respond to the unequal distribution of resources?

Section 4 People and the Environment
19. **Human/Environment Interaction** How can farmers restore the minerals in the soil?
20. **Human/Environment Interaction** What two problems can result from air pollution?

 **World Culture Regions**

Place Location Activity

On a separate sheet of paper, match the letters on the map with the numbered places listed below.

1. Latin America
2. North Africa, Southwest Asia, and Central Asia
3. Europe
4. Russia
5. East Asia
6. United States and Canada
7. Australia, Oceania, and Antarctica
8. Africa South of the Sahara

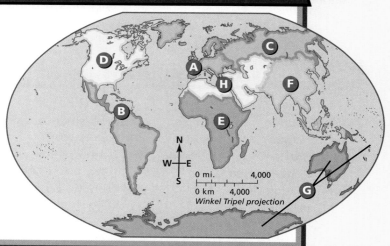

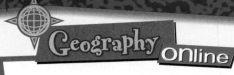

Self-Check Quiz Visit the *Geography: The World and Its People* Web site at gwip.glencoe.com and click on **Chapter 3— Self-Check Quizzes** to prepare for the Chapter Test.

Critical Thinking

21. **Making Predictions** In what ways do you think a company investing in a developing country could help the people there? How could that same company harm the culture?

22. **Sequencing Information** Make a chart like the one below, and list the ways you use electricity from the moment you wake up until you go to sleep. In the second column, write how you would perform the same activity if you had no electricity to rely on.

Activities With Electricity	Without Electricity

GeoJournal Activity

23. **Writing a Paragraph** Write a paragraph about the settlement of your community. Answer such questions as: Why did people originally settle in the area? How has the culture of your area changed?

Mental Mapping Activity

24. **Focusing on the Region** Draw a simple outline map of the United States. On your map, label the areas where the following activities take place:

- Commercial farming
- Manufacturing
- Raising livestock
- Fishing
- Obtaining oil

Technology Skills Activity

25. **Developing Multimedia Presentations** Research how your state's climate influences its culture, including tourist attractions, types of clothing, and the economy. Use your research to develop a commercial promoting your state.

Standardized Test Practice

Directions: Study the graph, then answer the following question.

Exports by Culture Region

Source: *Britannica Book of the Year*, 1999.

1. **According to the graph, how much do the United States and Canada export?**

   A $1,005,900,000,000

   B $1,005,900,000

   C $1,005,900

   D $1,005

Test-Taking Tip: In order to understand any type of graph, look carefully around the graph for keys that show how it is organized. On this bar graph, the numbers along the left side represent billions of dollars. Therefore, you need to multiply the number on the graph by 1,000,000,000 to get your answer.

◀ **Skier in Idaho's stretch of the Rocky Mountains**

Grocer in Chinatown, New York City

Farm on the Manitoba plains

NATIONAL
GEOGRAPHIC
SOCIETY

The United States and Canada

Which of the world's culture regions do you call home? It is probably the United States and Canada. If you look at a globe, you will see that the United States and Canada cover most of North America. These two nations share many of the same land-forms, including rugged mountains in the west, rounded mountains in the east, and rolling plains in the center.

NGS ONLINE
www.nationalgeographic.com/education

Focus on:
The United States and Canada

SPANNING MORE THAN 7 MILLION square miles (18 million sq. km), the United States and Canada cover much of North America. These huge countries share many of the same landscapes, climates, and natural resources.

The Land

The United States and Canada make up a region bordered by the very cold Arctic Ocean in the north and bathed by the Gulf of Mexico's warm currents in the south. The western coast faces the Pacific Ocean. Eastern shores are edged by the Atlantic.

Rugged mountains are found in the western part of each country. The Pacific ranges follow the coastline. Farther inland are the massive, jagged peaks of the Rocky Mountains. Relatively young as mountains go, the Rockies stretch more than 3,000 miles (4,828 km) from Alaska to the southwestern United States.

East of the Rockies are the wide and windswept Great Plains. This gently rolling landscape covers the central part of both the United States and Canada. In the United States, the Mississippi River—the largest river system in North America—flows through the heart of these plains.

The Appalachian range, much older than the Rockies, is the dominant landform in the eastern part of the region. East and south of the Appalachians' low, rounded peaks are coastal plains that end at the Atlantic shores.

The Climate

This region's vast size and varied landforms help give it great diversity in climate and vegetation. In the far northern parts of Alaska and Canada, amid the treeless tundra and dense evergreen forests, brief summers and bitterly cold winters prevail. The Pacific coast, from southern Alaska to northern California, has a mild, wet climate. Rain clouds blowing in from the ocean are blocked by the Pacific ranges. Robbed of moisture, the land immediately east of these mountains is dry.

Hot, humid summers and cold, snowy winters are the rule on the Great Plains. This humid continental climate extends from the plains across southeastern Canada and the northeastern United States. The southeastern states, however, enjoy much milder winters. The mildest of all are found on Florida's southern tip, the only part of the mainland that has a tropical climate.

Parachutist plunging toward the
Appalachian Mountains, West Virginia ▶

◀ Polar bear snoozing in the
Canadian Arctic

The Economy

The United States and Canada are prosperous countries. Abundant natural resources and plenty of skilled workers have been key ingredients in creating two of the most successful economies in the world. Both countries operate under the free enterprise system, in which individuals and groups—not the government—control businesses and industries.

The region's strong economy was built on agriculture, which remains important today. Fertile soil, numerous waterways, a favorable climate, and high-tech equipment have made the United States and Canada two of the world's top food producers. Livestock, grains, vegetables, and fruits all are raised by the region's farmers.

Rich oil, coal, and natural gas deposits occur in this region. So do deposits of valuable minerals, including copper, iron ore, nickel, silver, and gold. These energy sources and raw materials have made it possible for the United States and Canada to develop large industrial economies. Today, however, people are more likely to work in offices than in factories. Service industries such as banking, communications, entertainment, insurance, and health care employ most people in the region.

The People

The United States and Canada have a rich mix of cultures. Native Americans were the nations' first inhabitants. Centuries later, settlers from Europe arrived. Immigrants from Africa, Asia, Latin America, and almost every other part of the world eventually followed. Some came looking for religious or political freedom. Some came as enslaved laborers. Some came for a fresh start in these immense lands of boundless opportunity.

Today more than 310 million people call this region home. Thirty-one million of them live in Canada, while the remaining 281 million live in the United States. On either side of the border, most people live in urban areas. Toronto, Vancouver, and Montreal are among Canada's largest cities. In the United States, New York City, Los Angeles, and Chicago are the most populous cities.

Exploring the Region

1. **Which oceans border the region?**
2. **Why is the climate dry just east of the Pacific ranges?**
3. **What factors have helped make the region prosperous?**
4. **In which country do most of the region's people live?**

◀ **Worker in sterile gown manufacturing computer chips in Texas**

Inuit boys examining a Native American sculpture ▶

REGIONAL ATLAS

The United States and Canada

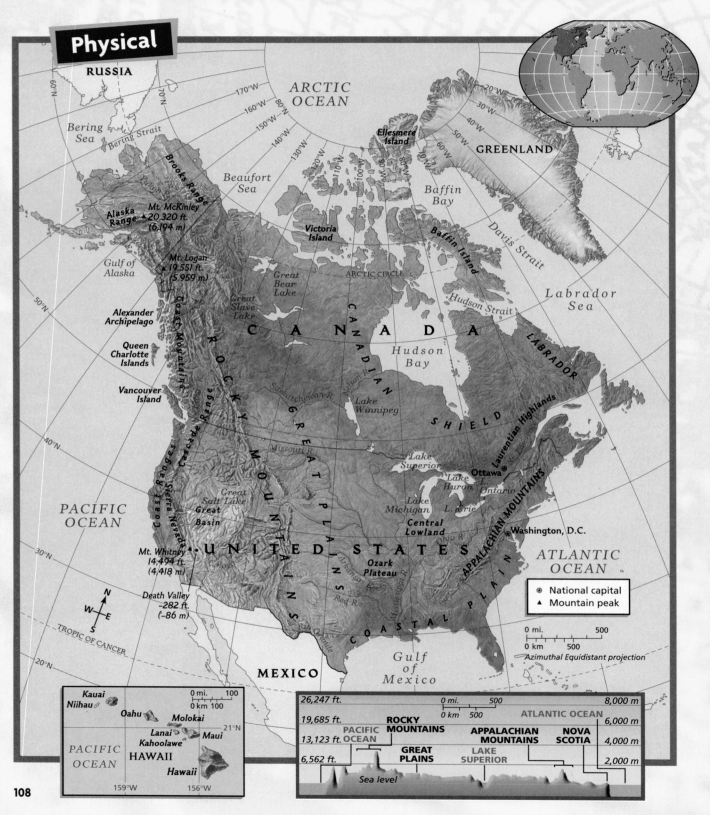

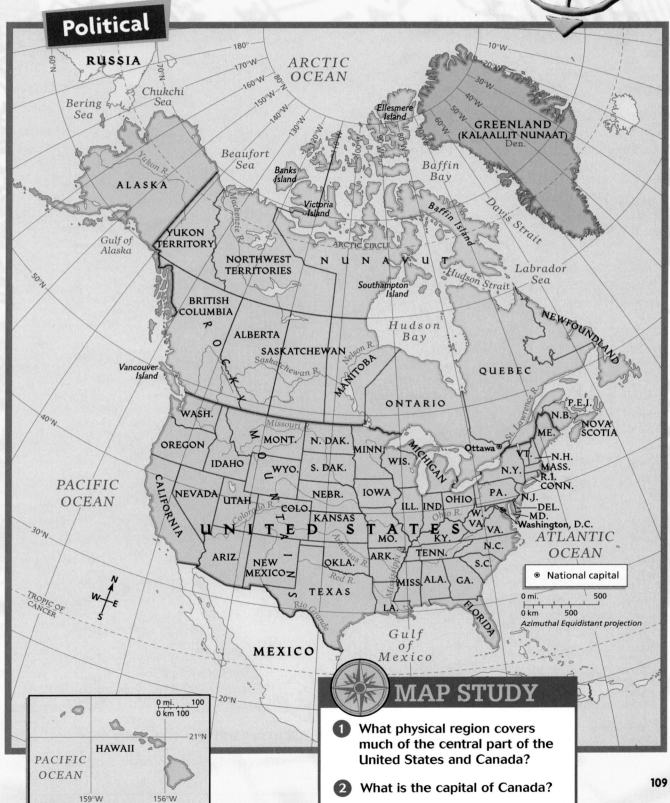

Political

RUSSIA

ARCTIC OCEAN

180°
170°W
160°W
150°W
140°W
130°W
120°W
10°W
20°W
30°W
40°W
50°W
60°W

60°N
70°N
80°N

Bering Sea

Chukchi Sea

Beaufort Sea

Ellesmere Island

GREENLAND (KALAALLIT NUNAAT) Den.

ALASKA

Yukon R.

Banks Island

Baffin Bay

Gulf of Alaska

Mackenzie R.

Victoria Island

Baffin Island

Davis Strait

YUKON TERRITORY

ARCTIC CIRCLE

NORTHWEST TERRITORIES

NUNAVUT

Labrador Sea

50°N

BRITISH COLUMBIA

Southampton Island

Hudson Strait

NEWFOUNDLAND

Vancouver Island

ALBERTA

SASKATCHEWAN

Saskatchewan R.

Nelson R.

MANITOBA

Hudson Bay

QUEBEC

40°N

PACIFIC OCEAN

WASH.

OREGON

IDAHO

MONT.

N. DAK.

MINN.

WIS.

MICHIGAN

ONTARIO

Ottawa ⊛

St. Lawrence R.

P.E.I.

N.B.

ME.

NOVA SCOTIA

VT.

N.H.

MASS.

N.Y.

R.I.

CONN.

Missouri R.

WYO.

S. DAK.

IOWA

PA.

N.J.

NEVADA

UTAH

Colorado R.

COLO.

NEBR.

ILL.

IND.

OHIO

W. VA.

DEL.

MD.

Washington, D.C.

CALIFORNIA

KANSAS

MO.

KY.

VA.

ATLANTIC OCEAN

ARIZ.

NEW MEXICO

OKLA.

Arkansas R.

ARK.

TENN.

N.C.

S.C.

UNITED STATES

ROCKY MOUNTAINS

Red R.

TEXAS

MISS.

ALA.

GA.

⊛ National capital

30°N

TROPIC OF CANCER

LA.

FLORIDA

0 mi.   500
0 km   500

Azimuthal Equidistant projection

Rio Grande

MEXICO

Gulf of Mexico

20°N

0 mi.   100
0 km   100

21°N

HAWAII

PACIFIC OCEAN

159°W   156°W

N W E S

MAP STUDY

1 What physical region covers much of the central part of the United States and Canada?

2 What is the capital of Canada?

REGIONAL ATLAS

The United States and Canada

Food Production

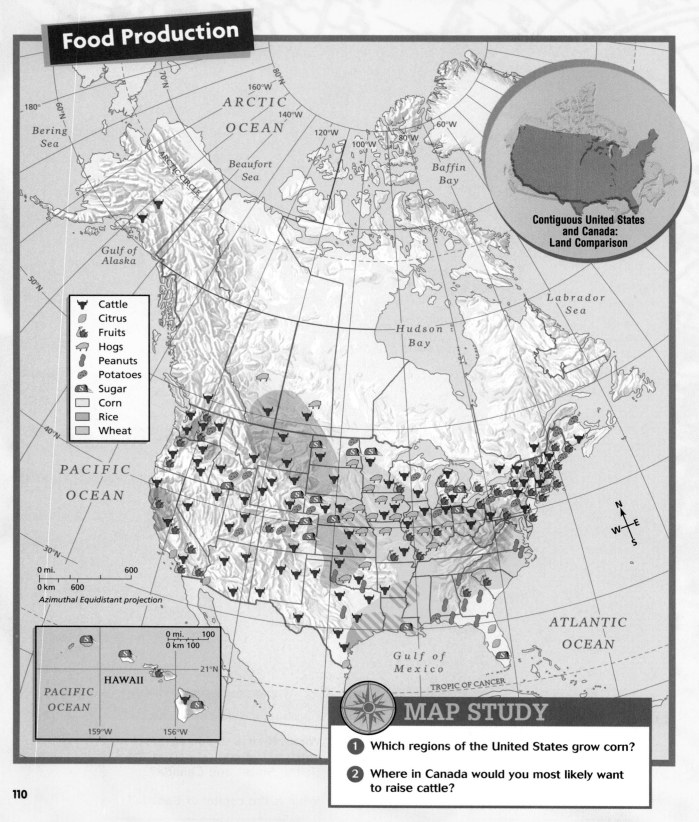

Contiguous United States and Canada: Land Comparison

Legend:
- Cattle
- Citrus
- Fruits
- Hogs
- Peanuts
- Potatoes
- Sugar
- Corn
- Rice
- Wheat

0 mi. 600
0 km 600
Azimuthal Equidistant projection

0 mi. 100
0 km 100
21°N
HAWAII
PACIFIC OCEAN
159°W 156°W

MAP STUDY

1. Which regions of the United States grow corn?

2. Where in Canada would you most likely want to raise cattle?

Geo Extremes

① HIGHEST POINT
Mount McKinley (Alaska)
20,320 ft. (6,194 m) high

② LOWEST POINT
Death Valley (California)
282 ft. (86 m)
below sea level

③ LONGEST RIVER
Mississippi-Missouri
(United States)
3,710 mi. (5,971 km) long

④ LARGEST LAKE
Lake Superior
31,700 sq. mi.
(82,103 sq. km)

⑤ LARGEST CANYON
Grand Canyon (Arizona)
277 mi. (446 km) long
1 mi. (1.6 km) deep

⑥ GREATEST TIDES
Bay of Fundy (Nova Scotia)
52 ft. (16 m)

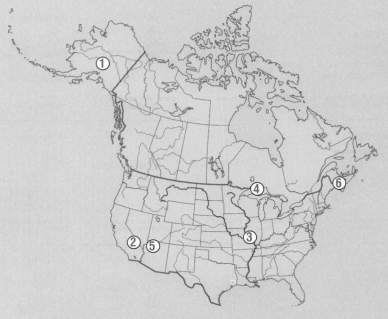

COMPARING POPULATION:
United States and Canada

UNITED STATES

CANADA

= 50,000,000

Source: *Population Reference Bureau*, 2000.

ETHNIC GROUPS:
United States and Canada

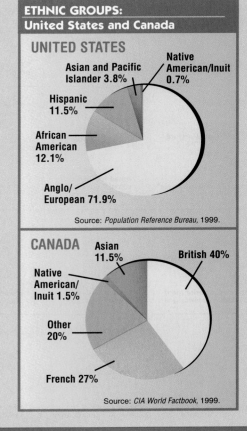

UNITED STATES

Asian and Pacific Islander 3.8%
Native American/Inuit 0.7%
Hispanic 11.5%
African American 12.1%
Anglo/European 71.9%

Source: *Population Reference Bureau*, 1999.

CANADA

Asian 11.5%
British 40%
Native American/Inuit 1.5%
Other 20%
French 27%

Source: *CIA World Factbook*, 1999.

GRAPHIC STUDY

① In what area of the United States do you find both the lowest point and largest canyon?

② How does the percentage of Native American/Inuit population in the United States compare with their percentage of the population in Canada?

Country Profiles

UNITED STATES

POPULATION:
281,000,000
75 per sq. mi.
29 per sq. km

LANGUAGE:
English

MAJOR EXPORT:
Machinery

CAPITAL:
Washington, D.C.

MAJOR IMPORT:
Crude Oil

LANDMASS:
3,717,796 sq. mi.
9,629,091 sq. km

Washington, D.C.

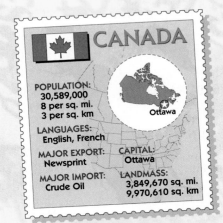

CANADA

POPULATION:
30,589,000
8 per sq. mi.
3 per sq. km

LANGUAGES:
English, French

MAJOR EXPORT:
Newsprint

CAPITAL:
Ottawa

MAJOR IMPORT:
Crude Oil

LANDMASS:
3,849,670 sq. mi.
9,970,610 sq. km

Ottawa

U.S. State Names: Meaning and Origin

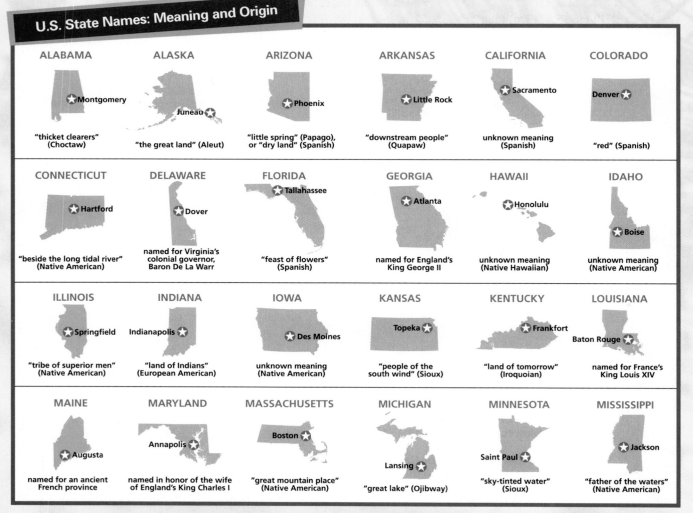

ALABAMA — Montgomery — "thicket clearers" (Choctaw)

ALASKA — Juneau — "the great land" (Aleut)

ARIZONA — Phoenix — "little spring" (Papago), or "dry land" (Spanish)

ARKANSAS — Little Rock — "downstream people" (Quapaw)

CALIFORNIA — Sacramento — unknown meaning (Spanish)

COLORADO — Denver — "red" (Spanish)

CONNECTICUT — Hartford — "beside the long tidal river" (Native American)

DELAWARE — Dover — named for Virginia's colonial governor, Baron De La Warr

FLORIDA — Tallahassee — "feast of flowers" (Spanish)

GEORGIA — Atlanta — named for England's King George II

HAWAII — Honolulu — unknown meaning (Native Hawaiian)

IDAHO — Boise — unknown meaning (Native American)

ILLINOIS — Springfield — "tribe of superior men" (Native American)

INDIANA — Indianapolis — "land of Indians" (European American)

IOWA — Des Moines — unknown meaning (Native American)

KANSAS — Topeka — "people of the south wind" (Sioux)

KENTUCKY — Frankfort — "land of tomorrow" (Iroquoian)

LOUISIANA — Baton Rouge — named for France's King Louis XIV

MAINE — Augusta — named for an ancient French province

MARYLAND — Annapolis — named in honor of the wife of England's King Charles I

MASSACHUSETTS — Boston — "great mountain place" (Native American)

MICHIGAN — Lansing — "great lake" (Ojibway)

MINNESOTA — Saint Paul — "sky-tinted water" (Sioux)

MISSISSIPPI — Jackson — "father of the waters" (Native American)

Countries, states, provinces, and flags not drawn to scale

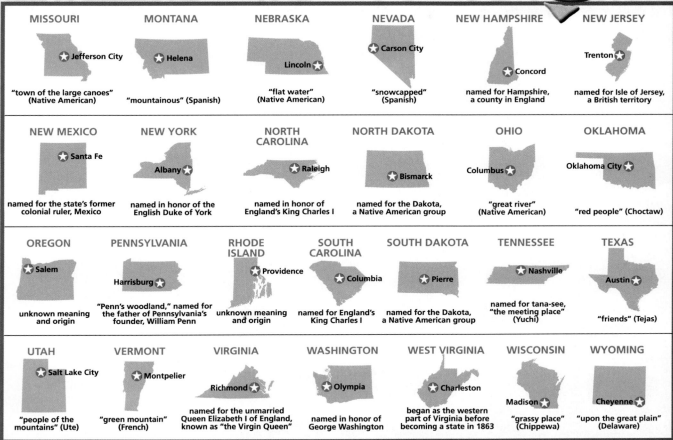

MISSOURI

Jefferson City

"town of the large canoes" (Native American)

MONTANA

Helena

"mountainous" (Spanish)

NEBRASKA

Lincoln

"flat water" (Native American)

NEVADA

Carson City

"snowcapped" (Spanish)

NEW HAMPSHIRE

Concord

named for Hampshire, a county in England

NEW JERSEY

Trenton

named for Isle of Jersey, a British territory

NEW MEXICO

Santa Fe

named for the state's former colonial ruler, Mexico

NEW YORK

Albany

named in honor of the English Duke of York

NORTH CAROLINA

Raleigh

named in honor of England's King Charles I

NORTH DAKOTA

Bismarck

named for the Dakota, a Native American group

OHIO

Columbus

"great river" (Native American)

OKLAHOMA

Oklahoma City

"red people" (Choctaw)

OREGON

Salem

unknown meaning and origin

PENNSYLVANIA

Harrisburg

"Penn's woodland," named for the father of Pennsylvania's founder, William Penn

RHODE ISLAND

Providence

unknown meaning and origin

SOUTH CAROLINA

Columbia

named for England's King Charles I

SOUTH DAKOTA

Pierre

named for the Dakota, a Native American group

TENNESSEE

Nashville

named for tana-see, "the meeting place" (Yuchi)

TEXAS

Austin

"friends" (Tejas)

UTAH

Salt Lake City

"people of the mountains" (Ute)

VERMONT

Montpelier

"green mountain" (French)

VIRGINIA

Richmond

named for the unmarried Queen Elizabeth I of England, known as "the Virgin Queen"

WASHINGTON

Olympia

named in honor of George Washington

WEST VIRGINIA

Charleston

began as the western part of Virginia before becoming a state in 1863

WISCONSIN

Madison

"grassy place" (Chippewa)

WYOMING

Cheyenne

"upon the great plain" (Delaware)

Canadian Province and Territory Names: Meaning and Origin

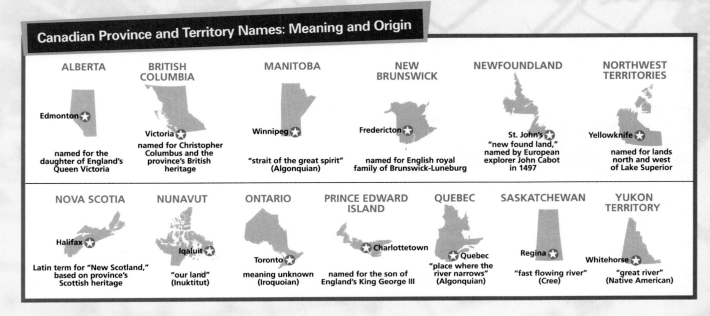

ALBERTA

Edmonton

named for the daughter of England's Queen Victoria

BRITISH COLUMBIA

Victoria

named for Christopher Columbus and the province's British heritage

MANITOBA

Winnipeg

"strait of the great spirit" (Algonquian)

NEW BRUNSWICK

Fredericton

named for English royal family of Brunswick-Luneburg

NEWFOUNDLAND

St. John's

"new found land," named by European explorer John Cabot in 1497

NORTHWEST TERRITORIES

Yellowknife

named for lands north and west of Lake Superior

NOVA SCOTIA

Halifax

Latin term for "New Scotland," based on province's Scottish heritage

NUNAVUT

Iqaluit

"our land" (Inuktitut)

ONTARIO

Toronto

meaning unknown (Iroquoian)

PRINCE EDWARD ISLAND

Charlottetown

named for the son of England's King George III

QUEBEC

Quebec

"place where the river narrows" (Algonquian)

SASKATCHEWAN

Regina

"fast flowing river" (Cree)

YUKON TERRITORY

Whitehorse

"great river" (Native American)

Chapter

4

The United States

The World and Its People ■ NATIONAL GEOGRAPHIC

To learn more about the people and places of the United States, view **The World and Its People Chapter 4** video.

Geography ONLINE

Chapter Overview Visit the **Geography: The World and Its People** Web site at gwip.glencoe.com and click on **Chapter 4– Chapter Overviews** to preview information about the United States.

A Vast, Scenic Land

Guide to Reading

Main Idea

The United States has a great variety of landforms and climates.

Terms to Know

- contiguous
- megalopolis
- coral reef

Places to Locate

- Atlantic Coastal Plain
- Gulf Coastal Plain
- Appalachian Mountains
- Central Lowland
- Mississippi River
- Great Lakes
- Great Plains
- Rocky Mountains
- Mt. McKinley

Reading Strategy

Make a chart like the one below. Fill in details about each of the seven physical regions of the United States.

Region	Details

NATIONAL GEOGRAPHIC

Exploring Our World

Who in the continental United States gets to see the sunrise first? The people in Maine are the first. As the earth rotates, the sun shines on an extremely varied land. It warms the valleys in the east, shimmers on the lakes in the north, and bakes the deserts in the southwest. In the far Pacific, the sun greets Hawaii's tropical beaches. Finally, the sun sets beyond Alaska in the north.

The United States stretches 2,807 miles (4,517 km) across the middle part of North America. The 48 states in this part of the country are **contiguous**, or joined together inside a common boundary. These states touch the Atlantic Ocean, the Gulf of Mexico, and the Pacific Ocean. Our neighbors are Canada to the north and Mexico to the south.

Two states lie apart from the other 48. Alaska—the largest state—lies in the northwestern portion of North America. Hawaii is in the Pacific Ocean, about 2,400 miles (3,862 km) southwest of California.

From Sea to Shining Sea

The United States ranks as the fourth-largest country in the world. Only Russia, Canada, and China are larger. Like a patchwork quilt, the United States has regional patterns of different landscapes. You can see swamps and deserts, tall mountains and flat plains.

The contiguous states have five main physical regions: the Coastal Plains, the Appalachian Mountains, the Interior Plains, the Mountains and Basins, and the Pacific Coast. Alaska and Hawaii each have their own unique set of physical landforms.

◄ Monument Valley, Utah

The Coastal Plains A broad lowland runs along the eastern and southeastern coasts of the United States. The eastern lowlands are called the **Atlantic Coastal Plain.** The lowlands in the southeast border the Gulf of Mexico and are called the **Gulf Coastal Plain.**

Find the Atlantic Coastal Plain on the map on page 117. Stretching from Massachusetts to Florida, it widens the farther south you go. Excellent harbors along the Atlantic Coastal Plain led to the growth of shipping ports like New York City. The soil in the northern part of this region tends to be thin and rocky.

Boston, New York City, Philadelphia, Baltimore, and Washington, D.C., all lie in the Atlantic Coastal Plain. These cities and their suburbs form an almost continuous line of settlement. Geographers call this kind of huge urban area a megalopolis. Find these cities on the map on page 133.

NATIONAL GEOGRAPHIC

The United States: Political

Albers Conic Equal-Area projection

- ⊛ National capital
- ⊙ State capital

Applying Map Skills

1. What is the national capital of the United States?

2. Which states do not lie within the 48 contiguous states?

Find NGS online map resources @
www.nationalgeographic.com/maps

Find the Gulf Coastal Plain on the map below. This plain is wider than the Atlantic plain. Soils in this region are better than those along the Atlantic coast. Texas and Louisiana both have rich deposits of oil and natural gas. The large cities of the Gulf Coastal Plain include Houston and New Orleans, which are shown on the map on page 133.

The Appalachian Mountains Along the western edge of the Atlantic Coastal Plain rise the **Appalachian** (A•puh•LAY•chuhn) **Mountains.** The second-longest range in North America, the Appalachians run almost 1,500 miles (2,414 km) from eastern Canada to Alabama. They are the oldest mountains on the continent. How can you tell? Their rounded peaks show their age. Erosion has worn them down over time. The highest peak, Mount Mitchell in North Carolina, reaches 6,684 feet (2,037 m).

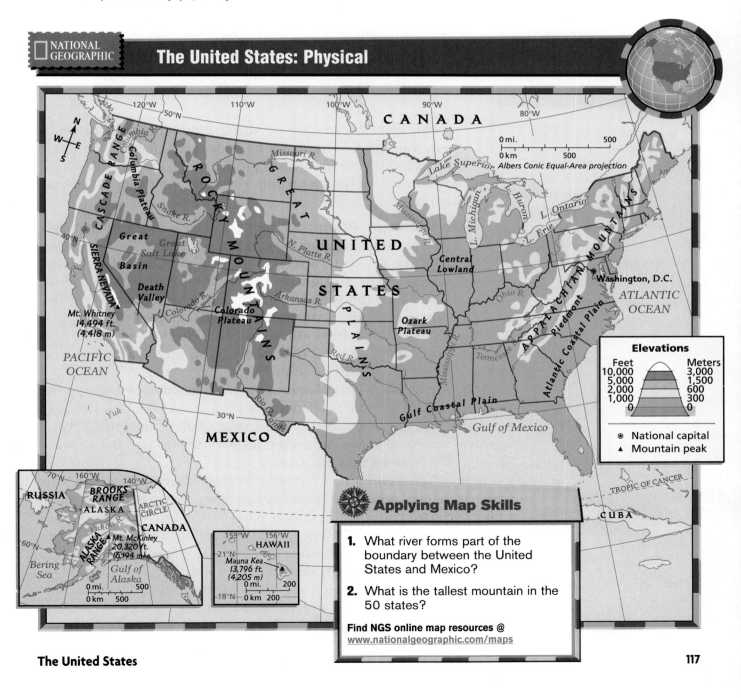

NATIONAL GEOGRAPHIC

The United States: Physical

Elevations

Feet	Meters
10,000	3,000
5,000	1,500
2,000	600
1,000	300
0	0

⊗ National capital
▲ Mountain peak

0 mi. 500
0 km 500
Albers Conic Equal-Area projection

Applying Map Skills

1. What river forms part of the boundary between the United States and Mexico?

2. What is the tallest mountain in the 50 states?

Find NGS online map resources @
www.nationalgeographic.com/maps

NATIONAL GEOGRAPHIC On Location

City and Country

The Interior Plains of the United States include industrial cities of the north, like Chicago (above), and the agricultural lands of the Great Plains, like this area in Texas (above right).

Region What river drains much of the Interior Plains?

On the eastern side of these mountains lies the hilly Piedmont area. The land here is very fertile. Farms in Pennsylvania's Piedmont area produce high crop yields. On the western side of the mountains, you find a system of valleys and high ridges.

The Interior Plains When you cross the Appalachians, heading west, you enter the vast Interior Plains. This region has two parts. The eastern section is called the **Central Lowland.** Here you will find grassy hills, rolling flatlands, and thick forests. The land is fertile, and farms are productive. This area also contains important waterways. The **Mississippi River** flows 2,340 miles (3,766 km) from Minnesota south to the Gulf of Mexico. Barges carry goods along the river. Many finish their journey in New Orleans, where the goods are shipped to other countries.

The **Great Lakes**—the largest group of freshwater lakes in the world—lie in the Central Lowland. Glaciers formed Lake Superior, Lake Michigan, Lake Huron, Lake Erie, and Lake Ontario millions of years ago. The waters of these connected lakes flow into the St. Lawrence River, which empties into the Atlantic Ocean.

West of the Central Lowland stretch the **Great Plains.** The landscape, blanketed with neat fields of grain and grassy pastures, takes on a checkerboard pattern. This region starts on the low western bank of the Mississippi River. Slowly the land rises to a height of about 6,500 feet (1,981 m) at the base of the Rocky Mountains. The Plains are about 500 miles (805 km) wide and stretch into Canada and down to the Mexican border. The rich grasslands of the Plains once provided food for millions of buffalo and the Native Americans who lived there. Today farmers grow grains and ranchers raise cattle.

Mountains and Basins West of the Great Plains rise the majestic **Rocky Mountains**—the longest mountain range in North America. This range begins in Alaska and runs all the way south to Mexico.

As you read in Chapter 1, the surface of the earth rides on huge sheets of rock called tectonic plates. Sometimes these plates collide with such force that they push the land up very high. That is how the Rocky Mountains were formed. The force of this collision has raised some peaks more than 14,000 feet (4,267 m). Running along these mountains is a ridge called the Continental Divide. This ridge separates rivers that flow west—toward the Pacific Ocean—from those that flow east—toward the Mississippi River. Many important rivers begin in the high, snowy peaks of the Rockies. The Rio Grande, Missouri, Platte, Arkansas, and Red Rivers all flow east. The Colorado, Snake, and Columbia Rivers flow west.

As you move west of the Rocky Mountains, you find three large plateaus. Find the northernmost one—the Columbia Plateau—on the map on page 117. South of it you see another plateau called the Great Basin, which includes the Great Salt Lake. The lake's high salt levels make it easy for you to float in it. The third plateau farther south is the Colorado Plateau. Here is where you find the brilliant orange and red rocks shown in the photograph on page 114. The Grand Canyon is also part of the magnificent scenery of this plateau.

The Pacific Coast Near the Pacific coast rise two other mountain ranges. Like the Rockies, the peaks in these ranges are very high. The Cascade Range reaches from Washington State south to California. Volcanoes formed these high peaks—and some of them still erupt. Along California's eastern side runs the Sierra Nevada. The name *Nevada* means "snow covered" in Spanish. Even in a place as far south as California, these high mountains remain covered with snow.

To the west of these Pacific ranges lie fertile valleys. The Willamette Valley in Oregon and the Central Valley in California both produce abundant crops. Many of the fruits and vegetables you eat may come from these valleys.

Alaska Mountain ranges form a semicircle over the northern, eastern, and south-central parts of Alaska. **Mt. McKinley**—the tallest mountain in North America—stands 20,320 feet (6,194 m) high in the Alaska Range. The northern part of the state borders on the frigid Arctic Ocean, and you can almost see Russia from Alaska's western shores. Most people in Alaska live along the southern coastal plain or in the central Yukon River valley.

Hawaii Eight large islands and more than 120 smaller islands make up Hawaii, our western state in the Pacific Ocean. Volcanoes on the ocean floor erupted and formed these islands. Some of the islands have coral reefs, formed by the skeletons of small sea animals. These structures lie just above or submerged just below the surface of the water.

✔ Reading Check How were the Rocky Mountains formed?

A Variety of Climates

Because the United States is such a large country, you probably expect it to have a variety of climates. You are right! Most of the country lies squarely in the middle latitude region—from 30°N to 60°N latitude. As you remember from Chapter 2, this part of the earth has the greatest variety of climates. With Alaska and Hawaii, our country also has high latitude and tropical climates.

Mid-Latitude Climates Look at the climate map below. As you can see, the area from the northern Great Plains to the northern Atlantic coast has a humid continental climate. Winters in this region are cold, and summers are hot and long. Rain falls on and off throughout the year. Snow often blankets the area during the winter—especially around the Great Lakes.

The southern part of the Great Plains and the southeastern states have a humid subtropical climate. Winters are milder than in the north, and summers are hot and humid. Being close to the Gulf of Mexico and the Caribbean Sea often means that summer months bring violent thunderstorms, hurricanes, and tornadoes.

NATIONAL GEOGRAPHIC

The United States: Climate

Tropical
- Tropical rain forest
- Tropical savanna

Dry
- Steppe
- Desert

Mid-Latitude
- Marine west coast
- Mediterranean
- Humid subtropical
- Humid continental

High Latitude
- Subarctic
- Tundra
- Highlands (climate varies with elevation)

Applying Map Skills

1. How many climate zones would you experience if you traveled directly from Miami to St. Louis to Denver to Los Angeles?

2. In which climate zone do you live?

Find NGS online map resources @
www.nationalgeographic.com/maps

As these views of a country lane in Vermont show, the humid continental climate region has four distinct seasons of the year.

Region How does a humid continental climate differ from a humid subtropical climate?

The Pacific coast from northern California up to Washington has a marine west coast climate. Temperatures are mild year-round, and winds from the Pacific bring plenty of rainfall. Huge forests of tall trees grow in these areas.

Southern California has a Mediterranean climate. People in this area enjoy dry, warm summers and mild, rainy winters. Fruits and vegetables, including grapes for wine, thrive in the Mediterranean climate of this region.

Dry Climates Much of the western Great Plains has a dry steppe climate. This region is dry much of the year, but enough rain falls to support the growth of thick grasses. You also find steppe climates on the plateaus west of the Rockies. Why are these areas dry? The Pacific mountain ranges block the humid ocean winds. Therefore, hot, dry air gets trapped in the plateaus and basins between the Pacific mountain ranges and the Rockies.

In the southwest, even less rain falls. This arid region has a desert climate, and temperatures climb very high. In southeastern California's Death Valley, the lowest point in the United States, temperatures soar as high as 125°F (52°C) in the summer. Turn to the vegetation map on page 66 of Chapter 2 and the photograph on page 69 to see what types of vegetation grow in the dry climate zones.

Hawaii

Lush tropical growth is found in Hawaii, our 50th state.

Location In what ocean is Hawaii located?

High Latitude Climates People in Alaska experience the cold climates of high latitude regions. The southern two-thirds of the state has a subarctic climate, with cool summers and freezing winters. Rainfall is heavy in coastal regions and lighter in the interior. Only small bushes can grow in the far north's tundra climate.

As you recall from Chapter 2, mountains have cool climates. Look at the climate map on page 120. Compare the highland climate areas to the physical map on page 117 to see which mountain ranges have highland climates.

Tropical Climates The southern tip of Florida is the only part of the contiguous United States with a tropical climate. This area has heavy rainfall and warm temperatures all year long. However, Hawaii is the wettest state in the nation. It lies within the tropics, yet cooling breezes from the ocean keep temperatures moderate. Average temperatures in Honolulu range from 72°F (22°C) in the cool months to 78°F (26°C) in the warm months.

✓ Reading Check Why are dry climates found in the western United States?

Section 1 Assessment

Defining Terms

1. **Define** contiguous, megalopolis, coral reef.

Recalling Facts

2. **Place** How does the United States rank in size among all the countries of the world?

3. **History** Which region once supported millions of buffalo and the Native Americans who depended on them?

4. **Place** What is the Continental Divide?

Critical Thinking

5. **Understanding Cause and Effect** What evidence indicates that the Appalachians are the oldest mountains in North America?

6. **Drawing Conclusions** What challenges do you think result from the distance between Alaska, Hawaii, and the other states?

Graphic Organizer

7. **Organizing Information** Create a diagram like this one to compare the Atlantic and Gulf Coastal Plains. In the overlapping area, write the characteristics that the two areas share. In the separate outer parts of the ovals, write the qualities that make each region unique.

Atlantic Coastal Plain   Gulf Coastal Plain

Applying Geography Skills

8. **Analyzing Maps** Look at the physical map on page 117 and the climate map on page 120. At what elevation does the city of Denver lie? What climate zone do you find just west of Denver?

An Economic Leader

Guide to Reading

Main Idea

The powerful United States economy runs on abundant resources and the hard work of Americans.

Terms to Know

- free enterprise system
- service industry
- navigable
- fossil fuels
- acid rain
- landfill
- recycling
- free trade

Places to Locate

- New York City
- Washington, D.C.
- Los Angeles

Reading Strategy

Complete a chart like this one. First, list the five economic regions of the United States. In the right column, list the economic activities carried out in each region.

Region	Economic Activities

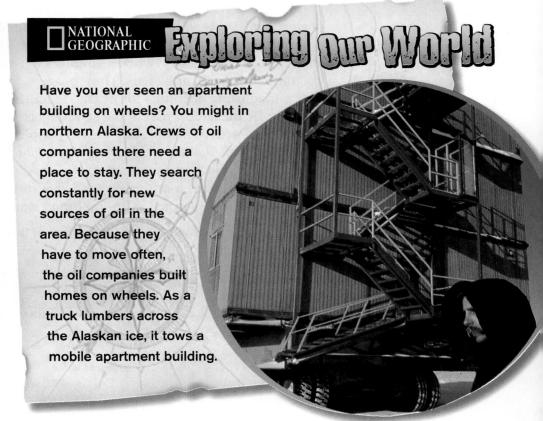

NATIONAL GEOGRAPHIC Exploring Our World

Have you ever seen an apartment building on wheels? You might in northern Alaska. Crews of oil companies there need a place to stay. They search constantly for new sources of oil in the area. Because they have to move often, the oil companies built homes on wheels. As a truck lumbers across the Alaskan ice, it tows a mobile apartment building.

The United States has a large, energetic, and growing economy. Fueling all of this economic activity is freedom. As you recall from Chapter 3, the free enterprise system is built on the idea that individual people have the right to run businesses to make a profit with limited interference from the government. Americans are free to start their own businesses and to keep the profits they earn after paying taxes. They are free to work in whatever jobs they want. This has helped create great economic success.

The World's Economic Leader

The United States is rich in resources and has a hardworking labor force. As a result, the country has built the world's largest economy—in terms of how much money is made from the sale of its goods and services. In fact, the American economy is larger than the next two largest economies—China's and Japan's—combined.

Farms in the United States produce about one-half of the world's corn and about one-tenth of its wheat. American farmers raise about 20 percent of the world's beef, pork, and lamb. The country exports more food than any other nation. Yet agriculture is only a small part of the American economy. Food makes up less than 2 percent of the value of all goods produced in the country.

The United States has rich mineral resources. About one-fifth of the world's coal and copper and one-tenth of the world's petroleum come from the United States. The country also has large amounts of iron ore, zinc, lead, silver, gold, and many other minerals. Mining, though, makes up little more than 1 percent of the nation's economy.

American factory workers build cars and airplanes. They make computers and appliances. They process foods and make medicines. Manufacturing accounts for nearly one-fifth of the American economy.

By far, the largest part of the economy is services. A **service industry** is a business that provides services to people instead of producing goods. Banking and finance are services. So is entertainment—and people all over the world buy American movies and CDs. The United States is a leader in tourism, another service industry. Computer-based, online services have also emerged as an important American service industry.

✓ Reading Check **What is the largest part of the United States economy?**

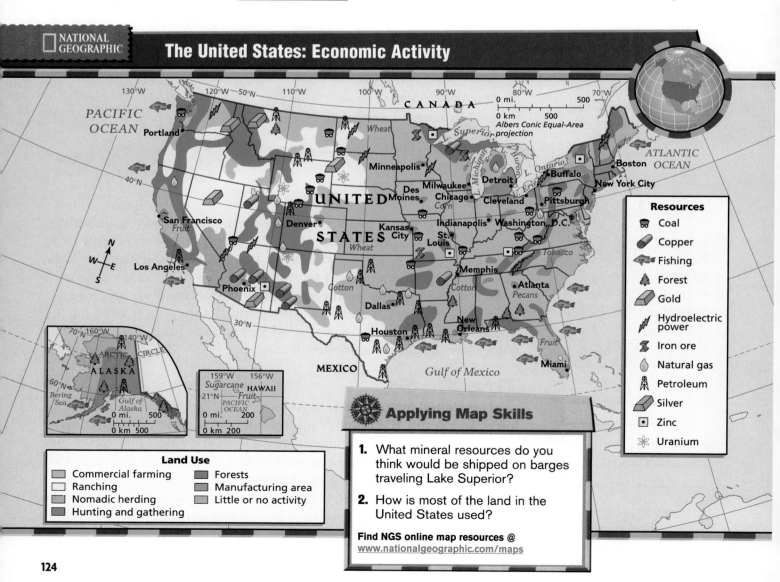

The United States: Economic Activity

NATIONAL GEOGRAPHIC

Resources
- Coal
- Copper
- Fishing
- Forest
- Gold
- Hydroelectric power
- Iron ore
- Natural gas
- Petroleum
- Silver
- Zinc
- Uranium

Land Use
- Commercial farming
- Ranching
- Nomadic herding
- Hunting and gathering
- Forests
- Manufacturing area
- Little or no activity

Applying Map Skills

1. What mineral resources do you think would be shipped on barges traveling Lake Superior?

2. How is most of the land in the United States used?

Find NGS online map resources @ www.nationalgeographic.com/maps

America's Economic Regions

Geographers group states together into five economic regions—the Northeast, the South, the Midwest, the Interior West, and the Pacific.

The Northeast Some farmers in central Pennsylvania and western New York grow grains and fruits. Yet as you read in Section 1, the rocky soil and steep hills in this region are a challenge to the farmers. The area boasts plenty of deep water ports and swiftly moving rivers, though. As a result, manufacturing, trade, and fishing are the heartbeat of this region. In fact, the Northeast was home to the first mills powered by running water and coal. Look at the economic activity map on page 124. As you can see, coal mining takes place in Pennsylvania and West Virginia.

With their deep, natural harbors, Boston, New York City, Philadelphia, and Baltimore all are important ports. Goods are shipped all over the world to and from these ports. These cities filled with skyscrapers are also important centers of banking, insurance, and finance. **New York City** is one of the financial capitals of the world. It is also a world center in the fashion, entertainment, and communications industries. Farther south, the nation's capital—**Washington, D.C.**—employs hundreds of thousands in government and tourism services.

The South With rich soil on most of the Coastal Plains, agriculture flourishes in the Southern states. Because of the region's warm, wet climate, farmers in Louisiana and Arkansas grow rice and sugarcane. Tobacco flourishes in Virginia and the Carolinas. In Florida you can sample the citrus fruits that farmers grow. Peanuts and pecans are found in Georgia. You also see cotton growing in several Southern states, including Texas. Texas, in fact, has more farms than any other state in the country. The people of Texas grow cotton, sorghum, and wheat and raise livestock as well.

The traditional image of the South as an agricultural region is changing, however. As you tour the South, you see expanding cities, growing industries, and diverse populations. New manufacturing centers have drawn new businesses and people to the South from the Northeast and elsewhere. Workers make textiles, electrical equipment, and airplane parts. Oil is found in Texas and Louisiana. As a result, they produce petroleum-based products.

Service industries are important in the South as well. Florida is a major tourist center. People come to enjoy amusement parks in Orlando, the Kennedy Space Center at Cape Canaveral, and the beautiful beaches on both coasts. Millions of people flock to New Orleans each year to taste its spicy food and hear its lively music. Houston, Dallas, Atlanta, and Miami are just a few of this region's major centers of business and finance.

The Midwest This part of the United States has been called "America's breadbasket." Miles and miles of grain and soybean fields greet you as you travel over flat land and fertile soil. In this farm belt, farmers grow corn, soybeans, oats, and wheat to feed animals and people all over the world. Dairy farms in the upper

Surf's Up!
Fourteen-year-old Shawn Kilgore lives on Florida's Captiva Island, along the Gulf of Mexico. "I really like warm weather," he says. "Who needs snow? My dog Sunny and I couldn't go surfing if we lived in Ohio where my cousins are." Shawn's parents manage a resort for tourists. "My mom and dad are always reminding me that we live in one of the world's richest countries. So my older sister and I volunteer to grocery-shop for people around here who can't do it themselves."

NATIONAL GEOGRAPHIC On Location

Farms

A farmer harvests alfalfa. America's economy was originally built on agriculture.

Region What economic region do you think this farm is in? How can you tell?

Midwest produce milk and cheese. However, technology has changed many farms from small family-owned acres to big businesses. The graph in the **Geography Handbook** on page 15 shows you the decrease in the number of farms over the past few decades.

Many of the region's rivers are **navigable,** or wide and deep enough to allow the passage of ships. As a result, many cities here are major ports—even though they are far from an ocean. Businesses in Cincinnati and Louisville send goods down the Ohio River. St. Louis and Memphis serve as centers of trade along the Mississippi River. Chicago's and Cleveland's industries ship goods through the Great Lakes and St. Lawrence Seaway to ports around the world.

Because of their abundant coal and iron resources, many cities in the Midwest are manufacturing centers. A complex network of railroads also helps the region's many industries. Detroit is called Motown (short for Motor Town) because the country's auto industry started and grew here. Other major industries include steel, heavy machinery, and auto parts.

The Interior West Magnificent landscapes greet visitors to this region. However, this area is short on an important resource—water. With its dry climate, the region discourages farming. Yet grasses thrive in much of the land, and where the land is irrigated, you find agriculture. Large areas are used for raising cattle and sheep. Ranches here may be huge—as large as 4,000 acres (1,619 ha). In the past, cowhands worked the range on horseback. Although they still use horses today, you are just as likely to see them driving a sturdy truck.

Look at the map on page 124. You see rich deposits of minerals and energy resources in the Interior West. The discovery of gold and silver in the mountains and riverbeds drew settlers here 150 years ago. Mining still plays an important role in the economy.

Many people work in service industries, too. Every year tourists travel to Denver, Salt Lake City, Albuquerque, and Phoenix. They use these cities as starting points for trips to sites such as Yellowstone National Park or the Grand Canyon. Some visit the ruins of ancient Native American settlements, such as those found at Mesa Verde in southwestern Colorado.

The Pacific The Pacific region includes the states on the western coast plus Alaska and Hawaii. The fertile valleys of California, Oregon, and Washington produce large amounts of food. As you learned in

Section 1, many of the fruits and vegetables you enjoy every day come from these states. Do you like pineapple? If so, it may have come from Hawaii. This state also grows sugarcane, coffee, and rice in its tropical climate and rich volcanic soil.

In this region, just like the Atlantic coast, fishing is a major industry. The states of Washington and Oregon draw many people to work in the lumber industry. Mineral resources are important in the Pacific region, too. California has gold, lead, and copper. Alaska has vast reserves of oil.

Factory workers in California and Washington make planes. The areas around San Francisco and Seattle are world-famous centers of research in computers and software. **Los Angeles** is the world capital of the movie industry. These states also attract millions of tourists who plan to visit California's redwood forests, Hawaii's tropical beaches, or the stunning glaciers of Alaska.

✓ Reading Check What goods are manufactured in the Pacific states?

Entering the Twenty-First Century

The American economy, although strong, faces challenges in the twenty-first century. One of these challenges is how to clean up pollution and trash. Americans burn fossil fuels—coal, oil, and natural gas—to power their factories and run their cars. Burning these fuels pollutes the air, endangering all who breathe it. The pollution also mixes with water vapor in the air to make acid rain, or rain containing high amounts of chemical pollutants. Acid rain damages trees and harms rivers and lakes.

The fast-paced American way of life creates another problem. People generate huge amounts of trash. Landfills, the areas where trash companies dump the waste they collect, grow higher and higher

NATIONAL GEOGRAPHIC On Location

Factories

A worker inspects computer components. Along with agriculture, America's economy is strong in technology, science, education, and medicine.

Place What areas in the United States are important software centers?

each year. Many communities now promote recycling, or reusing materials instead of throwing them out. Recycling cuts down on the amount of trash.

New Technology The ability to develop new technology has been a major source of strength for the American economy. Researchers work constantly to find new products to make people's lives easier, healthier—and more fun. Success has helped the country become a world leader in satellites, computers, medicine, and many other fields. Keeping our number one position will require just as much creative thinking and hard work. You will need to learn and use these new technologies to stay productive in your future jobs.

World Trade The United States leads the world in the value of all its imports and exports. Millions of Americans depend on trade for their jobs. American leaders have worked hard to promote free trade. Free trade means taking down trade barriers such as tariffs or quotas so that goods flow freely among countries. In 1993 the United States joined Mexico and Canada in the North American Free Trade Agreement (NAFTA). This agreement promised to remove all barriers to trade among those three countries.

 **Reading Check** How do factories and cars harm the environment in the United States?

Section 2 Assessment

Defining Terms

1. **Define** free enterprise system, service industry, navigable, fossil fuels, acid rain, landfill, recycling, free trade.

Recalling Facts

2. **Economics** Why is the Midwest called "America's breadbasket"?

3. **History** The discovery of which resources first brought settlers to the Interior West?

4. **Economics** What was the goal of the North American Free Trade Agreement (NAFTA)?

Critical Thinking

5. **Analyzing Information** Describe two characteristics of the United States that have helped it become a world leader.

6. **Understanding Cause and Effect** What reasons can you give for the economic changes taking place in the South?

Graphic Organizer

7. **Organizing Information** Draw a diagram like this one. Name one economic region of the United States in the center oval. In the outer ovals write one specific example of each subtopic under their headings.

Applying Geography Skills

8. **Analyzing Maps** Study the economic activity map on page 124 and the climate map on page 120. Then explain why there is little economic activity southwest of Phoenix, Arizona.

Geography Skill

Using Scale

Model cars or airplanes are like the real versions, except for size. Models are made to scale—for example, 1 inch on a model may stand for 1 foot in the real vehicle.

Learning the Skill

Scale is also used to represent size and distance on maps. For example, 1 inch on a map may represent 100 miles (161 km) on the earth's surface. On another map, 1 inch may represent 1,000 miles. A map scale is usually shown with a **scale bar.** This bar shows you how much real distance on the earth is shown by a measurement on the map. To use scale, follow these steps:

* On the scale bar, find the unit of measurement.
* Note the distance in miles or kilometers that the unit represents (1 inch = 5 miles? 1 inch = 100 miles?).
* Measure the distance between two points on the map.
* Multiply that number by the miles or kilometers each unit stands for.

Practicing the Skill

Look at the map to answer the following questions.

1. What is the scale of miles on this map?
2. What is the scale of kilometers on this map?
3. About how many inches long is the map in a direct line from the North Entrance to the South Entrance? How many miles is this measurement?

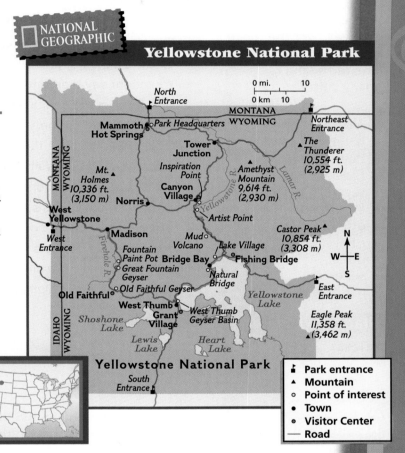

NATIONAL GEOGRAPHIC

Yellowstone National Park

0 mi. 10
0 km 10

North Entrance
MONTANA
WYOMING
Mammoth Hot Springs • ○ Park Headquarters
Northeast Entrance
Tower Junction •
▲ The Thunderer 10,554 ft. (2,925 m)
MONTANA / WYOMING
Mt. Holmes ▲ 10,336 ft. (3,150 m)
Inspiration Point
Amethyst Mountain ▲ 9,614 ft. (2,930 m)
Canyon Village ○
Norris •
West Yellowstone
Artist Point
West Entrance
Madison •
Mud ○ Volcano
Lake Village ○
Castor Peak ▲ 10,854 ft. (3,308 m)
Fountain Paint Pot ○
Great Fountain Geyser ○
Bridge Bay ○
Fishing Bridge •
Natural Bridge
Old Faithful Geyser ○
Old Faithful ○
West Thumb ○
Grant Village ○
West Thumb Geyser Basin
Yellowstone Lake
East Entrance
IDAHO / WYOMING
Shoshone Lake
Lewis Lake
Heart Lake
Eagle Peak 11,358 ft. ▲ (3,462 m)
Yellowstone National Park
South Entrance

N W E S

▴ Park entrance
▲ Mountain
○ Point of interest
• Town
⊙ Visitor Center
— Road

4. By road, about how far is the North Entrance from West Thumb?
5. How much farther by road is it from Old Faithful Geyser to the South Entrance than from the geyser to Bridge Bay?

Applying the Skill

Look at the economic activity map of the United States on page 124. About how many miles does 1 inch represent on the contiguous states? What is the scale of miles on the inset map of Hawaii?

GO TO

Practice key skills with **Glencoe Skillbuilder Interactive Workbook, Level 1.**

Section 3 The Americans

Section 3

Guide to Reading

Main Idea

The United States has attracted people from all over the world who have created a land of many cultures.

Terms to Know

- colony
- democracy
- federal republic
- secede
- immigrant
- ethnic group
- rural
- urban
- suburbs
- national park

Places to Locate

- Texas
- California
- Georgia
- Massachusetts

Reading Strategy

Create a diagram like this one. In each outer oval, write under the heading one fact about American society as it relates to the topic given.

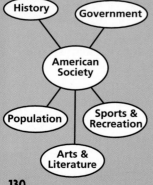

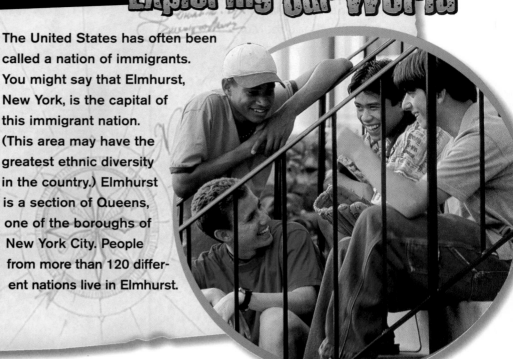

NATIONAL GEOGRAPHIC Exploring Our World

The United States has often been called a nation of immigrants. You might say that Elmhurst, New York, is the capital of this immigrant nation. (This area may have the greatest ethnic diversity in the country.) Elmhurst is a section of Queens, one of the boroughs of New York City. People from more than 120 different nations live in Elmhurst.

The United States is full of people from many different lands. What attracts people to the United States? One reason is the freedom that Americans enjoy. Economic opportunity is another reason. The United States gives people in many other lands hope for a better life.

A Rich History

The United States is a young country compared to many others. The country became independent a little more than 200 years ago. Before it became a country, however, the land's history stretched back thousands of years. Experts have long believed that the first people to settle in the Americas came from Asia. At the time, the earth's climate was much colder than it is now. Huge sheets of ice covered much of the Northern Hemisphere. As a result, sea levels were lower, and a land bridge connected Asia and Alaska. Herd animals crossed this bridge— and the people who hunted them followed.

These people fanned out over the Americas. Their descendants are today called Native Americans. Over time, they developed different ways of life using local resources. In the Northeast of the present-day

United States, the people hunted deer and fished. In the Southeast and Mississippi Valley, they grew corn and other crops. On the Great Plains, they hunted buffalo. In the dry Southwest, they irrigated the land to grow corn and beans. In the Northwest, they fished.

Around A.D. 1500, Europeans learned of the existence of the Americas. The raw materials they saw—forests, animal furs, and rich soils—soon led them to set up colonies, or overseas settlements tied to a parent country. The French built trading posts around the Great Lakes. The Spanish made communities in the area from **Florida** and **Texas** to **California.** British colonists settled along the Atlantic coast from **Georgia** to **Massachusetts.**

A Democratic Republic

By the mid-1700s, the people living in the British colonies had started to see themselves as Americans—different from the British. They resented certain actions of the British government, believing that these actions abused their rights. From 1775 to 1781, the new Americans fought a war that freed the colonies from British rule. With the help of France and Spain, they won the war and formed a new country—the United States of America.

In 1787 a group of leaders, including George Washington and James Madison, met to create a new form of government. They wrote the document called the Constitution of the United States that established the form of government still used today. Our government is based on the principle of democracy, a form of government in which the people rule. We have a *representative democracy,* in which voters choose leaders who make and enforce the laws.

The 13 former British colonies became the first 13 states. Each state had its own government, but people chosen by the voters of each state also served in a national government. In this way, the United States was a federal republic, or a government divided between national and state powers with a president who leads the national government. As you can see from the chart on page 132, the national government is divided into three branches.

A Period of Growth

From 1800 to 1900, the United States experienced tremendous growth. It grew from the 13 states along the Atlantic coast to include 45 states that reached to the Pacific Ocean. The population boomed as millions of people settled here from other lands. They cleared forests, farmed, and often fought with Native Americans who were being pushed out of the way. Farmers grew corn in the Midwest and cotton in the South. Some people mined gold and silver in the Rocky Mountains and California.

In the mid-1800s, the nation experienced a crisis. The South had built its economy on slavery. Hundreds of thousands of enslaved Africans had been forced to work in the agricultural South. Over time, slavery divided the country. In 1861 several Southern states seceded, or withdrew from the national government. For four years, the North and the South fought a bitter civil war. In the end, the Southern states were brought back into the Union, and slavery was abolished.

San Xavier del Bac

Is this Catholic church in Spain? In Mexico? No, this Spanish-style church, called San Xavier del Bac, stands near Tucson, Arizona. Settlers built the church in 1797, when the area was part of Spain's colonial empire. In fact, many Spanish settlements in the American Southwest were founded in the 1500s, long before the English Pilgrims sailed to the Americas on the *Mayflower.*

Analyzing the Diagram

The United States government has three main branches.

Place Which branch makes the laws?

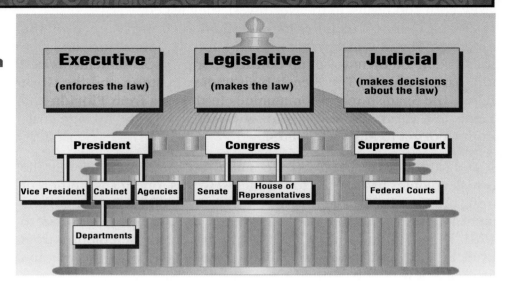

Executive (enforces the law)

Legislative (makes the law)

Judicial (makes decisions about the law)

President — Vice President, Cabinet, Agencies — Departments

Congress — Senate, House of Representatives

Supreme Court — Federal Courts

The Civil War did more than end slavery. It also launched the country into a period of great economic growth. Railroads crisscrossed the land, and factories sprang up, especially in the Northeast and Midwest. This economic growth attracted another great wave of immigrants, or people who move to a new country to make a permanent home.

A World Leader During the early 1900s, the United States became one of the leading economies in the world. Automobiles rolled off assembly lines, electricity became common, and other technologies—the telephone and radio, for example—entered daily life.

The world plunged into two World Wars in the 1910s and 1940s. The United States took part in these wars. Our country's leaders urged the world's people to fight for freedom and to defeat dangerous dictators. American factories built tanks and airplanes, while American soldiers helped win the wars.

After World War II, the United States enjoyed great influence around the world. American companies shipped their products to all continents. American leaders pushed for democracy and free enterprise in other countries. American culture spread around the globe.

At home, however, tensions existed among groups within American society. Many of the Americans who had fought in the two World Wars or had taken care of the home front were women, African Americans, and Hispanic Americans. After World War II, these groups wanted equal rights. Because the United States was often not tolerant toward minorities, many people, including such men as Martin Luther King, Jr., began to push for change. The poems on page 136 describe two views of Americans struggling to be accepted.

✓**Reading Check** How is our country a democracy and a federal republic?

One Out of Many

About 280 million people live in the United States, making it the third most populous country in the world. Compared with people in most other countries, Americans enjoy a high standard of living. Americans, on the average, can expect to live about 76 years. Medical advances have helped people achieve longer lives.

Because it is a nation of immigrants, the United States has a diverse population with various ethnic groups. An ethnic group is a group of people who share a culture, language, or history. Almost three-fourths of the people in our country are descended from European ethnic groups. African American ethnic groups form about 12 percent of the population. The fastest-growing ethnic groups are the Hispanic, who trace their heritage to the countries of Latin America and Spain.

Today many immigrants to the United States come from China, India, other Asian countries, and the Pacific islands. The smallest ethnic groups have lived in the country the longest—Native Americans who

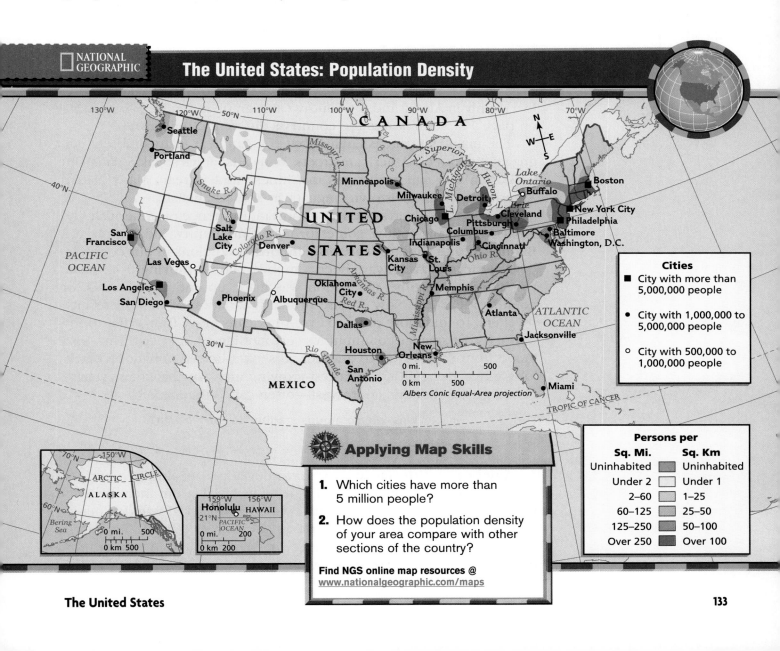

NATIONAL GEOGRAPHIC

The United States: Population Density

Cities
- ■ City with more than 5,000,000 people
- ● City with 1,000,000 to 5,000,000 people
- ○ City with 500,000 to 1,000,000 people

Albers Conic Equal-Area projection

Applying Map Skills

1. Which cities have more than 5 million people?

2. How does the population density of your area compare with other sections of the country?

Find NGS online map resources @
www.nationalgeographic.com/maps

Persons per	
Sq. Mi.	**Sq. Km**
Uninhabited	Uninhabited
Under 2	Under 1
2–60	1–25
60–125	25–50
125–250	50–100
Over 250	Over 100

Music

The music of North America stems from Native American, European, and African influences. This Native American flute was developed by peoples on the Great Plains. When you blow into the flute, the air turns upward and travels through an outside tube. Then the air goes back *into* the flute and passes over a sharp edge, which creates the sound. In addition to five or six playing holes, four "direction" holes are added. This allows the music to be heard to the north, south, east, and west.

Looking Closer **How does the flute player "amplify" his music in a natural way?**

GO TO

World Music: A Cultural Legacy
Hear music of this region on Disc 1, Track 4.

live mainly in Alaska and the Southwest. A graph in the **Regional Atlas** on page 111 summarizes the country's different ethnic groups.

The main language is English, but you can hear many different languages spoken on American streets. Many Hispanics speak Spanish. In California you can read signs in Chinese and Korean. Many Native Americans speak their own language as well as English.

Religion has always been an important influence on American life. Most Americans follow some form of Christianity. Judaism, Islam, Buddhism, and Hinduism are also important religions in our country.

Americans have always been a mobile people, moving from place to place. At one time, our nation was made up entirely of **rural,** or countryside, areas. Now we are primarily a nation of **urban,** or city, dwellers. To find more room to live, Americans move from cities to the **suburbs,** or smaller communities surrounding a larger city. They also move from one region to another to seek a better climate or better jobs. Since the 1970s, the fastest-growing areas in the country have been in the South and Southwest—often called the Sunbelt.

✓Reading Check **What ethnic groups are the fastest-growing?**

American Culture

American artists and writers have developed distinctly American styles. The earliest American artists used materials from their environments to create works of art. Native Americans carved wooden masks or made pottery from clay found in their areas. Artists like Maria Martinez carried on these traditions by turning Native American pottery into fine art. Later artists were attracted to the beauty of the

American land. Winslow Homer painted the stormy waters of the North Atlantic. Georgia O'Keeffe painted the colorful cliffs and deserts of the Southwest. Artists like Thomas Eakins painted scenes of city life.

Two themes are common to American literature. One theme focuses on the rich diversity of the people in the United States. The poetry of Langston Hughes and the novels of Toni Morrison portray the triumphs and sorrows of African Americans. The novels of Amy Tan examine the lives of Chinese Americans. Oscar Hijuelos and Sandra Cisneros write about the country's Hispanics.

A second theme focuses on the landscape and history of particular regions. Mark Twain's humorous books relate life along the Mississippi River in the mid-1800s. Nathaniel Hawthorne wrote about the people of New England. Willa Cather portrayed the struggles people faced in settling the Great Plains. William Faulkner examined life in the South.

Sports and Recreation Many Americans spend their leisure time at home, watching television, playing video games, or using a computer. Many also pursue active lives outdoors. They bike and hike, ski and skate, shoot baskets and kick soccer balls. Many enjoy spectator sports such as baseball and football. Stock-car races and rodeos also attract large crowds. Millions each year travel to national parks, or areas set aside to protect wilderness and wildlife and for recreation.

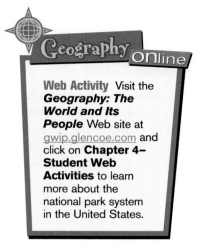

Geography Online

Web Activity Visit the **Geography: The World and Its People** Web site at gwip.glencoe.com and click on **Chapter 4– Student Web Activities** to learn more about the national park system in the United States.

√ Reading Check What are two common themes in American literature?

Section 3 Assessment

Defining Terms
1. Define colony, democracy, federal republic, secede, immigrant, ethnic group, rural, urban, suburbs, national park.

Recalling Facts
2. History What route do experts think the first Americans took to reach North America?

3. Government What document explains the form of government used in the United States?

4. Culture What theme do the works of Langston Hughes and Toni Morrison share?

Critical Thinking
5. Analyzing Information After World War II, what tensions existed within society?

6. Drawing Conclusions How does climate and culture influence the popularity of sports in your area?

Graphic Organizer
7. Organizing Information Draw a diagram like the one below. At the tops of the three arrows, complete the diagram by listing three reasons that Americans today are moving more frequently than ever.

↓ ↓ ↓

Americans are on the move.

Applying Geography Skills

8. Analyzing Maps According to the population density map on page 133, what are the two largest cities in the Pacific Northwest?

Making Connections

▲ Picking cotton near Dallas, Texas, 1907

Americans All

Native Americans and African Americans endured many years of injustice. Even so, the pride and determination of these Americans remained strong. Read the poems by Native American poet Simon J. Ortiz and African American poet Langston Hughes to see how they express these feelings.

Survival This Way
by Simon J. Ortiz (1941–)

Survival, I know how this way.
This way, I know.
It rains.
Mountains and canyons and plants
grow.
We travelled this way,
gauged our distance by stories
and loved our children.
We taught them
to love their births.
We told ourselves over and over
again, "We shall survive
this way."

"Survival This Way" by Simon J. Ortiz. Reprinted by permission of the author.

I, Too
by Langston Hughes (1902–1967)

I, too, sing America.

I am the darker brother.
They send me to eat in the kitchen
When company comes,
But I laugh,
And eat well,
And grow strong.

Tomorrow,
I'll be at the table
When company comes.
Nobody'll dare
Say to me,
"Eat in the kitchen,"
Then.

Besides,
They'll see how beautiful I am
And be ashamed—

I, too, am America.

"I, Too" from *Collected Poems* by Langston Hughes. Copyright © 1994 by the Estate of Langston Hughes. Reprinted by permission of Alfred A. Knopf, a Division of Random House, Inc.

▲ Native Americans on the Great Plains, 1891

Making the Connection

1. How does the poem "Survival This Way" tell how Native Americans feel about their children?

2. What does Langston Hughes mean by the phrase "I, too, sing America"?

3. **Making Comparisons** In what way do both poems convey a message of hope?

Section 1 — A Vast, Scenic Land

Terms to Know
contiguous
megalopolis
coral reef

Main Idea
The United States has a great variety of landforms and climates.
✓ **Region** The United States has five main physical regions: the Coastal Plains, the Appalachian Mountains, the Interior Plains, the Mountains and Basins region, and the Pacific Coast. Alaska and Hawaii make up two additional regions.
✓ **Place** Northeastern coastal lowlands have poor soil but are densely populated.
✓ **Economics** The Central Lowland is well suited to agriculture, as are western coastal valleys.
✓ **Place** The high Rocky Mountains and nearby basins have rich mineral deposits.

Section 2 — An Economic Leader

Terms to Know
free enterprise system
service industry
navigable
fossil fuels
acid rain
landfill
recycling
free trade

Main Idea
The powerful United States economy runs on abundant resources and the hard work of Americans.
✓ **Economics** Because of many natural resources and a hardworking labor force, the United States has the world's most productive economy.
✓ **Economics** Service industries contribute the most to the American economy, followed by manufacturing, agriculture, and mining.
✓ **Economics** The United States has five economic regions—the Northeast, the South, the Midwest, the Interior West, and the Pacific.
✓ **Economics** Creativity and hard work are needed to continue to develop new technologies and help the American economy grow.

Section 3 — The Americans

Terms to Know
colony
democracy
federal republic
secede
immigrant
ethnic group
rural
urban
suburbs
national park

Main Idea
The United States has attracted people from all over the world who have created a land of many cultures.
✓ **History** The American people are immigrants or the descendants of immigrants who came from all over the world.
✓ **Government** The United States has a representative democratic government with power shared by the states and the national government.
✓ **Culture** Ethnic groups in America are descendants of five main peoples: Europeans, Africans, Hispanics, Asians and Pacific Islanders, and Native Americans and Inuit.
✓ **Culture** American arts celebrate the country's ethnic and regional diversity.

Assessment and Activities

Using Key Terms

Match the terms in Part A with their definitions in Part B.

A.

1. contiguous
2. megalopolis
3. free enterprise system
4. fossil fuels
5. navigable
6. immigrant
7. colony
8. democracy
9. ethnic group
10. suburb

B.

a. oil, natural gas, and coal
b. smaller community surrounding a city
c. areas joined inside a common boundary
d. government by the people
e. limited government control over the economy
f. group of people who share a common culture, language, and history
g. a "supercity"
h. person who moves to a new country
i. wide and deep enough for ships to pass
j. overseas settlement tied to a parent country

Reviewing the Main Ideas

Section 1 A Vast, Scenic Land

11. **Region** What are the two parts of the Interior Plains?
12. **Place** Which part of the contiguous United States has a tropical climate?
13. **Region** Name four of the mountain ranges found in the United States.

Section 2 An Economic Leader

14. **Place** Name four of the mineral resources found in the United States.
15. **Economics** Name four of the South's agricultural products.
16. **Human/Environment Interaction** What is happening to America's landfills?

Section 3 The Americans

17. **History** Why did the American colonists want independence from Britain?
18. **Culture** Which two ethnic groups have grown the fastest in the United States?
19. **Culture** Which part of the United States did Georgia O'Keeffe show in her paintings?

 NATIONAL GEOGRAPHIC

The United States

Place Location Activity

On a separate sheet of paper, match the letters on the map with the numbered places listed below.

1. Rocky Mountains
2. Mississippi River
3. Appalachian Mountains
4. Washington, D.C.
5. Chicago
6. Lake Superior
7. Ohio River
8. Gulf of Mexico
9. Texas
10. Los Angeles

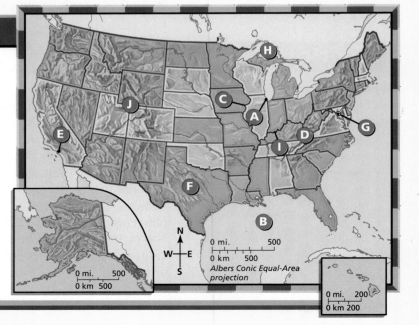

Self-Check Quiz Visit the *Geography: The World and Its People* Web site at gwip.glencoe.com and click on **Chapter 4– Self-Check Quizzes** to prepare for the Chapter Test.

Critical Thinking

20. **Understanding Cause and Effect** What physical features in the Midwest have affected the economy of that region?

21. **Categorizing Information** Draw a diagram like the one below. In each outer oval, write two facts about the United States under each heading.

GeoJournal Activity

22. **Writing a Paragraph** Write a paragraph describing the economic region in which you live. Explain what people in your area generally do for a living.

Mental Mapping Activity

23. **Focusing on the Region** Draw a simple outline map of the United States, then label the following:

- Appalachian Mountains
- Rocky Mountains
- Mississippi River
- Atlantic Ocean
- Great Plains
- Great Lakes
- Alaska
- Hawaii
- Pacific Ocean
- Gulf of Mexico

Technology Skills Activity

24. **Using a Spreadsheet** Choose one of the physical regions of the United States. List the names of the states in that region in a spreadsheet beginning with cell A2 and continuing down column A. List the populations of each state in column B, beginning with cell B2. Then make a bar graph that shows the populations of the states in that region.

Standardized Test Practice

Directions: Study the graph, then answer the following questions.

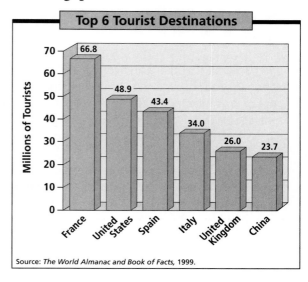

Top 6 Tourist Destinations

Source: *The World Almanac and Book of Facts,* 1999.

1. **According to the graph, about how many tourists visited the United States in 1999?**

   A 48.9

   B 66.8

   C 48,900

   D 48,900,000

2. **Which country on the graph had the least number of tourists?**

   F France

   G China

   H Spain

   J Italy

Test-Taking Tip: A common error when reading graphs is overlooking the information on the bottom and the side of the graph. Check the sides of the graph to see what the numbers mean.

TOO MUCH
Trash

Tons of Trash If you are an average American, you throw away about 4 pounds (2 kg) of trash each day. Not much, right? Think again. That is 1,460 pounds (663 kg) a year. By age 13, you have produced almost 10 tons (9 t) of trash!

Americans create more than a third of the world's trash—200 million tons (181 million t) each year. That is enough to fill a line of garbage trucks that would circle the earth 8 times!

What happens to trash?

State Recycling Rates
- 30% or greater
- 20–29%
- 10–19%
- Less than 10%
- Unavailable

Source: U.S. EPA Municipal Solid Waste Handbook—Internet Version.

- ⚙ Most ends up in landfills.

- ⚙ Some is burned in incinerators.

- ⚙ Some is dumped into lakes, rivers, and oceans.

All of these disposal methods create pollution and harm living things. When landfills fill up, new ones must be created. But sites for new landfills are getting hard to find. Would you want to live near one?

The Three R's Surprisingly, the solution to too much trash is simple. We need to produce less waste. How? By following the three R's—reduce, reuse, and recycle.

- ⚙ REDUCE the amount of trash you throw away each day.

- ⚙ REUSE products and containers.

- ⚙ RECYCLE some of your trash. About 80 percent of household trash can be recycled.

If we reduce, reuse, and recycle, we can win the war against trash.

Trash piles up at a landfill in New Jersey.

Making a Difference

Dig It! You are at the ball game. You toss part of your hot dog into a trash can. Eventually, the hot dog goes to a landfill. How long will it take the hot dog to decay in a landfill?

In 1973 archaeologist William Rathje began the Garbage Project. He wanted to "dig up" facts about the trash Americans throw away. Rathje and his University of Arizona students spent years studying garbage from landfills across the United States. His research results are full of surprises. In some landfills, team members found foods such as steaks and hot dogs that were 15 to 20 years old! Lack of air, light, and moisture prevents wastes from breaking down. Another surprise: About a third of the trash in landfills is paper. Yard waste, food scraps, plastics, construction materials, and furniture are some of the other items we throw away.

William Rathje

Recycling to the Max Linda Munn and her husband, Frank Schiavo, are teachers in California. They have not set out a curbside garbage can in more than 20 years. That is because they recycle or compost almost everything they use. They produce only about two handfuls of trash a week—and that goes to a recycling center, too.

A student recycles aluminum cans.

What Can You Do?

⊛ **Make Toys From Trash**
Create toys from discarded clean paper, cardboard, or plastic. Have a class contest and award prizes to students who reuse trash in the most creative ways.

⊛ **Campaign Against Waste**
Investigate products used every day. Which ones have too much packaging—layers of plastic or paper thrown away once the product is used? Can you think of ways to eliminate the excess? Identify companies that make these products and send them a letter or an e-mail outlining your packaging changes.

⊛ **Use the Internet**
Click the Games option at www.edf.org/Earth2Kids to learn more about recycling. Also check out the Environmental Protection Agency's kids' page at www.epa.gov/epaoswer/osw/kids.htm

Chapter 5 Canada

The World and Its People NATIONAL GEOGRAPHIC

To learn more about Canada's people and places, view *The World and Its People* **Chapter 5** video.

Geography ONLINE

Chapter Overview Visit the *Geography: The World and Its People* Web site at gwip.glencoe.com and click on **Chapter 5– Chapter Overviews** to preview information about Canada.

Landforms of the North

Guide to Reading

Main Idea

Canada is a vast country with many landforms and climates.

Terms to Know

- province
- glacier
- peninsula
- tundra
- prairie
- cordillera

Places to Locate

- Laurentian Highlands
- St. Lawrence River
- Hudson Bay
- Canadian Shield
- Interior Plains
- Rocky Mountains
- Coast Mountains

Reading Strategy

Create a chart like this one. In each space write the name of one of Canada's six physical regions and a fact about it.

Canada's Physical Regions

NATIONAL GEOGRAPHIC Exploring Our World

Have you ever seen up close and personal a lumbering, snarling grizzly bear? Many tourists come to Banff National Park in western Canada hoping to spot such a creature. Located in the Rocky Mountains, Banff is Canada's oldest, best-loved, and busiest national park. More than 4 million visitors a year are drawn to its spectacular mountain scenery.

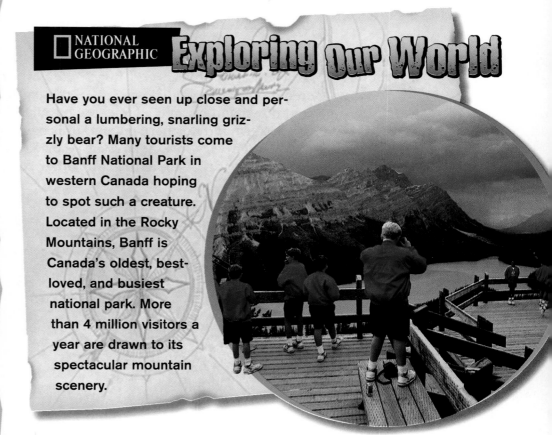

Vikings first landed their longboats on its eastern coast around A.D. 1000. Niagara Falls thunders in the southeast. Grizzly bears roam its western regions. What country are we describing? It is Canada.

From Atlantic to Pacific

The political map on page 144 shows you that Canada is located north of the contiguous United States. Between the two countries lies the world's longest undefended border, which stretches 5,522 miles (8,887 km) across North America. No military troops try to halt the thousands of people who cross this border every day. Like the United States, Canada has the Atlantic Ocean on its eastern coast and the Pacific Ocean on its western coast. The Arctic Ocean lies to the far north.

Canada is the world's second-largest country in land area. Only Russia is larger. Instead of being made up of states, Canada has 10 **provinces,** or regional political divisions. It also includes 3 territories.

Look at the map below to find the provinces of Newfoundland (NOO•fuhn•luhnd), Nova Scotia, New Brunswick, Prince Edward Island, Quebec (kwih•BEHK), Ontario, Manitoba, Saskatchewan (suh•SKA•chuh•wuhn), Alberta, and British Columbia. Now find the Yukon Territory and the Northwest Territories. In 1999 a third territory—Nunavut (NOO•nuh•vuht)—was carved out of part of the Northwest Territories. This area is the homeland of the Inuit, a Native American people.

✔ Reading Check What is the name of Canada's newest territory?

Canada's Landforms

Thousands of years ago, huge **glaciers,** or giant sheets of ice, covered most of Canada. The weight of these glaciers pushed much of the land down and created a large low basin. Highlands rose on the western, eastern, and northern edges of this basin. Water filled the land that was pushed very low. As a result, Canada today has many lakes and inland waterways—more than any other country in the world.

NATIONAL GEOGRAPHIC

Canada: Political

Applying Map Skills

1. What is Canada's national capital?

2. Name the Canadian provinces or territories that border Alaska.

Find NGS online map resources @ www.nationalgeographic.com/maps

⊛ National capital
⊚ Provincial capital
• Major city

0 mi. 500
0 km 500
Azimuthal Equidistant projection

Geographers divide Canada's vast water-studded land into six physical regions. These regions are the Eastern Highlands, the St. Lawrence and Great Lakes Lowlands, the Canadian Shield and Arctic Islands, the Interior Plains, the Rocky Mountains, and the Pacific Coast.

The Eastern Highlands Along Canada's southeastern Atlantic coast stretch the Appalachian Highlands. They continue farther south through the eastern United States as the Appalachian Mountains. In Canada's northeast, across the St. Lawrence River, lies another highland area—the **Laurentian** (law•REHN•chuhn) **Highlands.** Traveling through both of these parts of Canada, you see rolling hills and low mountains. The valleys between are dotted with tidy farms. Forests blanket much of the landscape. Many deepwater harbors nestle along the jagged, rocky coasts.

The St. Lawrence and Great Lakes Lowlands Cutting through the eastern highland areas are the lowlands of the **St. Lawrence River.** These lowlands continue west to the Great Lakes region. The St. Lawrence River and the Great Lakes form the major waterway

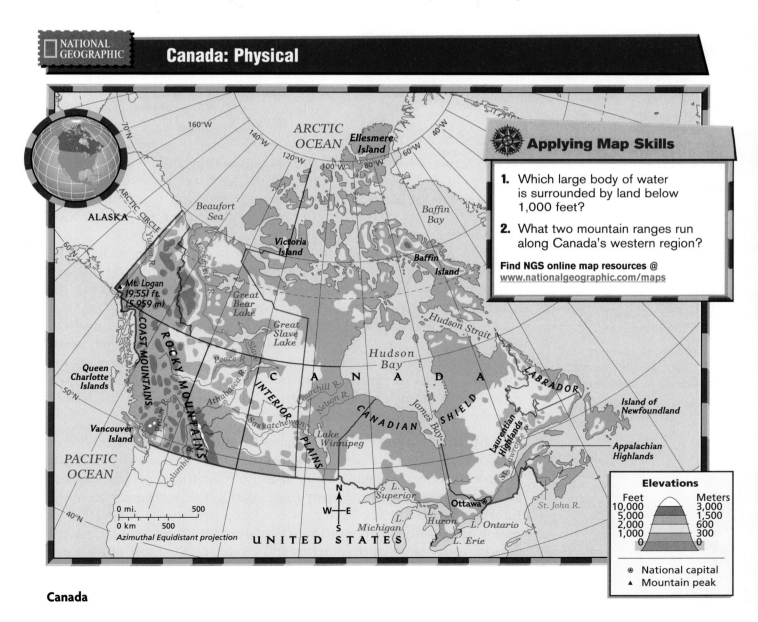

NATIONAL GEOGRAPHIC

Canada: Physical

Applying Map Skills

1. Which large body of water is surrounded by land below 1,000 feet?

2. What two mountain ranges run along Canada's western region?

Find NGS online map resources @
www.nationalgeographic.com/maps

Elevations

Feet		Meters
10,000		3,000
5,000		1,500
2,000		600
1,000		300
0		0

⊛ National capital
▲ Mountain peak

0 mi. 500
0 km 500
Azimuthal Equidistant projection

Canada

Horseshoe Falls, Canada

Canada's Horseshoe Falls—one of the two waterfalls that make up Niagara Falls—lies on the Niagara River, which flows from Lake Erie into Lake Ontario.

Place How might Niagara Falls benefit Canada's economy?

linking the Atlantic coast and central Canada. Huge, slow-moving barges carrying grain, ore, coal, and other resources course through this waterway that Canada shares with the United States.

Find the St. Lawrence River valley on the physical map on page 145. Now find the peninsula in Ontario that is bordered by Lake Ontario, Lake Erie, and Lake Huron. A **peninsula** is a piece of land with water on three sides. The St. Lawrence River valley and this peninsula have rich soil, good transportation routes, and sprawling urban centers. You find most of Canada's people, industries, and farms here.

The Canadian Shield and the Arctic Islands Wrapped around **Hudson Bay** is the huge, horseshoe-shaped region known as the **Canadian Shield.** It includes about half of Canada's land area. Hills worn down by erosion along with thousands of lakes carved by glaciers dot much of this region. Bare rock covers much of the land here. In other places there is enough thin soil to support a few hardy plants, such as mosses and shrubs. This soil cannot be used for farming, however. In the southern part, evergreen forests provide shelter and food for deer, elk, and moose. Much of the region is wilderness.

Deep within the Canadian Shield are iron ore, copper, nickel, gold, and uranium deposits. Because of the region's location and climate, few people live here. Many Canadians vacation along the lakes in the southern part of this region during the summer.

To the north lie the Arctic Islands. Ten large islands and hundreds of small ones lie almost entirely north of the Arctic Circle. Much of the land consists of tundra—vast rolling, treeless plains in which only the top few inches of ground thaw in summer. Only low-lying bushes, mosses, and sturdy grasses grow here. Glacial ice blankets the islands that are farthest north.

The Interior Plains Canada shares landforms with its neighbor to the south. The Great Plains of the United States become the **Interior Plains** of western Canada. Look at the physical map on page 145 to locate this region. You see many large lakes in the northern part of the Interior Plains. The southern part is a huge prairie—a rolling, inland grassy area with fertile soil. Herds of buffalo once roamed on the prairie. Today large cattle ranches and farms occupy most of the land.

The Rocky Mountains Another landform shared by Canada and the United States is the **Rocky Mountains,** part of an area called the cordillera (KAWR•duhl•YEHR•uh). A cordillera is a group of mountain ranges that run side by side. The Canadian Rockies are known for their scenic beauty and rich mineral resources. Tourists from around the world are drawn to this area, particularly to Banff and Jasper National Parks.

The Pacific Coast West of the Rockies you cross high plateaus until you reach the **Coast Mountains.** These mountains skirt Canada's Pacific shore and form another part of the cordillera. Like the Rockies, they cross into the United States.

A string of islands off Canada's west coast are actually partly underwater peaks of the Coast Mountains. The Inside Passage—a waterway that is 1,000 miles (1,609 km) long—separates these islands from the mainland Coast ranges. You can use the Inside Passage to travel by boat between the city of Seattle in Washington State and ports on Alaska's southeastern tail. Along the way, narrow fingers of the sea snake inland between steep mountainsides.

Near Canada's border with Alaska, several mountains soar more than 15,000 feet (4,572 m). Mount Logan—Canada's highest peak— even reaches 19,551 feet (5,959 m). Most of the people in the Pacific Coast region live along the coast and in river valleys that cut through the mountains.

✓ **Reading Check** What are two landforms that Canada shares with the United States?

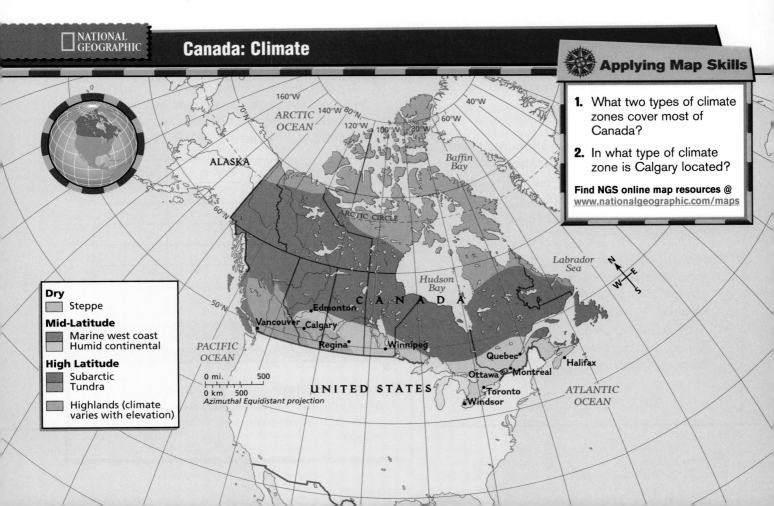

NATIONAL GEOGRAPHIC

Canada: Climate

Applying Map Skills

1. What two types of climate zones cover most of Canada?

2. In what type of climate zone is Calgary located?

Find NGS online map resources @ www.nationalgeographic.com/maps

Dry
■ Steppe

Mid-Latitude
■ Marine west coast
■ Humid continental

High Latitude
■ Subarctic
■ Tundra
■ Highlands (climate varies with elevation)

ARCTIC OCEAN

ALASKA

Baffin Bay

ARCTIC CIRCLE

Labrador Sea

Hudson Bay

CANADA

Edmonton
Vancouver • Calgary
PACIFIC OCEAN
Regina • Winnipeg
Quebec • Halifax
Ottawa • Montreal
UNITED STATES
Toronto
Windsor
ATLANTIC OCEAN

0 mi. 500
0 km 500
Azimuthal Equidistant projection

The Climate

One feature that Canada does not share with the United States is climate. The map on page 147 shows you that Canada generally has a cool or cold climate. In Canada's far north, people shiver in the cold tundra climate. Farther south, between 70°N and 50°N latitude, you find a subarctic climate with short, cool summers and long, cold winters.

Southeastern Canada has a humid continental climate. Winters can be long, cold, and snowy here. Summers are short and sometimes hot. As you might guess, most Canadians live in this area.

The humid continental climate continues west of the Great Lakes to the Interior Plains. The western part of the Plains has similar temperatures but less rainfall, making for a dry steppe climate. Rainfall here is low because of the rain shadow effect of the Rocky Mountains.

The southwestern Pacific coast, with its marine west coast climate, is the only area in Canada that has wet, mild winters. The Coast Mountains in this region cause warm winds from the Pacific Ocean to release moisture. The western, or windward, side of the mountains gets more rain and has warmer temperatures than any other part of Canada. In fact, British Columbia's capital—Victoria—is known for its well-kept gardens that bloom year-round.

✔ Reading Check Which part of Canada has wet, mild winters?

Section 1 Assessment

Defining Terms

1. Define province, glacier, peninsula, tundra, prairie, cordillera.

Recalling Facts

2. Region What is unusual about the border between Canada and the United States?

3. Place Name four of the mineral resources found in the Canadian Shield.

4. Region In what type of climate do most Canadians live?

Critical Thinking

5. Understanding Cause and Effect How did glaciers affect the physical landscape of Canada?

6. Analyzing Information What region in Canada has the most people and the most economic activity?

Graphic Organizer

7. Organizing Information Draw a chart like this one. For each climate type listed, write where in Canada that climate exists and one fact about the climate.

Climate	Region	Fact
Tundra		
Subarctic		
Humid Continental		
Steppe		
Marine West Coast		

Applying Geography Skills

8. Analyzing Maps Find the cities of Winnipeg, Manitoba, and St. John's, Newfoundland, on the political map on page 144. Which city is farther north?

Geography Skill

Reading a Physical Map

A map that shows the different heights of the land is called a **physical map.** Physical maps use colors and shading to also show *relief*–or how flat or rugged the land surface is. Colors are also used to show the land's *elevation*–or height above sea level. Green often shows the lowest elevations (closest to sea level). Yellows, oranges, browns, and reds usually mean higher elevations. Sometimes the highest areas, such as mountain peaks, are white.

Learning the Skill

To read a physical map, apply these steps:

- Read the map title to identify the region shown on the map.
- Use the map key to find the meaning of colors and symbols.
- Identify the areas of highest and lowest elevation on the map.
- Find important physical features, including mountains, rivers, and coastlines.
- Mentally map the actual shape of the land.

Practicing the Skill

Look at the map to answer the following:

1. What province is highlighted on the map?
2. What mountain ranges are labeled?

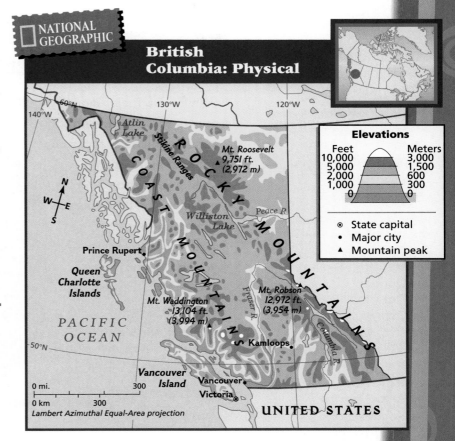

NATIONAL GEOGRAPHIC

British Columbia: Physical

Elevations

Feet		Meters
10,000		3,000
5,000		1,500
2,000		600
1,000		300
0		0

- ⊙ State capital
- • Major city
- ▲ Mountain peak

0 mi. 300
0 km 300
Lambert Azimuthal Equal-Area projection

3. What is the elevation of the green areas on the map (in feet and meters)?
4. What color on the map means 2,000–5,000 feet (600–1,500 m)?
5. Briefly describe the physical landscape of the area shown on the map, moving from west to east.

Applying the Skill

Look at the physical map of Canada on page 145. Describe the physical landscape of the country, moving from east to west.

GO TO

Practice key skills with **Glencoe Skillbuilder Interactive Workbook, Level 1.**

A Resource-Rich Country

Guide to Reading

Main Idea

Canada's economy benefits from rich natural resources, skilled workers, and close trading ties with the United States.

Terms to Know

- service industry
- newsprint
- secede
- acid rain

Places to Locate

- Newfoundland
- Maritime Provinces
- Quebec
- Ontario
- Ottawa
- St. Lawrence Seaway
- Prairie Provinces
- British Columbia
- Yukon Territory
- Northwest Territories
- Nunavut

Reading Strategy

Create a chart like this one and list Canada's six economic regions in the left column. In the right column, list the main economic activities in each region.

Economic Regions	Economic Activities

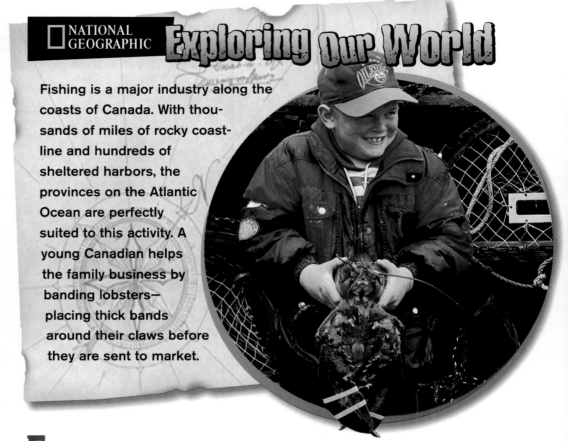

NATIONAL GEOGRAPHIC Exploring Our World

Fishing is a major industry along the coasts of Canada. With thousands of miles of rocky coastline and hundreds of sheltered harbors, the provinces on the Atlantic Ocean are perfectly suited to this activity. A young Canadian helps the family business by banding lobsters— placing thick bands around their claws before they are sent to market.

Fishing is Canada's oldest industry. Yet it is only one of many economic activities that Canadians carry on today. Productive farms, forests, mines, and businesses have turned Canada into one of the world's most advanced economies.

Canada's Economic Regions

Canada's economy is similar to that of the United States. Canada is known for fertile farmland, rich natural resources, and skilled workers. Manufacturing, farming, and service industries are the country's major economic activities. **Service industries** are businesses that provide services to people instead of producing goods. Fishing, mining, and lumbering are also important to the Canadian economy.

Canada, like the United States, has a free market economy in which people start and run businesses with limited government involvement. Canada's government, however, plays a more direct role in the

Canadian economy. For example, Canada's national and provincial governments provide health care for citizens. Broadcasting, transportation, and electric power companies are also government-owned. These public services might not have been available in Canada's remote areas without government support.

Geographers group Canada's provinces and territories into six economic regions: Newfoundland and the Maritime Provinces, Quebec, Ontario, the Prairie Provinces, British Columbia, and the North.

Newfoundland and the Maritime Provinces   Canada's easternmost economic region is formed by **Newfoundland** and the **Maritime Provinces.** The Maritime Provinces are Nova Scotia, New Brunswick, and Prince Edward Island. Fishing has long been a major economic activity here. The Grand Banks, off the coast of Newfoundland, is known as one of the best fishing grounds in the world. As a result of overfishing, however, fishing has been regulated. Today fewer Canadians make a living from the sea.

Farming is limited in this region. A short growing season and thin, rocky soil discourage large-scale agriculture. Yet small farms throughout the Maritime Provinces grow crops such as potatoes and apples.

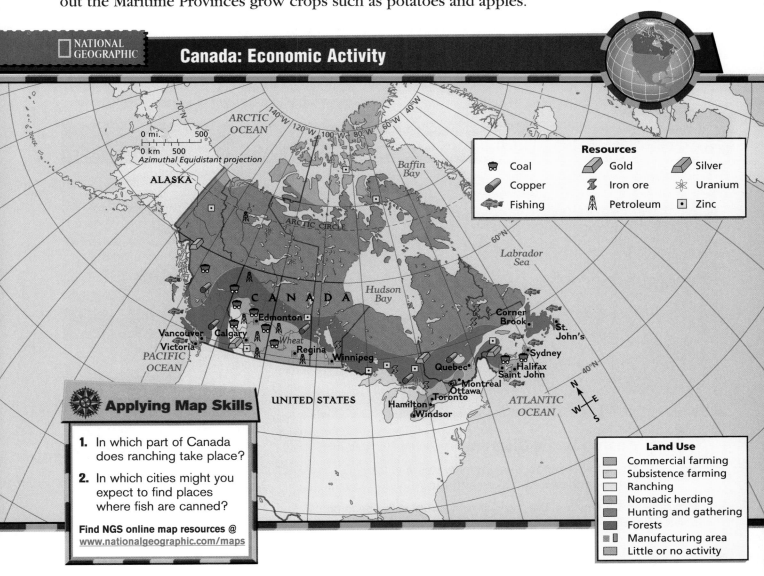

NATIONAL GEOGRAPHIC

Canada: Economic Activity

Resources

- Coal
- Copper
- Fishing
- Gold
- Iron ore
- Petroleum
- Silver
- Uranium
- Zinc

Applying Map Skills

1. In which part of Canada does ranching take place?

2. In which cities might you expect to find places where fish are canned?

Find NGS online map resources @ www.nationalgeographic.com/maps

Land Use

- Commercial farming
- Subsistence farming
- Ranching
- Nomadic herding
- Hunting and gathering
- Forests
- Manufacturing area
- Little or no activity

Manufacturing and mining provide jobs for most people in this region. Tourism is increasingly important, too. In 1997 Canada opened a bridge connecting Prince Edward Island to the mainland. Since then, the number of visitors to the island has doubled. Halifax is a major shipping center. Its harbor remains open in winter when ice closes most other eastern Canadian ports.

Quebec   Canada's largest province in land area is **Quebec.** It is home to almost 25 percent of Canadians. Most of Quebec's people live in cities in and around the St. Lawrence River valley. Montreal, an important port on the St. Lawrence River, is Canada's second-largest city and a major financial and industrial center. The city of Quebec, founded by the French in 1608, is the capital of the province. Many historic sites and a European charm make it popular with tourists.

Manufacturing and service industries are dominant in Quebec's economy. To the north, you find miners providing the province with iron ore, copper, and gold. Other economic activities in Quebec include agriculture and fishing.

Ontario   Canada's second-largest province is **Ontario,** but it has the most people and greatest wealth. Ontario is the leading industrial region of Canada. Its industries range from mining, timber, and transport equipment to oil, gas, and chemicals. The region produces more than half of Canada's manufactured goods.

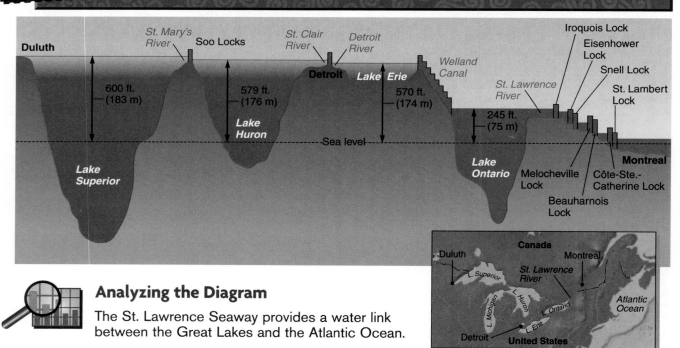

NATIONAL GEOGRAPHIC

St. Lawrence Seaway

Duluth

St. Mary's River — Soo Locks

600 ft. (183 m)

Lake Superior

St. Clair River — Detroit River

Detroit

579 ft. (176 m)

Lake Huron

Lake Erie

Welland Canal

570 ft. (174 m)

Sea level

St. Lawrence River

Iroquois Lock
Eisenhower Lock
Snell Lock
St. Lambert Lock

245 ft. (75 m)

Lake Ontario

Melocheville Lock

Beauharnois Lock

Côte-Ste.- Catherine Lock

Montreal

Canada
Duluth
L. Superior
L. Michigan
L. Huron
L. Ontario
L. Erie
St. Lawrence River
Montreal
Detroit
United States
Atlantic Ocean

Analyzing the Diagram

The St. Lawrence Seaway provides a water link between the Great Lakes and the Atlantic Ocean.

Movement On which of the lakes shown would ships become isolated from the Seaway if the lake level dropped sharply?

Southern Ontario also has fertile land and a growing season long enough for commercial farming. Farmers here grow grains, fruits, and vegetables and raise beef and dairy cattle.

Toronto, the capital of Ontario, is Canada's largest city. It is also the country's chief manufacturing, financial, and communications center. Most of Canada's large corporations have their headquarters here. **Ottawa,** the capital city of Canada, lies in Ontario near the border with Quebec. Many Canadians work in government offices in Ottawa.

Ontario borders the Great Lakes and the St. Lawrence River. To open the Great Lakes to ocean shipping, the United States and Canada built the **St. Lawrence Seaway.** Look at the diagram on page 152. It shows how a system of locks and canals allows ships to pass between the Great Lakes and the Atlantic Ocean. Turn to page 164 to learn more about the St. Lawrence Seaway.

The Prairie Provinces The **Prairie Provinces**—Manitoba, Saskatchewan, and Alberta—spread between the Canadian Shield and the Rocky Mountains. Farming and ranching are major economic activities here. Canada is one of the world's biggest producers of wheat, most of which is exported to Europe and Asia. Ranchers in Alberta raise most of Canada's cattle.

Some of the world's largest reserves of oil and natural gas are found in Alberta and Saskatchewan. These resources contribute to Canada's wealth. Huge pipelines carry the oil and gas to other parts of Canada and to the United States. Canada's fossil fuels and hydroelectric power make it the fifth-largest producer of energy in the world.

The Prairie Provinces have some of the fastest-growing cities in Canada. Calgary and Edmonton, both in Alberta, are leading oil and agricultural centers. Winnipeg, in Manitoba, is a major transportation and business center that links eastern and western Canada.

British Columbia Thick forests blanket much of **British Columbia.** As you might guess, timber, pulp, and paper industries provide much of British Columbia's income. The province helps make Canada the world's leading producer of newsprint, the type of paper used for printing newspapers. The mining of coal, copper, and lead also adds to the wealth of British Columbia.

Agriculture and fishing are strong economic activities in British Columbia. In river valleys between mountains, farmers raise cattle and poultry. They also grow fruits and vegetables. Fishing fleets sail out into the Pacific Ocean to catch salmon and other kinds of fish. Vancouver is a bustling trade center and the nation's main Pacific port.

NATIONAL GEOGRAPHIC **On Location**

Nighttime Harvest

At harvest time in southern Saskatchewan, the work goes on around the clock. The farms in the Prairie Provinces are large and depend on machinery.

Region What other economic activities take place in the Prairie Provinces?

The North   Geese, moose, bears, and caribou are the main inhabitants of many of Canada's northern regions. This area includes the **Yukon Territory,** the **Northwest Territories,** and **Nunavut.** The North accounts for 40 percent of Canada's land, but few people live here. Most are Inuit and other Native Americans who follow traditional hunting and gathering ways of life. The far north is Arctic wilderness. Many resources are thought to lie hidden beneath the frozen ground. The harsh climate and lack of roads make it very costly to mine these resources.

Web Activity Visit the *Geography: The World and Its People* Web site at gwip.glencoe.com and click on **Chapter 5– Student Web Activities** to learn more about Quebec's French culture.

Economic Issues   Canada's economy faces a number of issues. One issue concerns Quebec. About 80 percent of Quebec's population is French-speaking. The rest of Canada is largely English-speaking. Many people in Quebec want to secede, or withdraw, from Canada. If that happens, the province's economy would suffer.

Experts also worry about acid rain, or rain containing high amounts of chemical pollutants. This rain is damaging Canada's lakes and forests. Below-normal water levels on the Great Lakes are also causing concern.

A third issue involves the United States. About $1 billion worth of trade passes between Canada and the United States each day. In 1994 Canada, the United States, and Mexico created the North American Free Trade Agreement (NAFTA) to remove trade barriers among the three countries. Some Canadians fear that their economy is too dependent on the United States. They worry that the American economy is so large that it will dominate the partnership.

✓ Reading Check   Which economic region is Canada's main wheat-growing area?

Section 2 Assessment

Defining Terms
1. Define service industry, newsprint, secede, acid rain.

Recalling Facts
2. Economics What are the results of overfishing the Grand Banks?

3. Place Identify five facts about Toronto.

4. Economics Which province is the world's leading producer of newsprint?

Critical Thinking
5. Making Inferences Why is Vancouver a useful port for Canadian trade with Asian countries?

6. Drawing Conclusions Explain why some Canadians fear NAFTA.

Graphic Organizer
7. Organizing Information Draw a chart like this one. Then list each province or territory, what economic region it is in, what resources are found in it, and major cities, if any.

Province or Territory	Economic Region	Resources	Cities

Applying Geography Skills

8. Analyzing Maps Look at the economic activity map on page 151. Name the types of economic activity that take place north of the Arctic Circle.

The Canadians

Guide to Reading

Main Idea

Canadians of many different backgrounds live in towns and cities close to the United States border.

Terms to Know

- colony
- dominion
- parliamentary democracy
- prime minister
- bilingual

Places to Locate

- Quebec
- Montreal
- Nova Scotia
- New Brunswick
- Prince Edward Island
- Toronto

Reading Strategy

Draw a chart like this one and give at least two facts about Canada for each topic.

History		
Population		
Culture		

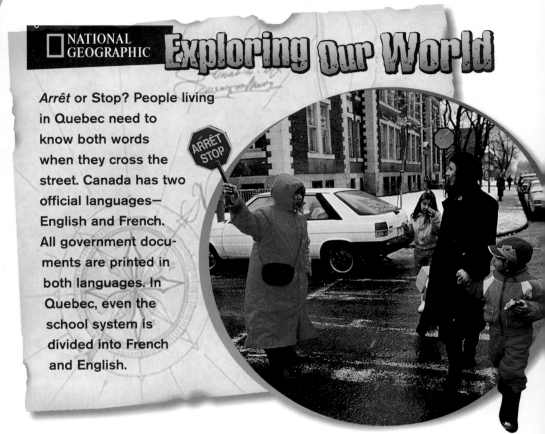

NATIONAL GEOGRAPHIC

Exploring Our World

Arrêt or Stop? People living in Quebec need to know both words when they cross the street. Canada has two official languages—English and French. All government documents are printed in both languages. In Quebec, even the school system is divided into French and English.

Like the United States, Canada's population is made up of many different cultures. The largest group of Canadians has a European heritage, but the country is home to people from all continents. Yet unlike the United States, Canada has had difficulty achieving a strong sense of being one nation. The country's vast distances and largely separate cultures have made Canadians feel more closely attached to their own region or culture than to Canada as a whole.

Canada's History

Canada's first peoples—the Inuit and other Native Americans—adjusted to the environment in a number of different ways. As a result, they developed a great variety of cultures. Some lived in coastal fishing villages. Others were hunters and gatherers constantly on the move. Still others founded permanent settlements near rivers where they raised crops on fertile land.

Thousands of years passed before European settlers arrived. The first Europeans in Canada were Viking explorers from northern Europe who landed in about A.D. 1000. They lived for a while on the

Clothing

The Inuit of the Canadian Arctic designed their clothes for protection from the harsh climate. Traditional clothing was made up of a caribou or sealskin parka, pants, mittens, and boots. In winter the Inuit wore their furs facing toward the skin. This created air pockets that trapped warm air close to the body. On top they wore another layer with the fur facing outward. The clothing flapped as the wearer moved, creating a breeze that kept the person from overheating while running or working.

Looking Closer How does traditional clothing protect the Inuit from the harsh climate?

Newfoundland coast but eventually left. In 1497 John Cabot, an Italian who was exploring for England, landed near Newfoundland. He claimed the area for England.

Then in 1534 French explorer Jacques Cartier sailed into the St. Lawrence River valley and claimed that area for France. French explorers, settlers, and missionaries soon followed and founded several cities. The most important were **Quebec** and **Montreal.** For almost 230 years, France ruled the area around the St. Lawrence River and the Great Lakes. This region was called New France.

During the 1600s and 1700s, England and France fought each other for territory around the globe. The British battled the French in several wars to win control of eastern North America. The last war, known in the United States as the French and Indian War, ended with a treaty signed in 1763. In that treaty, France gave Britain all of its land in Canada. During this time, European warfare and diseases were destroying the Native American cultures.

From Colony to Nation *Where* Europeans settled reflected their sharp cultural differences. For about 100 years, Great Britain held Canada as a colony. A colony is an overseas territory with ties to the parent country. The Atlantic coast and what is today Ontario became home to mostly English-speaking people. Between them lay the largely French-speaking region, which was called Quebec. The British kept Canada's English and French areas separate, with each region having its own colonial government.

Like the American colonists, Canadians wanted to have more control over their lives. In 1867 the British backed Canadian wishes by uniting the territories of Ontario, Quebec, **Nova Scotia,** and **New Brunswick** into one large nation known as the Dominion of Canada. As a dominion,

Canada had its own government to run local affairs. British officials in London, however, still controlled Canada's relations with other countries.

The new Canadian government promised continued protection for the French language and culture. Yet many English-speaking Canadians did not always keep to this promise. French speakers often claimed that they were treated unfairly because of their heritage. Canada often was torn apart by disputes between the two ethnic groups.

After Canada's founding, it gradually expanded its territory. In the late 1860s, Britain handed over to Canada a large area of land in western North America. In 1885 the Canadian government completed a railroad that crossed the continent from the Atlantic to the Pacific coast. This encouraged people from abroad and from other parts of Canada to settle in the west. On the vast expanse of prairies, settlers grew grain and raised cattle. Over the next 65 years, these lands—as well as the eastern territories of **Prince Edward Island** and Newfoundland—joined Canada as provinces.

During the 1900s, Canadians fought side by side with the British in two world wars. Canada's loyal support in these conflicts gradually led to the nation's full independence. In 1982 Canadians finally won the right to change their constitution without British approval. Today only one major link between Canada and Britain remains. The British king or queen still reigns as king or queen of Canada, but this is a ceremonial position with no real power.

Canada's Government The Canadians have a British-style parliamentary democracy. In a parliamentary democracy, voters elect representatives to a lawmaking body called Parliament. These representatives then choose an official named the prime minister to head the government. Since the British king or queen visits Canada only once in a while, a Canadian official called the governor-general carries out most of the government's ceremonial duties.

As you read in Section 2, some people in largely French-speaking Quebec want to pull out of Canada and form their own country. The majority of Quebec's people voted twice against a measure that would lead to independence. Those who favor the move still support independence, however. Only a free Quebec, they claim, can successfully protect French culture in a largely English-speaking North America. As a result, Canada's future as a united country is uncertain.

In recent years, Canada's Native Americans have strengthened their influence as well. In response, the Canadian government has given these peoples more control over land. In 1999 the new territory of Nunavut was created for the Inuit who live in the north. *Nunavut* means "our land" in one of the Inuit languages. The Inuit have formed their own government and also gained control over the minerals in the region.

✓ **Reading Check** What kind of government do Canadians have?

Time to Play
Fifteen-year-old Natalie Menard has been playing ice hockey since she was five years old. Winters are long in Quebec, so Natalie enjoys plenty of time on the ice. Most of her friends are of French descent, and all of their school classes are conducted in French. Natalie also enjoys visiting her cousin Angela, who lives in Toronto, Ontario. More than 6 miles (10 km) of covered walkways and underground tunnels in downtown Toronto connect subways with shops, offices, hotels, and restaurants. Natalie and Angela walk from place to place without even thinking of the weather.

Canada's People and Culture

Although Canada is large in land area, it has only about 30.6 million people. This compares with a population of about 280 million in the United States. Look at the population density map below. It shows you that most Canadians live within 100 miles (161 km) of the United States border.

Canada is a **bilingual** country, or a country that has two official languages—English and French. Yet a growing number of Canadians speak Ukrainian and Chinese. Because of heavy immigration and a high birthrate, Canada's population has grown rapidly since World War II. Today the population is extremely diverse, as you can see from the graph of Canada's ethnic groups on page 111. It also has a long history of religious diversity. Most Canadians are Roman Catholic or Protestant. Still, many Canadians follow Judaism, Buddhism, Hinduism, and Islam.

The Arts and Literature From poetry to novels to drama, Canadian authors write either in English or French about many subjects. The painter-writer Emily Carr portrayed the dark green forests and the Native American cultures of the Pacific Coast in her works. A group of artists, known together as the Group of Seven, used lavish colors and unusual shapes to portray the wild beauty of the Canadian north.

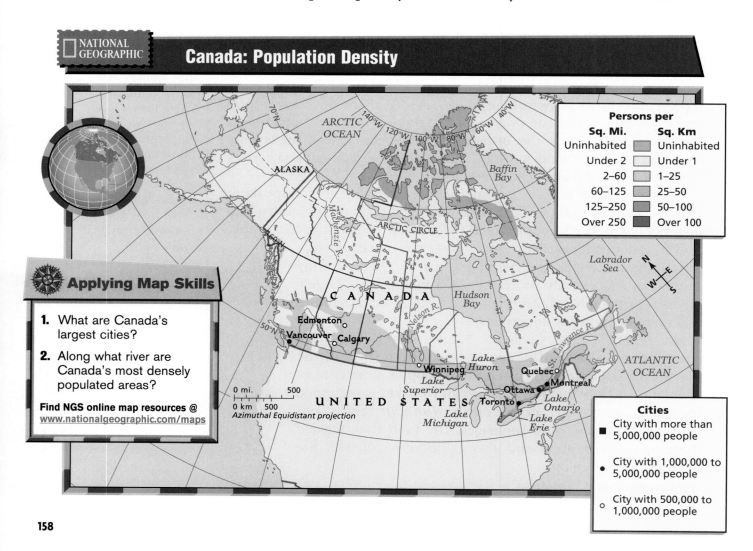

NATIONAL GEOGRAPHIC

Canada: Population Density

Persons per

Sq. Mi.		Sq. Km
Uninhabited		Uninhabited
Under 2		Under 1
2–60		1–25
60–125		25–50
125–250		50–100
Over 250		Over 100

Applying Map Skills

1. What are Canada's largest cities?

2. Along what river are Canada's most densely populated areas?

Find NGS online map resources @
www.nationalgeographic.com/maps

0 mi.   500
0 km   500
Azimuthal Equidistant projection

Cities

■ City with more than 5,000,000 people

● City with 1,000,000 to 5,000,000 people

○ City with 500,000 to 1,000,000 people

Food, Sports, and Recreation

Because Canada has such ethnic diversity, people here can enjoy a variety of tasty foods. People from many different groups have settled in cities like **Toronto.** You can walk down the street and sample the foods of Ukraine, Greece, Italy, the Caribbean, and China all in the same day.

Canadians enjoy a variety of activities, especially outdoor sports. You will find local parks and national parks crowded with people exercising and having fun. Many young Canadians enjoy playing ice hockey. They also take part in other winter sports, including skiing, skating, and snowboarding. During the summer, they might go sailing on Lake Ontario. Professional football and hockey are tremendous spectator sports. Many Canadian sports fans also flock to see the baseball games played in Toronto's large indoor stadium.

Reading Check How does Canada's population compare with that of the United States?

NATIONAL GEOGRAPHIC **On Location**

Toronto, Ontario

Toronto and its suburbs have well over 4 million people, making the area by far Canada's largest urban center.

Location On what body of water is Toronto located?

Section 3 Assessment

Defining Terms

1. **Define** colony, dominion, parliamentary democracy, prime minister, bilingual.

Recalling Facts

2. **History** Who were the first people to live in Canada?

3. **Government** What new territory was created in 1999 and what does its name mean?

4. **Movement** What led to the settlement of western Canada?

Critical Thinking

5. **Analyzing Information** What is the link between Canada and Great Britain?

6. **Summarizing Information** What are two reasons for Canada's population growth?

Graphic Organizer

7. **Organizing Information** Draw a diagram like this one, and list two examples under each heading in the outer ovals.

Food

Religion

Canada's Diversity

Sports

Language

Applying Geography Skills

8. **Analyzing Maps** Study the population density map on page 158. Which cities have populations between 500,000 and 1,000,000?

Snow and Ice Sculpting

While some art forms can exist almost anywhere, that is not the case for ice and snow sculptures. Both of these art forms find the support and weather they need at the city of Quebec's Winter Carnival. Each year more than 1 million people come to this two-week-long festival to enjoy the arts and to celebrate winter.

A Tradition in Ice Sculpture

Quebec held its first Winter Carnival in 1894 to lift the spirits of local people during the long winter months. A local artist carved likenesses of political leaders out of ice. Since then, ice and snow sculpture have become important features of the Winter Carnival. The centerpiece of the Winter Carnival is its glistening ice palace, home to *Bonhomme,* the snowman mascot of the Winter Carnival. Artists work for months creating the impressive castle, which is bathed in colored lights.

Snow Sculpture

The International Snow Sculpture competition is another annual tradition of the Quebec Winter Carnival. Teams of sculptors from about 20 countries compete to create the winning snow sculpture. Each team begins with a huge block of packed snow weighing between 30 and 40 tons (27 and 36 t). It reaches about 12 feet (4 m) high. Working just as they might with stone or wood, sculptors chisel figures into the snow. This time, however, their work could melt away with only a few days of warm weather. At the high point of the competition, during the Night of the Long Knives, the sculptors work day and night to finish their creations.

Carnival visitors can see other examples of ice and snow sculpture in front of shops and restaurants around the city. They can spend the night in a traditional snow igloo and slide down an icy toboggan run. Other activities include dogsled racing, a snow swim, and a canoe race down the ice-filled St. Lawrence River. With so many unique sights and activities, it is easy to understand how the Quebec Winter Carnival has become the world's largest winter festival.

◀ Quebec's Ice Palace

A giant snow sculpture ▶

Making the Connection

1. How did the tradition of snow and ice sculpture start at the Quebec Winter Carnival?

2. How is the art of the Quebec Winter Carnival dependent upon the weather?

3. **Sequencing Information** Describe how artists make snow sculptures in the competition.

Reading Review

Section 1 — Landforms of the North

Terms to Know
province
glacier
peninsula
tundra
prairie
cordillera

Main Idea
Canada is a vast country with many landforms and climates.

✓ Region Canada, the second-largest country in the world, is rich in natural resources.

✓ Region Canada has a variety of landforms, including mountains, lowlands, Arctic wilderness, and prairies.

✓ Economics The lowlands of the St. Lawrence River and the Great Lakes region have rich soil and good transportation routes.

✓ Region The Canadian Shield, covering half of Canada's land area, is rocky and unsuitable for farming but has mineral resources and water power.

✓ Region Most of Canada has a cool or cold climate. Milder temperatures are found in the southern part of the country and the southwest.

Section 2 — A Resource-Rich Economy

Terms to Know
service industry
newsprint
secede
acid rain

Main Idea
Canada's economy benefits from rich natural resources, skilled workers, and close trading ties with the United States.

✓ Economics Canada is one of the world's most economically advanced countries.

✓ Economics Most people in the east make a living in manufacturing, mining, and service industries. Some work in fishing and farming.

✓ Culture Quebec and Ontario have Canada's largest cities and most of its people.

✓ Economics Farming, ranching, oil, and forestry are major economic activities in the west.

✓ Economics Economic challenges include settling regional differences, working out Canada's relationship with the United States, and solving environmental problems.

Section 3 — The Canadians

Terms to Know
colony
dominion
parliamentary democracy
prime minister
bilingual

Main Idea
Canadians of many different backgrounds live in towns and cities close to the United States border.

✓ History Native Americans were the first Canadians. French and British settlers later built homes in Canada. Large numbers of immigrants have recently come from Asia and Eastern Europe.

✓ Government Canada's government is a parliamentary democracy headed by a prime minister.

✓ Culture Some people in French-speaking Quebec want to separate from the rest of Canada.

✓ Culture Canada's population is made up of many different ethnic groups. English and French are the country's two official languages.

Canada

Using Key Terms

Match the terms in Part A with their definitions in Part B.

A.
1. province
2. glacier
3. tundra
4. secede
5. newsprint
6. acid rain
7. peninsula
8. bilingual
9. prime minister
10. parliamentary democracy

B.
a. having or speaking two languages
b. giant sheet of ice
c. land surrounded by water on three sides
d. voters elect representatives to a lawmaking body called Parliament
e. government leader chosen by members of Parliament
f. vast, dry, treeless plain in the high latitudes
g. type of paper used for printing newspapers
h. regional political division
i. to withdraw
j. precipitation with large amounts of chemicals

Reviewing the Main Ideas

Section 1 Landforms of the North

11. **Location** What three oceans border Canada?
12. **Region** Name three natural resources found in the Interior Plains.
13. **Location** In which physical region do most Canadians live?

Section 2 A Resource-Rich Economy

14. **Government** Describe two ways in which Canada's government plays a role in the nation's economy.
15. **Economics** Why did the United States and Canada build the St. Lawrence Seaway?
16. **Economics** Name three economic activities found in British Columbia.
17. **Human/Environment Interaction** What are two problems affecting Canada's economy?

Section 3 The Canadians

18. **History** Who was the first explorer to claim part of Canada for France?
19. **Culture** Why do some of Quebec's people want independence from Canada?
20. **Culture** What are Canada's two official languages?

 **Canada**

Place Location Activity

On a separate sheet of paper, match the letters on the map with the numbered places listed below.

1. Hudson Bay
2. Nunavut
3. British Columbia
4. Ottawa
5. Quebec (province)
6. St. Lawrence River
7. Rocky Mountains
8. Winnipeg
9. Ontario
10. Nova Scotia

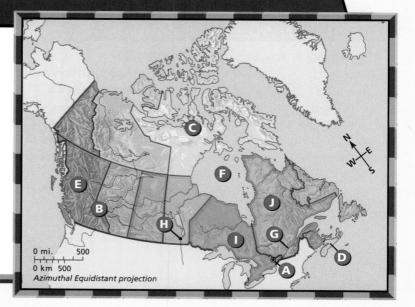

0 mi. 500
0 km 500
Azimuthal Equidistant projection

Critical Thinking

21. **Making Comparisons** Compare the overall climate of Canada with the overall climate of the United States.

22. **Analyzing Information** Why do most Canadians live near the United States border?

23. **Categorizing Information** Choose one of Canada's provinces or territories. Complete a chart like the one below with at least two facts or examples under the headings in each section.

Province or Territory	Landforms	Resources
	Major Cities	Products

GeoJournal Activity

24. **Writing a Speech** List what you know about how people can use forests and forest products. Then take the role of an environmental leader. Write a short speech explaining how Canada's forests are useful and why they need to be protected.

Mental Mapping Activity

25. **Focusing on the Region** Draw a simple outline map of Canada, then label the following on your map:

- Arctic Ocean
- Pacific Ocean
- Atlantic Ocean
- Rocky Mountains
- Hudson Bay
- Quebec (province)
- Ontario
- British Columbia
- Nunavut
- Ottawa

Technology Skills Activity

26. **Using the Internet** Access the Internet and search for information on the Inuit and the new territory of Nunavut. Create an illustrated time line that shows the steps leading to the creation of the new territory.

The **Princeton Review**

Standardized Test Practice

Directions: Study the graph, then answer the following questions.

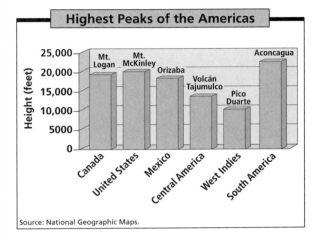

Source: National Geographic Maps.

1. **What is the tallest mountain in the Americas?**

   A Mt. Logan
   B Mt. McKinley
   C Orizaba
   D Aconcagua

2. **Which mountain in North America is the tallest?**

   F Mt. Logan
   G Mt. McKinley
   H Aconcagua
   J Orizaba

Test-Taking Tip: When you are not sure of an answer, use the process of elimination. On question 2, for example, you know that the United States and Canada are both in North America. You have not yet learned that Mexico is also part of North America. Even so, Mexico's highest peak is shorter than the highest peak in both Canada and the United States.

GEOGRAPHY & HISTORY

Iroquois greeting Cartier

THE ST. LAWRENCE:
A River of Trade

A cargo ship winds its way through the Thousand Islands (right), a group of more than 1,500 islands that lie in the St. Lawrence River near Lake Ontario.

Along with a system of lakes, locks, dams, and canals, the river forms the St. Lawrence Seaway, a busy highway for international trade. More than a thousand ships travel the waterway each year, hauling about 50 million tons (45 million t) of cargo. Yet the river was not always so easy to navigate. When French explorer Jacques Cartier sailed upriver in 1535, he thought he had found a shortcut to Asia. His dreams were ruined, however, when he was stopped by swift rapids near present-day Montreal. Many changes had to occur before the river would be tamed.

The Seaway's History

To the Iroquois Indians, who greeted Cartier upon his arrival, the St. Lawrence was "the river without end." In their birchbark canoes, they paddled far into the interior, bringing back beaver,

lynx, and other furs to trade with the Europeans. In 1608 Samuel de Champlain of France founded the first trading post on the river, later known as the city of Quebec.

Early explorers and settlers, however, could not get their boats through many parts of the St. Lawrence. In 1680 a canal was begun around the Montreal rapids. As more canals were built, the river became a major Canadian shipping route. In 1833 Canada opened the Welland Canal, which bypasses Niagara Falls and links Lake Erie with Lake Ontario.

By the early 1900s, Canada and the United States were planning to widen and deepen sections of the St. Lawrence to allow for larger ships. Construction of the St. Lawrence Seaway, as the waterway was called, began in 1954. The project called for building miles of canals, plus 15 locks, 3 dams, and several hydro-electric power stations. The

seaway opened in 1959. For the first time, ships could travel between the Atlantic Ocean and inland ports such as Chicago, Illinois, on Lake Michigan, and Duluth, Minnesota, on Lake Superior.

The Seaway Today

Today the St. Lawrence Seaway is still an important avenue of trade. Ships carry iron ore from Labrador to ports on the Great Lakes. The ships return with wheat and other grains. Many products travel from North American ports to countries overseas. In winter, thick ice closes sections of the seaway for months.

QUESTIONS

1 Why is the St. Lawrence Seaway important to the United States and Canada?

2 How did building the seaway help ships?

The busy St. Lawrence Seaway ▶

NATIONAL
GEOGRAPHIC
SOCIETY

The Great Lakes and St. Lawrence Seaway

N
W E
S

CANADA

•Duluth *Lake Superior*

Gulf of St. Lawrence

•Quebec

St. Lawrence R.

•Montreal

Lake Huron

Niagara Falls

Lake Michigan

•Chicago

Lake Ontario

ATLANTIC OCEAN

Lake Erie

0 mi.   500
0 km   500

UNITED STATES

Unit 3

Peruvian Indian woman and child

Spanish colonial architecture in Guatemala

NATIONAL
GEOGRAPHIC
SOCIETY

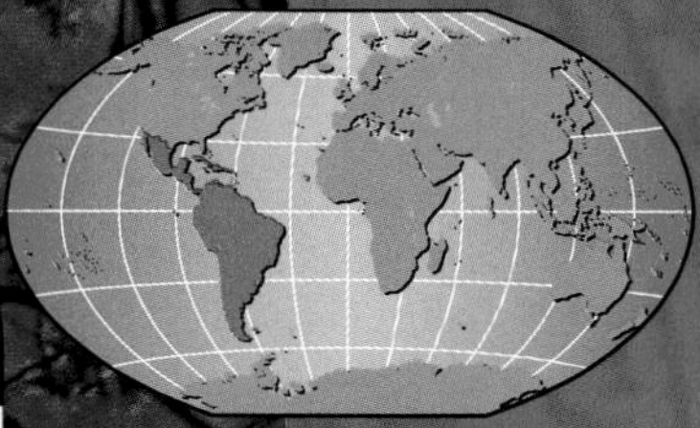

Latin America

W here can you find steamy tropical forests, frigid mountain peaks, thundering waterfalls, and peaceful island beaches? All of these contrasts can be found in Latin America—a huge part of the world made up of 33 nations on two continents. This region stretches from the Mexico–United States border, in North America, to the southernmost tip of South America.

NGS ONLINE
www.nationalgeographic.com/education

Spider monkey and Mayan ruins, Mexico ▲

Focus on:

Latin America

COMMON THREADS OF LANGUAGE AND RELIGION unite this region. Once claimed as European colonies, most Latin American countries still use either Spanish or Portuguese as the official language. These two languages are based on Latin, which is how the region gets its name. Most Latin Americans are Roman Catholic, another influence from colonial times.

The Land

Latin America stretches from the Rio Grande south to Tierra del Fuego, just 600 miles (966 km) from Antarctica's frozen shores. Three times larger than the continental United States, the region includes Mexico, Central America, the Caribbean islands, and South America.

Mountains are prominent features in many parts of Latin America. Some Caribbean islands are actually the exposed tops of ancient, submerged volcanoes. In Mexico, the branches of the Sierra Madre spread like welcoming arms to hug a central highland known as the Mexican Plateau. Mist-covered peaks stretch through the interior of Central America. Most impressive, however, are the Andes. The longest series of mountain ranges in the world, the lofty Andes follow the western coast of South America for 4,500 miles (7,242 km).

Narrow coastal plains line the edges of Mexico and Central America. South America has vast inland plains. These include the pampas of Argentina and the llanos of Colombia and Venezuela. The largest lowland area on this continent is the basin of the Amazon River, the longest river in the Western Hemisphere.

The Climate

Most of this region has a tropical climate. Daily showers drench the rain forests, which thrive in the lowlands. In Brazil the Amazon River and its tributaries snake through the largest area of rain forest regions, which covers roughly one-third of South America. Rain forests contain more species of plants and animals than any other ecosystem on Earth.

The climate tends to be drier and cooler at higher elevations and farther away from the Equator. Under these conditions, tall grasses and scattered trees flourish.

Drier still are parts of northern Mexico and southern Argentina. Here, rainfall is sparse and so is vegetation. Yet even these places are lush compared to the Atacama Desert, along Chile's coast. The barren Atacama is among the world's driest places.

Three-toed sloth in rain forest, Panama ▶

◀ Peaks of the Andes, Chile

The Economy

Latin America is rich in natural resources. Gold drew many of the first European conquerors. Copper, silver, iron ore, tin, and lead also are abundant in the region. Some Latin American countries are among the world's leading producers of oil and natural gas.

Agriculture plays an important role in the region's economy. Coffee, bananas, and sugarcane thrive in the moist, fertile lowlands. On higher ground, farmers raise grain and fruit, while cowhands known as gauchos drive huge herds of cattle across rolling grasslands.

Industrialization is increasing in Latin America. However, some countries are moving along this path more quickly than others. In recent years, Mexico, Brazil, and Chile have become major producers of manufactured goods. Some things have hindered industrial development in other parts of the region. These include shortages of money, skilled labor, and reliable transportation, along with geographic barriers such as rugged mountains and thick forests.

The People

Long before Europeans crossed the Atlantic Ocean, great Native American civilizations developed in Latin America. The Maya flourished in the Guatemalan and Honduran lowlands and across Mexico's Yucatán Peninsula. The central highlands of Mexico were the site of the Aztec Empire. In South America, the Inca established an empire that stretched from southern Colombia to central Chile.

Beginning in the 1500s, Spain and Portugal ruled most of Latin America. These European invaders destroyed the Native American civilizations. They also brought enslaved Africans to work alongside Native Americans on plantations.

Independence came for many Latin American countries in the early 1800s. These countries remain a cultural mixture—Native Americans, Europeans, Africans, and others all have left their mark.

Today most Latin Americans live in urban areas along the coasts of South America or in a band reaching from Mexico into Central America. Some of the largest cities in the world are in this region, including Mexico City, Rio de Janeiro, and Buenos Aires.

Exploring the Region

1. What is Latin America's longest series of mountain ranges?
2. What type of climate is found across most of the region?
3. What European countries once ruled Latin America?
4. Which Latin American countries are industrializing most rapidly?

◀ Mexican boy carrying decorated cross for religious celebration

UNIT 3

Rio de Janeiro, Brazil ▼

Latin America

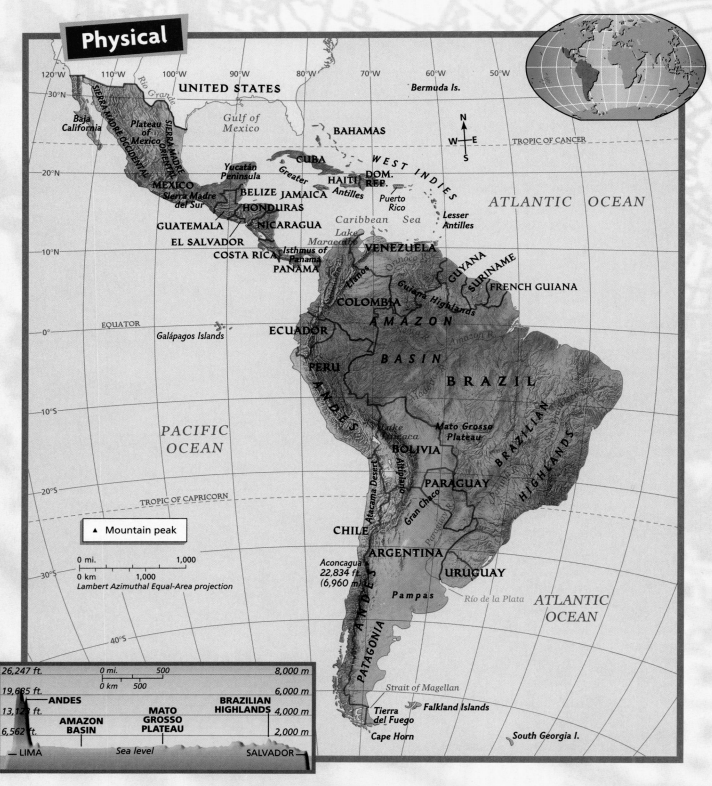

Physical

120°W 110°W 100°W 90°W 80°W 70°W 60°W 50°W

30°N

Rio Grande

UNITED STATES

Bermuda Is.

Baja California

SIERRA MADRE OCCIDENTAL

Plateau of Mexico

SIERRA MADRE ORIENTAL

Gulf of Mexico

BAHAMAS

N
W   E
S

TROPIC OF CANCER

20°N

Yucatán Peninsula

MEXICO

Sierra Madre del Sur

CUBA

Greater

HAITI

DOM. REP.

Antilles

BELIZE **JAMAICA**

Puerto Rico

HONDURAS

Lesser Antilles

W E S T   I N D I E S

ATLANTIC OCEAN

GUATEMALA

NICARAGUA

Caribbean   Sea

EL SALVADOR

Lake Maracaibo

10°N

COSTA RICA

Isthmus of Panama

VENEZUELA

Orinoco R.

GUYANA

SURINAME

PANAMA

Llanos

Guiana Highlands

FRENCH GUIANA

COLOMBIA

A M A Z O N

EQUATOR

Galápagos Islands

ECUADOR

0°

Negro R.

Amazon R.

B A S I N

PERU

B R A Z I L

Madeira R.

A
N
D
E
S

10°S

Lake Titicaca

Mato Grosso Plateau

São Francisco R.

PACIFIC OCEAN

Altiplano

BOLIVIA

B R A Z I L I A N

20°S

TROPIC OF CAPRICORN

Atacama Desert

PARAGUAY

H I G H L A N D S

Gran Chaco

Paraguay R.

Paraná R.

▲ Mountain peak

CHILE

0 mi.   1,000

0 km   1,000

Lambert Azimuthal Equal-Area projection

ARGENTINA

Aconcagua
22,834 ft.
(6,960 m)

URUGUAY

30°S

Pampas

Río de la Plata

ATLANTIC OCEAN

A
N
D
E
S

40°S

PATAGONIA

Strait of Magellan

26,247 ft.	0 mi.	500		8,000 m

0 km   500

19,685 ft.   6,000 m

ANDES

Tierra del Fuego

Falkland Islands

13,123 ft.

BRAZILIAN HIGHLANDS

4,000 m

MATO GROSSO PLATEAU

6,562 ft.

AMAZON BASIN

2,000 m

Cape Horn

South Georgia I.

— LIMA

Sea level

SALVADOR —

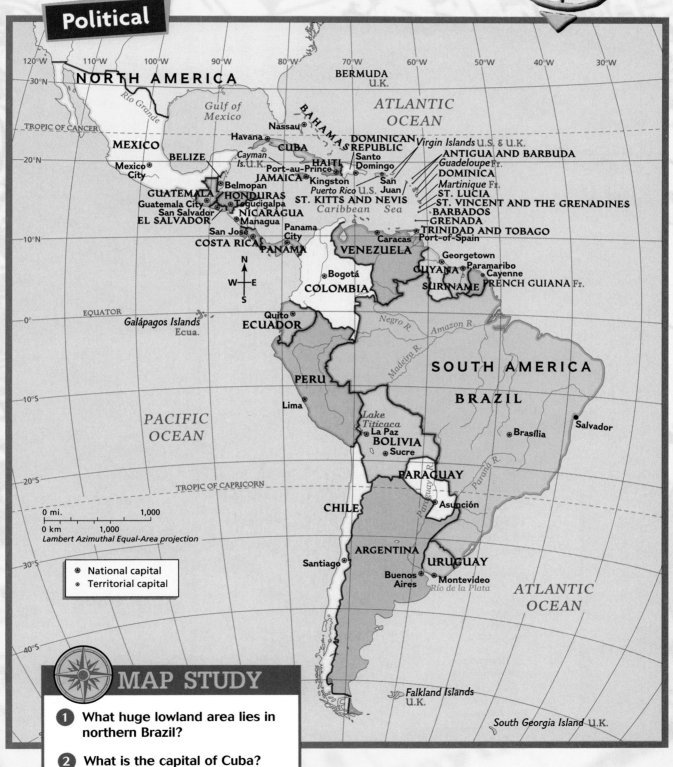

UNIT 3

Political

NORTH AMERICA

BERMUDA U.K.

Rio Grande

Gulf of Mexico

ATLANTIC OCEAN

TROPIC OF CANCER

Nassau

BAHAMAS

MEXICO

Havana

CUBA

DOMINICAN REPUBLIC

Virgin Islands U.S. & U.K.

Mexico City

BELIZE

Cayman Is. U.K.

Port-au-Prince

HAITI

Santo Domingo

ANTIGUA AND BARBUDA

Guadeloupe Fr.

DOMINICA

GUATEMALA

Belmopan

JAMAICA

Kingston

Puerto Rico U.S.

San Juan

Martinique Fr.

ST. LUCIA

HONDURAS

Guatemala City

Tegucigalpa

ST. KITTS AND NEVIS

ST. VINCENT AND THE GRENADINES

San Salvador

NICARAGUA

Caribbean Sea

BARBADOS

EL SALVADOR

Managua

GRENADA

San José

Panama City

TRINIDAD AND TOBAGO

COSTA RICA

Port-of-Spain

PANAMA

Caracas

Georgetown

N

W E

S

VENEZUELA

GUYANA

Paramaribo

Cayenne

Bogotá

SURINAME

FRENCH GUIANA Fr.

COLOMBIA

EQUATOR

Galápagos Islands Ecua.

Quito

ECUADOR

Negro R.

Amazon R.

PERU

SOUTH AMERICA

BRAZIL

Madeira R.

Lima

PACIFIC OCEAN

Lake Titicaca

La Paz

BOLIVIA

Sucre

Brasília

Salvador

PARAGUAY

Paraná R.

TROPIC OF CAPRICORN

CHILE

Paraguay R.

Asunción

0 mi.   1,000

0 km   1,000

Lambert Azimuthal Equal-Area projection

ARGENTINA

URUGUAY

⊗ National capital
• Territorial capital

Santiago

Buenos Aires

Montevideo

ATLANTIC OCEAN

Río de la Plata

Falkland Islands U.K.

MAP STUDY

1 What huge lowland area lies in northern Brazil?

2 What is the capital of Cuba?

173

Latin America

Urban Population Growth

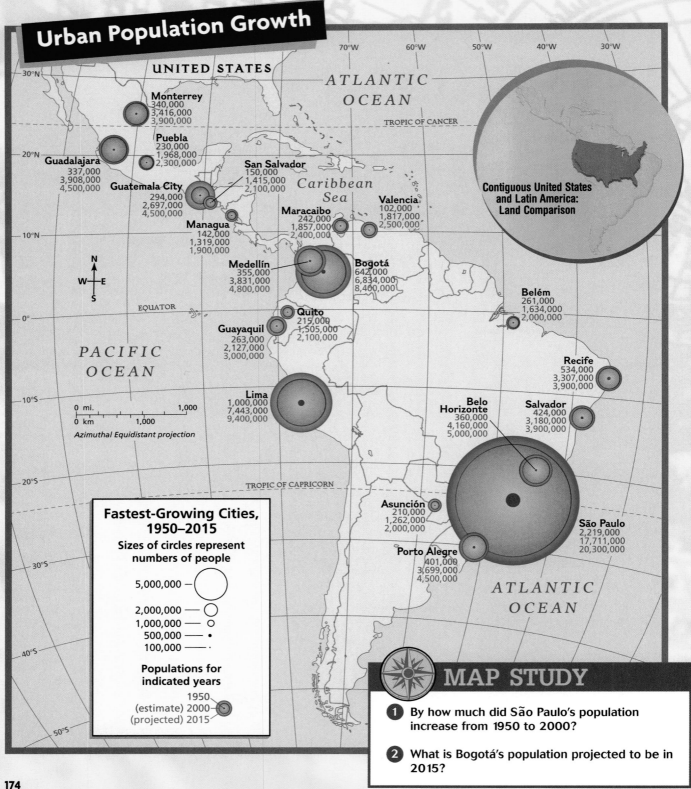

UNITED STATES

ATLANTIC OCEAN

TROPIC OF CANCER

Monterrey
340,000
3,416,000
3,900,000

Puebla
230,000
1,968,000
2,300,000

Guadalajara
337,000
3,908,000
4,500,000

San Salvador
150,000
1,415,000
2,100,000

Caribbean Sea

Guatemala City
294,000
2,697,000
4,500,000

Managua
142,000
1,319,000
1,900,000

Maracaibo
242,000
1,857,000
2,400,000

Valencia
102,000
1,817,000
2,500,000

Contiguous United States and Latin America: Land Comparison

N
W—E
S

Medellín
355,000
3,831,000
4,800,000

Bogotá
642,000
6,834,000
8,400,000

Belém
261,000
1,634,000
2,000,000

EQUATOR

Quito
215,000
1,505,000
2,100,000

Guayaquil
263,000
2,127,000
3,000,000

PACIFIC OCEAN

Recife
534,000
3,307,000
3,900,000

0 mi.   1,000
0 km   1,000

Azimuthal Equidistant projection

Lima
1,000,000
7,443,000
9,400,000

Belo Horizonte
360,000
4,160,000
5,000,000

Salvador
424,000
3,180,000
3,900,000

TROPIC OF CAPRICORN

Asunción
210,000
1,262,000
2,000,000

Fastest-Growing Cities, 1950–2015

Sizes of circles represent numbers of people

5,000,000 —

2,000,000 — ○

1,000,000 — ○

500,000 — •

100,000 — ·

São Paulo
2,219,000
17,711,000
20,300,000

Porto Alegre
401,000
3,699,000
4,500,000

ATLANTIC OCEAN

Populations for indicated years

1950
(estimate) 2000
(projected) 2015

MAP STUDY

1 By how much did São Paulo's population increase from 1950 to 2000?

2 What is Bogotá's population projected to be in 2015?

Geo Extremes

① **HIGHEST POINT**
Aconcagua (Argentina)
22,834 ft. (6,960 m) high

② **LOWEST POINT**
Valdés Peninsula (Argentina)
131 ft. (40 m) below sea level

③ **LONGEST RIVER**
Amazon River
(Brazil and Peru)
4,000 mi. (6,437 km) long

④ **LARGEST LAKE**
Lake Maracaibo (Venezuela)
5,217 sq. mi. (13,512 sq. km)

⑤ **HIGHEST LARGE
NAVIGABLE LAKE**
Lake Titicaca
(Peru and Bolivia)
12,500 ft. (3,810 m) high

⑥ **HIGHEST WATERFALL**
Angel Falls (Venezuela)
3,212 ft. (979 m) high

⑦ **DRIEST PLACE**
Atacama Desert (Chile)
rainfall barely measurable

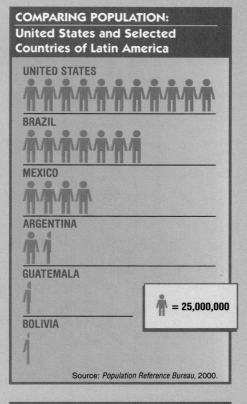

COMPARING POPULATION:
United States and Selected
Countries of Latin America

UNITED STATES

BRAZIL

MEXICO

ARGENTINA

GUATEMALA

👤 = 25,000,000

BOLIVIA

Source: *Population Reference Bureau*, 2000.

ETHNIC GROUPS:
Selected Countries of Latin America

ARGENTINA
| 97% | | 3% |

BOLIVIA
| 55% | 30% | 15% |

BRAZIL
6%
| 55% | 38% | 1% |

GUATEMALA
| 56% | 44% |

MEXICO
9%
| 60% | 30% | 1% |

☐ Black ☐ Indian ☐ African/European
☐ White ☐ Mestizo ☐ Other

Source: *CIA World Factbook*, 2000.

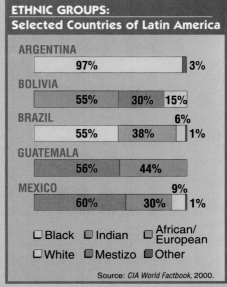

GRAPHIC STUDY

① What two Latin American "extremes" are found in Venezuela?

② What countries have a majority of mestizos (people of mixed European and Native American ancestry)?

Country Profiles

ANTIGUA and BARBUDA

POPULATION:
67,000
394 per sq. mi.
152 per sq. km

LANGUAGE:
English

MAJOR EXPORT:
Petroleum Products

MAJOR IMPORTS:
Foods and Livestock

CAPITAL:
St. John's

LANDMASS:
170 sq. mi.
440 sq. km

ARGENTINA

POPULATION:
36,568,000
34 per sq. mi.
13 per sq. km

LANGUAGE:
Spanish

MAJOR EXPORT:
Meat

MAJOR IMPORT:
Machinery

CAPITAL:
Buenos Aires

LANDMASS:
1,068,302 sq. mi.
2,766,889 sq. km

BAHAMAS

POPULATION:
301,000
56 per sq. mi.
22 per sq. km

LANGUAGES:
English, Creole

MAJOR EXPORT:
Pharmaceuticals

MAJOR IMPORT:
Foods

CAPITAL:
Nassau

LANDMASS:
5,382 sq. mi.
13,939 sq. km

BARBADOS

POPULATION:
269,000
1,620 per sq. mi.
626 per sq. km

LANGUAGE:
English

MAJOR EXPORT:
Sugar

MAJOR IMPORT:
Manufactured Goods

CAPITAL:
Bridgetown

LANDMASS:
166 sq. mi.
430 sq. km

BELIZE

POPULATION:
248,000
28 per sq. mi.
11 per sq. km

LANGUAGE:
English

MAJOR EXPORT:
Sugar

MAJOR IMPORT:
Machinery

CAPITAL:
Belmopan

LANDMASS:
8,867 sq. mi.
22,965 sq. km

BOLIVIA

POPULATION:
8,090,000
19 per sq. mi.
7 per sq. km

LANGUAGES:
Spanish, Quechua, Aymara

MAJOR EXPORT:
Metals

MAJOR IMPORT:
Machinery

CAPITALS:
La Paz, Sucre

LANDMASS:
424,164 sq. mi.
1,098,581 sq. km

BRAZIL

POPULATION:
167,988,000
51 per sq. mi.
20 per sq. km

LANGUAGE:
Portuguese

MAJOR EXPORT:
Iron Ore

MAJOR IMPORT:
Crude Oil

CAPITAL:
Brasília

LANDMASS:
3,286,488 sq. mi.
8,511,965 sq. km

CHILE

POPULATION:
15,018,000
51 per sq. mi.
20 per sq. km

LANGUAGE:
Spanish

MAJOR EXPORT:
Copper

MAJOR IMPORT:
Machinery

CAPITAL:
Santiago

LANDMASS:
292,135 sq. mi.
756,626 sq. km

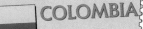

COLOMBIA

POPULATION:
38,581,000
88 per sq. mi.
34 per sq. km

LANGUAGE:
Spanish

MAJOR EXPORT:
Petroleum

MAJOR IMPORT:
Machinery

CAPITAL:
Bogotá

LANDMASS:
439,737 sq. mi.
1,138,914 sq. km

COSTA RICA

POPULATION:
3,594,000
182 per sq. mi.
70 per sq. km

LANGUAGE:
Spanish

MAJOR EXPORT:
Coffee

MAJOR IMPORT:
Raw Materials

CAPITAL:
San José

LANDMASS:
19,730 sq. mi.
51,100 sq. km

CUBA

POPULATION:
11,178,000
261 per sq. mi.
101 per sq. km

LANGUAGE:
Spanish

MAJOR EXPORT:
Sugar

MAJOR IMPORT:
Petroleum

CAPITAL:
Havana

LANDMASS:
42,804 sq. mi.
110,861 sq. km

Countries and flags not drawn to scale

DOMINICA

POPULATION:
71,000
245 per sq. mi.
95 per sq. km

LANGUAGES:
English, French

MAJOR EXPORT:
Bananas

MAJOR IMPORT:
Manufactured
Goods

Roseau

CAPITAL:
Roseau

LANDMASS:
290 sq. mi.
751 sq. km

DOMINICAN REPUBLIC

POPULATION:
8,299,000
441 per sq. mi.
170 per sq. km

LANGUAGE:
Spanish

MAJOR EXPORT:
Ferronickel

MAJOR IMPORT:
Foods

Santo Domingo

CAPITAL:
Santo Domingo

LANDMASS:
18,816 sq. mi.
48,734 sq. km

ECUADOR

POPULATION:
12,411,000
113 per sq. mi.
44 per sq. km

LANGUAGES:
Spanish, Quechua

MAJOR EXPORT:
Petroleum

MAJOR IMPORT:
Transport
Equipment

Quito

CAPITAL:
Quito

LANDMASS:
109,484 sq. mi.
283,561 sq. km

EL SALVADOR

POPULATION:
5,859,000
721 per sq. mi.
278 per sq. km

LANGUAGE:
Spanish

MAJOR EXPORT:
Coffee

MAJOR IMPORT:
Raw Materials

San Salvador

CAPITAL:
San Salvador

LANDMASS:
8,124 sq. mi.
21,041 sq. km

FRENCH GUIANA *

POPULATION:
185,000
5 per sq. mi.
2 per sq. km

LANGUAGE:
French

MAJOR EXPORT:
Shrimp

MAJOR IMPORT:
Foods

Cayenne

CAPITAL:
Cayenne

LANDMASS:
34,749 sq. mi.
89,999 sq. km

* Territory of France

GRENADA

POPULATION:
97,000
729 per sq. mi.
282 per sq. km

LANGUAGES:
English, French

MAJOR EXPORT:
Bananas

MAJOR IMPORT:
Foods

St. George's

CAPITAL:
St. George's

LANDMASS:
133 sq. mi.
344 sq. km

GUATEMALA

POPULATION:
12,336,000
293 per sq. mi.
113 per sq. km

LANGUAGES:
Spanish, Mayan
Languages

MAJOR EXPORT:
Coffee

MAJOR IMPORT:
Petroleum

Guatemala City

CAPITAL:
Guatemala City

LANDMASS:
42,042 sq. mi.
108,889 sq. km

GUYANA

POPULATION:
705,000
8 per sq. mi.
3 per sq. km

LANGUAGE:
English

MAJOR EXPORT:
Sugar

MAJOR IMPORT:
Manufactured
Goods

Georgetown

CAPITAL:
Georgetown

LANDMASS:
83,000 sq. mi.
214,969 sq. km

HAITI

POPULATION:
7,751,000
723 per sq. mi.
279 per sq. km

LANGUAGES:
French, Creole

MAJOR EXPORT:
Manufactured
Goods

MAJOR IMPORT:
Machinery

Port-au-Prince

CAPITAL:
Port-au-Prince

LANDMASS:
10,714 sq. mi.
27,750 sq. km

HONDURAS

POPULATION:
5,901,000
136 per sq. mi.
53 per sq. km

LANGUAGE:
Spanish

MAJOR EXPORT:
Bananas

MAJOR IMPORT:
Machinery

Tegucigalpa

CAPITAL:
Tegucigalpa

LANDMASS:
43,277 sq. mi.
112,088 sq. km

JAMAICA

POPULATION:
2,621,000
618 per sq. mi.
238 per sq. km

LANGUAGES:
English, Creole

MAJOR EXPORT:
Alumina

MAJOR IMPORT:
Machinery

Kingston

CAPITAL:
Kingston

LANDMASS:
4,244 sq. mi.
10,991 sq. km

REGIONAL ATLAS

Country Profiles

MEXICO

POPULATION:
99,734,000
132 per sq. mi.
51 per sq. km

LANGUAGES:
Spanish,
Native American
Languages

MAJOR EXPORT:
Crude Oil

MAJOR IMPORT:
Machinery

CAPITAL:
Mexico City

LANDMASS:
756,066 sq. mi.
1,958,201 sq. km

Mexico City

NICARAGUA

POPULATION:
4,952,000
99 per sq. mi.
38 per sq. km

LANGUAGE:
Spanish

MAJOR EXPORT:
Coffee

MAJOR IMPORT:
Manufactured
Goods

CAPITAL:
Managua

LANDMASS:
50,193 sq. mi.
129,999 sq. km

Managua

PANAMA

POPULATION:
2,809,000
94 per sq. mi.
36 per sq. km

LANGUAGE:
Spanish

MAJOR EXPORT:
Bananas

MAJOR IMPORT:
Machinery

CAPITAL:
Panama City

LANDMASS:
29,762 sq. mi.
77,082 sq. km

Panama City

PARAGUAY

POPULATION:
5,219,000
33 per sq. mi.
13 per sq. km

LANGUAGES:
Spanish, Guaraní

MAJOR EXPORT:
Cotton

MAJOR IMPORT:
Machinery

CAPITAL:
Asunción

LANDMASS:
157,048 sq. mi.
406,752 sq. km

Asunción

PERU

POPULATION:
26,624,000
54 per sq. mi.
21 per sq. km

LANGUAGES:
Spanish, Quechua,
Aymara

MAJOR EXPORT:
Copper

MAJOR IMPORT:
Machinery

CAPITAL:
Lima

LANDMASS:
496,225 sq. mi.
1,285,217 sq. km

Lima

PUERTO RICO*

POPULATION:
3,887,652
1,132 per sq. mi.
437 per sq. km

LANGUAGES:
Spanish, English

MAJOR EXPORT:
Pharmaceuticals

MAJOR IMPORT:
Chemical Products

CAPITAL:
San Juan

LANDMASS:
3,435 sq. mi.
8,897 sq. km

San Juan

* U.S. Commonwealth

ST. KITTS and NEVIS

POPULATION:
39,000
386 per sq. mi.
149 per sq. km

LANGUAGE:
English

MAJOR EXPORT:
Machinery

MAJOR IMPORT:
Electronic Goods

CAPITAL:
Basseterre

LANDMASS:
101 sq. mi.
261 sq. km

Basseterre

ST. LUCIA

POPULATION:
154,000
647 per sq. mi.
250 per sq. km

LANGUAGES:
English, French

MAJOR EXPORT:
Bananas

MAJOR IMPORT:
Foods

CAPITAL:
Castries

LANDMASS:
238 sq. mi.
617 sq. km

Castries

ST. VINCENT and the GRENADINES

POPULATION:
114,000
760 per sq. mi.
294 per sq. km

LANGUAGES:
English, French

MAJOR EXPORT:
Bananas

MAJOR IMPORT:
Foods

CAPITAL:
Kingstown

LANDMASS:
150 sq. mi.
388 sq. km

Kingstown

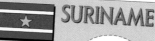

SURINAME

POPULATION:
431,000
7 per sq. mi.
3 per sq. km

LANGUAGE:
Dutch

MAJOR EXPORT:
Bauxite

MAJOR IMPORT:
Machinery

CAPITAL:
Paramaribo

LANDMASS:
63,037 sq. mi.
163,265 sq. km

Paramaribo

TRINIDAD and TOBAGO

POPULATION:
1,285,000
649 per sq. mi.
250 per sq. km

LANGUAGE:
English

MAJOR EXPORT:
Petroleum

MAJOR IMPORT:
Machinery

CAPITAL:
Port-of-Spain

LANDMASS:
1,981 sq. mi.
5,131 sq. km

Port-of-Spain

Countries and flags not drawn to scale

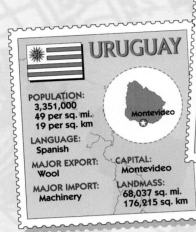

URUGUAY

POPULATION:
3,351,000
49 per sq. mi.
19 per sq. km

LANGUAGE:
Spanish

MAJOR EXPORT:
Wool

MAJOR IMPORT:
Machinery

CAPITAL:
Montevideo

LANDMASS:
68,037 sq. mi.
176,215 sq. km

Montevideo

VENEZUELA

POPULATION:
23,766,000
67 per sq. mi.
26 per sq. km

LANGUAGE:
Spanish

MAJOR EXPORT:
Petroleum

MAJOR IMPORT:
Raw Materials

CAPITAL:
Caracas

LANDMASS:
352,144 sq. mi.
912,050 sq. km

Caracas

VIRGIN ISLANDS*

POPULATION:
97,000
713 per sq. mi.
276 per sq. km

LANGUAGE:
English

MAJOR EXPORT:
Chemical Products

MAJOR IMPORT:
Crude Oil

CAPITAL:
Charlotte Amalie

LANDMASS:
136 sq. mi.
352 sq. km

Charlotte Amalie

* Territory of U.S.

Questions From Buzz Bee!

The following questions are taken from National Geographic GeoBees. Use your textbook, the Internet, and other library resources to find the answers.

1. In 1911 Hiram Bingham found the site of a major Incan city in the Peruvian Andes. Name this site.

2. Marajó Island, which is just south of the Equator and about the same size as Denmark, is bordered on the northwest by what river?

3. Cowhands who tend cattle on the grasslands of Argentina are part of the country's folklore. What are these cowhands called?

4. Name the world's most populous Spanish-speaking country.

5. Bananas grown on coastal lowlands north of Tegucigalpa are a major export of which Central American country?

The World and Its People | NATIONAL GEOGRAPHIC

To learn more about the people and places of Mexico, view *The World and Its People* **Chapter 6** video.

Geography Online

Chapter Overview Visit the *Geography: The World and Its People* Web site at gwip.glencoe.com and click on **Chapter 6— Chapter Overviews** to preview information about Mexico.

Mexico's Land and Economy

Guide to Reading

Main Idea

Mexico's mountainous landscape and varied climate create different economic regions.

Terms to Know

- land bridge
- peninsula
- latitude
- altitude
- hurricane
- vaquero
- *maquiladora*
- subsistence farm
- plantation

Places to Locate

- Pacific Ocean
- Gulf of Mexico
- Baja California
- Yucatán Peninsula
- Sierra Madre
- Mexico City
- Plateau of Mexico
- Monterrey
- Tijuana
- Ciudad Juárez
- Guadalajara

Reading Strategy

Make three charts like this one to show how Mexico's northern, central, and southern regions differ.

Region	→	
Landscape	→	
Climate	→	
Economy	→	

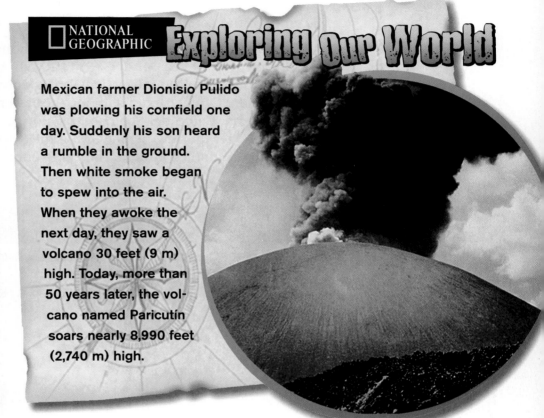

NATIONAL GEOGRAPHIC Exploring Our World

Mexican farmer Dionisio Pulido was plowing his cornfield one day. Suddenly his son heard a rumble in the ground. Then white smoke began to spew into the air. When they awoke the next day, they saw a volcano 30 feet (9 m) high. Today, more than 50 years later, the volcano named Paricutín soars nearly 8,990 feet (2,740 m) high.

Paricutín and other volcanoes are scattered throughout Mexico because the country sits where three plates in the earth's crust collide. Sometimes the movement of these plates can bring disastrous results. Hot magma, or melted rock, might shoot through a volcano. The ground might shift violently in an earthquake. Do you see why Native Americans once called Mexico "the land of the shaking earth"?

Bridging Two Continents

Mexico forms part of a land bridge, or narrow strip of land that joins two larger landmasses. This land bridge connects North America and South America. Look at the map on page 172. You can see that Mexico, part of North America, borders the southern United States. Farther south, Mexico reaches its narrowest point at the Isthmus of Tehuantepec (tay•WAHN•tah•PEHK). Here, only 140 miles (225 km) separate the **Pacific Ocean** from the **Gulf of Mexico**.

◀ The Lighthouse of Commerce stands guard over Monterrey and Saddle Mountain.

The Pacific Ocean borders Mexico on the west. Extending south along this western coast is **Baja** (BAH•hah) **California.** It is a long, thin peninsula, or piece of land with water on three sides. On Mexico's eastern side, the Gulf of Mexico and the Caribbean Sea border the shores. Between the Gulf and the Caribbean Sea is another peninsula—the **Yucatán** (YOO•kah•TAHN) **Peninsula.**

Mexico is a rugged land. If you saw it from space, you might think that the country looked like a crumpled piece of paper with deep folds. Towering mountain ranges and a huge, high plateau occupy the center of the country.

The Sierra Madre   Three different mountain ranges in Mexico make up the **Sierra Madre** (SYEHR•rah MAH•thray), or "mother range." The Sierra Madre Occidental runs down the western coast. The Sierra Madre Oriental extends along the eastern coast. The steep ridges of the Sierra Madre del Sur rise in southwestern Mexico. Because of the rugged terrain, few people live in the Sierra Madre. The mountains are rich in resources, though. They hold copper, zinc, silver, and timber.

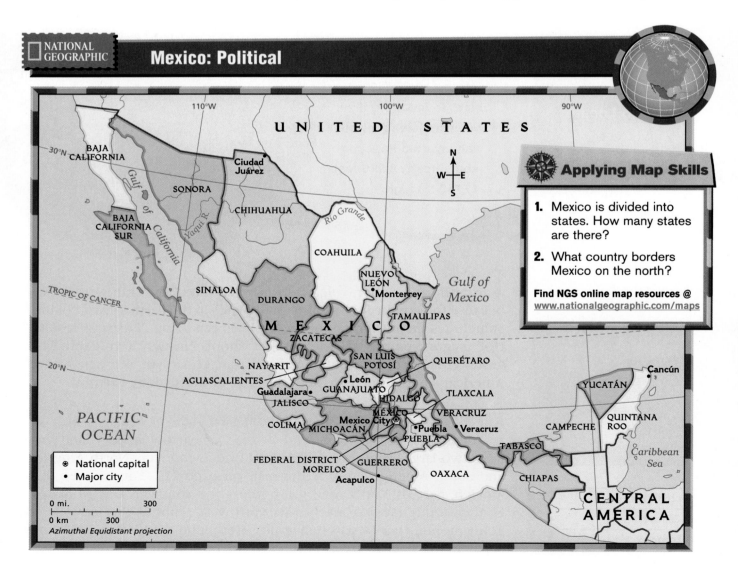

NATIONAL GEOGRAPHIC

Mexico: Political

Applying Map Skills

1. Mexico is divided into states. How many states are there?

2. What country borders Mexico on the north?

Find NGS online map resources @ www.nationalgeographic.com/maps

- ⊛ National capital
- • Major city

0 mi.   300
0 km   300
Azimuthal Equidistant projection

Many of Mexico's mountains are volcanoes. Popocatépetl (POH•puh•KAT•uh•PEHT•uhl), or "El Popo," as Mexicans call it, erupted violently centuries ago. In December 2000, El Popo erupted again, hurling molten rock into the sky. About 30,000 people from surrounding areas were forced to temporarily leave their homes. Tens of millions of people live 50 miles (80 km) or less from the mountain and could face even worse eruptions in the future.

Mexicans face another danger from the land. Earthquakes can destroy their cities and homes. A 1985 quake killed nearly 10,000 people in Mexico's capital, **Mexico City,** even though the earthquake's center was about 185 miles (298 km) away.

The Plateau of Mexico   The Sierra Madre surround the large, flat center of the country, the **Plateau of Mexico.** You find mostly deserts and grassy plains in the northern part of the Plateau. Broad, flat valleys that slice through the center hold many of the country's chief cities and most of its people. To the south, the Plateau steadily rises until it meets with the high, snowcapped mountains of the Sierra Madre del Sur.

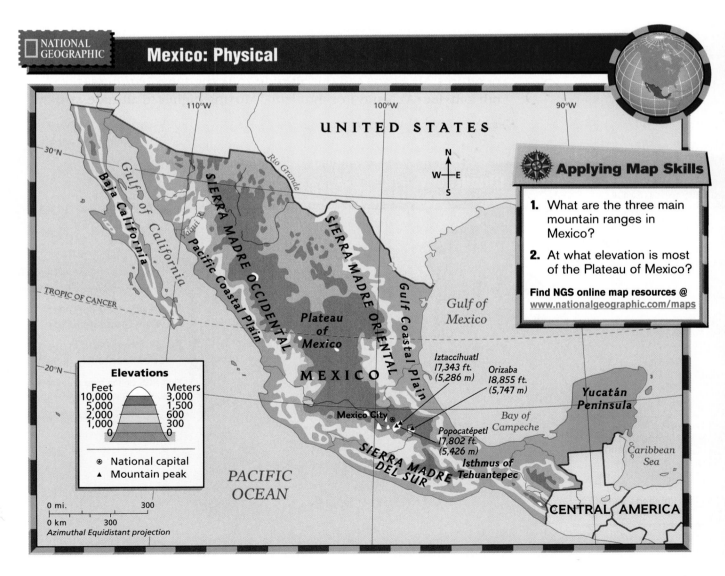

NATIONAL GEOGRAPHIC

Mexico: Physical

UNITED STATES

110°W   100°W   90°W

30°N

Gulf of California

Baja California

Rio Grande

Yaqui R.

SIERRA MADRE OCCIDENTAL

Pacific Coastal Plain

SIERRA MADRE ORIENTAL

Gulf Coastal Plain

TROPIC OF CANCER

Gulf of Mexico

Plateau of Mexico

MEXICO

20°N

Iztaccíhuatl 17,343 ft. (5,286 m)

Orizaba 18,855 ft. (5,747 m)

Yucatán Peninsula

Mexico City

Bay of Campeche

Popocatépetl 17,802 ft. (5,426 m)

Caribbean Sea

SIERRA MADRE DEL SUR

Isthmus of Tehuantepec

PACIFIC OCEAN

CENTRAL AMERICA

Elevations

Feet	Meters
10,000	3,000
5,000	1,500
2,000	600
1,000	300
0	0

⊛ National capital
▲ Mountain peak

0 mi.   300
0 km   300
Azimuthal Equidistant projection

Applying Map Skills

1. What are the three main mountain ranges in Mexico?

2. At what elevation is most of the Plateau of Mexico?

Find NGS online map resources @ www.nationalgeographic.com/maps

Coastal Lowlands Mexico's lowland plains squeeze between the mountains and the sea. The Pacific Coastal Plain begins with a hot, largely empty desert in the north. As you move farther south, better soil and rainfall allow ranching and farming along this plain. On the other side of the country, the Gulf Coastal Plain has more rain and fertile soil for growing crops and raising animals.

✓ **Reading Check** What two dangers from the land do Mexicans face?

Land of Many Climates

Mexico has many different climates. Why? As you read in Chapter 2, latitude—or location north or south of the Equator—affects temperature. The Tropic of Cancer, which cuts across the center of Mexico at 23½°N latitude, marks the northern edge of the tropics. Areas south of this line have warm temperatures throughout the year. Areas north of this line are warm in summer and cooler in winter.

Altitude, or height above sea level, affects temperature in Mexico as well. The higher up you go, the cooler the temperatures—even within the tropics. The diagram on page 185 shows that Mexico's mountains and plateau create three altitude zones. You could travel through all of these zones in a day's trip across the Sierra Madre.

Because they are near sea level, the coastal lowlands have high temperatures. Mexicans call this altitude zone the *tierra caliente* (tee•AY•rah kah•lee•AYN•tay), or "hot land." Moving higher in altitude, you find

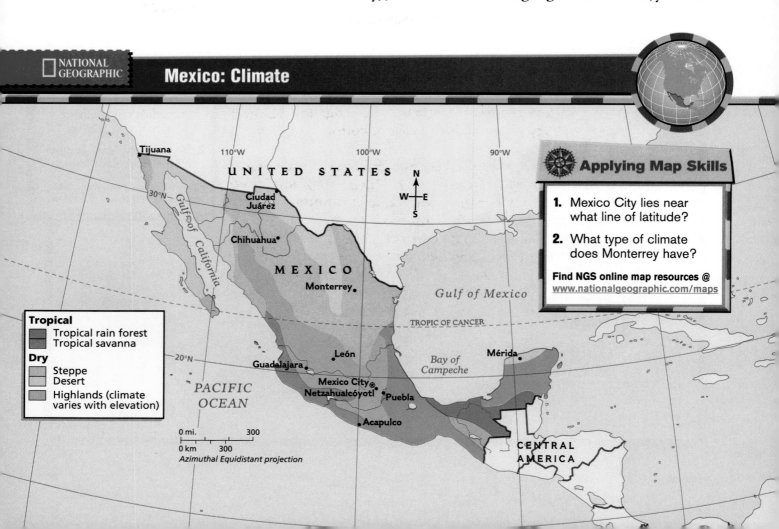

NATIONAL GEOGRAPHIC

Mexico: Climate

Applying Map Skills

1. Mexico City lies near what line of latitude?

2. What type of climate does Monterrey have?

Find NGS online map resources @ www.nationalgeographic.com/maps

Tropical
- Tropical rain forest
- Tropical savanna

Dry
- Steppe
- Desert
- Highlands (climate varies with elevation)

Tijuana
110°W
UNITED STATES
30°N
Gulf of California
Ciudad Juárez
Chihuahua
MEXICO
Monterrey
100°W
Gulf of Mexico
TROPIC OF CANCER
90°W
León
Guadalajara
Mérida
Bay of Campeche
20°N
PACIFIC OCEAN
Mexico City
Netzahualcóyotl
Puebla
Acapulco
CENTRAL AMERICA

0 mi. 300
0 km 300
Azimuthal Equidistant projection

Mexico's Altitude Zones

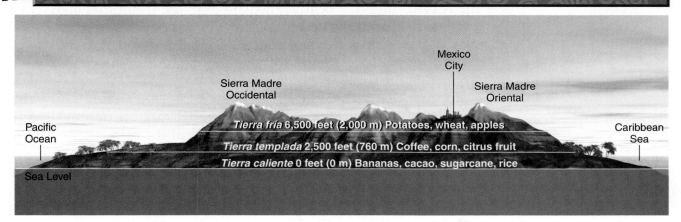

Mexico City

Sierra Madre Occidental

Sierra Madre Oriental

Pacific Ocean

Caribbean Sea

Tierra fría 6,500 feet (2,000 m) Potatoes, wheat, apples

Tierra templada 2,500 feet (760 m) Coffee, corn, citrus fruit

Tierra caliente 0 feet (0 m) Bananas, cacao, sugarcane, rice

Sea Level

the *tierra templada* (taym•PLAH•dah), or "temperate land." Here the climate is more moderate. In the highest zone, it is even cooler. This area is called the *tierra fría* (FREE•ah), or "cold land."

Rainfall varies throughout Mexico. Baja California and the north receive little precipitation. Other regions receive more, mostly in the summer and early fall. From June to October, hurricanes may hit Mexico. These fierce tropical storms with high winds and heavy rains form over the warm ocean waters and can strike Mexico with fury.

✓ Reading Check What is Mexico's warmest altitude zone?

Analyzing the Diagram

Mexico has zones of different climates that result from different altitudes.

Location In which altitude zone is Mexico City located?

Mexico's Economic Regions

Geography and climate give Mexico three distinct economic regions: the north, central Mexico, and the south. Large stretches of the north are too dry and rocky to grow crops without irrigation. Farmers there build canals to carry water to their fields of cotton, fruits, grains, and vegetables.

Did you know that the skills used by American cowhands began in Mexico? Cowhands called vaqueros (vah•KEHR•ohs) developed the tools and methods for herding, roping, and branding cattle. Vaqueros in northern Mexico carry on this work today.

Northern Mexico has seen an economic boom. **Monterrey,** Mexico's main producer of steel and cement, has long been a major industrial city. In this and other cities, many companies from the United States and elsewhere have built *maquiladoras* (mah•KEEL•ah•DOHR•as), or factories that assemble parts made in other countries. As a result, many Mexicans have flocked to cities such as **Tijuana** (tee•WAH•nah) and **Ciudad Juárez** (see•ooh•DAHD HWAH•rayz). *Maquiladoras* produce more than half of Mexico's manufacturing exports. Their growth, however, has raised fears about dangers to the environment and to workers' health and safety. Still, factory work and increased trade have improved life in Mexico's north.

More than half of Mexico's people live in the central region, the country's heartland. Why do they call this area home? The climate is one reason. Although central Mexico lies in the tropics, its high elevation keeps it from being hot and humid. Temperatures are mild, and the climate is pleasant year-round. A second reason is the fertile soil created by volcanic eruptions over the centuries. This allows good production for farming and ranching.

Large industrial cities such as Mexico City and **Guadalajara** also prosper in central Mexico. More than 18 million people live in Mexico City and its suburbs, making it one of the largest cities in the world.

The south is the poorest economic region of the country. The mountains towering in the center of this region have poor soil. Subsistence farms, or small plots where farmers grow only enough food to feed their families, are common here. In contrast, the coastal lowlands of this area have good soil and plentiful rain. Wealthy farmers grow sugarcane or bananas on plantations, large farms that raise a single crop for sale.

Both coasts also have beautiful beaches and a warm climate. Tourists from all over the world flock to such resorts as Acapulco and Puerto Vallarta on the Pacific coast and Cancún on the Yucatán Peninsula.

✓ Reading Check **How does the economic region of northern Mexico differ from that of southern Mexico?**

Assessment

Defining Terms

1. Define land bridge, peninsula, latitude, altitude, hurricane, vaquero, *maquiladora*, subsistence farm, plantation.

Recalling Facts

2. **Place** What bodies of water border Mexico?

3. **Location** What are the names of Mexico's three mountain ranges and where are they located?

4. **Economics** Why have many Mexicans moved to the cities of the north?

Critical Thinking

5. **Analyzing Information** Why is Mexico sometimes referred to as "the land of the shaking earth"?

6. **Understanding Cause and Effect** What two factors have the greatest effect on the climate of Mexico?

Graphic Organizer

7. **Organizing Information** Draw a diagram like this one, then list two facts that explain the large population growth of central Mexico.

```
──────────────▶  ┌─────────────┐
                 │    High     │
                 │ Population  │
                 │   Growth of │
──────────────▶  │Central Mexico│
                 └─────────────┘
```

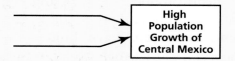

Applying Geography Skills

8. **Analyzing Maps** Refer to the physical map of Mexico on page 183. What is the elevation of the Yucatán Peninsula?

Technology Skill

Using the Internet

To learn more about almost any topic imaginable, use the **Internet**—a global network of computers. Many features, such as e-mail, interactive educational classes, and shopping services are offered on the Net.

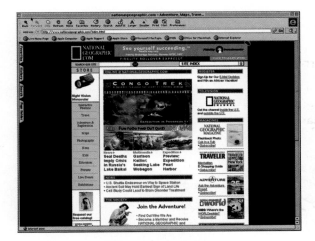

▲ The National Geographic Society's Web site

Learning the Skill

To get on the Internet, you need three things: (a) a personal computer, (b) a *modem*—or device that connects your computer to a telephone line, and (c) an account with an Internet service provider (ISP). An ISP is a company that enables you to log on to the Internet, usually for a fee.

After you are connected, the easiest way to access Internet sites is to use a "Web browser," a program that lets you view and explore information on the World Wide Web. The Web consists of many documents called "Web sites," each of which has its own address, or Uniform Resource Locator (URL). Many URLs start with the keystrokes *http://*

If you do not know the exact URL of a site, commercial "search engines" such as Yahoo! or AltaVista can help you find information. Type a subject or name into the "search" box, then press Enter. The search engine lists available sites that may have the information you are looking for.

Practicing the Skill

Follow these steps to learn how the Internet can help you find information about Mexico's Sierra Madre.

1. Log on to the Internet and access a search engine.
2. Search by typing "Sierra Madre" in the search box.
3. Scroll the list of Web sites that appears when the search is complete. Select a site to bring up and read or print.
4. If you get "lost" on the Internet, click on the back arrow key at the top of the screen until you find a familiar site.
5. Continue selecting sites until you have enough information to write a short report on minerals found in the Sierra Madre.

Applying the Skill

Follow the above steps to locate information about *maquiladoras.* Use the information you gather to create a chart or graph showing how the number of workers hired by *maquiladoras* has increased over the past 10 years.

Mexico's History and Government

Guide to Reading

Main Idea

Mexico's culture reflects a blend of its Native American and Spanish past.

Terms to Know

- hieroglyphics
- mural
- colony
- hacienda
- mestizo
- federal republic

Places to Locate

- Yucatán Peninsula
- Tenochtitlán
- Mexico City

Reading Strategy

Draw a chart like this one, then provide one example of how Native Americans and Europeans influenced Mexican culture.

Ethnic groups	Influence on Mexican Culture
Native American	
European	

NATIONAL GEOGRAPHIC

Exploring Our World

Thousands of people visit an ancient temple in Mexico's Yucatán Peninsula on the spring and fall equinoxes. On those two days, the setting sun casts a shadow on the stairs of the temple's north face. The area that is not shadowed looks like the ancient Native American god called Kukulcan, or the Feathered Serpent, going down the temple stairs.

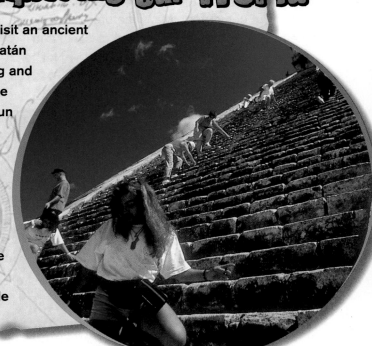

The first people to arrive in Mexico were the ancestors of today's Native Americans. Mexico's Native American heritage shapes the country's culture. So does Mexico's European heritage, brought by the Spaniards who conquered the area in the 1500s.

Native American Civilizations

Native Americans came to Mexico thousands of years ago. From about 1200 B.C. to the A.D. 1500s, these people built a series of brilliant, highly advanced civilizations on Mexican soil. Of these, the Mayan and Aztec civilizations are the best known. Look at the map on page 189 to see where the Mayan and Aztec civilizations thrived.

The Maya The Maya built a civilization in the rain forests of the **Yucatán Peninsula** and surrounding areas from about A.D. 250 to A.D. 900. Religion held Mayan society together. Mayan priests needed to measure time accurately to hold religious ceremonies at the correct moment. They studied the heavens and developed a calendar of 365 days.

The Maya were the first to grow corn. They also built huge stone temples in the shape of pyramids with steps. One of these structures, the temple of Kukulcan, showed careful planning. Each side of Kukulcan had 91 steps, totaling 364. The platform at the temple's top made one more step for a grand total of 365—just like the days in the year.

The Maya also developed hieroglyphics, a form of writing that uses signs and symbols. They had a complex number system. Artists decorated temples and tombs with elaborate murals, or wall paintings.

Around A.D. 900, Mayan civilization declined. Why? Historians do not know. Some suggest that the Maya overused the land and could not grow enough food. Others suggest that warfare or the spread of disease caused their decline. The Maya did not disappear, however. Their descendants still live in the same area and speak the Mayan language.

The Aztec Around A.D. 1200, a people called the Mexica moved into central Mexico from the north. The Spanish later called these people Aztec. The Aztec conquered a large empire in central Mexico. Their capital, **Tenochtitlán** (tay•NAWCH•teet•LAHN) was magnificent. Today **Mexico City**—Mexico's capital—stands on this ancient site.

Tenochtitlán was originally built on two islands in the middle of Lake Texcoco. Long causeways connected it to land. The city had huge stepped pyramids. Merchants traded gold, silver, and pottery in busy marketplaces. Farmers grew their crops in structures called "floating gardens," or rafts filled with mud. The rafts eventually sank to the lake bottom and piled up, forming fertile islands.

✓ **Reading Check** What two Native American cultures flourished in Mexico?

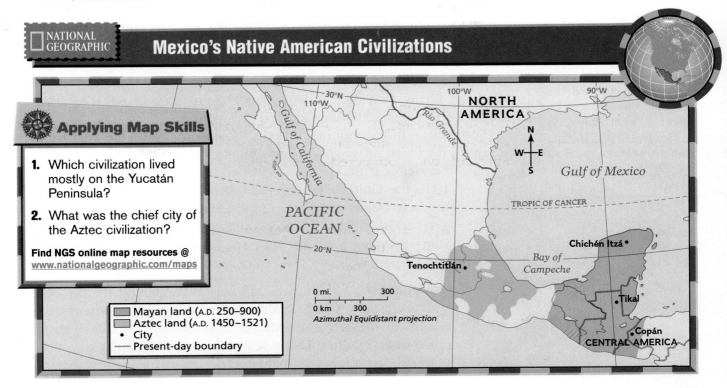

NATIONAL GEOGRAPHIC

Mexico's Native American Civilizations

Applying Map Skills

1. Which civilization lived mostly on the Yucatán Peninsula?

2. What was the chief city of the Aztec civilization?

Find NGS online map resources @ www.nationalgeographic.com/maps

Mayan land (A.D. 250–900)
Aztec land (A.D. 1450–1521)
• City
— Present-day boundary

0 mi. 300
0 km 300
Azimuthal Equidistant projection

NORTH AMERICA
Gulf of California
Rio Grande
Gulf of Mexico
TROPIC OF CANCER
PACIFIC OCEAN
Chichén Itzá
Tenochtitlán
Bay of Campeche
Tikal
Copán
CENTRAL AMERICA

Spanish Mexico

In 1519 Mexico's history changed dramatically. A Spanish army led by Hernán Cortés landed on Mexico's Gulf coast. He and about 600 soldiers marched to Tenochtitlán, which they heard was filled with gold. Some Native Americans who resented the harsh rule of the Aztec joined the Spanish. With their weapons, the Spaniards defeated the Aztec within two years.

Spain made Mexico a **colony,** or an overseas territory, because Mexico's rocky land held rich deposits of gold and silver. Many Spanish settlers came to live in Mexico. Some raised cattle on large ranches called **haciendas** (ah•see•AYN•dahs). Others started gold and silver mines. The Spaniards made Native Americans work on the ranches and in the mines. Thousands of Native Americans died from mistreatment. Many thousands more died of diseases like the common cold and smallpox caught from the Europeans. Spanish priests came to Mexico and in their own way tried to improve the lives of the Native Americans. Because of their work, many Native Americans accepted the priests' teachings. Today about 90 percent of Mexico's people follow the Roman Catholic religion.

A Mixing of Cultures Over time, the Spanish and Native American cultures mixed. Modern Mexicans reflect this blending of peoples. About 60 percent are **mestizos** (mays•TEE•zohs), having mixed Spanish and Native American heritage. Another 30 percent of the population are completely Native American, while the remaining 10 percent are descended from Europeans, especially the Spaniards.

✓ **Reading Check** Why was Mexico a valuable colony for Spain?

Independence and Revolution

The people of Mexico came to resent Spanish rule. In 1810 they rallied behind a Catholic priest, Miguel Hidalgo (mee•GAYL ee•DAHL•goh). He led an army of peasants in revolt. Spanish officials brought charges against Hidalgo and executed him, but the rebellion did not stop. Mexicans finally won their independence from Spain in 1821. In 1824 they set up a republic with an elected president.

Soon after independence, Mexico lost some valuable territory. Mexico's northern province of Texas fought for and won its own independence from Mexico and asked to join the United States. In 1846 the United States fought Mexico in a dispute over the southern boundary of Texas. In the treaty ending the war, Mexico gave up its claims to Texas. It was forced to sell to the United States what are today the states of California, Utah, and Nevada, and parts of Colorado, Wyoming, Arizona, and New Mexico.

Revolution For many decades, rich families, army officers, and Catholic Church leaders held most of the power and wealth in Mexico. Poor Mexicans grew angry. In 1910 their discontent exploded in a revolution. Emiliano Zapata, who commanded a rebel army, stated the goals of this revolution.

Digging for Treasure

Fourteen-year-old Benito is on a quest. He has joined his father and uncle in an archaeological dig, hoping to discover ruins from an ancient Aztec civilization. Although digging is hard and dirty work, Benito is excited to do it. His mother told him that he is a direct descendant of the Aztec. Perhaps he will be the one to find their fabled "city of gold." If he does not find riches, though, he has another plan. He hopes to become a famous baseball or soccer player someday. If he practices hard, he believes that he will be good enough to play in Mexico City's Azteca Stadium, which holds about 100,000 people.

He wanted to give to the poor "the lands, woods, and water that the landlords or bosses have taken from us." Zapata's forces swooped down and seized many large haciendas. They divided the land among the poor.

✓ Reading Check Who led the 1910 revolution against Spanish rule?

Mexico's Government

Mexico, like the United States, is a **federal republic** because power is divided between national and state governments, and a strong president leads the national government. Mexico's national government differs in that it has much more power than the states. The Mexican president, who can serve only one six-year term, also has more power than the legislative and judicial branches.

For many decades, one political party called the Institutional Revolutionary Party led Mexico. All the presidents and most other elected officials came from this party. In recent years, economic troubles and the people's lack of political power led to growing frustration. In the year 2000, the newly elected president of Mexico, Vicente Fox, came from a different political party—for the first time in 70 years.

✓ Reading Check What form of government does Mexico have?

Geography Online

Web Activity Visit the **Geography: The World and Its People** Web site at gwip.glencoe.com and click on **Chapter 6—Student Web Activities** to learn more about Mexico's history.

Section 2 Assessment

Defining Terms
1. **Define** hieroglyphics, mural, colony, hacienda, mestizo, federal republic.

Recalling Facts
2. **History** Describe three achievements of the ancient Maya.
3. **History** Which European country conquered and colonized Mexico?
4. **History** What were Emiliano Zapata's goals?

Critical Thinking
5. **Sequencing Information** Put the following events in the correct chronological order: Cortés conquers the Aztec, Hidalgo leads peasants in revolt, the Mexica move into central Mexico, Zapata leads a revolution.
6. **Understanding Cause and Effect** How did the arrival of Europeans affect the Native Americans in Mexico?

Graphic Organizer
7. **Organizing Information** Draw a diagram like this one. In one circle list facts about the Mexican government. In the other put information about the U.S. government. In the overlapping area write facts that are true about both.

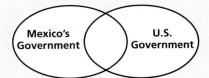

Mexico's Government / U.S. Government

Applying Geography Skills
8. **Analyzing Maps** Refer to the map of Mexico's Native American civilizations on page 189. Which Native American group settled the farthest south?

Making Connections

ART   SCIENCE   LITERATURE   TECHNOLOGY

The Aztec Calendar Stone

It is hard to imagine how a huge stone filled with carved figures can serve as a calendar. Known commonly as the Sun Stone, the Aztec calendar is full of both scientific and religious information.

History

In 1790 workers in the heart of the *zócalo,* or main square, of Mexico City uncovered a massive circular stone. Mexico City sits on top of Tenochtitlán, the ancient capital of the Aztec Empire. Some 300 years earlier, the Aztec at Tenochtitlán had carved the 25-ton (23-t) basalt rock calendar. Using stone tools, they created a monument that measured 12 feet (3.6 m) in diameter and 3 feet (.9 m) thick.

The face of the Aztec sun god appears at the center of the calendar stone. The sun god was thought to be one of the most important Aztec gods. Seven rings surround the sun god. In the closest ring are four squarelike spaces, each with a symbol that represents the four past ages of the world—the time that existed before humans appeared. Circling these symbols is a ring with signs representing the 20 days of the Aztec month.

Meaning of the Calendar

The Aztec calendar stone is actually two calendars in one. One calendar is a religious calendar based on a 260-day cycle. The Aztec believed that their lives depended on fulfilling their gods' demands. The calendar told Aztec priests when to make offerings and hold rituals for each god. It also divided the days among the gods. According to the Aztec view, this kept the universe in balance. An imbalance could lead to a power struggle among the gods and bring about the end of the world.

The second calendar is an agricultural calendar based on a 365-day solar cycle. The Aztec were very efficient farmers. They used this calendar to keep track of the seasons and ceremonies related to agricultural cycles.

Making the Connection

1. What does the Aztec agricultural calendar reveal about the scientific understanding of the Aztec?

2. Why was it important for the Aztec to divide the days among the gods?

3. **Making Comparisons** How do the two calendar systems of the Aztec differ?

◀ Today the Aztec calendar stone is displayed in the National Museum of Anthropology in Mexico City.

Guide to Reading

Main Idea

Mexicans enjoy a rich and lively culture but face many serious challenges.

Terms to Know

- plaza
- adobe
- industrialized
- service industry
- migrant worker
- smog

Places to Locate

- Gulf of Mexico
- Rio Grande

Reading Strategy

Draw a diagram like this one. In each of the smaller ovals, write an example of Mexican culture. Create as many smaller ovals as you need.

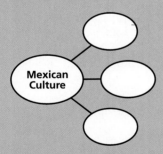

Mexican Culture

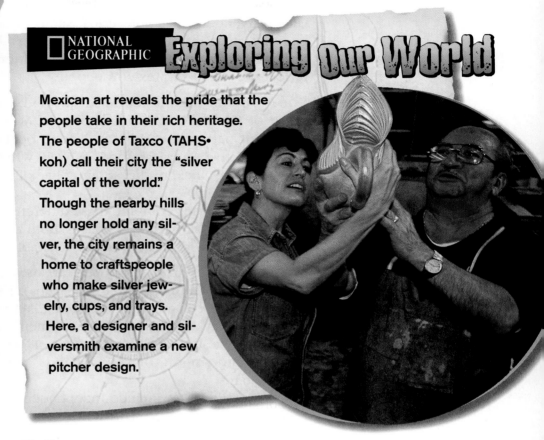

NATIONAL GEOGRAPHIC **Exploring Our World**

Mexican art reveals the pride that the people take in their rich heritage. The people of Taxco (TAHS•koh) call their city the "silver capital of the world." Though the nearby hills no longer hold any silver, the city remains a home to craftspeople who make silver jewelry, cups, and trays. Here, a designer and silversmith examine a new pitcher design.

Mexico—the third-largest country in size in Latin America, after Brazil and Argentina—has a large and dynamic population. More than 70 percent of all Mexicans live in the country's bustling cities.

Mexico's Cities and Villages

In the center of Mexico's cities, you often find large **plazas,** or public squares. Around each city's plaza stand important buildings such as a church and a government center. When you look at the buildings, you can see the architectural style of Spanish colonial times. Newer sections of the cities have a mix of towering glass office buildings and modern houses. In the poorer sections of town, people build small homes out of whatever materials they can find. This may include boards, sheet metal, or even cardboard.

Rural villages also have central plazas. Streets lead from the plazas to residential areas. Many homes are made of **adobe** (uh•DOH•bee), or sun-dried clay bricks. The roofs might be made of straw or of colored tile, in the Spanish style.

✓ Reading Check What do you find in the center of Mexico's cities and villages?

Mexican Culture

Mexican artists and writers have created many national treasures. In the early 1900s, Mexican painters produced beautiful murals—just as Native American painters had done centuries before. Among the most famous of these mural painters were José Clemente Orozco and Diego Rivera. Rivera's wife Frida Kahlo became well known for her paintings that revealed her inner feelings. Modern writers such as Carlos Fuentes and Octavio Paz have written poems and stories that reflect the values of Mexico's people.

Food   If you have tasted Mexican food, you know that it is a rich blend of flavors. Corn—first grown in Mexico—continues to be an important part of the Mexican diet. Chocolate, tomatoes, and chilies were all Native American foods as well. When the Spanish came, they brought beef, chicken, cheese, and olive oil, which Mexicans added to their cooking.

Today Mexicans combine these different cooking traditions in popular foods like tacos and enchiladas. Both dishes combine a flat bread called a tortilla with meat or beans, vegetables, cheese, and spicy chilies.

Celebrations   Throughout the year, Mexicans enjoy several special celebrations called fiestas (fee•EHS•tuhs). These special days include parades, fireworks, music, and dancing. Mariachi (MAHR•ee•AH•chee) bands may play such traditional instruments as the violin, guitar, horn, and bass at fiestas. More likely, however, you will hear the fast-paced rhythms and singing of Latino bands, which have influenced the United States.

National holidays include Independence Day (September 16) and Cinco de Mayo (May 5). This holiday celebrates the day in 1862 that Mexicans defeated an invading French army in battle. November 2 is a

Art

Mexican artist Diego Rivera is one of the most famous mural painters of the twentieth century. He believed that art belonged to the people. In Mexico City, Rivera's murals line the courtyard of the Ministry of Education building and cover the walls of the National Palace. With their characteristically vivid colors and distinctive style, Rivera's murals tell the story of the work, culture, and history of the Mexican people.

Looking Closer How did Rivera's work support his belief that art belongs to the people?

Mexico Through the Centuries ▶

A skilled seamstress (left) makes clothing in a *maquiladora* in northern Mexico. On September 16, parades celebrate the women and men who helped win Mexican independence (above).

Place What is the purpose of fiestas?

special religious celebration called the Day of the Dead. On this day, families gather in cemeteries where they honor their departed loved ones by laying down food and flowers.

✓ Reading Check What are some important celebrations in Mexico?

Mexico's Economy Today

With many resources and workers, Mexico has a growing economy. Did you know that Mexico's economy ranks among the top 12 in the world? As in the past, agriculture is still important. Farmers raise food to feed people at home—and also to ship around the world. Corn, beans, wheat, and rice are the main crops grown for food. Exports include coffee, cotton, vegetables, fruit, livestock, and tobacco.

In recent years, Mexico has **industrialized,** or changed its economy to rely less on farming and more on manufacturing. Factories in Mexico now make cars, consumer goods, and steel. The labels on your clothing may even say "Made in Mexico."

Mexico has large deposits of petroleum and natural gas in the **Gulf of Mexico** and along the southern coast. As a result, Mexico numbers among the world's major oil-producing nations.

Mexico is also home to important service industries. **Service industries,** you recall, are businesses that provide services to people rather than produce goods. Banking is a major service industry in Mexico, as is tourism.

Mexico

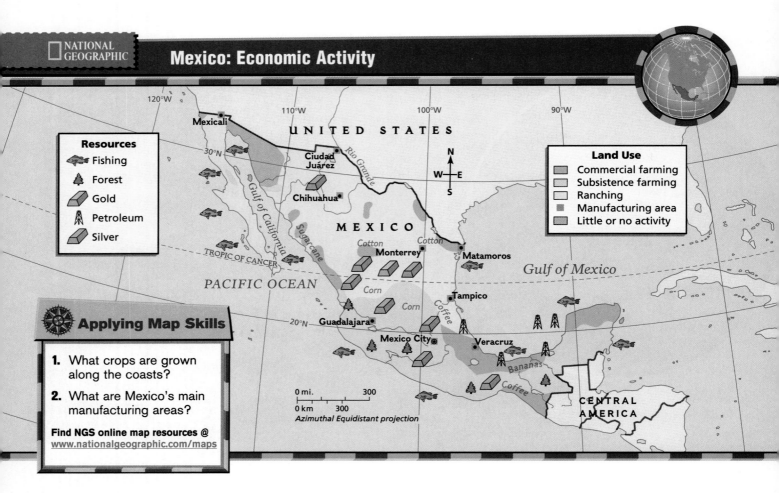

Resources
- Fishing
- Forest
- Gold
- Petroleum
- Silver

Land Use
- Commercial farming
- Subsistence farming
- Ranching
- Manufacturing area
- Little or no activity

UNITED STATES

MEXICO

Gulf of Mexico

PACIFIC OCEAN

CENTRAL AMERICA

Mexicali · Ciudad Juárez · Chihuahua · Monterrey · Matamoros · Tampico · Guadalajara · Mexico City · Veracruz

Cotton · Cotton · Corn · Corn · Sugarcane · Coffee · Bananas · Coffee

Gulf of California · Rio Grande · TROPIC OF CANCER

0 mi. 300
0 km 300
Azimuthal Equidistant projection

Applying Map Skills

1. What crops are grown along the coasts?

2. What are Mexico's main manufacturing areas?

Find NGS online map resources @
www.nationalgeographic.com/maps

Free Trade   To promote economic growth, the government of Mexico has moved to break down barriers to trade with other countries. In the mid-1990s, Mexico, Canada, and the United States signed the North American Free Trade Agreement, or NAFTA. Mexico's leaders hoped that free trade would encourage companies to open factories in Mexico, creating jobs. In 1999 Mexico signed a similar agreement with many countries in Europe.

✓ Reading Check Where are most of Mexico's oil fields located?

Mexico's Challenges

Mexico has tried to use its resources to improve the lives of its people. These actions have had strong effects on Mexican life—and created some challenges for the future.

Population   Mexico's population has increased rapidly in recent decades. Because many people have moved to the cities to find jobs, the cities have grown quickly. Many people have been forced to take jobs that pay low wages. As a result, thousands of people crowd together in poor sections of the cities.

Those Mexicans who cannot find any work in their country may become migrant workers. These are people who travel from place to place when extra workers are needed to plant or harvest crops.

They legally and sometimes illegally cross Mexico's long border to work in the United States. Though the pay is low, the migrant workers can earn more in the United States than in Mexico.

Foreign Investment and Foreign Debt   For many decades, the Mexican government refused to let foreign companies build factories in Mexico. Leaders feared that the companies would take their profits to their own country, draining money out of Mexico. In the 1990s, the government changed this policy. Mexican officials were still concerned that money would be lost, but they hoped that the new factories would create jobs for more Mexicans.

To help its economy grow, Mexico borrowed money from foreign banks. The government then had to use any money it earned in taxes to pay back the loans. As a result, Mexico's leaders did not have enough funds to spend on the Mexican people when the economy began to struggle. Many Mexicans grew angry. Yet if the government did not make the loan payments, banks would refuse to lend more money for future plans. Because there are still loans to be repaid, Mexicans will face this situation for many years.

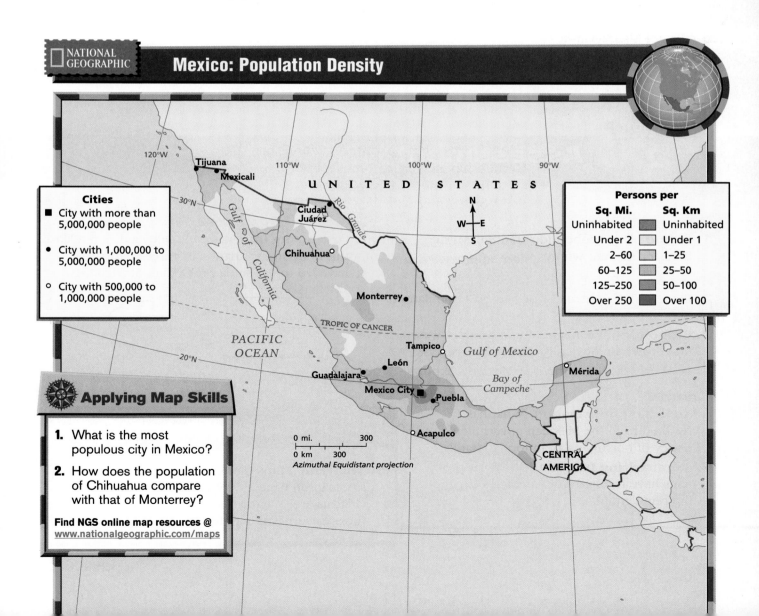

NATIONAL GEOGRAPHIC

Mexico: Population Density

Cities

■ City with more than 5,000,000 people

● City with 1,000,000 to 5,000,000 people

○ City with 500,000 to 1,000,000 people

Persons per		
Sq. Mi.		**Sq. Km**
Uninhabited		Uninhabited
Under 2		Under 1
2–60		1–25
60–125		25–50
125–250		50–100
Over 250		Over 100

Applying Map Skills

1. What is the most populous city in Mexico?

2. How does the population of Chihuahua compare with that of Monterrey?

Find NGS online map resources @
www.nationalgeographic.com/maps

Mexico City

Hazy smog fills the sky over Mexico City.

Human/Environment Interaction How has the landscape contributed to air pollution in Mexico City?

Pollution As Mexico's population boomed, its cities grew very large. At the same time, the economy industrialized. Both of these changes contributed to rising pollution in Mexico.

The mountains that surround Mexico City trap the exhaust fumes from hundreds of thousands of cars. People wake each day to a thick haze of fog and chemicals called smog. Many people wear masks when they leave their homes to go to work or school. In northern Mexico, many factories release dangerous chemicals into the air or water. One environmental group says that the **Rio Grande** is now one of the most polluted rivers in North America.

✓ Reading Check **What challenges does Mexico face?**

Section 3 Assessment

Defining Terms
1. Define plaza, adobe, industrialized, service industry, migrant worker, smog.

Recalling Facts
2. **Place** What percentage of Mexico's population lives in urban areas?
3. **Economics** What agricultural product was first grown in Mexico by Native Americans?
4. **Economics** Name three of Mexico's exports.

Critical Thinking
5. **Analyzing Information** What has resulted from the Mexican government's policy of borrowing from foreign banks?
6. **Summarizing Information** What problems have resulted from Mexico's expanding population?

Graphic Organizer
7. **Organizing Information** Draw a diagram like this one. On the arrows list three factors that have led to the smog problem of Mexico City.

Smog in Mexico City

Applying Geography Skills

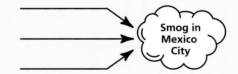

8. **Analyzing Maps** Look at the population density map on page 197. What is the population of Guadalajara? What is the population density of the area surrounding Mérida?

Reading Review

Section 1 | Mexico's Land and Economy

Terms to Know

land bridge
peninsula
latitude
altitude
hurricane
vaquero
maquiladora
subsistence farm
plantation

Main Idea

Mexico's mountainous landscape and varied climate create different economic regions.

✓ **Location** Mexico connects North and South America.

✓ **Region** Mexico has three mountain ranges, a large central plateau, and coastal lowlands.

✓ **Location** Much of Mexico lies in the tropics, but the climate in some areas is cool because of high elevation.

✓ **Region** Landforms and climate combine to create three economic zones in Mexico.

Section 2 | Mexico's History and Government

Terms to Know

hieroglyphics
mural
colony
hacienda
mestizo
federal republic

Main Idea

Mexico's culture reflects a blend of its Native American and Spanish past.

✓ **History** Mexico's Native American civilizations—the Maya and Aztec—made many contributions to Mexico's culture.

✓ **Culture** Mexico's people reflect the country's Native American and Spanish roots. The main language is Spanish, and the main religion is Catholicism.

✓ **History** The Spanish ruled Mexico from the 1500s to 1821, when Mexico won its independence.

✓ **Government** Mexico is a federal republic with a strong central government and a system of state governments.

Section 3 | Mexico Today

Terms to Know

plaza
adobe
industrialized
service industry
migrant worker
smog

Main Idea

Mexicans enjoy a rich and lively culture but face many serious challenges.

✓ **Culture** More than 70 percent of Mexicans live in cities today.

✓ **Economics** Farming is still important in Mexico, but manufacturing, service industries, and oil refining play larger roles in the economy.

✓ **Culture** Challenges facing Mexico include problems caused by population growth, foreign investment and debt, and pollution.

Assessment and Activities

 Using Key Terms

Match the terms in Part A with their definitions in Part B.

A.
1. altitude
2. hurricane
3. vaquero
4. *maquiladora*
5. mestizo
6. adobe
7. plaza
8. smog
9. mural
10. subsistence farm

B.
a. factories that assemble parts from other countries
b. cowhand
c. sun-dried clay bricks
d. wall painting
e. height above sea level
f. fog mixed with smoke
g. produces only enough to support a family's needs
h. fierce tropical storm
i. public square
j. a person of Native American and Spanish ancestry

Reviewing the Main Ideas

Section 1 Mexico's Land and Economy
11. **Location** How does Mexico's latitude affect its climate?
12. **Place** Name three mineral resources found in the mountains of Mexico.
13. **Movement** How have *maquiladoras* affected northern Mexico's cities?

Section 2 Mexico's History and Government
14. **History** What was the capital of the ancient Aztec?
15. **Movement** What effects did Spanish conquest have on Native Americans?
16. **History** When did Mexico win its independence from Spain?

Section 3 Mexico Today
17. **Culture** Who are two of Mexico's famous mural painters?
18. **Culture** What does Cinco de Mayo celebrate?

 **Mexico**

Place Location Activity

On a separate sheet of paper, match the letters on the map with the numbered places listed below.

1. Sierra Madre Occidental
2. Mexico City
3. Plateau of Mexico
4. Yucatán Peninsula
5. Baja California
6. Rio Grande
7. Gulf of Mexico
8. Guadalajara
9. Monterrey
10. Sierra Madre del Sur

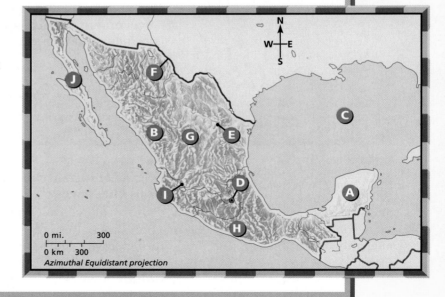

0 mi. 300
0 km 300
Azimuthal Equidistant projection

Critical Thinking

19. **Understanding Cause and Effect** Why have Mexico's leaders encouraged free trade agreements with other countries?

20. **Sequencing Information** Choose eight events and their dates from Mexico's history and place them in the correct order on a time line like the one below.

GeoJournal Activity

21. **Writing a Brochure** In this chapter, you learned about Mexico's economic and environmental challenges. Research the economy of your city or town. Include the types of industries, trading partners, and environmental effects of the industries. Share the information you gather by making a brochure. Include a map of your area.

Mental Mapping Activity

22. **Focusing on the Region** Draw a simple outline map of Mexico, then label the following:

- Pacific Ocean
- Gulf of Mexico
- Yucatán Peninsula
- Baja California
- Sierra Madre Occidental
- Sierra Madre Oriental
- Sierra Madre del Sur
- Plateau of Mexico
- Mexico City
- Rio Grande

Technology Skills Activity

23. **Developing a Multimedia Presentation** Imagine that you work for Mexico's Economic Development Office. Create a multimedia presentation to present to a group of foreign investors. Use a software application such as PowerPoint® to showcase positive features like climate, resources, and labor supply.

The Princeton Review

Standardized Test Practice

Directions: Read the paragraph below, then answer the question that follows.

The Aztec civilization was organized into classes. At the top was the emperor. His power came from his control of the army and the religious beliefs of the people. Next came the nobles, followed by commoners. Commoners included priests, merchants, and artists. Below commoners were the serfs, or workers who farmed the nobles' fields. Slaves, the lowest class, included criminals and people in debt, as well as female and children prisoners of war. Male prisoners of war were sacrificed to the Aztec gods. The Aztec believed that live human sacrifices were needed to keep the gods pleased and to prevent floods and other disasters.

1. **Which of the following statements is an opinion about the information given above?**

   F The Aztec civilization was organized into classes.

   G Male prisoners of war were sacrificed to the Aztec gods.

   H Slaves included children.

   J The Aztec should not have sacrificed people to the gods.

Test-Taking Tip: This question asks you to identify an opinion. An opinion is a person's belief. It is not a proven fact (such as answer F). Opinions often contain subjective words, like *easier,* or *best,* or *should.*

201

GeoLAB ACTIVITY

Built on Solid Ground?

1 Background

The earth's surface seems solid and stable. The outer layer of our planet, however, is split into large pieces called *plates*. Plates are like pieces of a puzzle—an enormous puzzle in which the pieces slowly move together, apart, and past one another. Where the edges of plates bump together, as they do in the region around Mexico, earthquakes can occur. The type of land on which buildings are constructed helps determine how much damage an earthquake will cause. If you lived in a part of the world where earthquakes occur, where would you want to build your house?

2 Materials

- 2 widemouthed clear jars
- 1 funnel
- 1 measuring cup full of water
- sand (enough to fill half of one jar)
- gravel (enough to fill half of one jar)
- 2 rocks (rocks should be larger than the gravel but small enough to fit inside the jars)

Earthquake Damage in Mexico City

Believe It or Not!

The incredible power of earthquakes was seen in the western states of Montana and Wyoming in 1959. An earthquake vibrated 500,000 square miles of wilderness area, ripping off enormous chunks of 8,257-foot Mount Jackson in Yellowstone National Park. A new lake was created, and 160 geysers roared to life.

3 What to Do

1. Fill one jar halfway with gravel. Put one rock on top of the gravel.
2. Fill the second jar halfway with sand. Push the funnel into the edge of the sand so that it is standing upright against the side of the jar. Then put one rock on top of the sand.
3. Slowly pour water from the measuring cup into the funnel. Add water until it is just below the level of the top of the sand. Remove the funnel.
4. Compare the two jars. Make a sketch of each one.
5. Now make an "earthquake" by giving the table or jars several quick shakes.
6. Compare the two jars again, and make a second sketch of each one.

Sand mixed with water represents loosely packed soil, which is found in much of the United States. ▼

4 LAB ACTIVITY REPORT

1. Describe what you saw when you added water to the sand through the funnel.
2. Compare the sketches you made before and after you shook the jar holding the gravel.
3. In which jar did the rock move most during the "earthquake"?
4. **Drawing Conclusions** On which of these surfaces would it be safest to build a house? Why?

5 Extending the Lab

Activity

Using construction paper and tape, make a model of a building you think could withstand an earthquake. Consider such details as the number of walls; how wide and tall the building is; and the size, number, and type of support columns between floors. When your model is completed, place it on a table. See how well your model withstands an "earthquake" by shaking the table in different ways and with differing amounts of force.

The World and Its People

NATIONAL GEOGRAPHIC

To learn more about the people and places of Central America and the West Indies, view *The World and Its People* **Chapter 7** video.

Geography oNLine

Chapter Overview Visit the *Geography: The World and Its People* Web site at gwip.glencoe.com and click on **Chapter 7— Chapter Overviews** to preview information about Central America and the West Indies.

Guide to Reading

Main Idea

Rich soil and a warm climate make Central America largely a farming region.

Terms to Know

- isthmus
- hurricane
- plantation
- subsistence farm
- canopy
- eco-tourist
- literacy rate
- republic
- parliamentary democracy

Places to Locate

- Belize
- Guatemala
- El Salvador
- Honduras
- Nicaragua
- Costa Rica
- Panama

Reading Strategy

Make a chart like this one, listing each country in Central America and key facts about each country.

Country	Key Facts

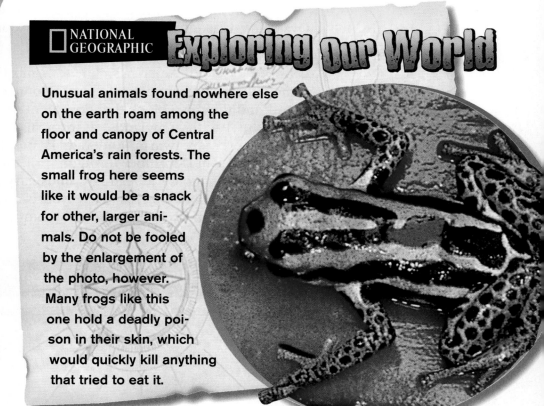

NATIONAL GEOGRAPHIC Exploring Our World

Unusual animals found nowhere else on the earth roam among the floor and canopy of Central America's rain forests. The small frog here seems like it would be a snack for other, larger animals. Do not be fooled by the enlargement of the photo, however. Many frogs like this one hold a deadly poison in their skin, which would quickly kill anything that tried to eat it.

Home to many different animals and plants, Central America is an **isthmus,** or a narrow piece of land that links two larger areas of land— North America and South America. Most of the countries on the isthmus have two coastlines—one on the Pacific Ocean and one on the Caribbean Sea. This narrow region (actually part of North America) stretches more than 1,000 miles (1,609 km) from Mexico southeast to the South American continent. Seven countries make up Central America: **Belize, Guatemala, El Salvador, Honduras, Nicaragua, Costa Rica,** and **Panama.** In Panama, the isthmus is only about 37 miles (60 km) wide.

A Rugged Land

Like Mexico, Central America sits where plates in the earth's crust meet. The collision of these plates produces volcanoes and earthquakes in the region. The Central Highlands, which curve like a backbone through inland Central America, are actually a chain of volcanic mountains. Because of their ruggedness, the Central Highlands are difficult to cross. This causes serious problems for transportation and has also kept many of the region's people isolated from one another. The volcanoes of the Central Highlands do bring some benefits to farmers, though. Volcanic material has made the soil very fertile.

The map on page 207 shows you coastal plains on either side of the Central Highlands. These plains are named for the bodies of water they meet—the Pacific Lowlands on the west and the Caribbean Lowlands on the east. The Pacific Lowlands in Nicaragua hold several freshwater lakes. Oval-shaped Lake Nicaragua covers 3,100 square miles (8,029 sq. km), making it Central America's largest lake.

Climate   Central America's climate is mostly tropical, but there are some differences from place to place. Mountainous areas are cool year-round, and they also block the movement of winds and moisture.

In the Pacific Lowlands, a tropical savanna climate prevails. Temperatures are warm, and rain is plentiful from May through November. From December through April, the climate is hot and drier.

In contrast, the Caribbean Lowlands have a hot, tropical rain forest climate throughout the year. Here you can expect about 100 inches (254 cm) of rain each year. Breezes from the Caribbean Sea provide

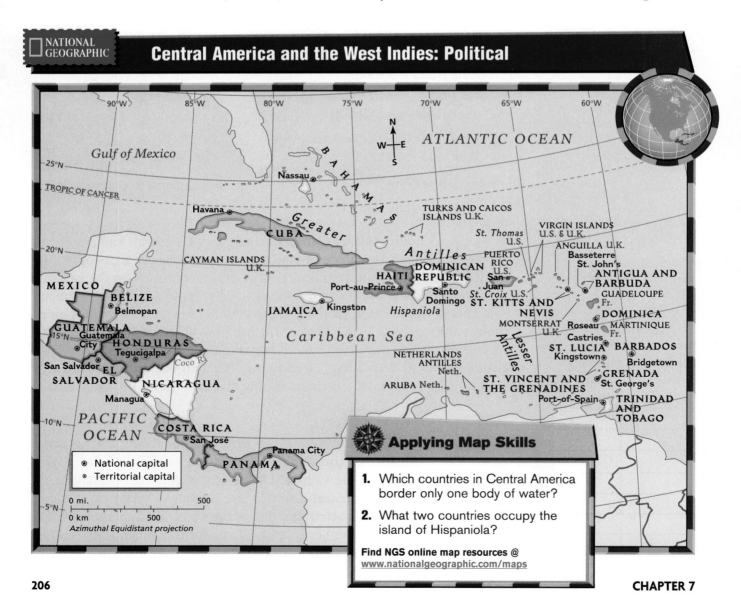

NATIONAL GEOGRAPHIC

Central America and the West Indies: Political

National capital
Territorial capital

0 mi.   500
0 km   500
Azimuthal Equidistant projection

Applying Map Skills

1. Which countries in Central America border only one body of water?

2. What two countries occupy the island of Hispaniola?

Find NGS online map resources @
www.nationalgeographic.com/maps

some cooling relief. These breezes can be replaced by deadly hurricanes during the summer and fall, though. **Hurricanes** are fierce storms with winds of more than 74 miles (119 km) per hour.

✓ Reading Check How have the volcanoes in Central America been helpful?

The Economy

The economies of the Central American countries depend on farming and harvesting wood from their rain forests. Central America has two kinds of farms. Wealthy people and companies own **plantations**—commercial farms that grow crops for sale. Major crops include coffee, bananas, cotton, and sugarcane. Plantations export their harvest to the United States and other parts of the world. Farmers in Guatemala and Costa Rica also grow flowers and ornamental plants for export.

Many farms in Central America are not plantations but **subsistence farms,** or small plots of land where poor farmers grow only enough food to feed their families. Subsistence farmers typically raise livestock and grow corn, beans, and rice.

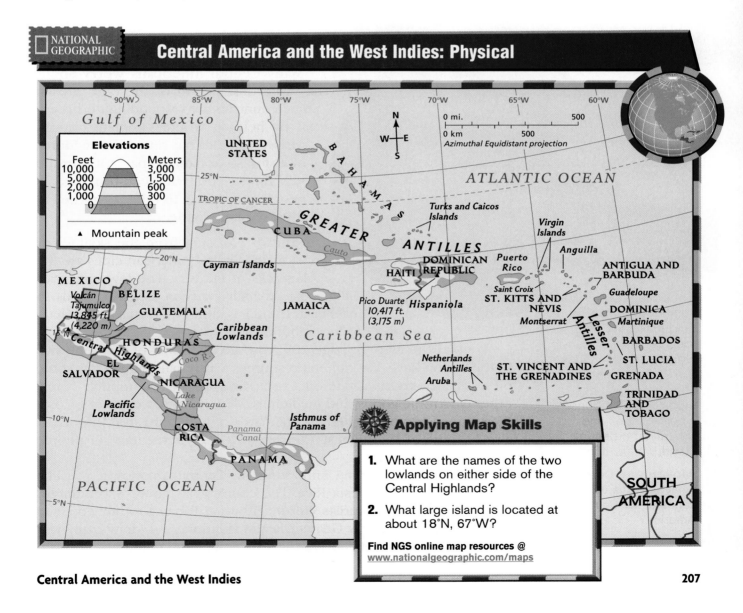

NATIONAL GEOGRAPHIC

Central America and the West Indies: Physical

Applying Map Skills

1. What are the names of the two lowlands on either side of the Central Highlands?

2. What large island is located at about 18°N, 67°W?

Find NGS online map resources @
www.nationalgeographic.com/maps

Economic Highs and Lows

San José, Costa Rica's capital (top), has shopping malls and fast-food chains like many North American cities. In 1998 Hurricane Mitch caused massive mudslides that buried whole villages and destroyed crops in Honduras (bottom).

Movement During what seasons do hurricanes strike Central America?

Rain Forests Under Central America's green canopy, or topmost layer of the rain forest that shades the forest floor, valuable resources can be found. The dense forests offer expensive woods—mahogany and rosewood, for example. Unusual animal and plant species also thrive here. Scientists research the plants to develop new medicines.

Both local and foreign-owned companies have set up large-scale operations in the rain forests. Lumber companies cut down and export the valuable trees. Other companies and local farmers also cut or burn the trees to clear land for farming. Without trees to hold the soil in place, heavy rains wash it and its nutrients away. As a result, the land becomes poor just a few years after being cleared. The businesses and farmers then move on, clearing trees from another piece of land.

Many Central Americans worry about the rapid destruction of the rain forests. Some countries are responding to this crisis by helping workers replant cleared areas. Costa Rica has set aside one-fourth of its forests as national parks. It uses the rain forests to attract eco-tourists, or people who travel to other countries to enjoy natural wonders.

Industry Missing from the skylines of most major Central American cities are the smokestacks of industry. The few industries that exist generally focus on preparing foods. In Guatemala, Honduras, and Nicaragua, some factories produce clothing for export.

Look at the map on page 209. You see that most of the countries in the region have few mineral resources. Guatemala, which has some oil reserves, exports crude oil. Costa Rica produces computer chips, other electronic goods, and medicines. With its varied economy, Costa Rica enjoys one of the highest standards of living in Latin America. It also has one of the highest literacy rates, or percentage of people who can read and write.

Tourism is of growing importance in Central America. If you like bird-watching, go to Costa Rica. The country has about 850 different kinds of birds. Many tourists visit neighboring Panama to buy goods in its large tax-free markets. Guatemala and Honduras also draw many visitors to the magnificent ruins of their ancient Mayan culture.

The Panama Canal   The economy in Panama is based on farming—as the economy is throughout Central America—but Panama also earns money from its canal. The Panama Canal stretches across the narrow Isthmus of Panama. Ships pay a fee to use the canal to shorten travel time between the Atlantic and Pacific Oceans. Turn to page 212 to see how the canal works.

The United States built the canal and owned it for more than 80 years. On December 31, 1999, Panama was given control of the canal. Panama hopes to use this waterway to build its economy. Nearly half of Panama's 2.8 million people live and work in the canal area.

✓**Reading Check** **What are the major crops grown on Central America's plantations?**

The History and People of Central America

Native Americans settled Central America thousands of years ago. The Maya flourished in the rain forests of the north from about A.D. 250 to A.D. 900. Look at the Native American civilizations map on

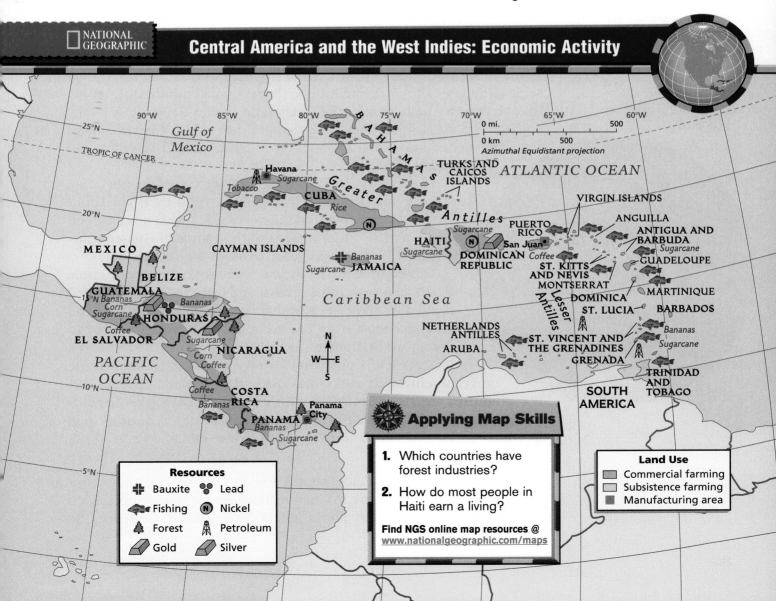

NATIONAL GEOGRAPHIC

Central America and the West Indies: Economic Activity

Applying Map Skills

1. Which countries have forest industries?

2. How do most people in Haiti earn a living?

Find NGS online map resources @
www.nationalgeographic.com/maps

Resources
- ✚ Bauxite
- 🐟 Fishing
- 🌲 Forest
- Gold
- Lead
- Ⓝ Nickel
- Petroleum
- Silver

Land Use
- Commercial farming
- Subsistence farming
- Manufacturing area

page 189. In Tikal (tee•KAHL), Guatemala, and Copán (koh•PAHN), Honduras, the Maya created impressive temples and sculptures. Then they mysteriously left their cities. Many of their descendants still live in the area today.

In the 1500s, Spaniards established settlements in Central America. For the next 300 years, Spanish landowners forced Native Americans to work on plantations. The two cultures gradually blended. Native Americans started to speak the Spanish language and follow the Roman Catholic faith. Native American traditions, in turn, affected the Spanish settlers.

One Central American country has a different history. The area that today is Belize was a British colony from the 1600s to the late 1900s. Early British settlers brought enslaved Africans to cut trees for their valuable woods. In Belize, African and British influences are strong.

Most Central American countries gained independence from Spain by 1821. The two exceptions are Panama and Belize. Panama was part of the South American country of Colombia for decades. In 1903 the United States helped Panama win its independence in exchange for the right to build the Panama Canal. Belize was the last Central American country to gain independence. It ceased to be a British colony in 1981.

After Independence   Most Central American countries faced constant strife after they became independent. A small number of people in each country held most of the wealth and power. Rebel movements arose as poor farmers fought for changes that would give them land and better lives. Civil wars raged in Nicaragua, El Salvador, and Guatemala as recently as the 1980s and 1990s.

In Guatemala, government military forces fought rebel groups living in the highlands from 1960 to 1996. About 150,000 people died, and the civil war severely weakened Guatemala's economy. Tens of thousands of Guatemalans left the country to look for work in the United States.

In contrast, Costa Ricans have enjoyed peace. A stable democratic government rules, and the country has avoided conflict for most of its history. As a result of these peaceful relations, the country has no army—only a police force to maintain law and order.

Today each country in Central America has a democratic government, with voters choosing government officials. Six countries are also republics with elected presidents as head of the government. Belize is a British-style parliamentary democracy, in which an elected legislature chooses a prime minister to head the government.

Daily Life   Nearly 36 million people live in Central America. About one-third of this number live in Guatemala, the most heavily populated country in the region. Only about 250,000 people live in Belize, the

teen Scene

What a Catch!

The deep blue waters of Lake Nicaragua are home to the world's only freshwater sharks and swordfish. Now the lake holds one less swordfish. Amadeo Robelo, who lives in Granada, Nicaragua, just spent three hours battling the powerful fish. Amadeo enjoys fishing with his father on weekends. After years of fighting in Nicaragua's civil war, his father wants to see the peace last and spend time with his children. He also wants Amadeo to become part of Nicaragua's middle class—something new in a region where you are either one of the few with wealth or one of the many who live in poverty.

least populous country. Spanish is the official language throughout the region, except for English-speaking Belize. Many Central Americans also speak Native American languages, such as Mayan. Guatemala's population, for instance, is largely Native American and has 21 different Native American languages. Most Central Americans follow the Roman Catholic religion.

About 50 percent of all Central Americans live on farms or in small villages. At least one major city, usually the capital, is densely populated in each country. Guatemala's capital, Guatemala City, ranks highest in population with about 2,200,000 people. People living in urban areas hold manufacturing or service industry jobs, or they work on farms outside the cities. Those living in coastal areas may harvest shrimp, lobster, and other seafood to sell in city markets or for export.

Whether rural or urban, most people enjoy a major celebration called Carnival. This festival comes before Lent, a solemn period of prayer and soul-searching before the Christian celebration of Easter. During Carnival—and at other times—bands play salsa, a mixture of Latin American popular music, jazz, and rock. Do you like baseball? It is a national sport in Nicaragua and very popular in Panama, too. Most people throughout the region also enjoy *fútbol,* or soccer.

Reading Check Why is the government of Belize different from that of other countries in Central America?

Assessment

Defining Terms
1. **Define** isthmus, hurricane, plantation, subsistence farm, canopy, eco-tourist, literacy rate, republic, parliamentary democracy.

Recalling Facts
2. **Location** Why does Central America experience earthquakes and volcanoes?
3. **Culture** What is the major religion and language of Central America?
4. **Place** Which country in Central America is the most heavily populated? The most sparsely populated?

Critical Thinking
5. **Making Comparisons** How have the differences in government stability affected the citizens of Guatemala and Costa Rica?

6. **Analyzing Cause and Effect** How do the Central Highlands affect climate?

Graphic Organizer
7. **Organizing Information** Draw a diagram like this one. On the lines list the major products and industries of Central America.

Major products and industries

Applying Geography Skills

8. **Analyzing Maps** Refer to the political map on page 206. Which countries of Central America border Mexico? Which border the Pacific Ocean?

The Panama Canal Locks

Before the Panama Canal was built, it took 60 days for a ship to sail around the southern tip of South America. Now about 35 oceangoing vessels can travel each day through the canal in about eight hours.

Digging the Canal

The first attempts to build a canal across Panama were begun in 1881 by a private French company. Huge expenses, poor planning, and the effects of diseases such as malaria and yellow fever stopped construction. In 1904 the United States government took over. Doctors had recently learned that bites from infected mosquitoes caused malaria and yellow fever. Workers drained swamps and cleared brush to remove the mosquitoes' breeding grounds. Then the digging began. The canal's course ran through hills of soft volcanic soil. Massive landslides regularly occurred before the 50-mile (80-km) canal was completed in 1914.

An Engineering Masterpiece

To move ships through the canal, engineers designed three sets of locks—the largest concrete structures on the earth. They allow ships to move from one water level to another by changing the amount of water in the locks. Together, the locks can raise or lower ships about 85 feet (26 m)—the height of a seven-story building. The diagram below shows you how these locks work.

Making the Connection

1. Why was a canal through Panama desirable?
2. What function do locks perform?
3. **Understanding Cause and Effect** How did medical advances affect the construction of the Panama Canal?

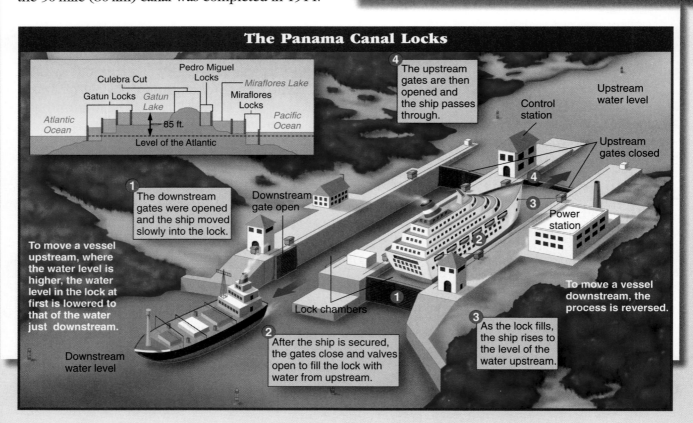

The Panama Canal Locks

Culebra Cut · Pedro Miguel Locks · Miraflores Lake · Gatun Locks · Gatun Lake · Miraflores Locks · Atlantic Ocean · Pacific Ocean · 85 ft. · Level of the Atlantic

4. The upstream gates are then opened and the ship passes through.

Control station

Upstream water level

Upstream gates closed

1. The downstream gates were opened and the ship moved slowly into the lock.

Downstream gate open

Power station

To move a vessel upstream, where the water level is higher, the water level in the lock at first is lowered to that of the water just downstream.

Lock chambers

To move a vessel downstream, the process is reversed.

2. After the ship is secured, the gates close and valves open to fill the lock with water from upstream.

3. As the lock fills, the ship rises to the level of the water upstream.

Downstream water level

Guide to Reading

Main Idea

The islands of the West Indies rely on tourism to support their economies.

Terms to Know

- archipelago
- bauxite
- colony
- communist state
- cooperative
- embargo
- free trade zone
- commonwealth

Places to Locate

- Caribbean Sea
- Greater Antilles
- Cuba
- Jamaica
- Hispaniola
- Puerto Rico
- Lesser Antilles
- Haiti
- Dominican Republic

Reading Strategy

Draw a diagram like this one. In each circle, list a country in the West Indies and features that are unique to it. Where the circles overlap, list features that are similar to both countries.

Country 1   Country 2

NATIONAL GEOGRAPHIC **Exploring Our World**

The warm waters of the Caribbean Sea lure millions of tourists to the West Indies every year. Some tourists go scuba diving so they can see the colorful fish that swim in the islands' clear waters. Others shop at the local stores, buying hand-crafted goods. This diver uses a metal detector to look for objects from a Spanish ship that sank in the 1600s.

Several **archipelagos** (AHR•kuh•PEH•luh•GOHS), or groups of islands, dot the **Caribbean Sea.** These island groups are known as the West Indies. East of Florida are the Bahamas, an archipelago of nearly 700 islands. South of Florida you find the **Greater Antilles** (an•TIH•leez). This group includes the large islands of **Cuba, Jamaica, Hispaniola** (HIHS•puhn•YOH•luh), and **Puerto Rico.** To the southeast are smaller islands called the **Lesser Antilles.**

Mountaintop Islands

When you look at the islands of the West Indies, you are really look-ing at the tops of mountains. Many West Indian islands are part of an underwater chain of mountains formed by volcanoes. A typical vol-canic island has central highlands ringed by coastal plains. The vol-canic soil in the highlands is rich.

Other islands are limestone mountains pushed up from the ocean floor by pressures under the earth's crust. Limestone islands generally are flatter than volcanic islands. The sandy soil found on many lime-stone islands is not good for farming.

The islands of the West Indies vary in size. Cuba, the largest island, is slightly smaller than the state of Pennsylvania. Among the smallest islands is Montserrat, slightly smaller than the city of Miami, Florida.

Climate The West Indies lie in the tropics. Most islands have a fairly constant tropical savanna climate. Sea and wind, more than elevation, affect the climate here. Cool northeast breezes sweep across the Caribbean Sea. They take on the temperature of the cooler water beneath them. When the winds blow onshore, they keep temperatures warm and pleasant, from 70°F to 85°F (21°C to 30°C).

Caribbean breezes sometimes bring gentle rains, but not *always.* For half the year, hurricanes threaten the West Indies. The word "hurricane" comes from the Taíno, an early Native American people who lived on the islands. They worshiped a god of storms named Hurakan.

✓ Reading Check **What formed the islands of the West Indies?**

The Economy of the West Indies

Tourism and farming are the most important economic activities in the West Indies. A warm, sunny climate and beautiful beaches attract millions of tourists each year. Tourism is the region's major industry. Airlines and cruise ships make regular stops at different islands.

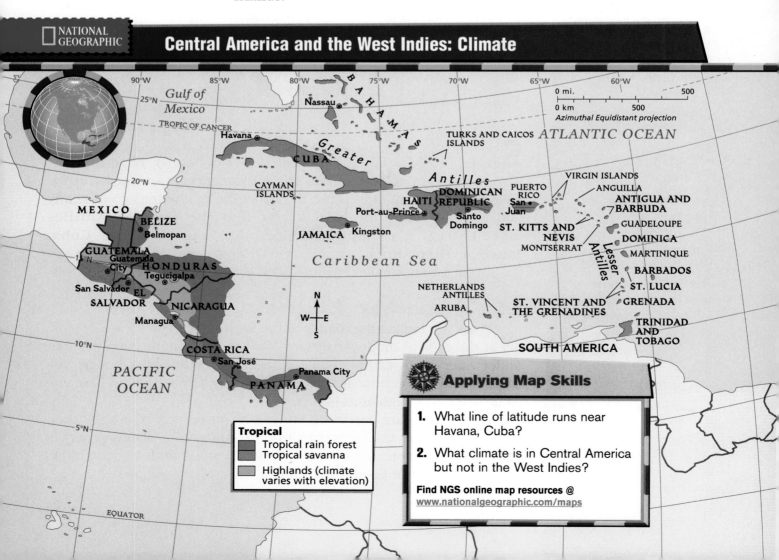

NATIONAL GEOGRAPHIC

Central America and the West Indies: Climate

Tropical
- Tropical rain forest
- Tropical savanna
- Highlands (climate varies with elevation)

Applying Map Skills

1. What line of latitude runs near Havana, Cuba?

2. What climate is in Central America but not in the West Indies?

Find NGS online map resources @
www.nationalgeographic.com/maps

Wealthy landowners grow crops such as sugarcane, bananas, coffee, and tobacco for export. Many laborers work on the plantations that grow these commercial crops. Some areas are used for subsistence farming. People may own small plots of land or rent it from someone else. They grow rice and beans, which are basic parts of the diet in this region, as well as fruits and vegetables. Puerto Rico has dairy and livestock industries as well.

The Caribbean islands face an economic danger by depending on one crop. If the crop fails, no income is earned. If too much of the crop is produced worldwide, overall prices fall and the economy is in serious trouble. Another economic challenge is that many West Indians cannot find jobs when the growing season is over, or they have to work for low pay. Workers often pack up and head to other West Indian islands, Central America, or the United States.

Mining, Manufacturing, and Trade Most of the islands do not have large amounts of minerals, although several islands have some resources. Jamaica, for example, mines **bauxite,** a mineral used to make aluminum. The country of Trinidad and Tobago exports oil products. In Puerto Rico, companies make chemicals and machinery. Haiti and the Dominican Republic have textile factories where workers make cloth. Workers in Jamaica also make clothing.

Several islands have banking and financial industries. The Panama Canal has also increased business in the region. Since the 1500s, trading ships have sailed to the West Indies. Turn to page 224 to learn more about this trade, known as the Columbian Exchange.

√ Reading Check **What is the major industry in the West Indies?**

History and Culture

When Christopher Columbus reached the island of San Salvador—now part of the Bahamas—in 1492, who met him? It was a Native American group—the Taíno. The Taíno and other Native Americans lived in the islands long before the coming of Europeans.

The Spaniards established the first permanent European settlement in the Western Hemisphere in 1496. That settlement is now the city of Santo Domingo, capital of the Dominican Republic. During the next 200 years, the Spaniards, the English, the French, and the Dutch also founded **colonies,** or overseas settlements, on many of the islands. They found the soil and climate perfect for growing sugarcane.

By the mid-1600s, most Native Americans had died from European diseases and harsh treatment. The Europeans then brought enslaved Africans to work on sugar plantations. When the slave trade ended in the early 1800s, plantation owners still in need of workers brought them from Asia, particularly India. The Asians agreed to work a set number of years in return for free travel to the West Indies.

Independence During the 1800s and 1900s, many Caribbean islands won their freedom from European rule. The first to become independent were the larger island countries, such as Haiti, the

Bee Hummingbird
How small is this bird? The bee hummingbird of Cuba measures only 2 inches (5.1 cm) from head to tail. That is small enough to make it the tiniest bird in the world. People walking through the swamps where it lives sometimes mistake the bird for an insect. The bird's wings move so fast—80 beats per second—that the human eye cannot see them. At two grams, the bee hummingbird weighs less than a penny.

Dominican Republic, and Cuba. Later, smaller islands such as Barbados and Grenada became countries. Many countries—like Haiti and the Dominican Republic—are republics. Others—like Jamaica and the Bahamas—are British-style parliamentary democracies.

Cuba is the only country in the Western Hemisphere with a government based on the ideas of communism. In a communist state, government leaders have strong control of the economy and society as a whole.

Some Caribbean islands still are not independent. Two large islands—Martinique and Guadeloupe—have ties to France. Puerto Rico and some of the Virgin Islands are linked to the United States. Other small islands in the Lesser Antilles are owned by the British or the Dutch.

Daily Life Although many people in Central America have Native American ancestors, the peoples of the West Indies have African or mixed African and European ancestry. Large Asian populations live in Trinidad and Jamaica as well.

More than 37 million people live in the West Indies. Cuba, with about 11.2 million people, has the largest population in the region. Saint Kitts and Nevis has only about 40,000 people. Most people speak a European language and follow the Roman Catholic or Protestant faiths.

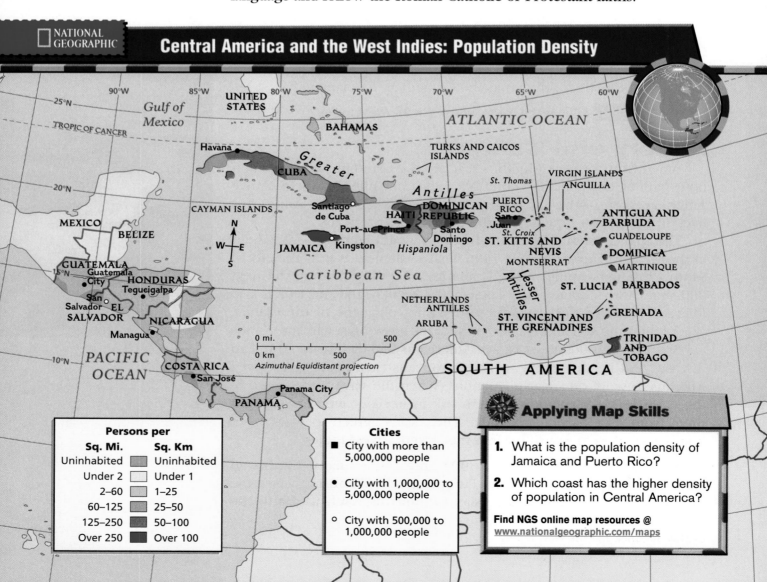

NATIONAL GEOGRAPHIC

Central America and the West Indies: Population Density

Persons per

Sq. Mi.	Sq. Km
Uninhabited	Uninhabited
Under 2	Under 1
2–60	1–25
60–125	25–50
125–250	50–100
Over 250	Over 100

Cities

■ City with more than 5,000,000 people

● City with 1,000,000 to 5,000,000 people

○ City with 500,000 to 1,000,000 people

Applying Map Skills

1. What is the population density of Jamaica and Puerto Rico?

2. Which coast has the higher density of population in Central America?

Find NGS online map resources @ www.nationalgeographic.com/maps

Schoolgirls walk past vast sugar plantations on Barbados that European countries started in the colonial period (left). A steel-drum band entertains tourists in Trinidad (above).

Region What attracts so many tourists to the islands of the West Indies?

About 60 percent of West Indians live in cities and villages. The other 40 percent live and work in the countryside. Many islanders have jobs in the hotels or restaurants that serve the tourist industry.

If you visit the Caribbean, you are likely to hear lively music. The bell-like tones of the steel drum, developed in Trinidad, are part of the rich musical heritage of the region. Enslaved Africans created a kind of music called calypso. Jamaica's reggae music combines African rhythms and American popular music. Cuban salsa blends African rhythms, Spanish styles, and jazz.

On several islands, you will hear a different sound—the crack of a baseball bat. People in Puerto Rico, the Dominican Republic, and Cuba have a passion for baseball. Soccer is another popular sport.

✓ Reading Check **Where was the first permanent European settlement in the West Indies?**

Island Profiles

The islands of the West Indies share many similarities, but they also have differences. Some of these differences can be seen in Cuba, Jamaica, Haiti, the Dominican Republic, and Puerto Rico.

Cuba One of the world's top sugar producers, Cuba lies about 90 miles (145 km) south of Florida. Most farmers work on **cooperatives,** or farms owned and operated by the government. In addition to

Central America and the West Indies

growing sugarcane, they grow coffee, tobacco, rice, and fruits. In Havana, Cuba's capital and the largest city in the West Indies, workers make food products, cigars, and household goods.

Cuba is a communist state led by a dictator, President Fidel Castro. Cuba won its independence from Spain in 1898. The country had a democratic government, although military leaders sometimes seized power. In 1959 Castro led a revolution that took control of the government. He set up a communist state and turned to the Soviet Union for support. When he seized property belonging to American companies, the United States government responded. It put in place an embargo, or a ban on trade, against Cuba.

Cuba relied on aid from the Soviet Union. When the Soviet Union broke apart in 1989, it stopped giving economic support to the island. The Cuban economy is struggling, and many Cubans live in poverty.

Jamaica   What languages would you hear if you visited Jamaica? Officially, English is the spoken language. Yet many Jamaicans speak Creole, which is a mixture of African languages and English, French, and Spanish. Creole is a musical-sounding language with a characteristic Jamaican rhythm. Almost all of Jamaica's 2.6 million people are of African or mixed African and European backgrounds. A small number are of Asian ancestry.

Once a British colony, Jamaica became independent in 1962. Kingston is the country's capital and largest city. With misty blue mountains and sunny beaches, Jamaica attracts thousands of tourists every year. The country is also one of the world's leading producers of bauxite. The main export crops are sugar, coffee, and bananas.

Haiti   On the western half of the island of Hispaniola, you will find the country of **Haiti.** Led by a formerly enslaved man, Francois-Dominique Toussaint-Louverture, Haiti fought for and won its independence from France in 1804. It was the second independent republic in the Western Hemisphere (after the United States). It became the first nation in the history of the world to be founded by formerly enslaved persons. About 95 percent of Haiti's 7.8 million people are of African ancestry. Civil war has left Haiti's economy in ruins, and most Haitians are poor. Coffee and sugar, the main export crops, are shipped through Port-au-Prince, the country's capital.

Dominican Republic   The **Dominican Republic** shares the island of Hispaniola with Haiti, filling the eastern part. Though they share the same island, the two countries have different histories and little contact. Haiti was a French colony. The Dominican Republic was settled by Spaniards, who brought enslaved Africans to work on sugar plantations. Sugar is still an important crop to the Dominicans. Tourism is growing, too, and many Dominicans sell goods in the country's free trade zone. Free trade zones are areas where people can buy goods from other countries without paying taxes.

The government of the Dominican Republic hopes to build up the country's electrical power so the economy can grow more quickly.

Poverty remains a problem, though. As a result, many Dominicans have left the country. Thousands have come to the United States looking for work.

Puerto Rico To be or not to be a state in the United States, that is the question that Puerto Ricans ask themselves every few years. The last time they voted on the question, they said no. How did Puerto Rico become part of the United States? The island was a Spanish colony from 1508 to 1898. After the Spanish-American War in 1898, the United States won control of Puerto Rico. Since 1952 the island has been a **commonwealth,** or a partly self-governing territory, under American protection. By law, Puerto Ricans are citizens of the United States. They can come and go from the island to the United States as they wish. Today nearly 3 million Puerto Ricans live in the United States. The island itself holds about 3.9 million.

Puerto Rico has a high standard of living compared to most other Caribbean islands. It also boasts more industry, with factories producing chemicals, machinery, clothing, and other products. San Juan, the capital and largest city, is home to about 1 million people. In rural areas, farmers grow sugarcane and coffee. Puerto Rico makes more money from tourism than any country in the region.

Geography ONline

Web Activity Visit the **Geography: The World and Its People** Web site at gwip.glencoe.com and click on **Chapter 7— Student Web Activities** to learn more about Puerto Rico.

✓ Reading Check What is a commonwealth?

Section 2 Assessment

Defining Terms
1. **Define** archipelago, bauxite, colony, communist state, cooperative, embargo, free trade zone, commonwealth.

Recalling Facts
2. **Region** What three archipelagos make up the West Indies?
3. **History** Name four groups who have influenced the culture of the Caribbean region.
4. **Government** How is Cuba different from every other country in the Western Hemisphere?

Critical Thinking
5. **Drawing Conclusions** Describe two reasons Puerto Ricans might be satisfied remaining a commonwealth, rather than becoming a state.

6. **Making Predictions** What is the danger of a country depending on only one crop?

Graphic Organizer
7. **Organizing Information** Complete a chart like the one below with facts about Haiti and the Dominican Republic.

Country	Haiti	Dominican Republic
Colonized by		
Products		
Problems		

Applying Geography Skills

8. **Analyzing Maps** Refer to the population density map on page 216. What cities in the West Indies have more than 1 million people?

Geography Skill

Interpreting an Elevation Profile

You have learned that differences in land elevation are often shown on physical or relief maps. Another way to show elevation is on **elevation profiles.** When you view a person's profile, you see a side view. An elevation profile is a diagram that shows a side view of the landforms in an area.

Learning the Skill

Suppose you could slice right through a country from top to bottom and could look at the inside, or *cross section.* The cross section, or elevation profile, below pictures the island of Jamaica. It shows how far Jamaica's landforms extend below or above sea level.

Follow these steps to understand an elevation profile:

- Read the title of the profile to find out what country you are viewing.
- Look at the line of latitude written along the bottom of the profile. On a separate map, find the country and where this line of latitude runs through it.
- Look at the measurements along the sides of the profile. Note where sea level is located and the height in feet or meters.
- Now read the labels on the profile to identify the heights of the different landforms shown along with their elevation.
- Compare the highest and lowest points.

Practicing the Skill

Use the elevation profile below to answer the following questions.

1. At what elevation is Kingston?
2. What are the highest mountains, and where are they located?
3. Where are the lowest regions?
4. Along what line of latitude was this cross section taken?

Applying the Skill

Turn to page 17 in the **Geography Handbook.** Use the elevation profile of Africa to answer questions 2–4 above about *that* continent.

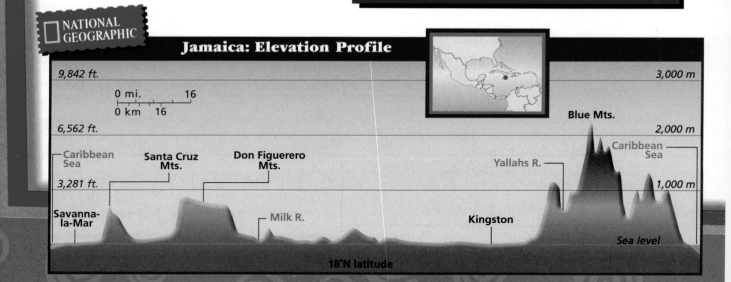

NATIONAL GEOGRAPHIC

Jamaica: Elevation Profile

9,842 ft.   3,000 m

0 mi.   16
0 km   16

6,562 ft.   2,000 m

Blue Mts.

Caribbean Sea   Santa Cruz Mts.   Don Figuerero Mts.   Yallahs R.   Caribbean Sea

3,281 ft.   1,000 m

Savanna-la-Mar   Milk R.   Kingston

Sea level

18°N latitude

Section 1 | Central America

Terms to Know

isthmus
hurricane
plantation
subsistence farm
canopy
eco-tourist
literacy rate
republic
parliamentary democracy

Main Idea

Rich soil and a warm climate make Central America largely a farming region.

✓Region Central America includes seven countries: Belize, Guatemala, Honduras, El Salvador, Nicaragua, Costa Rica, and Panama.

✓Region Volcanic mountains run down the center of Central America with coastal lowlands on either side.

✓Economics Most people in the region farm—either on plantations or subsistence farms.

✓Culture Most countries in Central America have a blend of Native American and Spanish cultures.

Section 2 | The West Indies

Terms to Know

archipelago
bauxite
colony
communist state
cooperative
embargo
free trade zone
commonwealth

Main Idea

The islands of the West Indies rely on tourism to support their economies.

✓Region The West Indies include three different island groups: the Bahamas, the Greater Antilles, and the Lesser Antilles.

✓History Most of the islands were at one time colonies of European countries.

✓Economics Farming and tourism are the major economic activities in the West Indies.

✓Culture The cultures of the West Indies mix Native American, European, African, and Asian influences.

✓Government Most governments in the West Indies are democratic, but a dictator rules Communist Cuba.

◀ **The Panama Canal**

Assessment and Activities

 ## Using Key Terms

Match the terms in Part A with their definitions in Part B.

A.

1. isthmus
2. literacy rate
3. plantation
4. eco-tourist
5. archipelago
6. bauxite
7. commonwealth
8. embargo
9. free trade zone
10. republic

B.

a. large commercial farm
b. mineral ore from which aluminum is made
c. ban on trade
d. narrow piece of land connecting two larger pieces of land
e. area where people can buy goods from other countries without paying taxes
f. a person who travels to another country to enjoy its natural wonders
g. country with an elected president
h. percentage of adults who can read and write
i. partly self-governing territory
j. group of islands

 ## Reviewing the Main Ideas

Section 1 Central America

11. **Region** What seven countries make up Central America?
12. **Place** How do the climates of the Pacific Lowlands and the Caribbean Lowlands differ?
13. **History** In what Central American countries did the Maya live?
14. **Culture** What percentage of Central Americans live in cities and small villages?

Section 2 The West Indies

15. **Economics** What two activities form the basis of the West Indian economies?
16. **Region** Many of the Caribbean islands were formed by what type of tectonic activity?
17. **Movement** Why do many West Indians move to other countries?
18. **History** What was the first nation in the world to be founded by formerly enslaved people?
19. **History** When did the United States gain control of Puerto Rico?

 NATIONAL GEOGRAPHIC | **Central America and the West Indies**

Place Location Activity

On a separate sheet of paper, match the letters on the map with the numbered places listed below.

1. Guatemala
2. Caribbean Sea
3. Cuba
4. Puerto Rico
5. Costa Rica
6. Panama
7. Bahamas
8. Haiti
9. Jamaica
10. Honduras

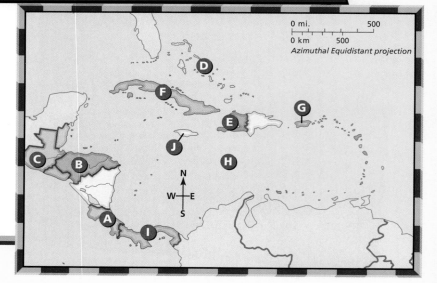

Critical Thinking

20. **Analyzing Information** Explain why Cuba's location is an important factor in the United States's relationship with that nation.

21. **Categorizing Information** Create a diagram like this with details about the people, history, and economy of a country in Chapter 7.

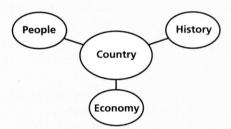

GeoJournal Activity

22. **Writing an Itinerary** Write an itinerary, or travel plan, for a cruise through the Caribbean. Your ship should make five stops. Include a map showing the route and descriptions of the sites and activities at each stop.

Mental Mapping Activity

23. **Focusing on the Region** Draw an outline map of Central America and the West Indies, then label the following:

- Pacific Ocean
- Caribbean Sea
- Guatemala
- Panama
- Cuba
- Puerto Rico
- Lesser Antilles
- Bahamas

Technology Skills Activity

24. **Building a Database** Create a database about Central America. Your database should have a record for each country. Each record should have a field for the following: population, per capita GDP, and capital city. Sort the records from largest to smallest for per capita GDP. What generalizations can you make based on these data?

The Princeton Review

Standardized Test Practice

Directions: Study the graph, then answer the question that follows.

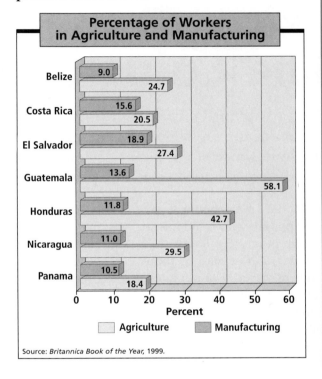

Percentage of Workers in Agriculture and Manufacturing

Country	Agriculture	Manufacturing
Belize	9.0	24.7
Costa Rica	15.6	20.5
El Salvador	18.9	27.4
Guatemala	13.6	58.1
Honduras	11.8	42.7
Nicaragua	11.0	29.5
Panama	10.5	18.4

Source: *Britannica Book of the Year,* 1999.

1. Of the countries shown on the graph, which has the lowest percentage of workers in manufacturing?
   A Belize
   B Honduras
   C Nicaragua
   D Panama

Test-Taking Tip: The process of elimination can be helpful here. First look at the key to see which of the bars represents manufacturing. Then look at the four answer choices and compare their percentages.

Columbus with King Ferdinand and Queen Isabella of Spain

The Columbian Exchange

The next time you eat a french fry, think about the long history of the lowly potato. The story begins high in the Andes mountain ranges of Bolivia and Peru (right), where thousands of years ago potatoes grew wild.

By the 1400s the Inca, an early people who ruled a vast empire in western South America, had developed thousands of varieties of potatoes. Yet how potatoes got from such a faraway time and place to be part of our everyday diet is a story that began even before the Inca.

Two Separate Worlds

Before the 1400s, people living in the world's Eastern Hemisphere were unknown to those living in the Western Hemisphere. This changed on October 12, 1492, when explorer Christopher Columbus, who had sailed from Spain, landed in the Bahamas in the Americas. Believing he had reached the Indies of Asia, Columbus named the people on the islands "Indians" and claimed the land for Spain. Columbus returned to the Americas the following year, bringing more than a thousand men in 17 ships. With his second trip, Columbus began what became known as "the Columbian exchange"—an exchange of people, animals, plants, and even diseases between the two hemispheres.

For Better and for Worse

The Europeans brought many new things to the Americas. Columbus brought horses, which helped the Native Americans with labor, hunting, and transportation. European farm animals such as sheep, pigs, and cattle created new sources of income. Explorers brought crops—oats, wheat, rye, and barley—that eventually covered North America's Great Plains. The sugarcane brought by Europeans flourished on plantations in Central and South America.

Some parts of the exchange were disastrous, however. Europeans brought diseases that killed millions of Native Americans. Plantation owners put enslaved Africans to work in their fields.

From the Americas, explorers returned home with a wide variety of plants. Spanish sailors carried potatoes to Europe. Nutritious and easy to grow, the potato became one of Europe's most important foods. (European immigrants then brought the potato to North America.) Corn from the Americas fed European cattle and pigs. Tobacco grown there became as valuable as gold. Peanuts, tomatoes, hot peppers, and cacao seeds (from which chocolate is made) changed the landscapes, eating habits, and cooking styles in Europe, Asia, and Africa.

QUESTIONS

1 What is "the Columbian exchange"?

2 Exchanges continue today. What are some present-day exchanges among the world's hemispheres?

Women in Peru tend a potato field. ▶

The Spread of Plants and Animals

NORTH AMERICA

EUROPE

PACIFIC OCEAN

ATLANTIC OCEAN

AFRICA

SOUTH AMERICA

N
W   E
S

Potatoes
Tobacco
Corn
Sugarcane
Cotton
Cacao
Chili peppers
Tomatoes
Horses and other livestock

0 mi.   4,000
0 km   4,000
Mercator projection

Brazil and Its Neighbors

The World and Its People
NATIONAL GEOGRAPHIC

To learn more about the people and places of Brazil and its neighbors, view **The World and Its People Chapter 8** video.

Geography ONLINE

Chapter Overview Visit the **Geography: The World and Its People** Web site at gwip.glencoe.com and click on **Chapter 8— Chapter Overviews** to preview information about Brazil and its neighbors.

Brazil

Guide to Reading

Main Idea

Brazil is a large country with many resources, a lively culture, and serious economic challenges.

Terms to Know

- basin
- *selva*
- escarpment
- favela
- inflation
- republic

Places to Locate

- Amazon River
- Paraná River
- São Francisco River
- Brazilian Highlands
- Brasília
- Rio de Janeiro
- São Paulo

Reading Strategy

Create a chart like the one below and fill in at least one key fact about Brazil in each category.

Brazil	
Land	
Climate	
History	
Economy	
Government	
People	

NATIONAL GEOGRAPHIC Exploring Our World

Some of the world's largest fresh-water fish swim in the mighty Amazon River in Brazil. Called pirarucu (pih•RAHR•uh•KEW), these fish can grow up to 15 feet (4.6 m) long. What a catch! The people who catch these huge fish often make the fish scales into souvenir key chains for tourists.

Like the pirarucu fish, Brazil is BIG. With a land area of 3,286,488 square miles (8,511,964 sq. km), Brazil is the fifth-largest country in the world and the largest in South America. In fact, Brazil makes up almost half of South America. It borders every South American country except Chile and Ecuador.

Brazil's Land and Climate

Because Brazil covers such a large area, it has many different types of landforms. The map on page 229 shows you that Brazil has narrow coastal plains, highland areas, and lowland river valleys.

The **Amazon River** is the world's second-longest river, winding almost 4,000 miles (6,437 km) from the Andes mountain ranges to the Atlantic Ocean. The river has such a powerful current at its mouth that it carries soil 60 miles (97 km) out to sea. On its journey to the Atlantic, the Amazon drains water from a wide, flat basin. A **basin** is a low area surrounded by higher land. In the Amazon Basin, heavy rain falls in the summer and autumn. Rainfall can reach as much as 120 inches (305 cm) a year. These rains support the growth of thick tropical rain forests, which Brazilians call *selvas.* Turn to page 252 to learn more about this rain forest, which covers one-third of Brazil.

Brazil has lowlands along the **Paraná River** and the **São Francisco River.** Both rivers begin in southeast Brazil. The Paraná flows to the southwest and the São Francisco to the northeast.

In addition to lowlands, Brazil has highlands. The **Brazilian Highlands** cover about half of the country, including much of the east and south. The highlands drop sharply to the Atlantic Ocean. This drop is called the Great Escarpment. An escarpment, as you remember, is a steep cliff between higher and lower land.

NATIONAL GEOGRAPHIC

Brazil and Its Neighbors: Political

Applying Map Skills

1. What is the capital of Brazil?

2. What country in this region is a territory of France?

Find NGS online map resources @ www.nationalgeographic.com/maps

0 mi. 800
0 km 800
Azimuthal Equidistant projection

- ⊛ National capital
- ⊙ Other capital
- • Major city

Varied Climate   Different climates exist in Brazil due to the country's huge size and varied landforms. Traveling in the Amazon Basin, you would feel the steamy temperatures and heavy rains of a tropical rain forest. In the highlands and on much of the coast, you would enjoy a tropical savanna climate, with hot temperatures and wet and dry seasons. In the south, you would have the warm temperatures of a humid subtropical zone. The capital, **Brasília,** is closer to the Equator than the coastal city of **Rio de Janeiro** (REE•oh DAY zhuh•NEHR•oh) but has milder temperatures. Why? Brasília is located at a higher altitude.

✓ **Reading Check** **What causes Brazil to have different climates?**

NATIONAL GEOGRAPHIC

Brazil and Its Neighbors: Physical

0 mi.   800
0 km   800
Azimuthal Equidistant projection

ATLANTIC OCEAN

VENEZUELA
GUYANA
SURINAME
FRENCH GUIANA
Mt. Roraima
9,094 ft.
(2,772 m)
Guiana Highlands
Llanos
Lake Maracaibo
Orinoco R.
EQUATOR
Amazon
Negro R.
Amazon R.
Amazon R.
Basin
Madeira R.
Xingu R.
Tocantins R.
São Francisco R.

SOUTH AMERICA

BRAZIL

Mato Grosso Plateau

Brazilian Highlands

PACIFIC OCEAN

Gran Chaco
Paraguay R.
Iguazú Falls

PARAGUAY

▲ Pico da Bandeira
9,482 ft.
(2,890 m)

TROPIC OF CAPRICORN
Mt. Ojos del Salado
22,572 ft.
(6,880 m)

Aconcagua
22,834 ft.
(6,960 m)

Mt. Tupungato
22,310 ft.
(6,800 m)

A N D E S

ARGENTINA

URUGUAY

Paraná R.
Uruguay R.

Pampas

Río de la Plata

Elevations

Feet		Meters
10,000		3,000
5,000		1,500
2,000		600
1,000		300
0		0

▲ Mountain peak

Patagonia

N
W—E
S

Strait of Magellan

Falkland Is.

Cape Horn

South Georgia I.

Applying Map Skills

1. Which area of Brazil—the north or the south—has the highest elevation?

2. Name two rivers that flow into the Amazon River.

Find NGS online map resources @ www.nationalgeographic.com/maps

Brazil's Economy

How do Brazilians earn a living? Agriculture, mining, and forestry have been important for centuries. Today Brazil's economy is diverse and productive, yet the country still faces serious economic challenges.

The North   For centuries the Amazon Basin was a mysterious region whose secrets were guarded by the Native Americans living there. This began to change in the mid-1800s. World demand skyrocketed for the rubber harvested from the basin's trees, and new settlers streamed to Brazil's interior. In recent years, government-built roads have brought even more people to the Amazon Basin. Mining companies dig for minerals such as bauxite, tin, and iron ore. Logging companies harvest mahogany and other woods from the rain forest. Farmers use the cleared land to grow soybeans and tobacco and to graze cattle.

Farmers along the Atlantic coastal plain grow tobacco, sugarcane, bananas, and cacao for export. Several large port cities—including Recife and Salvador—are located along the coast. Farther inland, thinly populated plateaus and highlands receive less than 40 inches (102 cm) of rain a year. Farmers in this area must use irrigation to bring water to their fields of beans, corn, and cotton.

The South   In the southern part of Brazil are rich mineral resources and fertile farmland. This region boasts one of the largest iron-ore deposits in the world, and the highlands are perfect for growing coffee. As the graph below shows, Brazil produces far more coffee than other countries. Brazil also exports large amounts of another popular breakfast drink—orange juice. Rice and bananas flourish as well.

The towering forests of the north are matched by the towering skyscrapers of the southeast. Here stand Brazil's major cities and centers of industry. Tourists flock to Rio de Janeiro, which has more than

Leading Coffee-Producing Countries

Analyzing the Graph

Brazil's highlands have the right soil and climate to grow coffee.

Region Which leading coffee-producing countries are in Latin America?

 *Textbook* update

Visit gwip.glencoe.com and click on **Chapter 8— Textbook Updates.**

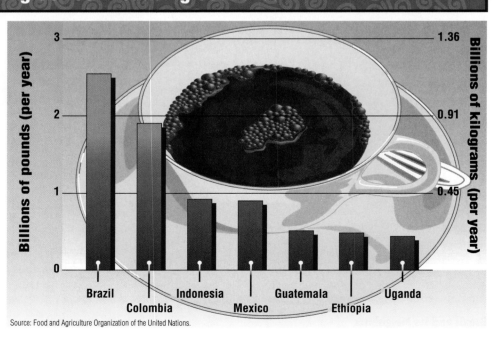

Source: Food and Agriculture Organization of the United Nations.

10 million people. If you visited Rio, you might enjoy the city's beautiful coastline, sandy beaches, and annual celebration of the holiday called Carnival. **São Paulo** (SAH•oh POW•loh), home to more than 16.5 million people, is one of the fastest-growing urban areas in the world. It is also Brazil's leading trade and industrial center. Vehicles, electronic products, steel, and more are manufactured here.

Farther south along the coast are more large cities, such as Curitiba (KUR•uh•TEE•buh) and Porto Alegre—each humming with industry. As you move west from the coast, you find cattle grazing on grasslands. Brazil—the leading cattle producer in South America—exports beef all over the world.

Brazil's Economic Challenges   Brazil's economy has brought wealth to many Brazilians and built a large and strong middle class. Yet as many as one-fifth of Brazil's people live in extreme poverty. Many Brazilian cities are surrounded by favelas, or slum areas. Thousands of poor people move here looking for work in the urban factories. They live in crude shacks with neither running water nor sewage systems. City governments have tried to clean up these areas, but people continue to settle here because they have no money to pay for housing. Many children as young as 10 go to work to help earn money.

To increase jobs and products for export, the government has encouraged mining, logging, and farming in the rain forest. However, these activities cause problems. First, they damage the land. Mining often results in dangerous levels of mercury building up in streams and lakes. Logging and the deliberate burning of trees to clear land for farming destroy large amounts of forest. In addition, these activities threaten the Native Americans who live in the rain forest. As more people settle in the Amazon Basin, Native Americans find it difficult to follow traditional ways of life. In the 1990s, the government announced plans to set aside 10 percent of the Amazon forest as parks. Another 10 percent will be set aside for native peoples.

√Reading Check   **What city is Brazil's leading trade and industrial center?**

Brazil's History and Culture

With about 168 million people, Brazil has the largest population of all Latin American nations. Unlike most of Latin America, Brazil's culture is largely Portuguese rather than Spanish. The Portuguese were the first and largest European group to colonize Brazil. Today Brazilians are of European, African, Native American, Asian, or mixed ancestry. Almost all of them speak a Brazilian form of Portuguese, which includes many words from Native American and African languages. Most of the population follow the Roman Catholic religion, but many Brazilians combine Catholicism with beliefs and practices from African and Native American religions.

Influence of History   Native Americans were the first people to live in Brazil. In the 1500s, the Portuguese took control of fertile coastal areas. They forced Native Americans to work on large plantations that

Rio de Janeiro

A huge statue of Christ overlooks Rio de Janeiro (right). Crowds of people in Rio de Janeiro celebrate Carnival wearing brightly colored costumes (above).

Place What groups make up Brazil's population?

grew tobacco and, later, sugarcane. Many Native Americans died from disease or overwork. To replace them, early Portuguese settlers brought people from Africa and enslaved them. Slavery finally was banned in 1888, but Africans remained in Brazil, most of them living in the northeastern part of the country. Over the years, African traditions have influenced Brazilian religion, music, dance, and food.

In the 1980s, economic problems created hard times for Brazil's people. Inflation soared. **Inflation** is an overall increase in the prices of goods across the entire economy. In the early 1990s, prices rose at a rate of 7,000 percent a year. This means that something costing $1 in January would cost about $70 in December. A new president ordered economic reforms that quickly slowed the rate of price increases, and Brazil's economy began to recover.

Moving to the Cities Look at the population map on page 246. You see that much of Brazil is sparsely populated. Millions of people have moved from rural areas to coastal cities in hopes of finding better jobs. The government has tried to encourage people to move from crowded coastal areas to less populated inland areas. Highways now crisscross the country and reach many once remote regions. In 1960 Brazil moved its capital from coastal Rio de Janeiro 600 miles (966 km) inland to the newly built city of Brasília. With more than 1.7 million people, Brasília is a modern and rapidly growing city.

The Government Brazil declared independence from Portugal in 1822. At first the new nation was an empire, with emperors ruling from 1822 to 1889. Like other Latin American countries, Brazil then experienced periods of military dictatorship. Today Brazil is a democratic republic, where people elect a president and other leaders. In Brazil, though, citizens cannot choose whether to vote or not vote. People from ages 18 to 70 are required by law to vote. Brazil has more than a dozen political parties—not just two main ones, as in the United States.

The national government of Brazil is much stronger than its 26 state governments. Brazil's president has more power over the country than an American president does in the United States.

Leisure Time Brazilians live for soccer, which they call *fútbol.* Every village has a soccer field, and the larger cities have stadiums. Maracana Stadium in Rio de Janeiro seats 220,000 fans.

Brazil is also famous for Carnival. This festival is celebrated just before the beginning of Lent, the Christian holy season that comes before Easter. The most spectacular Carnival is held each year in Rio de Janeiro. The celebration includes Brazilian music and showy parades.

Brazil has one of the largest television networks in the world. This network produces prime-time soap operas called *telenovelas.* These programs are wildly popular in Brazil—and viewers in more than 60 other nations enjoy them too.

✓ Reading Check Why do most Brazilians speak Portuguese?

Assessment

Defining Terms

1. Define basin, *selva,* escarpment, favela, inflation, republic.

Recalling Facts

2. Location Where is the source of the Amazon River located?

3. Economics What resources attract companies to the Amazon Basin?

4. Culture What is the major religion of Brazil?

Critical Thinking

5. Making Comparisons How does the economy in northern Brazil differ from the economy in southern Brazil?

6. Summarizing Information What economic challenges face Brazilians?

Graphic Organizer

7. Organizing Information Draw a diagram like this one. Beside the left arrow, write the cause of the government action. On the right, list results of this action.

> Government action: Government encouraged mining, logging, and farming in the rain forest.

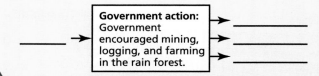

Applying Geography Skills

8. Analyzing Maps and Graphs Look at the maps on page 229 and page 238, and the graph on page 230. At what elevation is most coffee in Brazil grown? How much coffee does Brazil produce in a year?

Critical Thinking Skill

Sequencing and Categorizing Information

Sequencing means placing facts in the order in which they occurred. *Categorizing* means organizing information into groups of related facts and ideas. Both actions help you deal with large quantities of information in an understandable way.

Learning the Skill

Follow these steps to learn sequencing and categorizing skills:

- Look for dates or clue words that provide you with a chronological order: *in 2004, the late 1990s, first, then, finally, after the Great Depression,* and so on.
- If the sequence of events is not important, you may want to categorize the information instead. Categories might include economic activities or cultural traits.
- List these characteristics, or categories, as the headings on a chart.
- As you read, fill in details under the proper category on the chart.

Practicing the Skill

Read the sentences below, then answer the questions that follow.

After Brazil's independence from Portugal in 1823, a bill was presented to build a new capital named Brasília. More than 100 years later, in 1955, a planning committee chose the site for the new capital. The first streets were paved in 1958. On April 20, 1960, the festivities to officially "open" the new capital started at 4:00 P.M.

Brasília has both positive and negative aspects. The positive include virtually no air pollution, no threat of natural disasters, many green areas, and a pleasant climate. The negative aspects of the capital include very high housing prices, inefficient public transportation, few parking spaces, and long distances between the various government buildings.

1. What information can be organized sequentially?
2. What categories can you use to organize the information? What facts could be placed under each category?

Applying the Skill

Find two newspaper or magazine articles about Brazil or another South American country. Sequence or categorize the information on note cards or in a chart.

GO TO

Practice key skills with **Glencoe Skillbuilder Interactive Workbook, Level 1.**

Brasília

Argentina

Guide to Reading

Main Idea

Argentina is a large nation with diverse landforms, a strong economy, and a vibrant culture.

Terms to Know

- tannin
- *estancia*
- gaucho

Places to Locate

- Andes
- Aconcagua
- Patagonia
- pampas
- Buenos Aires
- Río de la Plata
- Gran Chaco

Reading Strategy

Draw a time line like this one and list at least five key dates and events in Argentina's history.

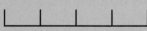

NATIONAL GEOGRAPHIC Exploring Our World

You might not expect to find much on the cold, windswept plateau of Patagonia in southern Argentina. The region is dry and bleak—but certainly not empty. Millions of sheep are raised here. This shepherd and sheepdogs watch the flock on a pasture that holds the world's largest sheep farm. More than 60,000 sheep graze here.

South of Brazil lies Argentina. It shares a long border with Chile to the west, touches Bolivia and Paraguay to the north, and borders on Brazil and Uruguay to the northeast. Argentina's eastern side is a long coast along the Atlantic Ocean. Its southern tip reaches almost to the continent of Antarctica.

Mountains, Plateaus, and Plains

Argentina's land area of 1,068,302 square miles (2,766,889 sq. km) makes it South America's second-largest country, after Brazil. Argentina is about the size of the United States east of the Mississippi River. Within this vast area, you can find mountains, deserts, plains, and forests.

The **Andes** tower over the western part of Argentina. Snowcapped peaks and clear blue lakes draw tourists for skiing and hiking. **Aconcagua** (AH•kohn•KAH•gwah) soars to a height of 22,834 feet (6,960 m), making it the highest mountain in the Western Hemisphere. On the eastern edge of the Andes are foothills. Through this area flow mountain streams, which farmers use to grow grains, cotton, and grapes.

Tropical

Tropical rain forest
Tropical savanna

Dry

Steppe
Desert

Mid-Latitude

Marine west coast
Humid subtropical

Highlands (climate
varies with elevation)

Applying Map Skills

1. What types of climates does Argentina have?

2. In which climate zone is the capital of Argentina?

Find NGS online map resources @ www.nationalgeographic.com/maps

South and east of the Andes lies a dry, windswept plateau called **Patagonia.** Most of Patagonia gets little rain and has poor soil. As a result, sheep raising is the only major economic activity. Find Patagonia on the physical map on page 229.

North of Patagonia, in the center of Argentina, are vast treeless plains known as the **pampas.** They stretch from the Andes foothills to the Atlantic coast. Argentina's economy depends on this region's fertile soil and mild climate. Similar to the Great Plains of the United States, the pampas are home to farmers who grow grains and ranchers who raise livestock. More than two-thirds of the country's people live in this region.

Buenos Aires, Argentina's capital and largest city, lies in the area where the pampas meet the **Río de la Plata.** The Río de la Plata is not really a river but a funnel-shaped bay that enters the Atlantic Ocean. It is formed by the Paraná and Uruguay Rivers.

In the northwest, heavy summer rains help great forests grow in the area called the **Gran Chaco.** The few people who live here practice subsistence farming and harvest quebracho (kay•BRAH•choh) trees. These hardwood trees produce tannin, a substance used in processing leather. To the east, hot, humid grasslands lie between the Uruguay and Paraná Rivers. Farmers graze livestock and grow crops on this fertile soil.

√ **Reading Check** Why are the pampas an important region of Argentina?

Argentina's Economy

Argentina has vast grasslands, so it is not surprising that Argentina's economy depends on farming and ranching. The country's major farm products include beef, sugarcane, wheat, soybeans, and corn. Most of these products are grown throughout the pampas and the northeastern part of the country. Huge *estancias* (ay•STAHN•see•ahs), or ranches, cover the pampas. Owners of these ranches hire gauchos (GOW•chohs), or cowhands, to take care of the livestock. Gauchos are the national symbol of Argentina, admired for their independence and horse-riding skills. The livestock that the gauchos herd and tend are a vital part of the country's economy.

Beef and food products are Argentina's chief exports. Turn to page 240 to read more about gaucho life on the pampas.

Manufacturing   Argentina is one of the most industrialized countries in South America. As you can see from the map on page 238, most of the country's factories are in or near Buenos Aires. Argentina's leading manufactured goods are food products, automobiles, chemicals, textiles, books, and magazines.

Petroleum is Argentina's most valuable mineral resource. The country's major oil fields are in Patagonia and the Andes. Other minerals—zinc, iron, copper, tin, and uranium—are mined in the Andes as well.

✓ Reading Check   What are Argentina's chief exports?

Argentina's People

Look at the population density map on page 246. Note that the most densely settled area is in and around Buenos Aires. Many people also live on the pampas. Why is this? The Andean region and Patagonia have harsh climates, so settlement in those areas is sparse. Buenos Aires and the pampas have a mild—although sometimes humid—climate that makes these areas more attractive to settlers.

Early History   Before the arrival of the Europeans, Native Americans farmed and hunted on the land. In the late 1500s, the Spaniards settled

NATIONAL GEOGRAPHIC On Location

Buenos Aires

This broad street in Buenos Aires is the Avenida 9 de Julio—or Avenue of the Ninth of July. Nearly 450 feet (137 m) wide, the street is named in honor of the day Argentina won independence from Spain.

Place Who led Argentina's fight for freedom from Spain?

NATIONAL GEOGRAPHIC

Brazil and Its Neighbors: Economic Activity

Resources

✚	Bauxite	⬥	Iron ore
⬛	Coal	◗	Manganese
◖	Copper	⬦	Natural gas
▽	Diamonds	⚲	Petroleum
🐟	Fishing	▱	Silver
🌲	Forest	▼	Tin
▭	Gold	✳	Uranium
		⊡	Zinc

Land Use

- ▨ Commercial farming
- ▧ Subsistence farming
- ☐ Ranching
- ▨ Forests
- ☐ Little or no activity
- ■ Manufacturing area

Applying Map Skills

1. What agricultural activities take place throughout Argentina?

2. In what countries is gold mined?

Find NGS online map resources @ www.nationalgeographic.com/maps

---

the area that is now Buenos Aires. By 1800 Buenos Aires was a flourishing port. Unfortunately, most of the Native Americans in the region had died from disease or were killed by Europeans.

In 1816 a general named José de San Martín led Argentina in its fight for freedom from Spain. After independence, the country was torn apart by conflict between the national government in Buenos Aires and various outlying regions. By the mid-1850s, a strong national government had emerged. Argentina entered a time of prosperity that lasted until the end of the 1800s. The growth of livestock raising and other industries brought wealth to some of Argentina's people. Still, because it was hard to make a living, many poor farmers migrated to the cities.

Political Unrest During the first half of the 1900s, Argentina entered a period of political unrest. Elected leaders governed poorly, the economy suffered, and the military took over. One of these military leaders, Juan Perón, became a dictator in the late 1940s. With his popular wife Eva at his side, Perón tried to improve the economy and give more help to workers. His crackdown on freedom of speech and the press made people unhappy, however. In 1955 a revolt drove Perón from power, and democracy returned.

Military officers again took control of Argentina in the 1970s. They ruled harshly, and political violence resulted in the deaths of many people. In 1982 Argentina suffered defeat in a war with the United Kingdom for control of the Falkland Islands. The Falklands, known in Argentina as the Malvinas, lie in the Atlantic Ocean off the coast of Argentina. Argentina's loss forced the military to step down, and elected leaders regained control of the government.

Today Argentina is a democratic republic. A powerful elected president leads the nation for a four-year term. A legislature with two houses makes the laws. As in the United States, nine judges appointed by the president sit on the country's highest court.

The country is also divided into 23 provinces and the federal district—which is Buenos Aires, the capital. The national government is much stronger than the provincial governments.

Argentina's Culture About 85 percent of Argentina's people are of European ancestry. During the late 1800s, immigrants in large numbers came to Argentina from Spain and Italy. Their arrival greatly influenced Argentina's society and culture. European ways of life are stronger in Argentina today than in most other Latin American countries.

The official language of Argentina is Spanish, although the language includes many Italian words. Most people are Roman Catholic. About 80 percent of Argentina's people live in cities and towns. Buenos Aires and its suburbs hold more than 12 million people. Buenos Aires has wide streets, European-style buildings, and shops and theaters. Its citizens call themselves *porteños* (pohr•TAY•nyohs), which means "people of the port." They have a passion for the national dance of Argentina, the tango.

✓ **Reading Check** Why does Argentina have a strong European culture?

I Protest!
Argentina's government wants to cut spending for schools, so Guillermo Larzabal intends to protest with his teachers in front of the congress building in Buenos Aires. "This is exciting!" he says. "I'm usually busy going to school and practicing *fútbol*. During the summer, from December to February, I spend all day kicking a *fútbol* with my friends. Sometimes our parents let us walk along the Avenida 9 de Julio. We watch the people and eat *helado*—ice cream."

Section 2 Assessment

Defining Terms

1. Define tannin, *estancia,* gaucho.

Recalling Facts

2. **Location** Near what other continent is South America's Argentina located?

3. **Region** Describe two ways in which the pampas are similar to the Great Plains of the United States.

4. **Economics** List four of Argentina's mineral resources.

Critical Thinking

5. **Analyzing Cause and Effect** Which of Juan Perón's policies led to his removal from office?

6. **Making Comparisons** How do the economic activities in Patagonia, the pampas, and the Gran Chaco differ?

Graphic Organizer

7. **Organizing Information** Draw a diagram like this one, then write one fact about Argentina in each section of the "pie."

Argentina

Applying Geography Skills

8. **Analyzing Maps** Study the physical map on page 229. What body of water separates the southern tip of Argentina from the mainland? What body of water borders Argentina on the east?

Poetry on the Pampas

As you learned in Section 2, gauchos herd cattle on the pampas. In 1872 José Hernández wrote the epic poem *El Gaucho Martín Fierro*. The poem tells the story of Martín Fierro, who recalls his life as a gaucho on the pampas. The following lines were translated from the poem.

El Gaucho Martín Fierro
by José Hernández (1834–1886)

A son am I of the rolling plain,
  A gaucho born and bred;
    For me the whole great world is small,
    Believe me, my heart can hold it all;
The snake strikes not at my passing foot,
  The sun burns not my head.

. . . .

Ah, my mind goes back and I see again
  The gaucho I knew of old;
    He picked his mount, and was ready aye,
    To sing or fight, and for work or play,
And even the poorest one was rich
  In the things not bought with gold.

The neediest gaucho in the land,
  That had least of goods and gear,
    Could show a troop of a single strain,
    And rode with a silver-studded rein,
The plains were brown with the grazing herds,
  And everywhere was cheer.

And when the time of the branding came,
  It did one good to see
    How the hand was quick and the eye
      was true,
    When the steers they threw with the
      long lassoo [lasso],
And the merry band that the years have swept
  Like leaves from the autumn tree.

Excerpt from *The Gaucho Martin Fierro*, adapted from the Spanish and rendered into English verse by Walter Owen. Copyright © 1936 by Farrar & Rinehart. Reprinted by permission of Henry Holt and Company, LLC.

Gauchos on Argentina's pampas ▲

Making the Connection

1. How does the poet describe the land on which the gaucho lives?

2. How can you tell from the poem that a gaucho is often on the move?

3. **Drawing Conclusions** What evidence does the poem give that the gaucho's way of life was a proud and happy one?

Caribbean South America

Guide to Reading

Main Idea

Each nation in Caribbean South America has a unique culture, depending on who settled the country.

Terms to Know

- llanos
- hydroelectric power
- altitude
- caudillo

Places to Locate

- Venezuela
- Lake Maracaibo
- Caracas
- Orinoco River
- Guiana Highlands
- Guyana
- Suriname
- French Guiana

Reading Strategy

Create a chart like this one, then list the names of the groups of people who live in each land.

Country	Groups
Venezuela	
Guyana	
French Guiana	
Suriname	

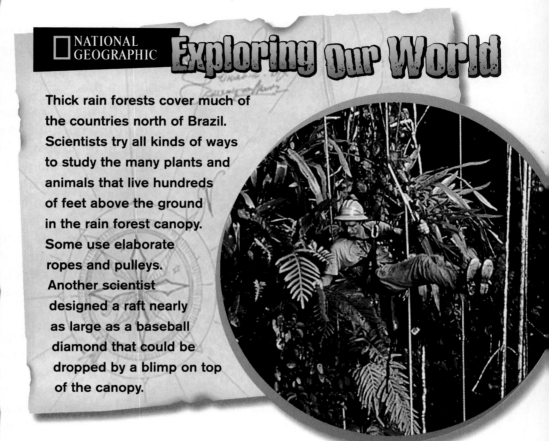

NATIONAL GEOGRAPHIC **Exploring Our World**

Thick rain forests cover much of the countries north of Brazil. Scientists try all kinds of ways to study the many plants and animals that live hundreds of feet above the ground in the rain forest canopy. Some use elaborate ropes and pulleys. Another scientist designed a raft nearly as large as a baseball diamond that could be dropped by a blimp on top of the canopy.

Four lands make up Caribbean South America: Venezuela, Guyana, Suriname, and French Guiana. They all border Brazil and either the Caribbean Sea or the Atlantic Ocean.

Venezuela: An Oil-Rich Land

Venezuela (VEH•nuh•ZWAY•luh) is the westernmost country of Caribbean South America. In the northwest lie the lowland coastal areas surrounding **Lake Maracaibo** (MAH•rah•KY•boh), the largest lake in South America. Swamps fill much of this area, and few people live here. The great number of towering oil wells, however, gives you a clue that rich oil fields lie under the lake and along its shores. Venezuela has more oil reserves than any other country in the Americas.

The Andean highlands begin south of the lake and are part of the Andes mountain ranges. This area includes most of the nation's cities,

Angel Falls

Angel Falls—the highest waterfall in the world at 3,212 feet (979 m)—roars over a cliff in Venezuela. It would take 11 football fields stacked end-to-end to reach the top.

Place What is one of the rivers that provides Venezuela with hydroelectric power?

including **Caracas** (kah•RAH•kahs), the capital and largest city.

East of the highlands, you see grassy plains known as the **llanos** (LAH•nohs). The llanos have many ranches, farms, and oil fields. Venezuela's most important river—the **Orinoco**—flows across the llanos. This river is a valuable source of **hydroelectric power,** or water-generated electricity, for Venezuela's cities.

South and east of the llanos rise the **Guiana Highlands,** deeply cut by rivers. Angel Falls— the world's highest waterfall—spills over a bluff in this region.

The Climate Because it is close to the Equator, Venezuela has a mostly tropical climate. In the Guiana Highlands to the south, you enter a steamy rain forest. In the grassy llanos, a tropical savanna climate brings you hot temperatures but less rain. Much of the Caribbean coast is dry, with some areas receiving only 16 inches (41 cm) of rainfall yearly. As in Mexico, temperatures in Venezuela differ with **altitude,** or height above sea level. In the Andean highlands you are usually warm in the daytime but cool at night.

The Economy Venezuelans once depended on crops such as coffee and cacao to earn a living. Since the 1920s, petroleum has changed the country's economy. Venezuela is a world leader in oil production and one of the chief suppliers of oil to the United States. Because the government owns the oil industry, oil provides nearly half of the government's income. Iron ore, limestone, bauxite, gold, diamonds, and emeralds also are mined. Factories make steel, chemicals, and food products. About 10 percent of the people farm, growing sugarcane and bananas or raising cattle.

History and Government Originally settled by Native Americans, Venezuela became a Spanish colony in the early 1500s. The Spanish gave the country its name. With its many rivers, the land in South America reminded Spanish explorers of Venice, Italy, which is full of canals. They named the area *Venezuela,* which means "Little Venice."

In the early 1800s, rebellion swept across the Spanish colonial empire. Simón Bolívar (see•MOHN boh•LEE•vahr)—born in Venezuela—became one of the leaders of this revolt. He and his soldiers freed Venezuela and neighboring regions from Spanish rule. In 1830 Venezuela became independent.

During most of the 1800s and 1900s, the country was governed by military rulers called **caudillos** (kow•THEE•yohz). Their rule was often harsh. Since 1958, Venezuela has been a democracy led by a president and a two-house legislature.

Rising oil prices during the 1970s brought more money to the country. The middle class grew, and many people prospered. When oil prices fell in the 1990s, the country suffered. The government did not have the money to give the poor and unemployed the services they needed. In 1998 Venezuelans showed their impatience with the government. They elected a former military leader, Hugo Chavez, as president. Chavez proposed major changes to the country's constitution and economy. In December 1999, the Venezuelan people voted to accept the new constitution.

The People   Most of Venezuela's 23.8 million people have a mix of European, African, and Native American backgrounds. Spanish is the major language of the country, and the major religion is Roman Catholicism. Nearly 90 percent of Venezuelans live in cities. More than 3.6 million people live in Caracas, which holds towering skyscrapers surrounded by mountains.

✔ **Reading Check** What product changed Venezuela's economy?

The Guianas

Caribbean South America also includes the countries of **Guyana** (gy•AH•nuh) and **Suriname** (SUR•uh•NAH•muh) and the territory of **French Guiana** (gee•A•nuh). Guyana was a British colony called British Guiana. Suriname, once a colony of the Netherlands, was called Dutch Guiana. As a result, these three lands are called "the Guianas."

The Guianas have similar landforms. Highlands in the interiors are covered by thick rain forests. As you move toward the Caribbean coast, the land descends to low coastal plains. The climate is hot and tropical. Two rainy seasons bring heavy rain, which causes the growth of rain forests. Most people live on the coastal plains because of the cooling ocean winds. The map on page 238 shows that sugarcane grows in Guyana and French Guiana, while rice and bananas flourish in Suriname. Many people also earn their living mining gold and bauxite.

Guyana   In the early 1600s, the Dutch were the first Europeans to settle in Guyana. They forced Native Americans and Africans to work on tobacco, coffee, and cotton farms and, later, on sugarcane plantations. Great Britain won possession of the Dutch colonies in the early 1800s and ended slavery. Still needing workers, the British paid Indians from Asia to move here. Today people from India make up most of Guyana's population. Another one-third are of African ancestry. Small numbers of Native Americans and Europeans also live here. Christianity and Hinduism are the chief religions. Most people speak English. Georgetown, the capital, is the major city.

Guyana won its independence from Britain in 1966. Guyana remains a very poor country, however, and depends on aid from the United Kingdom.

Roping a Capybara
Capybaras are the world's largest rodents. They may grow to be 2 feet tall and 4 feet long, and weigh more than 100 pounds. Found in Central and South America, the capybara (KA•pih•BAR•uh) lives along rivers and lakes, and eats vegetation. Here, a gaucho ropes a dog-sized capybara in Venezuela. Some Venezuelans eat capybara during the Easter season.

Suriname The British were the first Europeans to settle Suriname, but the Dutch gained control in 1667. As in Guyana, the Dutch brought enslaved Africans to work on large sugar plantations. Because of harsh treatment, many Africans fled into the isolated interior of the country. Their descendants still live there today. Later the Dutch hired workers from the Asian lands of India and Indonesia.

Asians form a large part of Suriname's population. About half of Suriname's people practice Christianity. The rest follow Hinduism or Islam. The main language is Dutch. Paramaribo (PAH•rah•MAH•ree•boh) is the capital and chief port. In 1975 Suriname won its independence from the Dutch. The country is poor, so it still relies on Dutch aid.

French Guiana French Guiana became a colony of France in the 1600s and remains one today. The country is headed by a French official called a *prefect,* who lives in the capital, Cayenne (ky•EHN). The French government provides jobs and aid to many of French Guiana's people.

Most people in French Guiana are of African or mixed African and European ancestry. They speak French and are Roman Catholic. In Cayenne, you see sidewalk cafés, police in French uniforms, and shoppers using francs, the French currency—just as you would in Paris, France. You also see local influences, such as Carnival, Native American woodcarving, and Caribbean music and dance.

✓ Reading Check **What European countries influenced the development of Guyana, Suriname, and French Guiana?**

Assessment

Defining Terms
1. **Define** llanos, hydroelectric power, altitude, caudillo.

Recalling Facts
2. **Place** What is the largest lake in South America?
3. **History** Who was Simón Bolívar?
4. **Economics** Name four of Venezuela's mineral resources.

Critical Thinking
5. **Drawing Conclusions** Why is Hinduism one of the major religions of Guyana?
6. **Understanding Cause and Effect** Why does a drop in oil prices hurt Venezuela's poor and unemployed?

Graphic Organizer
7. **Organizing Information** Draw a diagram like this one. In the top box, under the heading list similarities about the Guianas. In the smaller boxes, under the headings write facts about each country that show their differences.

```
              The Guianas
        ┌─────────┼─────────┐
     Guyana   Suriname    French
                          Guiana
```

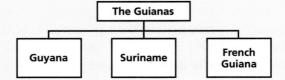

Applying Geography Skills

8. **Analyzing Maps** Look at the population density map on page 246. What is the population of Caracas? What is the population density of the area surrounding Caracas?

244

Uruguay and Paraguay

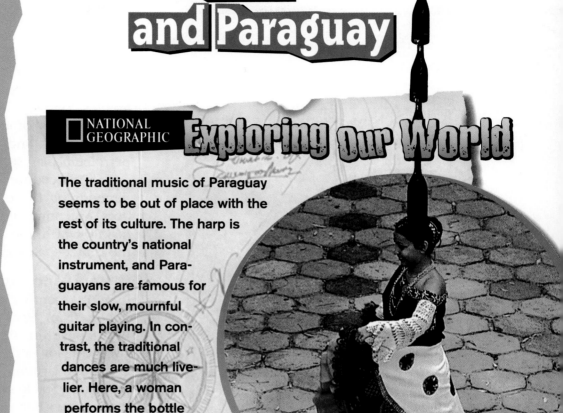

NATIONAL GEOGRAPHIC

Exploring Our World

The traditional music of Paraguay seems to be out of place with the rest of its culture. The harp is the country's national instrument, and Paraguayans are famous for their slow, mournful guitar playing. In contrast, the traditional dances are much livelier. Here, a woman performs the bottle dance—a difficult feat even though the bottles are attached to one another.

In the 1800s, European settlers built ranches on the vast plains of Uruguay and Paraguay, Brazil's southern neighbors. Raising livestock and manufacturing products from livestock form the economic base of these two countries.

Uruguay

Uruguay shares a border with southeastern Brazil. It touches water on the other three sides, with the Atlantic Ocean on the east, the Uruguay River on the west, and the Río de la Plata on the south. If you traveled in Uruguay, you would see grass blanketing its rolling, hilly plains. The hills descend to a low, narrow, fertile plain along the Atlantic coast.

You would enjoy the country's generally mild climate. Temperatures are moderate and rarely fall below freezing. Sometimes, though, a cold and violent wind rises in Argentina and blows north.

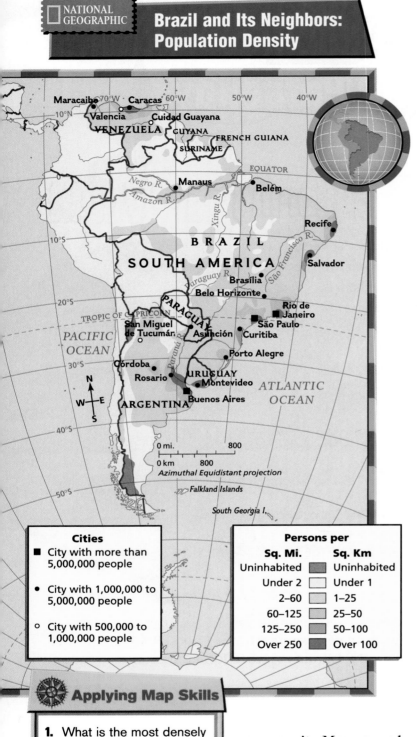

Cities

■ City with more than 5,000,000 people

● City with 1,000,000 to 5,000,000 people

○ City with 500,000 to 1,000,000 people

Persons per	
Sq. Mi.	**Sq. Km**
Uninhabited	Uninhabited
Under 2	Under 1
2–60	1–25
60–125	25–50
125–250	50–100
Over 250	Over 100

Applying Map Skills

1. What is the most densely populated area of Uruguay?

2. What cities in this region have more than 5 million people?

Find NGS online map resources @ www.nationalgeographic.com/maps

The Economy Uruguay's grasslands support sheep and cattle raising, the country's main economic activities. In fact, both Uruguay and Paraguay are covered with plains that are perfect for raising livestock. Cowhands are as common here as they are in our state of Texas. Animal products—meat, wool, and hides—top all exports. As you might expect, the major industries—textiles, footwear, and leather goods—use the products of the vast animal herds. Making tires and cement and refining petroleum play smaller roles in the economy. The economic health of Uruguay depends on the economies of its neighbors, Argentina and Brazil. Uruguay trades mainly with these countries but also with the United States and the European Union.

History and Government Uruguay joined Argentina in declaring independence from Spain in 1811. Later the Portuguese in Brazil took control of Uruguay. Not until 1828, with Argentina's help, did Uruguay win complete independence. Stable civilian governments led the country for most of its history. For many years, the government spent more on education than it did on defense. Uruguay's prosperity allowed it to become a welfare state, a country that uses tax money to support people who are sick, needy, jobless, or retired. Falling world demand for meat and wool and high government spending hurt the economy in the 1950s and 1960s, though. In addition, because of political troubles, military leaders took control of the government.

In 1985 democratic government returned to Uruguay. New leaders tried various economic changes to bring back prosperity. Many people resisted efforts to cut government spending, however. They feared losing their welfare benefits and suffering a lower standard of living. Reform efforts continue at a slow pace.

The People Most of Uruguay's 3.4 million people are of European ancestry. Only a small number have African or mixed ancestry. Very few Native Americans remain in the country. Spanish is the official language, and the Roman Catholic faith is the major religion. More

Itaipu Dam

About 40,000 workers labored to build Paraguay's Itaipu Dam. Brazil funded its construction. In return, Brazil pays low prices for the electricity it buys from Paraguay.

Human/Environment Interaction
How do both Paraguay and Brazil benefit from this arrangement?

than 90 percent of the people live in cities and towns. Nearly 1.4 million Uruguayans live in the coastal city of **Montevideo** (MAHN•tuh•vuh•DAY•oh), the country's capital.

Many people in rural areas own or rent small farms. Others work as gauchos on huge ranches. Legends about the gauchos have inspired Uruguay's folk music and literature. Uruguayans still flock to see gaucho rodeos.

✓ **Reading Check** What is a welfare state?

Paraguay

Paraguay sits near the center of South America. Paraguay is landlocked because it has no seacoast. The **Paraguay River** divides Paraguay into east and west regions. Rolling hills, forests, rivers, and a humid subtropical climate make the eastern region a pleasant place to live. Most of Paraguay's people live here.

To the west is a plains region called the **Gran Chaco,** which you may remember is shared with Argentina. This area is home to less than 5 percent of Paraguay's people. Droughts in winter and floods in summer discourage people from living there. In Gran Chaco's tropical savanna climate, grasses, palm trees, and quebracho trees thrive. Quebracho trees provide tannin, a chemical used to process leather.

The Economy Forestry and farming are the major economic activities in Paraguay. Large cattle ranches cover much of the country. Most farmers, however, use small plots of land to grow grains, soybeans, and cassava. Cassava roots can be ground up and eaten or used to make tapioca. They can also be sliced and fried just like potatoes.

Paraguay also exports electricity. The country has the world's largest hydroelectric power generator at the Itaipu (ee•TY•poo) Dam, on the **Paraná River.** Paraguay sells nearly 90 percent of the electricity it produces to neighboring countries.

History and Government   A Native American group called the Guaraní were the first people to live in Paraguay. By the 1500s, settlers and Roman Catholic missionaries from Spain arrived in the country. Paraguay was a colony of Spain until it became independent in 1811.

Paraguay has had a troubled history. Several wars in the 1800s left the nation very poor. Leaders had to invite investors from Argentina and Brazil to buy land to help the economy recover. Foreigners still own much of Paraguay's land. In addition, military leaders ruled harshly, punishing those who criticized them. In the 1990s, the country adopted a new constitution. Today the people of Paraguay elect their leaders.

The People   Paraguayans today are mostly of mixed Guaraní and Spanish ancestry. Both Spanish and Guaraní are official languages, but more people speak Guaraní. Most people practice the Roman Catholic faith. About one-half of the people live in cities. **Asunción** (ah•SOON•see•OHN) is the capital and largest city.

Paraguayan arts are influenced by Guaraní culture. Guaraní lace is Paraguay's most famous handicraft. Like people in Uruguay, the people of Paraguay enjoy meat dishes and sip *yerba maté,* a tealike drink.

Web Activity Visit the *Geography: The World and Its People* Web site at gwip.glencoe.com and click on **Chapter 8— Student Web Activities** to learn more about Paraguay.

✓ Reading Check **What are the main economic activities in Paraguay?**

Section 4 Assessment

Defining Terms
1. **Define** welfare state, landlocked.

Recalling Facts
2. **Economics** List three of Uruguay's major products.
3. **Culture** What percentage of Uruguay's people live in cities and towns?
4. **Human/Environment Interaction** What is the significance of the Itaipu Dam?

Critical Thinking
5. **Making Predictions** How can being landlocked affect a country's economy?
6. **Making Comparisons** How do Uruguay and Paraguay differ in regard to their Native American populations?

Graphic Organizer
7. **Organizing Information** Draw a diagram like this one, then on the lines list facts about the two regions of Paraguay.

Eastern Region — Paraguay — Gran Chaco

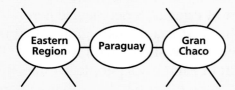

Applying Geography Skills

8. **Analyzing Maps** Refer to the population density map on page 246. About how many people live in Uruguay's capital city?

Reading Review

Section 1 | Brazil

Terms to Know
basin
selva
escarpment
favela
inflation
republic

Main Idea
Brazil is a large country with many resources, a lively culture, and serious economic challenges.

✓ Economics Brazil's prosperous economy includes mining, forestry, growing coffee and sugarcane, and manufacturing goods.

✓ Economics Brazil is trying to reduce its number of poor people and balance the use of resources with the preservation of its rain forests.

✓ Culture Most Brazilians are of mixed Portuguese, African, and Native American ancestry. Most live along the Atlantic coast and the Amazon River.

Section 2 | Argentina

Terms to Know
tannin
estancia
gaucho

Main Idea
Argentina is a large nation with diverse landforms, a strong economy, and a vibrant culture.

✓ Region Few people live in Argentina's Andes region, Patagonia, or Gran Chaco. The most populous area is the vast grassland called the pampas.

✓ Economics Argentina's strong economy includes mining, farming, ranching, and manufacturing.

✓ Culture Argentina's capital, Buenos Aires, is a huge city with European style.

Section 3 | Caribbean South America

Terms to Know
llanos
hydroelectric power
altitude
caudillo

Main Idea
Each nation in Caribbean South America has a unique culture, depending in part on who settled the country.

✓ Culture Most Venezuelans are of mixed European, African, and Native American ancestry. Most live in cities in the central highlands.

✓ Culture Guyana and Suriname have large numbers of people descended from workers who were brought from Africa and Asia.

✓ Government French Guiana is a territory of France.

Section 4 | Uruguay and Paraguay

Terms to Know
welfare state
landlocked

Main Idea
Farming and ranching form the economic base of Uruguay and Paraguay.

✓ Place Uruguay and Paraguay have large areas of grass-covered plains that support ranching and industries that depend on raising livestock.

✓ Economics Paraguay grows grains, soybeans, and cassava, and the country exports electricity.

Brazil and Its Neighbors

Chapter 8

Assessment and Activities

Using Key Terms

Match the terms in Part A with their definitions in Part B.

A.

1. basin
2. *estancia*
3. escarpment
4. caudillo
5. landlocked
6. gaucho
7. *selva*
8. inflation
9. llanos
10. tannin

B.

a. steep cliff separating two flat land surfaces, one higher than the other
b. cowhand
c. overall increase in prices across an entire economy
d. military dictator
e. large, grassy plains region of Latin America
f. a country that has no border on a sea or an ocean
g. broad, flat lowland surrounded by higher land
h. substance from tree bark used in turning hides into leather
i. rain forest in Brazil
j. large ranch in Argentina

Reviewing the Main Ideas

Section 1 Brazil

11. **Place** Describe the Amazon River.
12. **Location** Why are Brazil's inland areas sparsely populated?
13. **Government** What are the voting requirements in Brazil?

Section 2 Argentina

14. **Place** What is Argentina's capital?
15. **Place** How is Patagonia different from the pampas?
16. **Culture** Why does Argentina have only a small percentage of Native Americans?

Section 3 Caribbean South America

17. **Economics** Which of Venezuela's resources is its main source of income?
18. **Culture** Where do most of the people of the Guianas live?

Section 4 Uruguay and Paraguay

19. **Culture** What are the major language and religion of Uruguay?
20. **Economics** What are the major economic activities of Paraguay?

NATIONAL GEOGRAPHIC

Brazil and Its Neighbors

Place Location Activity

On a separate sheet of paper, match the letters on the map with the numbered places listed below.

1. Brazil
2. Amazon River
3. Argentina
4. Rio de Janeiro
5. Paraguay
6. Orinoco River
7. Río de la Plata
8. Venezuela
9. Brasília
10. Suriname

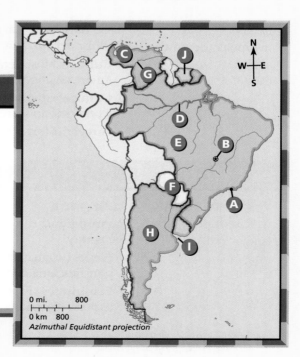

0 mi. 800
0 km 800
Azimuthal Equidistant projection

Self-Check Quiz Visit the *Geography: The World and Its People* Web site at gwip.glencoe.com and click on **Chapter 8— Self-Check Quizzes** to prepare for the Chapter Test.

Critical Thinking

21. **Analyzing Information** What facts support the statement "Argentina is one of the most industrialized countries in South America"?

22. **Identifying Points of View** In a chart like the one below, identify arguments for and against the cutting down of the rain forest.

Cutting Down the Rain Forest	
For	Against

GeoJournal Activity

23. **Writing a Research Report** Research the way coffee is grown, processed, and transported to the market and consumer. Write a report on your findings. Include a map of the areas where coffee is grown, a diagram with information about how it is grown and harvested, and a flowchart that shows how the coffee gets from harvest to consumer.

Mental Mapping Activity

24. **Focusing on the Region** Draw a simple outline map of South America, then label the following:

- Caribbean Sea
- Atlantic Ocean
- Pacific Ocean
- Amazon River
- Venezuela
- Brazil
- Argentina
- Brazilian Highlands
- Río de la Plata
- Guiana Highlands

Technology Skills Activity

25. **Using the Internet** Conduct a search for information about the Amazon rain forest and create an annotated bibliography of five useful Web sites. Your bibliography should contain the Web address, a brief summary of the information found on the site, and whether or not you feel the site is useful.

The Princeton Review

Standardized Test Practice

Directions: Read the passage below and answer the question that follows.

The Amazon Basin is a gigantic system of rivers and rain forests, covering half of Brazil and extending into neighboring countries. Much of the Amazon is still unexplored, and the rain forest holds many secrets. Some of the animals found here include the jaguar, tapir, spider monkey, sloth, river dolphin, and boa constrictor. Forest birds include toucans, parrots, hummingbirds, and hawks. More than 1,800 species of butterflies and 200 species of mosquitoes give you an idea about the insect population. In addition, the fish—such as piranha, pirarucu, and electric eel—are very unusual. Biologists cannot identify much of the catch found in markets.

1. **Based on this passage, which of the following generalizations is most accurate?**

   F The Amazon rain forest covers about one-third of the South American continent.

   G Native Americans living in the Amazon rain forest are losing their traditional way of life.

   H The Amazon Basin is huge, and its rain forests hold thousands of animal species.

   J The Amazon Basin is located only in Brazil.

Test-Taking Tip: This question asks you to make a generalization about the Amazon Basin. A *generalization* is a broad statement. Look for facts and the main idea *in the passage* to support your answer. Do not rely only on your memory. The main idea can help you eliminate answers that do not fit. Also, look for the statement that is true AND that is covered in the paragraph.

VANISHING
Rain Forests

WEST INDIES

CENTRAL
AMERICA

SOUTH
AMERICA

☐ Rain forests

Rain Forest Riches Imagine never tasting chocolate. Think about never eating a banana, chewing gum, or munching cashews. If there were no rain forests, we would have none of these foods. We also would not have many of the drugs used to treat malaria, multiple sclerosis, and leukemia. In fact, rain forest plants provide a fourth of the world's medicines.

Millions of kinds of plants and animals live in rain forests—more than half of all species on Earth. Scientists have studied only a fraction of these species. So no one really knows what new foods, medicines, or animals are there, just waiting to be discovered.

Rain Forest Destruction Yet we may never know. Why? Because a chunk of rain forest the size of two football fields vanishes every second! The forests are being destroyed for many reasons.

- Loggers cut trees and sell the lumber worldwide.

- Ranchers and farmers clear land for cattle and crops.

- Miners level acres of forest to get at valuable minerals.

People are trying to find ways to use rain forests without destroying them. Changing farming practices and developing different forest industries are possible solutions. However, time is running out. Can we afford to lose rain forests and all their treasures?

Male golden toad

Settlers clear trees for a home in the rain forest.

Making a Difference

Discovering New Monkeys How would it feel to discover an animal that no one knew existed? Dutch scientist Marc van Roosmalen knows. He recently discovered a new species of monkey (photo, at right) in Brazil.

Van Roosmalen runs an orphanage for abandoned monkeys. One day, a man showed up with a tiny monkey van Roosmalen had never seen before. He spent about a year tracking down a wild population of the monkeys deep in the Amazon rain forest. Of some 250 kinds of monkeys known worldwide, about 80 live in Brazil. At least 7 new species have been discovered since 1990.

New species *Callithrix humilis*, a dwarf marmoset

Rain Forest Field Trip With help from the Children's Environmental Trust Foundation, students from Millbrook, New York, traveled to Peru's Yarapa River region, deep in the Amazon rain forest. Students studied the forest from platforms built in the canopy, and they soared among the tall trees using ropes. The students met rain forest creatures at night, went birdwatching at dawn, and swam in the Yarapa River—home to crocodiles called caimans.

Back in Millbrook, the students educate others about saving rain forests. They also raise money to help support a Peruvian zoo that protects rain forest animals.

Millbrook student traps insects for study.

What Can You Do?

Write a Note
Write to your government representatives and encourage them to support plans that help save rain forests.

Check Out Your Community
What environmental problems face your community? What can you do to help solve the problems? For example, does your community have problems with water pollution or water shortages? What steps does your community take to make sure you have clean water to drink?

Use the Internet
Learn more about efforts to save rain forests. Check out the Rainforest Action Network at www.ran.org or search National Geographic's Web site at www.nationalgeographic.com

253

Chapter 9

The Andean Countries

The World and Its People NATIONAL GEOGRAPHIC

To learn more about the people and places of the Andean countries, view *The World and Its People* **Chapter 9** video.

Geography ONLINE

Chapter Overview Visit the **Geography: The World and Its People** Web site at gwip.glencoe.com and click on **Chapter 9— Chapter Overviews** to preview information about the Andean countries.

Guide to Reading

Main Idea

Although it has many resources, Colombia faces political and economic turmoil.

Terms to Know

- cordillera
- llanos
- cash crop
- mestizo
- republic
- campesino

Places to Locate

- Colombia
- Andes
- Magdalena River
- Bogotá
- Cartagena
- Medellín
- Cali

Reading Strategy

Create a chart like the one below and list advantages that Colombia enjoys in the left column. In the right column, list the challenges that it faces.

Colombia	
Advantages	Challenges

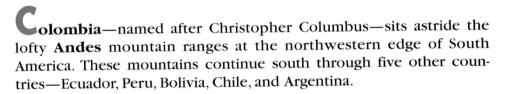

NATIONAL GEOGRAPHIC Exploring Our World

In a thin vein of black shale, a miner in Colombia spots a glistening green stone. He is not the first Colombian to mine the precious gemstones we call emeralds. The Colombian mine called Muzo has been producing top-quality emeralds for a thousand years. Early Native American rulers would offer these gems—more rare than diamonds—to their gods.

Colombia—named after Christopher Columbus—sits astride the lofty **Andes** mountain ranges at the northwestern edge of South America. These mountains continue south through five other countries—Ecuador, Peru, Bolivia, Chile, and Argentina.

Colombia's Land and Climate

Colombia—almost three times larger than Montana—has coasts on both the Caribbean Sea and the Pacific Ocean. The Andes sweep through the western part of Colombia. Here they become a cordillera (KAWR•duhl•YEHR•uh), or a group of mountain ranges that run side by side. Colombia's major river, the **Magdalena River,** flows between the central and eastern Andes to the Caribbean Sea. Nearly 80 percent of Colombia's people live in the valleys and highland plateaus of the Andes. **Bogotá** (BOH•goh•TAH), Colombia's capital and largest city, is located on a high Andean plateau.

Busy ports, such as **Cartagena** (KAHR•tah•HAY•nah), handle Colombia's trade on the Caribbean coast. In the west, thick forests spread over the lowlands along the Pacific coast. Few people live there.

◀ Mount Cotopaxi overlooks Quito, Ecuador

Tropical rain forests spread across the southeastern plain into the Amazon Basin. Only a few Native American groups live in this hot, steamy region. In the northeast you find hot grasslands called the **llanos.** Here, ranchers drive their cattle across the rolling plains.

Colombia lies entirely within the tropics. Temperatures in places are very hot, and heavy rains fall along the coasts and in the interior plains. In the high elevations of the Andes, however, temperatures are very cool for a tropical area. Bogotá lies at 8,355 feet (2,547 m) above sea level. High temperatures average only 67°F (19°C).

✓ **Reading Check** **Where do most of Colombia's people live?**

NATIONAL GEOGRAPHIC

The Andean Countries: Political

Applying Map Skills

1. What bodies of water does Colombia border?

2. What country has a name that sounds like "Equator"?

Find NGS online map resources @ www.nationalgeographic.com/maps

Colombia's Economy

Colombia has many natural resources. The mountains hold valuable minerals and precious stones, and Colombia has more coal than any other country in South America. Second only to Brazil in its potential hydroelectric power, Colombia also has large petroleum reserves in the lowlands. In addition, the country is a major supplier of gold and the world's number one source of emeralds.

In regard to manufacturing, factories in Colombia produce a variety of products. Workers make clothing, leather goods, food products, paper, chemicals, and iron and steel products.

NATIONAL GEOGRAPHIC

The Andean Countries: Physical

Caribbean Sea
ATLANTIC OCEAN
Cauca R.
Magdalena R.
Llanos
Guaviare R.
COLOMBIA
Japurá R.
EQUATOR
Galápagos Islands
ECUADOR
PERU
Marañón R.
Ucayali R.
SOUTH AMERICA
Mt. Huascarán 22,205 ft. (6,768 m)
A N D E S
Lake Titicaca
BOLIVIA
PACIFIC OCEAN
Altiplano
Atacama Desert
TROPIC OF CAPRICORN
Mt. Ojos del Salado 22,572 ft. (6,880 m)
CHILE
Aconcagua 22,834 ft. (6,960 m)
Mt. Tupungato 22,310 ft. (6,800 m)

0 mi. 800
0 km 800
Azimuthal Equidistant projection

N W E S

Strait of Magellan

Tierra del Fuego
Cape Horn

Elevations
Feet		Meters
10,000		3,000
5,000		1,500
2,000		600
1,000		300
0		0

▲ Mountain peak

Applying Map Skills

1. What are the inland plains of Colombia called?

2. What physical feature runs through all of these countries?

Find NGS online map resources @
www.nationalgeographic.com/maps

257

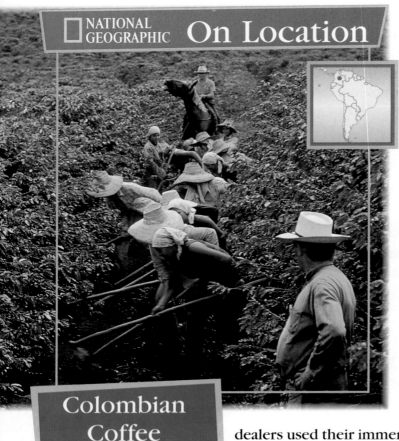

Colombian Coffee

Many historians believe that coffee was "discovered" in Ethiopia, Africa. Eventually Spanish missionaries brought the first coffee plants to Colombia.

Region What other crops does Colombia export?

Agriculture The coastal regions and the highlands have good soil for growing crops. In fact, the differences in land elevation allow Colombians to grow a variety of crops. Coffee is the country's major **cash crop**—a product sold for export. Colombian coffee—grown on large plantations and small farms—is known all over the world for its rich flavor.

Did you know that the average American eats about 28 pounds (13 kg) of bananas a year? Colombia exports bananas as well as cacao, sugarcane, rice, and cotton. Huge herds of cattle roam large *estancias* in the llanos. The rain forests of the eastern plains also supply a valuable resource—lumber.

Economic Challenges Despite many natural resources, Colombia faces economic challenges. In the 1980s, drug dealers became a major force in Colombia. The dealers paid farmers more to grow coca leaves—which are used to make the drug cocaine—than the farmers could earn growing coffee. At the same time, the drug dealers used their immense profits to build private armies. They threatened—and even killed—government officials who tried to stop them.

The government of Colombia has stepped up its efforts to break the power of the drug dealers. It has had some success, but drug dealers continue to flourish. In addition, the government has tried to persuade thousands of farmers to switch back to growing other crops.

Reading Check What crop has been a problem in Colombia? Why?

Colombia's History and People

About 38.6 million people live in Colombia. Nearly all Colombians are **mestizos** (meh•STEE•zohs), meaning they have mixed European and Native American or African backgrounds. Most speak Spanish and follow the Roman Catholic faith.

In 1810 Colombia was one of the first Spanish colonies in the Americas to declare independence. Simón Bolívar, whom you read about in Chapter 8, led this struggle for independence. In 1819 Colombia became part of New Granada, an independent country that included Venezuela, Ecuador, and Panama. Later, these other regions broke away and became separate countries.

Colombia today is a **republic** with an elected president. Political violence has scarred the country's history, though. During the late 1800s alone, Colombia suffered through more than 50 revolts and 8 civil wars. Fighting broke out again in 1948. About 300,000 people died in this conflict, which ended in the late 1960s.

To prevent further turmoil, the two main political parties agreed to govern the country together. Efforts were made to improve the lives of poor farmers by giving them more land. Factories and industrial jobs opened up. Still, a wide gap between rich and poor remained, causing further troubles.

In the 1960s, groups of rebels in the countryside began fighting the government. This latest civil war has lasted more than 35 years and left more than 100,000 people dead. In late 1999, more than 13 million Colombians joined in a massive protest across the country, urging an end to the fighting.

A Diverse Culture   Colombia has a rapidly growing urban population. Colombian farmers, or **campesinos,** and their families have journeyed to cities to look for work or to flee the fighting in the countryside. Thirty cities have more than 100,000 people each. The largest ones are Bogotá, **Medellín** (MAY•thay•YEEN), and **Cali** (KAH•lee).

You can see Colombia's Spanish, Native American, and African heritages reflected in its culture. Native American skills in weaving and pottery date back before the arrival of Columbus. Caribbean African rhythms blend with Spanish-influenced music. The Colombian writer Gabriel García Márquez is one of Latin America's most famous authors. His novels blend legends and fantasy with events of everyday life.

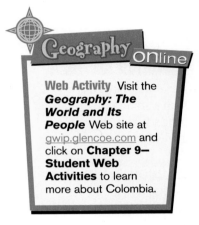

Geography Online

Web Activity Visit the **Geography: The World and Its People** Web site at gwip.glencoe.com and click on **Chapter 9— Student Web Activities** to learn more about Colombia.

✓Reading Check **What is a mestizo?**

Section 1 Assessment

Defining Terms

1. Define cordillera, llanos, cash crop, mestizo, republic, campesino.

Recalling Facts

2. Economics Colombia is the world's number one source of what mineral?

3. Movement What are two reasons campesinos and their families have moved to the cities?

4. Economics What are four agricultural products of Colombia?

Critical Thinking

5. Analyzing Cause and Effect Why does Bogotá, which is located in the tropics, have an average temperature of only 67°F (19°C)?

6. Drawing Conclusions Why do you think it is so difficult for Colombian farmers to stop growing coca?

Graphic Organizer

7. Organizing Information Draw a time line like this one, then put the following events and their dates in the correct order on it: Massive protest held by more than 13 million Colombians, Native Americans settle the region, Colombia declares independence from Spain, Colombia suffers 50 revolts and 8 civil wars, Colombia becomes part of New Granada.

├────────┼────────┼────────┼────────┤

Applying Geography Skills

8. Analyzing Maps Study the physical map on page 257. What rivers run through Colombia? Which side of the country is mountainous?

Peru and Ecuador

Guide to Reading

Main Idea

Peru and Ecuador share similar landscapes, climates, and history.

Terms to Know

- altiplano
- navigable
- foothill
- subsistence farm
- empire

Places to Locate

- Peru
- Ecuador
- Lake Titicaca
- Lima
- Galápagos Islands
- Guayaquil
- Quito

Reading Strategy

Draw two ovals like these. Under each heading list facts about Peru and Ecuador in the outer parts of the ovals. Where the ovals overlap, write facts that apply to both countries.

NATIONAL GEOGRAPHIC Exploring Our World

They built thousands of miles of roads. They built a city on mountain peaks and were expert bridge builders. The Inca accomplished these feats in western South America during the 1400s and 1500s. The ruins of their ancient city of Machu Picchu (MAH•choo PEEK•choo), built nearly 8,000 feet (2,438m) high in the Andes, were not even known to modern people until 1911.

Peru and **Ecuador** lie along the Pacific coast of South America west of Brazil and south of Colombia. The Andes form the spine of these countries. *Peru*—a Native American word that means "land of abundance"—is rich in mineral resources.

Peru

Dry deserts, snowcapped mountains, and hot, humid rain forests greet you in Peru. Most of Peru's farms and cities lie on a narrow coastal strip of plains and deserts. The cold Peru Current in the Pacific Ocean keeps temperatures here fairly mild even though the area is very near the Equator. Find the Peru Current on the map on page 58.

The Andes, with their highland valleys and plateaus, sweep through the center of Peru. South-central Peru contains a large highland plateau called the **altiplano.** Here you can see **Lake Titicaca** (TEE•tee•KAH•kah), the highest navigable lake in the world. **Navigable** means that a body of water is wide and deep enough to allow ships to travel in it. The altiplano continues south into Bolivia, which you will learn about in Section 3.

East of the Andes you descend to the foothills and flat plains of the Amazon Basin. **Foothills** are the low hills at the base of a mountain range. Rainfall is plentiful here, and temperatures remain high throughout the year. Thick rain forests cover almost all of the plains area.

Peru's Economy Peru's economy relies on a variety of natural resources. The Andes contain many minerals, including copper, silver, gold, and iron ore. Peru's biggest export is copper from mines in the south. The second-largest export—fish—comes from the Peru Current, the cool Pacific Ocean current that parallels the coast.

About a third of Peru's people farm the land. Some grow sugarcane, cotton, and coffee for export. Like Colombia, Peru grows coca leaves. Most people, however, work on **subsistence farms,** where they grow enough food to meet their family's needs. The chief crops are rice, plantains (a kind of banana), and corn. Do you enjoy eating potatoes? Thank the Native Americans of the Andes. Thousands of years ago, they were the first to grow this food. Potatoes are still Peru's main food crop, and farmers grow hundreds of varieties in different colors and shapes.

From Empire to Republic During the 1400s, a Native American people called the Inca had a powerful civilization in the area that is now Peru. Their **empire,** or group of lands under one ruler, stretched more than 2,500 miles (4,023 km) along the Andes from northern Ecuador to central Chile.

The Incan emperor developed a complex system of tax collection, courts, military posts, trade inspections, and work rules. Officials kept records by using quipu, or rope with knotted cords of different lengths and colors. Each knot meant a different item or number. Work crews built irrigation systems, roads, and suspension bridges that linked the regions of the empire to Cuzco, the capital city of the Inca. You can still see the remains of magnificent fortresses and buildings erected centuries ago by skilled Incan builders. The photograph on page 260 shows the ruins of one of the Inca's most famous cities—Machu Picchu.

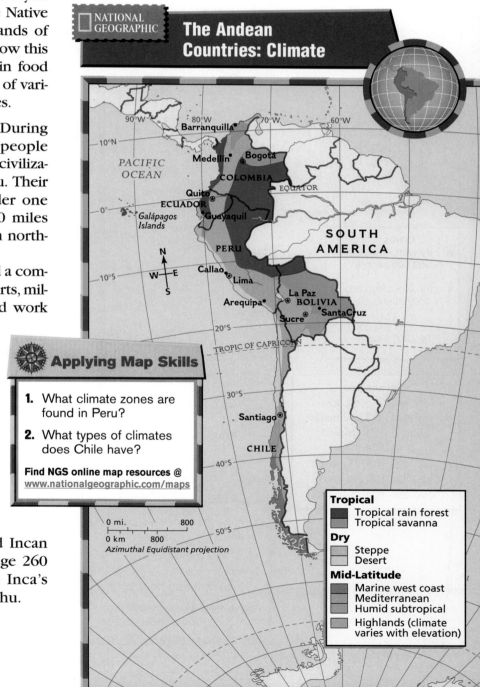

NATIONAL GEOGRAPHIC

The Andean Countries: Climate

Applying Map Skills

1. What climate zones are found in Peru?

2. What types of climates does Chile have?

Find NGS online map resources @ www.nationalgeographic.com/maps

0 mi. 800
0 km 800
Azimuthal Equidistant projection

Tropical
- Tropical rain forest
- Tropical savanna

Dry
- Steppe
- Desert

Mid-Latitude
- Marine west coast
- Mediterranean
- Humid subtropical
- Highlands (climate varies with elevation)

In the early 1500s, Spaniards arrived in Peru, craving the gold and silver found here. They defeated the Inca and made Peru a Spanish territory. Peru gained its freedom from Spain in the 1820s. After independence, Peru fought wars with neighboring Chile and Ecuador over land.

Peru is now a republic with an elected president. In recent years, the country's economy has grown very rapidly. Many of Peru's people, however, still live in poverty and cannot find steady jobs.

Peru's Culture Peru has 26.6 million people, making it the fourth-most populous country in South America. The people of Peru live mostly in cities or towns in or near the plain along the Pacific coast. **Lima** (LEE•mah), with more than 6 million people, is the capital and largest city. In recent years, many people from the countryside have moved to Lima in search of work. Because of this sudden rise in population, the city has become overcrowded, noisy, and polluted. Many people live in barrios, or very poor neighborhoods.

About half of Peru's people are Native American. In fact, Peru has one of the largest Native American populations in the Western Hemisphere. Many live in the Andean highlands or the eastern rain forests where they follow a traditional way of life. Most of them blend the Catholic faith, Peru's main religion, with beliefs of their ancestors.

Peruvians also include many people of mixed or European ancestry. People of Asian heritage form a small but important part of the population. In the 1990s, Alberto Fujimori (FOO•jee•MAW•ree), a Peruvian of Japanese ancestry, was Peru's president for 10 years.

Spanish is Peru's official language, but about 70 Native American languages also are spoken. You can hear the sounds of Quechua (KEH•chuh•wuh), the ancient language of the Inca, in many Native American villages.

✓ **Reading Check** Who built a huge empire centered in Peru?

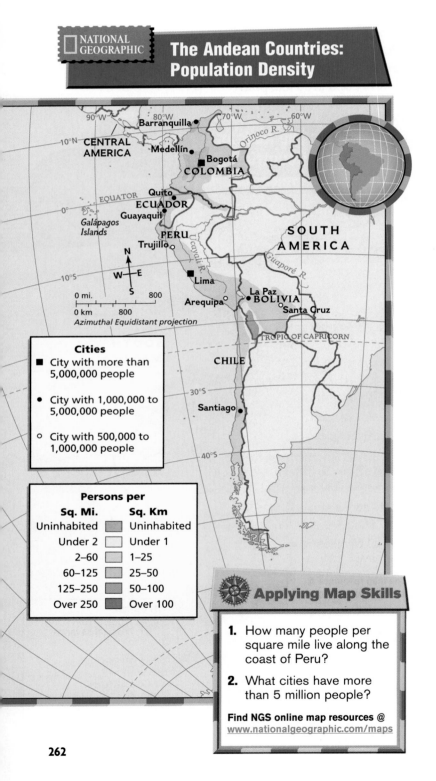

NATIONAL GEOGRAPHIC

The Andean Countries: Population Density

Cities

- ■ City with more than 5,000,000 people
- ● City with 1,000,000 to 5,000,000 people
- ○ City with 500,000 to 1,000,000 people

Persons per	
Sq. Mi.	**Sq. Km**
Uninhabited	Uninhabited
Under 2	Under 1
2–60	1–25
60–125	25–50
125–250	50–100
Over 250	Over 100

0 mi. 800
0 km 800
Azimuthal Equidistant projection

Applying Map Skills

1. How many people per square mile live along the coast of Peru?

2. What cities have more than 5 million people?

Find NGS online map resources @ www.nationalgeographic.com/maps

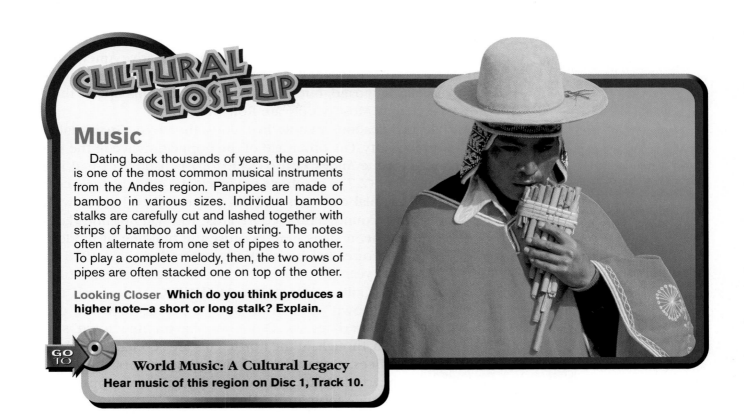

Music

Dating back thousands of years, the panpipe is one of the most common musical instruments from the Andes region. Panpipes are made of bamboo in various sizes. Individual bamboo stalks are carefully cut and lashed together with strips of bamboo and woolen string. The notes often alternate from one set of pipes to another. To play a complete melody, then, the two rows of pipes are often stacked one on top of the other.

Looking Closer Which do you think produces a higher note—a short or long stalk? Explain.

GO TO

World Music: A Cultural Legacy
Hear music of this region on Disc 1, Track 10.

Ecuador

Ecuador is one of the smallest countries in South America. In fact, the entire country is no bigger than the state of Nevada in the United States. Can you guess how it got its name? *Ecuador* is the Spanish word for "Equator," which runs right through Ecuador. Find the country on the map on page 256. Located between Colombia and Peru, Ecuador looks like a grinning face gazing westward across the Pacific Ocean at the **Galápagos Islands.** Owned by Ecuador since 1832, these scattered islands are known for their rich plant and animal life. Turn to page 265 to learn more about the unique Galápagos Islands.

Ecuador's land and climate are similar to Peru's. Swamps and fertile plains stretch along Ecuador's Pacific coast. The Peru Current in the Pacific Ocean keeps coastal temperatures mild. The Andes run through the center of the country. The higher you climb up these mountains, the colder the climate gets. In contrast, hot, humid rain forests cover the lowlands of eastern Ecuador.

Ecuador's Economy   Agriculture is Ecuador's most important economic activity. Bananas, cacao, coffee, rice, sugarcane, and other export crops grow in the coastal lowlands. Here you will find **Guayaquil** (GWY•ah•KEEL), Ecuador's most important port city. Farther inland, farms in the Andean highlands grow coffee, beans, corn, potatoes, and wheat. The eastern lowlands yield petroleum, Ecuador's major mineral export.

Andean Dig

Why is this man upside down? He is trying to recover a hard-to-reach frozen mummy in the Andes. This archaeological site lies more than 22,000 feet (6,706 m) above sea level. In 1999 scientists found the frozen bodies of three young Native Americans who were sacrificed 500 years ago to persuade the gods to bring good weather.

Ecuador's People Of Ecuador's 12.4 million people, about 40 percent are mestizos and another 40 percent are Native Americans. African Americans are a small part of the population. Spanish is Ecuador's official language, but many Native Americans speak their traditional languages. Most people are Roman Catholic.

About half of Ecuador's people live along the coast. Guayaquil is the most populous city. The other half of the population live in the valleys and plateaus of the Andes. **Quito** (KEE•toh), Ecuador's capital, lies more than 9,000 feet (2,743 m) above sea level. The city's historic center has Spanish colonial churches and old whitewashed houses with red-tiled roofs built around central courtyards. You will not find flashing neon signs here because the building of modern structures has been strictly controlled since 1978. In that year, the United Nations Educational, Scientific, and Cultural Organization (UNESCO) declared the "old town" section of Quito a protected world cultural heritage site. Quito does have a "new town" section, though, in the north. This area contains modern offices, embassies, and shopping centers. From the heart of Quito, you can see several snowcapped volcanoes looming in the distance. Very few people live in the lowlands of the east.

✓Reading Check   **Why are Ecuador's eastern lowlands important economically?**

Section 2 Assessment

Defining Terms

1. **Define** altiplano, navigable, foothill, subsistence farm, empire.

Recalling Facts

2. **Place** Describe the foothills east of the Andes.
3. **History** Who were the first people to grow potatoes?
4. **Economics** What is Ecuador's major mineral export?

Critical Thinking

5. **Analyzing Information** Why is Peru's name, which means "land of abundance," appropriate? Why is it also inappropriate?
6. **Analyzing Cause and Effect** What effect does the Peru Current have on the coastal areas of Peru and Ecuador?

Graphic Organizer

7. **Organizing Information** Draw two diagrams like this one. Under each heading list facts about each category for Peru and Ecuador.

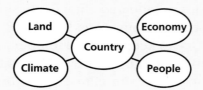

Applying Geography Skills

8. **Analyzing Maps** Turn to the climate map on page 261. In what climate zone is Quito, Ecuador, located? Lima, the capital of Peru, is located in which climate zone?

Making Connections

ART   SCIENCE   LITERATURE   TECHNOLOGY

The Galápagos Islands

The Galápagos Islands are located in the eastern Pacific Ocean about 600 miles (966 km) west of mainland Ecuador. Since 1959 about 95 percent of the islands has been maintained as a national park.

History of Exploration

From the first documented visit to the Galápagos Islands in 1535, people have commented on the islands' unusual wildlife. Sailors, including pirates and whalers, stopped on the islands to collect water and to trap the huge *galápagos*, or tortoises, found on the islands. Sailors valued the tortoises as a source of fresh meat because the giant tortoises could live on ships for months without food or water.

Charles Darwin

The most famous visitor to the Galápagos Islands was Charles Darwin, a scientist from England. He was studying animals all over the world. In 1835 Darwin spent five weeks visiting four of the biggest islands in the Galápagos. He carefully studied the volcanic landscape and the plant and animal life that he saw. He took notes on the differences between animals such as finches, mockingbirds, and iguanas from island to island. Darwin believed that these differences showed how populations of the same species change to fit their environment.

A Fragile Environment

Today the Galápagos Islands are still prized for their amazing variety of animal and plant life. Many of the species found here exist nowhere else on the earth. For instance, the Marine iguana that lives here is the only seagoing lizard in the world.

Unfortunately, years of contact between the islands and humans have had serious effects. Three of the 14 types of tortoises are extinct, and others are seriously threatened. Populations of goats, pigs, dogs, rats, and some types of plants, brought by visitors, have grown so large that they threaten the survival of native plants and animals. Demand for exotic marine life, including sharks and sea cucumbers, has led to overfishing. The government of Ecuador, along with environmentalists worldwide, is now working to protect the islands.

▲ Giant Galápagos tortoise

Making the Connection

1. Why did sailors long ago stop at the islands?
2. What did Darwin observe about the islands?
3. **Drawing Conclusions** Why are environmentalists and the government of Ecuador working to protect the Galápagos Islands?

Bolivia and Chile

Guide to Reading

Main Idea

Bolivia and Chile share the Andes, but their economies and people are different.

Terms to Know

- landlocked
- sodium nitrate

Places to Locate

- Bolivia
- Chile
- Sucre
- La Paz
- Atacama Desert
- Santiago
- Cape Horn

Reading Strategy

Create a chart like the one below. In each row, write at least one fact about Bolivia and one about Chile.

	Bolivia	Chile
Land		
Climate		
Economy		
People		

■ NATIONAL GEOGRAPHIC **Exploring Our World**

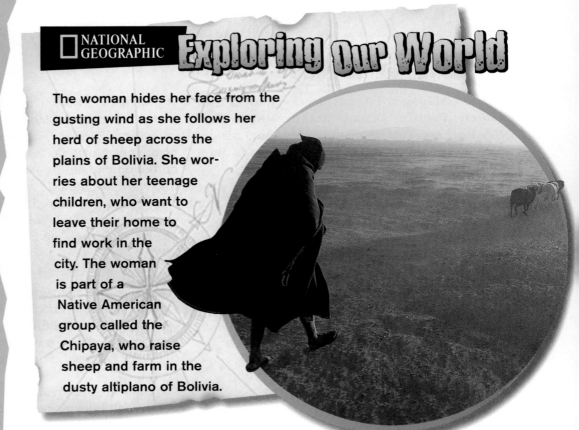

The woman hides her face from the gusting wind as she follows her herd of sheep across the plains of Bolivia. She worries about her teenage children, who want to leave their home to find work in the city. The woman is part of a Native American group called the Chipaya, who raise sheep and farm in the dusty altiplano of Bolivia.

At first glance, **Bolivia** and **Chile** seem very different. Bolivia lacks a seacoast, while Chile has a long coastline on the Pacific Ocean. The Andes, however, affect the climate and cultures of both countries.

Bolivia

Bolivia lies near the center of South America. It is a landlocked country, having no land that touches a sea or an ocean. Bolivia also is the highest and most isolated country in Latin America. Why? The Andes dominate Bolivia's landscape. Look at the physical map on page 257. You see that in western Bolivia, the Andes surround a high plateau called the altiplano. Because of the high elevation, the altiplano has a cool climate. Unless you were born in this area, you would find that the cold, thin air at 12,000 feet (3,658 m) makes it difficult to breathe. Few trees grow on the altiplano, and most of the land is too dry to farm. Still, the vast majority of Bolivians live on this high plateau. Those areas that have water have been farmed for many centuries.

Bolivia also has lowland plains and tropical rain forests. A vast lowland plain spreads over northern and eastern Bolivia. Tropical rain forests cover the northern end of the plain. Grasslands and swamps sweep across the rest of the plain. Most of this area has a hot, humid climate. South-central Bolivia, however, has gently sloping hills and broad valleys. The land is more fertile here, and many farms dot this region.

Bolivia's Economy   Bolivia is rich in minerals such as tin, silver, and zinc. Miners remove these minerals from high in the Andes. Workers in the eastern lowlands draw out gold, petroleum, and natural gas.

Still, Bolivia is a poor country. About two-thirds of the people live in poverty. Throughout the highlands, many villagers practice subsistence farming. They struggle to grow wheat, potatoes, and barley. At higher elevations, herders raise animals such as alpacas and llamas for wool and for carrying goods. In the south, farmers plant soybeans, a growing export. Timber is another important export.

Bolivian leaders hope that some new projects will expand the economy. Peru and Bolivia are building a highway to the Pacific coast. Bolivia also made an agreement with Brazil that calls for a pipeline to carry natural gas to Brazil. Increased trade from these projects is expected to spark economic growth in Bolivia.

Bolivia's People   What is unusual about Bolivia's capital? There is not just one capital city, but two. The official capital is **Sucre** (SOO•kray). The administrative capital and largest city is **La Paz** (lah•PAHZ). Both capital cities are located in the altiplano. La Paz—at 12,000 feet (3,658 m)—is the highest capital city in the world.

Most of Bolivia's more than 8 million people live in the Andean highlands. About half are of Native American ancestry, and another 40 percent are mestizos. In the cities, most people follow modern ways of living. In the country, you may hear traditional sounds—music played with flutelike instruments.

✓Reading Check   What is the altiplano?

NATIONAL GEOGRAPHIC

The Andean Countries: Economic Activity

Land Use
- Commercial farming
- Subsistence farming
- Ranching
- ■ Manufacturing area
- Little or no activity

Resources

🪨 Coal		🟡 Lead	
Copper		N Nitrates	
🐟 Fishing		Natural gas	
🌲 Forest		Petroleum	
▽ Gems		Silver	
Gold		▼ Tin	
Iron ore		⊡ Zinc	

0 mi.   800
0 km   800
Azimuthal Equidistant projection

Applying Map Skills

1. What types of land use are found in Bolivia?

2. What energy resources does Colombia have?

Find NGS online map resources @ www.nationalgeographic.com/maps

NATIONAL GEOGRAPHIC On Location

Chile's Contrasts

Chile has a wide variety of climates and landforms. The moderate capital city of Santiago in central Chile (above) contrasts sharply with the icy southern region (right).

Location What group of islands lies at the southern tip of Chile?

Chile

Find Chile on the map on page 256. It is almost twice the size of California. Though its average width is only 110 miles (177 km), Chile stretches 2,652 miles (4,268 km) along the Pacific Ocean.

About 80 percent of Chile's land is mountainous. The high Andes run along Chile's border with Bolivia and Argentina. Except in the altiplano area of Chile's north, very few Chileans live in the Andes.

From north to south, you find different regions in Chile. The first is a dry region known as the **Atacama Desert.** It is one of the driest places on the earth. Why? This area is in the rain shadow of the Andes. Winds from the Atlantic Ocean bring precipitation to regions east of the Andes, but they carry no moisture past them. In addition, the cold Peru Current in the Pacific Ocean does not evaporate as much as a warm current does. As a result, only dry air hits the coast.

A steppe climate zone lies just north of **Santiago,** Chile's capital. It receives some rainfall and has moderate temperatures. Most of Chile's people live in a central region called the Central Valley. With a mild Mediterranean climate, the fertile valleys here have the largest concentration of cities, industries, and farms.

The lake region, also known as "the south," has a marine west coast climate that supports thick forests. Chile's far south is a stormy, windswept region of snowcapped volcanoes, thick forests, and huge glaciers. The Strait of Magellan separates mainland Chile from a group of islands known as Tierra del Fuego (FWAY•goh)—or "Land of Fire." This region is shared by both Chile and Argentina. Cold ocean waters batter the rugged coast around **Cape Horn,** the southernmost point of South America.

Chile's Economy   In recent years, Chile has enjoyed high economic growth, and the number of people living below the poverty line has fallen by half. Mining forms the backbone of Chile's economy. The Atacama region is rich in minerals. Chile ranks as the world's leading copper producer. The country also mines and exports gold, silver, iron ore, and sodium nitrate—used as a fertilizer and in explosives.

Agriculture is also a major economic activity. Farmers produce wheat, corn, beans, sugar, and potatoes. The grapes and apples you eat in winter may come from Chile's summer harvest. Many people also raise cattle, sheep, and other livestock.

Chile has factories that process fish and other foods. Other workers make wood products, iron, steel, vehicles, cement, and textiles. Service industries such as banking and tourism also thrive.

Chile's Culture   Of the 15 million people in Chile, most are mestizos. A large minority are of European descent, and some Native American groups live in the altiplano and "the south." Nearly all the people speak Spanish, and most are Roman Catholic. Some 80 percent of Chile's population live in urban areas. Chile has been a democratic republic since the end of strict military rule in 1989.

✓ Reading Check   **What percentage of Chileans live in urban areas?**

Assessment

Defining Terms

1. **Define** landlocked, sodium nitrate.

Recalling Facts

2. **Economics** What percentage of Bolivia's population lives in poverty?
3. **Geography** What makes La Paz unique?
4. **Economics** Chile is the world's leading producer of what mineral?

Critical Thinking

5. **Analyzing Cause and Effect** Why is the Atacama Desert one of the driest places?
6. **Making Comparisons** What are differences and similarities between the economies of Bolivia and Chile?

Graphic Organizer

7. **Organizing Information** Draw a diagram like this one. Under each arrow list supporting facts for the main idea given.

> **Main Idea: Bolivia is rich in minerals but is still a poor country.**
> ↑   ↑   ↑   ↑

Applying Geography Skills

8. **Analyzing Maps** Study the physical map on page 257. The southernmost tip of South America is part of what country?

Technology Skill

Using a Database

An electronic **database** is a collection of data—names, facts, and statistics—that is stored in a file on the computer. Databases are useful for organizing large amounts of information. The information in a database can be sorted and presented in different ways.

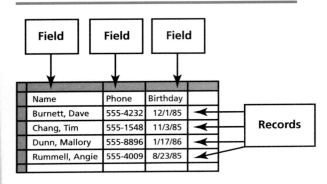

▲ Using a database can help organize statistics, names and addresses, and even baseball card collections.

Learning the Skill

The database organizes information in categories called *fields*. For example, a database of your friends might include the fields **Name, Address, Telephone Number,** and **Birthday.** Each person you enter into the database is called a *record*. After entering the records, you might create a list sorted by birthdays or use the records to create address labels. Together, all the records make up the database.

Geographers use databases for many purposes. They often have large amounts of data that they need to analyze. For example, a geographer might want to compare and contrast certain information about the Andean countries. A database would be a good place to sort and compare information about the land, economies, and people of these countries.

Practicing the Skill

Follow these steps to build a database about the Andean countries.

1. Determine what facts you want to include in your database and research to collect that information.
2. Follow the instructions in the database that you are using to set up fields. Then enter each item of data in its assigned field.
3. Determine how you want to organize the facts in the database—chronologically by the date, alphabetically, or by some other category.
4. Follow the instructions in your computer program to sort the information.
5. Check that all the information in your database is correct. If necessary, add, delete, or change information or fields.

Applying the Skill

Research and build a database that organizes information about an Andean country of your choice. Explain why the database is organized the way it is.

Section 1 — Colombia

Terms to Know
cordillera
llanos
cash crop
mestizo
republic
campesino

Main Idea
Although it has many resources, Colombia faces political and economic turmoil.

✓ Place Colombia, in the northwestern corner of South America, has central highlands, two coastal lowlands, and an interior plain with few people.

✓ Economics Colombia is rich in coal, hydroelectric power, gold, and emeralds.

✓ Government The government of Colombia is struggling to combat the power of drug dealers who make huge fortunes from selling cocaine, which comes from the coca plant.

✓ History Civil war has led many people to leave the countryside for Colombia's cities.

Section 2 — Peru and Ecuador

Terms to Know
altiplano
navigable
foothill
subsistence farm
empire

Main Idea
Peru and Ecuador share similar landscapes, climates, and history.

✓ Place Peru and Ecuador have narrow coastal plains with mild temperatures; the high Andes in the center; and hot, rainy lowlands in the interior.

✓ Economics Peru's main exports are copper and fish. Many people farm. Ecuador's economy is focused on agriculture.

✓ Culture Most people in Peru and Ecuador are Native Americans or mestizos.

▲ Dancers in Peru

Section 3 — Bolivia and Chile

Terms to Know
landlocked
sodium nitrate

Main Idea
Bolivia and Chile share the Andes, but their economies and people are different.

✓ Human/Environment Interaction Bolivia is a poor country consisting mainly of the towering Andes and a high plateau that is difficult to farm.

✓ Region In addition to its mountains, Chile has five different regions. Most people live in the fertile Central Valley.

✓ Economics Chile has a diverse economy that includes mining—especially copper and sodium nitrate—farming, and manufacturing.

Using Key Terms

Match the terms in Part A with their definitions in Part B.

A.

1. cordillera
2. campesino
3. cash crop
4. altiplano
5. navigable
6. foothill
7. empire
8. sodium nitrate
9. landlocked
10. mestizo

B.

a. person of mixed Native American and European ancestry
b. crop grown to be sold, often for export
c. mineral used in making fertilizer
d. group of lands under one ruler
e. group of mountain ranges that run side by side
f. a body of water wide and deep enough for ships to pass through
g. a country that has no land on a sea or an ocean
h. low hills at the base of a mountain range
i. farmer in Colombia
j. high plateau region of the Andes

 ## Reviewing the Main Ideas

Section 1 Colombia

11. **Location** Colombia borders which two major bodies of water?
12. **Place** List four of Colombia's natural resources.
13. **Culture** What is the heritage of most of Colombia's people?

Section 2 Peru and Ecuador

14. **Place** What is the highest navigable lake in the world?
15. **History** What ancient Native American civilization of the Andean countries lived in Peru?
16. **Place** Who owns the Galápagos Islands?

Section 3 Bolivia and Chile

17. **Location** Why is it difficult for visitors to breathe when visiting the altiplano?
18. **Economics** What types of projects does Bolivia have planned that it hopes will expand its economy?
19. **Government** What type of government does Chile have?

 NATIONAL GEOGRAPHIC   **The Andean Countries**

Place Location Activity

On a separate sheet of paper, match the letters on the map with the numbered places listed below.

1. Colombia
2. Peru
3. Chile
4. Andes
5. Lake Titicaca
6. Quito
7. Bogotá
8. Strait of Magellan
9. Lima
10. Bolivia

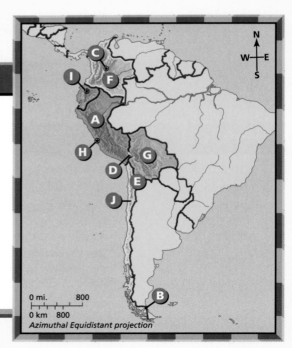

0 mi.   800
0 km   800
Azimuthal Equidistant projection

Self-Check Quiz Visit the *Geography: The World and Its People* Web site at gwip.glencoe.com and click on **Chapter 9—Self-Check Quizzes** to prepare for the Chapter Test.

Critical Thinking

20. **Making Inferences** Why are Native Americans who live in the Andean highlands more likely to follow a traditional way of life than those who live in the cities?

21. **Analyzing Cause and Effect** On a diagram like the one below, list factors that have led to political violence during Colombia's history.

```
┌──────────┐
│ ———————— │──────▶  Political Violence
│ ———————— │         in Colombia
│ ———————— │
└──────────┘
```

GeoJournal Activity

22. **Writing About Cities** Research one of the major cities of the Andean countries. Prepare a report that includes a map, fact bank, pictures, and a T-shirt design that shows a famous landmark of the city.

Mental Mapping Activity

23. **Focusing on the Region** Draw a simple outline map of South America, then label the following:

- Pacific Ocean
- Andes
- Atacama Desert
- Strait of Magellan
- Chile
- Peru
- Colombia
- Galápagos Islands
- Lake Titicaca
- Ecuador

Technology Skills Activity

24. **Building a Database** Create a fact sheet about the Andean countries by building a database. Create fields for such categories as physical features, natural resources, capital cities, population, and type of government. When you have entered data for each field, print your fact sheet.

Standardized Test Practice

Directions: Read the paragraphs below, then answer the question that follows.

Simón Bolívar, an aristocrat from Venezuela, led many of South America's lands to independence. He believed in equality and saw liberty as "the only object worth a man's life." Called "the Liberator," Bolívar devoted his life to freedom for Latin Americans.

Bolívar was the son of a rich family in New Granada, or what is today Colombia, Venezuela, Panama, and Ecuador. In 1805 he went to Europe. There, he learned about the French Revolution and its ideas of democracy. He returned home, vowing to free his people from Spanish rule. In 1810 Bolívar started a revolt against the Spaniards in Venezuela. Spanish officials soon crushed the movement, but Bolívar escaped and trained an army. During the next 20 years, Bolívar and his forces won freedom for the present-day countries of Venezuela, Colombia, Panama, Bolivia, and Ecuador.

1. **What is the main idea of the paragraphs above?**

   A Bolívar was the son of a rich family.

   B Bolívar traveled to Europe and learned about democracy.

   C Simón Bolívar was called "the Liberator."

   D Bolívar devoted his life to freedom for Latin Americans.

Test-Taking Tip: This question asks you to find the main idea, or to make a generalization. Most of the answer choices provide specific details, not a general idea. Which of the answers is more of a general statement?

Unit 4

Woman in Hungary
creating folk art

Ancient ruins in
Delphi, Greece

Europe

Y ou have learned about the Americas. Now let us spin the globe and travel to the Eastern Hemisphere. The first region you will learn about is Europe—relatively small as continents go, but rich in history and culture. Like the United States, most nations in Europe are industrialized and have high standards of living. Unlike the United States, however, the people of Europe do not share a common language or government.

▲ The Louvre museum, Paris, France

NGS ONLINE
www.nationalgeographic.com/education

275

Focus on:

Europe

BOTH A CONTINENT and a region, Europe has a wide range of cultures—and a history of conflict among its people. Recently, connections in trade, communication, and transportation have helped to create greater unity among European nations.

The Land

Jutting westward from Asia, Europe is a great peninsula that breaks into smaller peninsulas and is bordered by several large islands. Europe's long, jagged coastline is washed by many bodies of water, including the Arctic and Atlantic Oceans, and the North, Baltic, and Mediterranean Seas. Deep bays and well-protected inlets shelter fine harbors.

Mountains sweep across much of the continent. Those in the British Isles and large parts of northern Europe are low and rounded. Higher and more rugged are the Pyrenees, between France and Spain, and the Carpathians, in eastern Europe. The snow-capped Alps are Europe's highest mountains, towering over the central and southern parts of the continent.

Curving around these mountain ranges are broad, fertile plains. In the north, the North European Plain stretches from the British Isles to the Russian border. Cities, towns, and farms dot the gently rolling landscape.

For centuries, Europe's rivers have provided links between coastal ports and inland population centers. In western Europe, the Rhine flows northwest from the Alps until it empties into the North Sea. The Danube winds through eastern Europe, bound for the Black Sea.

The Climate

Despite its northern location, Europe enjoys a relatively mild climate. The secret lies in the region's closeness to the Atlantic. An ocean current known as the North Atlantic Current brings warm waters and winds to bathe Europe's western shores. As a result, northwestern Europe enjoys mild temperatures all year, along with plentiful rainfall. Farther south, countries along the Mediterranean Sea have hot, dry summers and mild winters. The region's northernmost countries have longer, colder winters than their southern neighbors. Winters are also cold in Europe's interior, which lies far from the influence of the Atlantic.

The vegetation varies from one climate zone to another. In Scandinavia's far north, you would find mostly mosses and small shrubs blanketing a tundra-like landscape. In northwestern and eastern Europe, grasslands and forests cover the rolling land. Farther south, drought-resistant shrubs and small trees cover rugged hills.

UNIT

4

Village at the foot of the Alps, Switzerland ▼

◄ Fisherman mending nets in Malta

277

The Economy

An abundance of key natural resources, waterways, and ports has helped make Europe a global economic power. Agriculture, manufacturing, and service industries dominate the region's economies.

Some of the most productive farmland in the world can be found on the European continent. From the fertile black soil, farmers gather bountiful harvests of grains, fruits, and vegetables. Cattle and sheep graze through lush European pastures.

Vast reserves of oil and natural gas lie offshore. Rich deposits of iron ore, coal, and other minerals have provided the raw materials for heavy industry and manufacturing. Europe was the birthplace of the Industrial Revolution, which transformed the region from an agricultural society into an industrial one. Today, countries such as France, Germany, Italy, Poland, and the United Kingdom rank among the world's top manufacturing centers.

The People

After Asia, Europe is the most densely populated continent on the earth. In some European countries, such as Sweden, most people belong to the same ethnic group. The populations of other countries, however, are made up of several ethnic groups. Some ethnic groups live together peacefully. Other groups often face tension and conflict.

Europeans enjoy a rich cultural heritage that stretches back thousands of years. Walk through the heart of any large European city and you might see ancient Roman ruins, Gothic cathedrals built in the Middle Ages, and sculptures created by Renaissance masters.

Throughout their long history, Europeans have explored and settled other lands. They have spread their culture to every part of the globe. Competition among European nations in the past century led to two world wars and a bitter division into Communist and non-Communist areas. Setting aside their differences, many European nations have recently joined the European Union. They are moving into the new century as a united economic force.

Exploring the Region

1. **What bodies of water border Europe?**
2. **Why is Europe's climate relatively mild?**
3. **What has helped make Europe a global economic power?**
4. **How did European culture spread to other parts of the world?**

◀ **Cargo lining the docks of Rotterdam, a port city in the Netherlands**

Children by road signs in Ireland ▼

the
BALLyvaughan
inn
RESTAURANT

CARR CHALADH
51 KILLIMER
CAR FERRY

AILLWEE CAVE
OPEN TO VISITORS

T LIOS DÚIN BHEARNA 10
69 LISDOONVARNA

GREGANS
3M. CASTLE HOTEL

BÓTHAR COIS FAIRRGE
COAST ROAD

THAIR NA TRA
LIOS DÚIN BHEARNA
LISDOONVARNA 19
VIA COAST ROAD

CEANN BÓIRNE
BLACK HEAD 6 L54

IONAD IASCAIREACHTA km
SHORE ANGLING
CENTRE 16

MONK'S 500
ON T R Mtrs.

REGIONAL ATLAS

Europe

Physical

Mountain peak

0 mi. 500
0 km 500
Lambert Azimuthal Equal-Area projection

ARCTIC CIRCLE

ICELAND

60°N

Faroe Is.

Shetland Is.

Orkney Is.

MERIDIAN OF GREENWICH (LONDON)

Norwegian Sea

N O R W A Y

S C A N D I N A V I A

S W E D E N

FINLAND

Baltic Sea

ESTONIA

LATVIA

LITHUANIA

RUSSIA

RUSSIA

ATLANTIC OCEAN

50°N

UNITED KINGDOM

IRELAND

British Isles

North Sea

Thames R.

Jutland

DENMARK

NETH.

GERMANY

Elbe R.

POLAND

BELARUS

N O R T H   E U R O P E A N   P L A I N

BELG.

LUX.

Loire R.

Seine R.

FRANCE

LIECH.

SWITZ.

CZECH REP.

Carpathian Mountains

SLOVAKIA

AUSTRIA

UKRAINE

Dnieper R.

MOLDOVA

Bay of Biscay

A L P S

Mt. Blanc 15,771 ft. (4,807 m)

Matterhorn 14,690 ft. (4,478 m)

SLOV.

Hungarian Plain

HUNGARY

ROMANIA

40°N

ANDORRA

Pyrenees

Ebro R.

MONACO

SAN MARINO

Corsica

CROATIA

BOSN. & HERZG.

YUG.

Danube R.

Black Sea

PORTUGAL

Douro R.

Tagus R.

S P A I N

I B E R I A N   P E N I N S U L A

Apennines

Adriatic Sea

Balkan Peninsula

BULGARIA

MACED.

Sardinia

ITALY

ALBANIA

Strait of Gibraltar

Sicily

GREECE

Aegean Sea

M e d i t e r r a n e a n   S e a

MALTA

Crete

CYPRUS

20°W 10°W 0° 10°E 20°E 30°E

26,247 ft. — 8,000 m
0 mi. 500
0 km 500
19,685 ft. — 6,000 m
PYRENEES   ALPS
13,123 ft. — 4,000 m
6,562 ft. — 2,000 m
— LISBON   Sea level   WARSAW —

Political

ARCTIC CIRCLE

Jan Mayen
Nor.

MERIDIAN OF GREENWICH
(LONDON)

20°W   10°W   0°   10°E   20°E   30°E   40°E   50°E

60°N

• Reykjavík
ICELAND

Faroe Islands
Den.

Rockall ·
U.K.

Norwegian Sea

N O R W A Y

S W E D E N

Gulf of Bothnia

F I N L A N D

• Helsinki

R U S S I A

Oslo ⊛

Stockholm ⊛

⊛ Tallinn
ESTONIA

**ATLANTIC
OCEAN**

SCOTLAND

Skagerrak

*North
Sea*

Baltic Sea

LATVIA
⊛ Rīga

N. IRE.
UNITED

IRELAND
Dublin ⊛

*Irish
Sea*

DENMARK
Copenhagen ⊛

RUSSIA

LITHUANIA
⊛ Vilnius

KINGDOM
WALES

*Celtic
Sea*

ENGLAND

London ⊛

NETH.
⊛ Amsterdam

Elbe R.

Berlin ⊛

POLAND
Warsaw ⊛

• Minsk

BELARUS

⊛ National capital

Brussels ⊛
BELG.
LUX.

GERMANY

Oder R.

Prague ⊛
CZECH REP.

Kiev ⊛

Dnieper R.

0 mi.   400
0 km   400
Lambert Azimuthal
Equal-Area projection

Seine R.

• Paris

Rhine R.

SLOVAKIA

Vienna ⊛ ⊛ Bratislava

Dniester R.

U K R A I N E

*Bay of
Biscay*

FRANCE

⊛ Bern
SWITZ.

LIECH.

AUSTRIA

⊛ Budapest

MOLDOVA
⊛ Chişinău

*Sea of
Azov*

Rhône R.

SLOV.
Ljubljana ⊛

HUNGARY

⊛ Zagreb

ROMANIA

40°N

PORTUGAL

ANDORRA

MONACO

*Corsica
Fr.*

**SAN
MARINO**

CROATIA

**BOSN. &
HERZG.**

VOJVODINA

⊛ Belgrade

Bucharest •

Danube R.

Black Sea

Lisbon •

⊛ Madrid

ITALY

*Adriatic
Sea*

Sarajevo ⊛

S
E
R
B
I
A

YUG.

Europe–Asia
boundary

SPAIN

VATICAN CITY
(Within Rome)

• Rome

MONTENEGRO

KOSOVO

BULGARIA
⊛ Sofia

Bosporus

*Strait of
Gibraltar*

• GIBRALTAR
U.K.

Balearic Is.
Sp.

*Sardinia
It.*

*Tyrrhenian
Sea*

Tirana ⊛
ALBANIA

⊛ Skopje
MACED.

TURKEY

M e d i t e r

Sicily

*Ionian
Sea*

G
R
E
E
C
E

*Aegean
Sea*

Dardanelles

Valletta ⊛
MALTA

*r a n e
a n*

⊛ Athens

Nicosia ⊛

Crete

CYPRUS

Sea

MAP STUDY

1 What body of water lies between
Scandinavia and Poland?

2 What is the capital of the United
Kingdom?

REGIONAL ATLAS

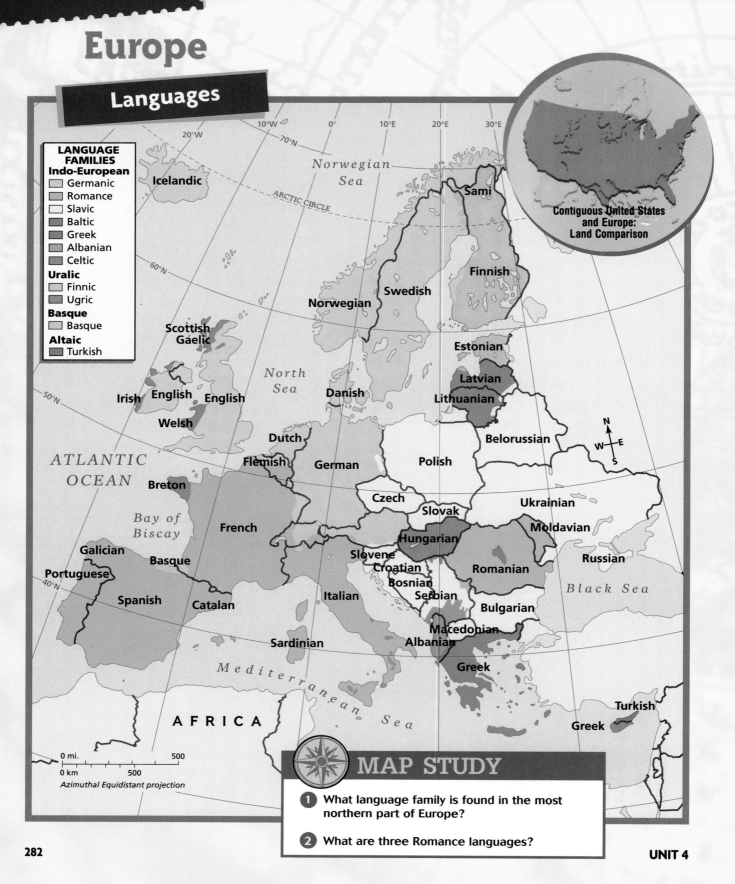

Europe

Languages

LANGUAGE FAMILIES
Indo-European
- Germanic
- Romance
- Slavic
- Baltic
- Greek
- Albanian
- Celtic

Uralic
- Finnic
- Ugric

Basque
- Basque

Altaic
- Turkish

Icelandic

Norwegian Sea

ARCTIC CIRCLE

Sami

Finnish

Swedish

Norwegian

Estonian

Latvian

Lithuanian

North Sea

Scottish Gaelic

Danish

Belorussian

Irish English English

Welsh

Dutch

Polish

ATLANTIC OCEAN

Flemish

German

Czech

Ukrainian

Breton

Slovak

Bay of Biscay

French

Hungarian

Moldavian

Galician

Slovene

Russian

Basque

Croatian

Romanian

Portuguese

Bosnian

Spanish

Italian

Serbian

Catalan

Bulgarian

Black Sea

Macedonian

Sardinian

Albanian

Greek

Mediterranean Sea

AFRICA

Turkish

Greek

Contiguous United States and Europe: Land Comparison

0 mi. 500
0 km 500
Azimuthal Equidistant projection

20°W   10°W   0°   10°E   20°E   30°E
70°N
60°N
50°N
40°N

MAP STUDY

1 What language family is found in the most northern part of Europe?

2 What are three Romance languages?

UNIT 4

Geo Extremes

① HIGHEST POINT
Mont Blanc (France and Italy)
15,771 ft. (4,807 m) high

② LOWEST POINT
Nieuwerkerk aan
den IJssel (Netherlands)
22 ft. (7 m) below sea level

③ LONGEST RIVER
Danube (central Europe)
1,776 mi. (2,858 km) long

④ LARGEST LAKE
Lake Vänern (Sweden)
2,156 sq. mi. (5,584 sq. km)

⑤ HIGHEST WATERFALL
Mardalsfossen,
Southern (Norway)
2,149 ft. (655 m) high

⑥ LARGEST ISLAND
Great Britain
84,210 sq. mi.
(218,103 sq. km)

COMPARING POPULATION:
United States and Selected Countries of Europe

UNITED STATES

GERMANY

UKRAINE

SPAIN

BELGIUM

👤 = 25,000,000

Source: *Population Reference Bureau,* 2000.

RELIGIONS:
Selected Countries of Europe

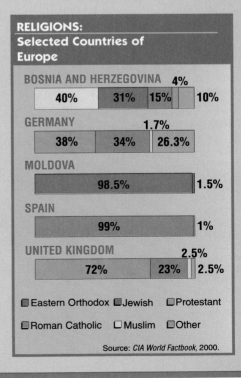

BOSNIA AND HERZEGOVINA			4%	
40%	31%	15%		10%
GERMANY		1.7%		
38%	34%	26.3%		
MOLDOVA				
98.5%			1.5%	
SPAIN				
99%			1%	
UNITED KINGDOM		2.5%		
72%		23%	2.5%	

☐ Eastern Orthodox ☐ Jewish ☐ Protestant
☐ Roman Catholic ☐ Muslim ☐ Other

Source: *CIA World Factbook,* 2000.

GRAPHIC STUDY

① What two countries share the highest point in Europe?

② Roughly what is the population of Germany? What percentage of the population is Protestant?

REGIONAL ATLAS

Country Profiles

ALBANIA

POPULATION:
3,460,000
312 per sq. mi.
120 per sq. km

LANGUAGE:
Albanian

MAJOR EXPORT:
Asphalt

MAJOR IMPORT:
Machinery

CAPITAL:
Tirana

LANDMASS:
11,100 sq. mi.
28,748 sq. km

ANDORRA

POPULATION:
66,000
377 per sq. mi.
146 per sq. km

LANGUAGES:
Catalan, French, Spanish

MAJOR EXPORT:
Electricity

MAJOR IMPORT:
Manufactured Goods

CAPITAL:
Andorra la Vella

LANDMASS:
175 sq. mi.
453 sq. km

AUSTRIA

POPULATION:
8,087,000
250 per sq. mi.
96 per sq. km

LANGUAGE:
German

MAJOR EXPORT:
Machinery

MAJOR IMPORT:
Petroleum

CAPITAL:
Vienna

LANDMASS:
32,377 sq. mi.
83,856 sq. km

BELARUS

POPULATION:
10,167,000
127 per sq. mi.
49 per sq. km

LANGUAGES:
Belarussian, Russian

MAJOR EXPORT:
Machinery

MAJOR IMPORT:
Fuels

CAPITAL:
Minsk

LANDMASS:
80,154 sq. mi.
207,598 sq. km

BELGIUM

POPULATION:
10,225,000
868 per sq. mi.
335 per sq. km

LANGUAGES:
Flemish, French

MAJOR EXPORTS:
Iron and Steel

MAJOR IMPORT:
Fuels

CAPITAL:
Brussels

LANDMASS:
11,783 sq. mi.
30,518 sq. km

BOSNIA and HERZEGOVINA

POPULATION:
3,839,000
194 per sq. mi.
75 per sq. km

LANGUAGE:
Serbo-Croatian

MAJOR EXPORT:
N/A

MAJOR IMPORT:
N/A

CAPITAL:
Sarajevo

LANDMASS:
19,741 sq. mi.
51,129 sq. km

BULGARIA

POPULATION:
8,188,000
191 per sq. mi.
74 per sq. km

LANGUAGE:
Bulgarian

MAJOR EXPORT:
Machinery

MAJOR IMPORT:
Fuels

CAPITAL:
Sofia

LANDMASS:
42,823 sq. mi.
110,912 sq. km

CROATIA

POPULATION:
4,600,000
211 per sq. mi.
81 per sq. km

LANGUAGE:
Serbo-Croatian

MAJOR EXPORT:
Transport Equipment

MAJOR IMPORT:
Machinery

CAPITAL:
Zagreb

LANDMASS:
21,829 sq. mi.
56,538 sq. km

CYPRUS

POPULATION:
875,000
384 per sq. mi.
148 per sq. km

LANGUAGES:
Greek, Turkish

MAJOR EXPORT:
Citrus Fruits

MAJOR IMPORT:
Manufactured Goods

CAPITAL:
Nicosia

LANDMASS:
2,277 sq. mi.
5,897 sq. km

CZECH REPUBLIC

POPULATION:
10,284,000
338 per sq. mi.
130 per sq. km

LANGUAGES:
Czech, Slovak

MAJOR EXPORT:
Machinery

MAJOR IMPORT:
Crude Oil

CAPITAL:
Prague

LANDMASS:
30,450 sq. mi.
78,864 sq. km

DENMARK

POPULATION:
5,325,000
320 per sq. mi.
124 per sq. km

LANGUAGE:
Danish

MAJOR EXPORT:
Machinery

MAJOR IMPORT:
Machinery

CAPITAL:
Copenhagen

LANDMASS:
16,638 sq. mi.
43,092 sq. km

ESTONIA

POPULATION:
1,441,000
83 per sq. mi.
32 per sq. km

LANGUAGE:
Estonian

MAJOR EXPORT:
Textiles

MAJOR IMPORT:
Machinery

CAPITAL:
Tallinn

LANDMASS:
17,413 sq. mi.
45,099 sq. km

Countries and flags not drawn to scale

FINLAND

POPULATION:
5,170,000
40 per sq. mi.
15 per sq. km

LANGUAGES:
Finnish, Swedish

MAJOR EXPORT:
Paper

MAJOR IMPORT:
Foods

CAPITAL:
Helsinki

LANDMASS:
130,558 sq. mi.
338,145 sq. km

Helsinki

FRANCE

POPULATION:
59,067,000
281 per sq. mi.
109 per sq. km

LANGUAGE:
French

MAJOR EXPORT:
Machinery

MAJOR IMPORT:
Crude Oil

CAPITAL:
Paris

LANDMASS:
210,026 sq. mi.
543,965 sq. km

Paris

GERMANY

POPULATION:
81,950,000
594 per sq. mi.
230 per sq. km

LANGUAGE:
German

MAJOR EXPORT:
Machinery

MAJOR IMPORT:
Machinery

CAPITAL:
Berlin

LANDMASS:
137,857 sq. mi.
357,046 sq. km

Berlin

GREECE

POPULATION:
10,539,000
207 per sq. mi.
80 per sq. km

LANGUAGE:
Greek

MAJOR EXPORT:
Foods

MAJOR IMPORT:
Machinery

CAPITAL:
Athens

LANDMASS:
50,962 sq. mi.
131,990 sq. km

Athens

HUNGARY

POPULATION:
10,076,000
281 per sq. mi.
108 per sq. km

LANGUAGE:
Hungarian

MAJOR EXPORT:
Machinery

MAJOR IMPORT:
Crude Oil

CAPITAL:
Budapest

LANDMASS:
35,919 sq. mi.
93,030 sq. km

Budapest

ICELAND

POPULATION:
277,000
7 per sq. mi.
3 per sq. km

LANGUAGE:
Icelandic

MAJOR EXPORT:
Fish

MAJOR IMPORT:
Machinery

CAPITAL:
Reykjavik

LANDMASS:
39,769 sq. mi.
103,001 sq. km

Reykjavik

IRELAND

POPULATION:
3,734,000
138 per sq. mi.
53 per sq. km

LANGUAGES:
English, Irish Gaelic

MAJOR EXPORT:
Chemical
Products

MAJOR IMPORT:
Foods

CAPITAL:
Dublin

LANDMASS:
27,137 sq. mi.
70,284 sq. km

Dublin

ITALY

POPULATION:
57,717,000
496 per sq. mi.
192 per sq. km

LANGUAGE:
Italian

MAJOR EXPORT:
Metals

MAJOR IMPORT:
Machinery

CAPITAL:
Rome

LANDMASS:
116,324 sq. mi.
301,277 sq. km

Rome

LATVIA

POPULATION:
2,430,000
97 per sq. mi.
38 per sq. km

LANGUAGES:
Latvian, Russian

MAJOR EXPORT:
Wood

MAJOR IMPORT:
Fuels

CAPITAL:
Rīga

LANDMASS:
24,942 sq. mi.
64,599 sq. km

Rīga

LIECHTENSTEIN

POPULATION:
32,000
516 per sq. mi.
200 per sq. km

LANGUAGE:
German

MAJOR EXPORT:
Machinery

MAJOR IMPORT:
Machinery

CAPITAL:
Vaduz

LANDMASS:
62 sq. mi.
160 sq. km

Vaduz

LITHUANIA

POPULATION:
3,700,000
147 per sq. mi.
57 per sq. km

LANGUAGES:
Lithuanian, Polish,
Russian

MAJOR EXPORTS:
Foods and
Livestock

MAJOR IMPORT:
Minerals

CAPITAL:
Vilnius

LANDMASS:
25,174 sq. mi.
65,200 sq. km

Vilnius

LUXEMBOURG

POPULATION:
432,000
433 per sq. mi.
167 per sq. km

LANGUAGES:
Luxembourgian,
German, French

MAJOR EXPORT:
Steel Products

MAJOR IMPORT:
Minerals

CAPITAL:
Luxembourg

LANDMASS:
998 sq. mi.
2,586 sq. km

Luxembourg

Country Profiles

MACEDONIA, Former Yugoslav Republic of

POPULATION:
2,019,000
203 per sq. mi.
79 per sq. km

LANGUAGES:
Macedonian,
Albanian

MAJOR EXPORT:
Manufactured
Goods

MAJOR IMPORT:
Fuels

CAPITAL:
Skopje

LANDMASS:
9,928 sq. mi.
25,713 sq. km

MALTA

POPULATION:
380,000
3,115 per sq. mi.
1,203 per sq. km

LANGUAGES:
Maltese, English

MAJOR EXPORT:
Machinery

MAJOR IMPORT:
Foods

CAPITAL:
Valletta

LANDMASS:
122 sq. mi.
316 sq. km

MOLDOVA

POPULATION:
4,284,000
324 per sq. mi.
126 per sq. km

LANGUAGES:
Moldovan, Russian

MAJOR EXPORT:
Foods

MAJOR IMPORT:
Petroleum

CAPITAL:
Chișinău

LANDMASS:
13,217 sq. mi.
33,999 sq. km

MONACO

POPULATION:
33,000
55,000 per sq. mi.
17,368 per sq. km

LANGUAGE:
French

MAJOR EXPORT:
N/A

MAJOR IMPORT:
N/A

CAPITAL:
Monaco

LANDMASS:
0.6 sq. mi.
1.9 sq. km

NETHERLANDS

POPULATION:
15,799,000
986 per sq. mi.
381 per sq. km

LANGUAGE:
Dutch

MAJOR EXPORT:
Manufactured
Goods

MAJOR IMPORT:
Raw Materials

CAPITAL:
Amsterdam

LANDMASS:
16,023 sq. mi.
41,499 sq. km

NORWAY

POPULATION:
4,462,000
36 per sq. mi.
14 per sq. km

LANGUAGE:
Norwegian

MAJOR EXPORT:
Petroleum

MAJOR IMPORT:
Machinery

CAPITAL:
Oslo

LANDMASS:
125,182 sq. mi.
324,220 sq. km

POLAND

POPULATION:
38,674,000
320 per sq. mi.
124 per sq. km

LANGUAGE:
Polish

MAJOR EXPORT:
Manufactured
Goods

MAJOR IMPORT:
Machinery

CAPITAL:
Warsaw

LANDMASS:
120,725 sq. mi.
312,677 sq. km

PORTUGAL

POPULATION:
9,992,000
280 per sq. mi.
108 per sq. km

LANGUAGE:
Portuguese

MAJOR EXPORT:
Clothing

MAJOR IMPORT:
Machinery

CAPITAL:
Lisbon

LANDMASS:
35,672 sq. mi.
92,389 sq. km

ROMANIA

POPULATION:
22,460,000
245 per sq. mi.
95 per sq. km

LANGUAGES:
Romanian, Hungarian,
German

MAJOR EXPORT:
Textiles

MAJOR IMPORT:
Fuels

CAPITAL:
Bucharest

LANDMASS:
91,699 sq. mi.
237,499 sq. km

SAN MARINO

POPULATION:
26,000
1,083 per sq. mi.
426 per sq. km

LANGUAGE:
Italian

MAJOR EXPORT:
Building Stone

MAJOR IMPORT:
Manufactured
Goods

CAPITAL:
San Marino

LANDMASS:
24 sq. mi.
61 sq. km

SLOVAKIA

POPULATION:
5,401,000
285 per sq. mi.
110 per sq. km

LANGUAGES:
Slovak, Hungarian

MAJOR EXPORT:
Transport
Equipment

MAJOR IMPORT:
Machinery

CAPITAL:
Bratislava

LANDMASS:
18,921 sq. mi.
49,006 sq. km

SLOVENIA

POPULATION:
1,978,000
253 per sq. mi.
98 per sq. km

LANGUAGES:
Slovene,
Serbo-Croatian

MAJOR EXPORT:
Transport
Equipment

MAJOR IMPORT:
Machinery

CAPITAL:
Ljubljana

LANDMASS:
7,819 sq. mi.
20,251 sq. km

Countries and flags not drawn to scale

SPAIN

POPULATION:
39,418,000
202 per sq. mi.
78 per sq. km

LANGUAGES:
Spanish, Catalan,
Galician, Basque

MAJOR EXPORTS:
Cars and Trucks

CAPITAL:
Madrid

MAJOR IMPORT:
Machinery

LANDMASS:
194,897 sq. mi.
504,782 sq. km

SWEDEN

POPULATION:
8,856,000
51 per sq. mi.
20 per sq. km

LANGUAGE:
Swedish

MAJOR EXPORT:
Paper Products

CAPITAL:
Stockholm

MAJOR IMPORT:
Crude Oil

LANDMASS:
173,732 sq. mi.
449,964 sq. km

SWITZERLAND

POPULATION:
7,119,000
447 per sq. mi.
172 per sq. km

LANGUAGES:
German, French,
Italian, Romansch

MAJOR EXPORT:
Precision
Instruments

CAPITAL:
Bern

MAJOR IMPORT:
Machinery

LANDMASS:
15,941 sq. mi.
41,288 sq. km

UKRAINE

POPULATION:
49,910,000
214 per sq. mi.
83 per sq. km

LANGUAGES:
Ukrainian, Russian

MAJOR EXPORT:
Metals

CAPITAL:
Kiev

MAJOR IMPORT:
Machinery

LANDMASS:
233,206 sq. mi.
604,001 sq. km

UNITED KINGDOM

POPULATION:
59,364,000
630 per sq. mi.
243 per sq. km

LANGUAGES:
English, Welsh,
Scottish Gaelic

MAJOR EXPORT:
Manufactured
Goods

CAPITAL:
London

MAJOR IMPORT:
Foods

LANDMASS:
94,248 sq. mi.
244,101 sq. km

VATICAN CITY

POPULATION:
1,000

LANGUAGES:
Italian, Latin

MAJOR EXPORT:
N/A

CAPITAL:
N/A

MAJOR IMPORT:
N/A

LANDMASS:
0.2 sq. mi.
0.4 sq. km

YUGOSLAVIA
(Serbia and Montenegro)

POPULATION:
10,646,000
270 per sq. mi.
104 per sq. km

LANGUAGES:
Serbo-Croatian,
Albanian

MAJOR EXPORT:
Manufactured
Goods

CAPITAL:
Belgrade

MAJOR IMPORT:
Machinery

LANDMASS:
39,450 sq. mi.
102,173 sq. km

Questions From Buzz Bee!

The following questions are taken from National Geographic GeoBees. Use your textbook, the Internet, and other library resources to find the answers.

1 Which European capital city located on the Seine River is known as the City of Light?

2 Which country leased Hong Kong from China for 99 years?

3 Slovakia and which other present-day central European country became independent in 1993?

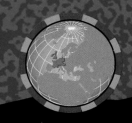

The World and Its People
NATIONAL GEOGRAPHIC

To learn more about the people and places of Western Europe, view **The World and Its People Chapter 10** video.

Geography Online

Chapter Overview Visit the **Geography: The World and Its People** Web site at gwip.glencoe.com and click on **Chapter 10— Chapter Overviews** to preview information about Western Europe.

The United Kingdom

Guide to Reading

Main Idea

The United Kingdom is a major industrial country that once ruled a vast global empire.

Terms to Know

- moor
- loch
- currency
- parliamentary democracy
- constitutional monarchy
- devolution

Places to Locate

- United Kingdom
- England
- Wales
- Scotland
- Great Britain
- Northern Ireland
- Ireland
- London
- Thames River

Reading Strategy

Create a diagram like this one, filling in the names of the four regions that make up the United Kingdom and one fact about each.

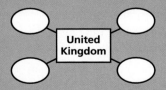

United Kingdom

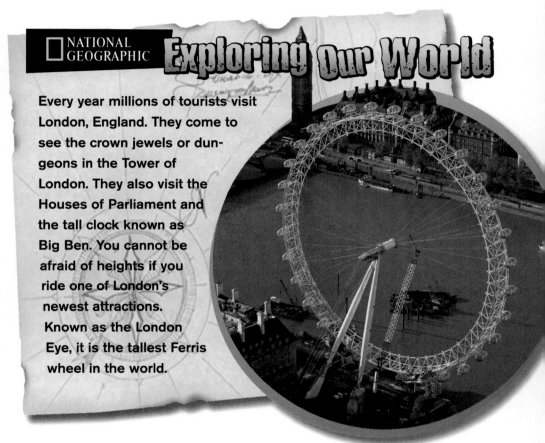

NATIONAL GEOGRAPHIC Exploring Our World

Every year millions of tourists visit London, England. They come to see the crown jewels or dungeons in the Tower of London. They also visit the Houses of Parliament and the tall clock known as Big Ben. You cannot be afraid of heights if you ride one of London's newest attractions. Known as the London Eye, it is the tallest Ferris wheel in the world.

About the size of Oregon, the **United Kingdom** lies just northwest of the continent of Europe. Four regions make up this country. **England, Wales,** and **Scotland** are located on the island of **Great Britain.** The United Kingdom's fourth region—**Northern Ireland**—shares a different island with the Republic of **Ireland.**

The Land and Climate

Find the English Channel on the map on page 291. If you entered Great Britain from this body of water, you would be standing on rolling plains covered by a patchwork of fertile fields and meadows. Soon, however, vast urban areas—such as **London** and Birmingham—would come into view. This is a reminder that the south and east of Great Britain contain the most crowded areas in the United Kingdom.

As you travel west and north—in northern England, Wales, and Scotland—highland areas come into view. Here you see rugged hills and low mountain ranges. You also cross **moors**—treeless, windy highland areas that have damp ground. Water is never far away. Northwest England holds the beautiful blue waters of the Lake District. Farther north, in Scotland, narrow bays called **lochs** cut into the highland coasts and reach far inland. A lowland area in central Scotland contains two large cities—Glasgow and Edinburgh—and most of Scotland's people, farmland, and industry. Across the Irish Sea is Northern Ireland, with its landscape of gentle mountains, valleys, and fertile lowlands. Belfast is the region's major city and port.

NATIONAL GEOGRAPHIC

Western Europe: Political

National capital ⊛
Major city •

0 mi. 200
0 km 200
Azimuthal Equidistant projection

Applying Map Skills

1. What four regions make up the United Kingdom?

2. What capital in Western Europe is farthest east?

Find NGS online map resources @
www.nationalgeographic.com/maps

The climate map on page 297 shows you that the United Kingdom has a mild climate even though it lies as far north as Canada. Why is this so? The North Atlantic Current carries warm waters from the Caribbean Sea and Gulf of Mexico to Great Britain. Winds blowing over these waters warm the United Kingdom in winter and cool it in summer.

The ocean winds also bring plenty of rain, which is good for agriculture but not for tourists. If you visit the United Kingdom, take your raincoat. Clouds fill the sky more than half the days of the year.

✓ Reading Check Why is the United Kingdom's climate mild even though it lies as far north as Canada?

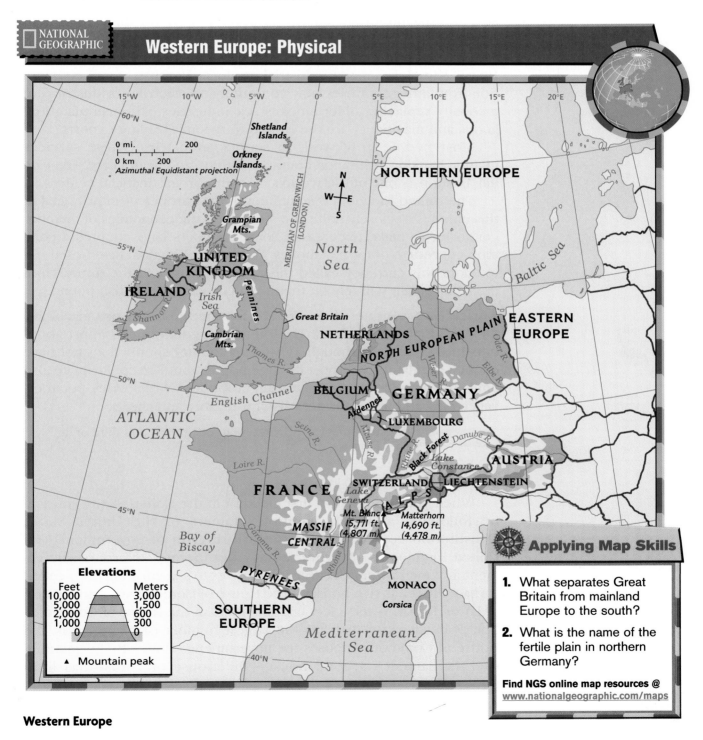

NATIONAL GEOGRAPHIC

Western Europe: Physical

Applying Map Skills

1. What separates Great Britain from mainland Europe to the south?

2. What is the name of the fertile plain in northern Germany?

Find NGS online map resources @ www.nationalgeographic.com/maps

Western Europe

The Economy

More than 200 years ago, inventors and scientists in the United Kingdom sparked the Industrial Revolution. Fuel-powered machines in factories began producing more goods than ever before. The Industrial Revolution helped make the United Kingdom the world's leading economic power during the 1800s. Although the country's economic influence declined during the 1900s, the United Kingdom is still a major industrial and trading country.

Resources and Manufacturing The United Kingdom is rich in energy resources. It pumps oil and natural gas from under the North Sea. Another important energy resource is coal. All three fuels help power the country and are exported to other countries.

Heavy machinery, ships, textiles, and automobiles were once the United Kingdom's major industrial products. Because of stiff competition from other countries, new computer and electronic industries are gradually replacing older smokestack industries. Still, manufactured goods and machinery are the United Kingdom's leading exports.

Most people now work in service industries. These services include banking, insurance, communications, and health care. The capital, London, is one of the world's most important financial centers.

The United Kingdom is a member of the European Union. A goal of this union is to closely unite and strengthen the economies of member countries. To make trade easier, some members have adopted a common currency, or form of money. The United Kingdom has chosen *not* to adopt this currency, called the euro, but it may in the future. Until then, the British will rely on their own currency, the British pound.

Agriculture Farmers here are efficient because they use machines. As a result, yields are high. The United Kingdom must import about one-third of its food, though. Why? The country simply does not have enough farmland to support its large population. The cool climate also limits the growing season. The main crops are wheat, barley, potatoes, vegetables, and fruits. Farmers also raise cattle and sheep.

✓ Reading Check **Why must the United Kingdom import one-third of its food?**

The Government

When you vote in your first election in the United States, you will be following a British tradition. Framers of the United States Constitution copied parts of the British form of government. The United Kingdom is a **parliamentary democracy,** a form of government in which voters elect representatives to a lawmaking body called Parliament. It has two houses—the House of Commons and the House of Lords. British voters elect 651 members of the House of Commons. The political party that has the largest number of members in the House of Commons chooses the government's leader, the prime minister. Parliament's second house—the House of Lords—has little power. Most members of this House are nobles who have inherited their titles. The House of Lords cannot block any laws that the House of Commons wants to pass, but it can help in revising laws.

The United Kingdom's government is also a **constitutional monarchy**, in which a queen or king is the official head of state. The monarch represents the country at public events but has little power. The prime minister and other elected officials make government decisions.

Devolution   In the late 1990s, the British government began a new policy called devolution. Devolution involves the transfer of certain powers from the central government to regional governments. New regional governments were created for Scotland, Wales, and Northern Ireland. The goals of devolution are to give the regions more power over their own affairs and to help them preserve their cultures.

√ Reading Check   **How is the prime minister chosen?**

History and Culture

More than 59 million people live in the United Kingdom. Many people in the United States and other parts of the world have ancestors from here. The British people speak English, although you can hear two older languages—Welsh and Scottish Gaelic—spoken in outlying areas of the country. Most people are Protestant Christians, although immigrants from former British colonies practice Islam and other religions.

The British people are descendants of various groups. Among early arrivals were the Celts (KEHLTS), who sailed to Great Britain and Ireland from the European mainland around 500 B.C. Their descendants live today in Scotland, Wales, Ireland, and in the southwest corner of England. The Romans invaded Great Britain in 55 B.C. In the A.D. 400s and 500s, the Anglos, Saxons, and Jutes came from mainland Europe. In 1066 the Normans from France conquered England.

In the late 1500s, England became a major European power. In 1707 England, Wales, and Scotland united to form Great Britain. When Ireland joined in 1801, the country's name became the United Kingdom. Traders, soldiers, and settlers set sail to distant lands and won control of large areas throughout the world. By the mid-1800s, the United Kingdom governed the world's largest overseas empire. The United States, Canada, India, and Australia all once belonged to it.

In the 1900s, the United Kingdom fought and was victorious in the two World Wars. During World War II, the nation's leader was Prime Minister Winston Churchill. His rousing speeches gave the British people

NATIONAL GEOGRAPHIC **On Location**

Loch Ness

Castle ruins overlook Loch Ness in Scotland.

Movement Who were among the first people to settle in Great Britain?

courage during the war's darkest days. Despite victory, the United Kingdom was weakened by both conflicts. After World War II, nearly all of the countries in the British Empire became independent nations.

Culture   The United Kingdom is a densely populated country. About 90 percent of the people live in cities and towns. With more than 7 million people, the city of London is one of Europe's most heavily populated cities.

Ancient and modern meet in London. In the heart of the city is Westminster Abbey, where Britain's kings and queens have been crowned since 1066. Farther down the **Thames** (TEHMZ) **River,** you can see the Tower of London. The crown jewels that belong to the British monarchs have been kept there since 1303. People get around London by riding colorful double-decker buses or by using the subway, which they call "the Tube."

What do the Beatles, Led Zeppelin, and Ewan MacGregor have in common? All of these performers come from the United Kingdom. For centuries, the people of the United Kingdom have left their mark on world culture. British writers include William Shakespeare, Charles Dickens, Charlotte and Emily Brontë, and Jane Austen. Visitors flock to the West End, London's famous theater district, to see the plays of Shakespeare or the modern musicals of Andrew Lloyd Webber.

✓ Reading Check   **What percentage of the United Kingdom's people live in towns and cities?**

Assessment

Defining Terms
1. **Define** moor, loch, currency, parliamentary democracy, constitutional monarchy, devolution.

Recalling Facts
2. **Place** What is the largest city in the United Kingdom?
3. **Economics** What resources does the United Kingdom take from under the North Sea?
4. **History** How did Winston Churchill encourage the people of Britain during World War II?

Critical Thinking
5. **Analyzing Information** What are three facts that support the statement "Ancient and modern meet in London"?
6. **Understanding Cause and Effect** As a result of devolution, regional governments

were created in which parts of the United Kingdom?

Graphic Organizer
7. **Organizing Information** Complete a chart like the one below by writing two facts about the United Kingdom under each category heading.

The United Kingdom		
Climate	Resources	Products
Government	History	Culture

Applying Geography Skills

8. **Analyzing Maps** Look at the physical map on page 291. Name the three mountain ranges found on the island of Great Britain.

Making Connections

ART SCIENCE LITERATURE TECHNOLOGY

Stonehenge

Stonehenge, one of the world's best-known and most puzzling ancient monuments, stands in southern England.

History of Stonehenge

The most noticeable part of Stonehenge is its huge stones set up in four circular patterns. A circular ditch and mound form a border around the site. Shallow dirt holes also circle the stones.

Stonehenge was built over a period of more than 2,000 years. The earliest construction, that of the circular ditch and mound, probably began about 3100 B.C. The outer ring of large pillars, topped with horizontal rocks, was built about 2000 B.C. An inner ring of stone pillars also supports horizontal stones.

There was no local source of stone, so workers carried it from an area that was about 20 miles (32 km) north. The stones are huge—up to 30 feet (9 m) long and 50 tons (45,359 kg) in weight. Before setting the stones in place, workers smoothed and shaped them. They carved joints into the stones so that they would fit together perfectly. Then the builders probably used levers and wooden supports to raise the blocks into position.

About 500 years later, builders added the third and fourth rings of stones. This time they used bluestone, which an earlier group of people had transported 240 miles (386 km) from the Preseli Mountains of Wales.

What Does It Mean?

Although much is known about when people built Stonehenge, experts do not agree who built it. Early theories suggested that an ancient group known as Druids or the Romans built the monument. Now archaeologists believe that the monument was completed long before either of these groups came to the area.

▲ Stonehenge

An even greater mystery is *why* Stonehenge was built. Most experts agree that Stonehenge was probably used as a place of worship. Some believe that the series of holes, stones, and archways were used as a calendar. By lining up particular holes and stones, people could note the summer and winter solstices. They could also keep track of the months. Some scientists think that early people used the site to predict solar and lunar eclipses.

Making the Connection

1. About how old is Stonehenge?
2. From where did the stones used at Stonehenge come?
3. **Sequencing Information** Describe the order in which Stonehenge was built.

The Republic of Ireland

Guide to Reading

Main Idea

Once a war-torn, agricultural country, the Republic of Ireland is now enjoying economic growth.

Terms to Know

- peat
- bog

Places to Locate

- Ireland
- Shannon River
- Dublin

Reading Strategy

Create a chart like this one and fill in at least one key fact about the Republic of Ireland in each category.

Ireland	
Land	
Climate	
Economy	
People	

NATIONAL GEOGRAPHIC **Exploring Our World**

On Ireland's Laytown Beach, hooves pound the packed sand as spectators cheer on their favorite horse. Ireland's fascination with fast horses dates back to about 500 B.C. That is when the Celts—who loved chariot racing—first reached the island. Ireland's green pastures are rich in calcium, helping horses build strong bones.

The way of life in **Ireland** (officially called the Republic of Ireland) differs from that in the United Kingdom. One key difference is in religion. More than 90 percent of Ireland's people are Catholic. Another difference is in the size of its population. Only 3.7 million people live in Ireland, compared to more than 59 million in the United Kingdom.

The Emerald Isle

Surrounded by the blue waters of the Atlantic Ocean and the Irish Sea, Ireland's lush green meadows and tree-covered hills stand out. As a result, it is called the Emerald Isle.

In western Ireland, the land rises to rocky cliffs that overlook the sea. At Ireland's center lies a wide, rolling plain covered with forests and farmland. Much of the area is rich in peat, or wet ground with decaying plants, which can be dried and used for fuel. Peat is dug from bogs, or low swampy lands. The **Shannon River** flows southwest through the plain. **Dublin,** Ireland's capital and largest city, lies in the east.

No part of Ireland is more than 70 miles (113 km) from the sea. As a result, Ireland is warmed by moist winds that blow over the North Atlantic Current. These winds, along with frequent rain and mist, keep Ireland's landscape green all year. Because the island is small and the mountains are not high, the entire island has the same climate.

√Reading Check **Why is Ireland called the Emerald Isle?**

Agriculture and Industry

If you look at the map on page 302, you will see that Ireland has few mineral resources. The country does have rich soil and pasture-land, however. In the mid-1800s, Irish farmers grew potatoes as their main food. Then disaster struck. Too much rain allowed a disease to spread among the plants. Potatoes began to rot in the fields. The Irish—with little else to eat—suffered terribly. A million or more died during what was known as the Potato Famine. An equal number moved to other countries, especially the United States. Ireland's farmers still grow potatoes, but they also grow barley, wheat, sugar beets, and turnips. They raise beef and dairy cattle as well.

Although farming is still important, manufacturing employs more people and contributes more to the country's economy. Ireland has attracted foreign companies to set up operations. At the same time,

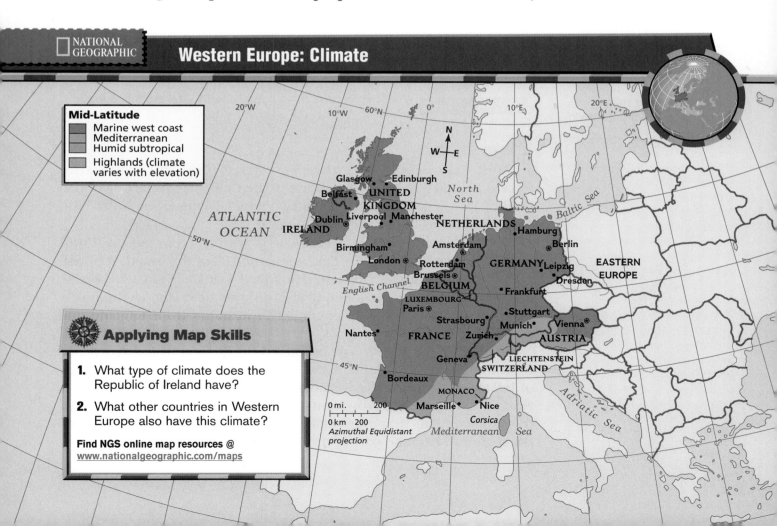

NATIONAL GEOGRAPHIC

Western Europe: Climate

Mid-Latitude
- Marine west coast
- Mediterranean
- Humid subtropical
- Highlands (climate varies with elevation)

Applying Map Skills

1. What type of climate does the Republic of Ireland have?

2. What other countries in Western Europe also have this climate?

Find NGS online map resources @
www.nationalgeographic.com/maps

Rural and Urban Life

Bright green meadows greet people who venture into Ireland's countryside (right). In Dublin, a sidewalk artist shows her own bright colors (above).

Place What religion do most Irish practice?

Ireland joined the European Union so it could market its products more widely.

With this economic boom, the Irish now work in many different manufacturing industries. Some have jobs processing foods and beverages. Others make textiles, clothing, pharmaceuticals, and computer equipment.

✓Reading Check **What two factors have contributed to Ireland's economic growth?**

The Irish

Most Irish people trace their ancestry to the Celts who settled Ireland around 500 B.C. Today a form of the Celtic language, called Irish Gaelic, and English are Ireland's two official languages. Most Irish speak English as their everyday language.

Political conflict has marked Ireland's history. From the A.D. 1100s to the early 1900s, English officials governed all of the island. The Irish people resisted British rule and demanded that their largely Catholic country become independent. British officials held land in the northern part of Ireland and encouraged Protestants from England and Scotland to move there.

In 1922 the southern part of Ireland finally won its independence from the United Kingdom. In 1949 it became known as the Republic of Ireland. The northern part remained within the United Kingdom as Northern Ireland. The Protestants who lived in Northern Ireland liked the arrangement. However, many Catholics who lived there wanted the region to become part of the Republic of Ireland.

The Northern Ireland Conflict Disagreement over Northern Ireland exploded into violence in the 1960s through the 1990s. Thousands died in ongoing attacks by Protestants or Catholics on people of the other faith. In 1998 officials of the United Kingdom and the Republic of Ireland met with leaders of the different sides in Northern Ireland. They all signed an agreement to end the political violence. The agreement gave Northern Ireland its own elected assembly. The assembly—with both Catholic and Protestant members—has the power to govern Northern Ireland. At the same time, the Republic of Ireland dropped the claim that it had the right to rule Northern Ireland and agreed to cooperate with the new assembly. The next year, the new government took office. Although disagreements continue to occur, many people hope that the new peace will last.

Irish Life and Culture In the past, most Irish lived in the countryside, but today Ireland is an urban nation. About 58 percent of Ireland's people live in cities and towns. Nearly one-third live in and around Dublin, the capital. Life often centers on the neighborhood church.

You may think that rock groups like the Cranberries, the Pogues, and U2 are Ireland's most famous cultural export. Irish music, both modern and traditional, is performed around the world. Of all the arts, however, the Irish have had the greatest influence on literature. Playwright George Bernard Shaw, poet William Butler Yeats, and novelist James Joyce are some of the country's best-known writers.

Reading Check **What are the two official languages of Ireland?**

Time to Perform!
Ann McBriarty is carrying on a long-held tradition of Irish step dancing. She has learned the four basic step dances: the reel, the light jig, the slip jig, and the hornpipe. Ann wears the costume of her dancing school now. She says, "After I reach a certain level, I may wear a dress that is different from the other dancers."

Section 2 Assessment

Defining Terms
1. **Define** peat, bog.

Recalling Facts
2. **Place** What is the major influence on Ireland's climate?
3. **Economics** Why did Ireland join the European Union?
4. **History** Who were the Celts?

Critical Thinking
5. **Understanding Cause and Effect** What event in the mid-1800s led many Irish people to move to the United States?
6. **Drawing Conclusions** What conclusions can be drawn from the fact that no part of Ireland is more than 70 miles from the sea?

Graphic Organizer
7. **Organizing Information** Place the following events and their dates in order on a time line like the one below: (a) Celts arrive in Ireland; (b) Officials sign peace agreement; (c) British officials encourage Protestants to settle in Northern Ireland; (d) Southern Ireland becomes known as the Republic of Ireland.

Applying Geography Skills

8. **Analyzing Maps** Turn to the physical map on page 291. What bodies of water surround Ireland?

France

Guide to Reading

Main Idea

A center of European culture, France is also a major agricultural and manufacturing country.

Terms to Know

- navigable
- republic

Places to Locate

- France
- Pyrenees
- Alps
- Massif Central
- Rhône River
- North European Plain
- Seine River
- Paris
- Rhine River

Reading Strategy

Draw a diagram like this. Then list three French agricultural and manufacturing products in the small ovals.

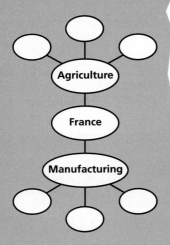

Agriculture

France

Manufacturing

NATIONAL GEOGRAPHIC Exploring Our World

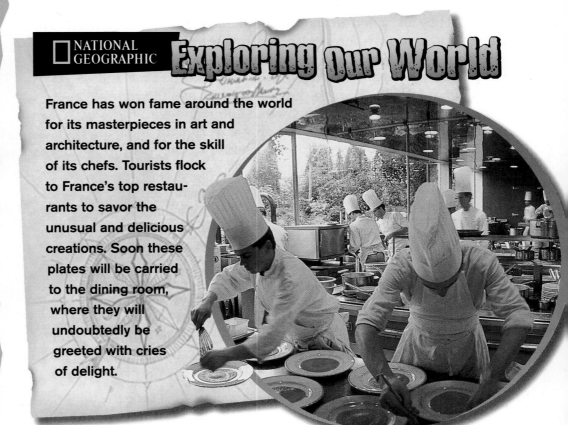

France has won fame around the world for its masterpieces in art and architecture, and for the skill of its chefs. Tourists flock to France's top restaurants to savor the unusual and delicious creations. Soon these plates will be carried to the dining room, where they will undoubtedly be greeted with cries of delight.

France and its neighbors in Western Europe rank as major economic and cultural centers of the world. For centuries, these countries struggled with one another for power and land. In the first half of the 1900s, warfare among them tore apart the European continent. Since the end of World War II, however, these countries have set aside their differences. Joined in economic partnership in the European Union, they look forward to a peaceful and prosperous twenty-first century.

France's Land and Climate

The largest country in Western Europe, **France** is slightly smaller than the state of Texas. On its southwestern and eastern borders, France touches several European countries. Look at the map on page 281 to see which seven countries are France's neighbors.

The map also shows that seas wash the other half of France's borders. If you like to swim, beaches beckon on the Mediterranean Sea, the Atlantic Ocean, and the English Channel. Since the mid-1990s, a tunnel—called the Chunnel—runs under the English Channel. People and goods now move rapidly between France and Great Britain.

Mountains and Highlands In the southwest, the **Pyrenees** (PIHR•uh•NEEZ) mountain range separates France from Spain. In the southeast, the **Alps** divide France from Italy and Switzerland. The soaring, icy peaks of the French Alps include Mont Blanc. At 15,771 feet (4,807 m), Mont Blanc is one of the highest mountains in Europe. Another highland region—the **Massif Central** (ma•SEEF sahn•TRAHL)—rises in south-central France. The long **Rhône** (ROHN) **River** flows through a wide valley that separates the Massif from the Alps.

Plains and Rivers Most of northern France is part of the vast **North European Plain.** The rich soil in this flat lowland area makes France a major agricultural country. In many French towns, you can find open-air markets displaying an abundance of fresh farm produce.

A network of rivers connects the different regions of France. Most of these rivers are navigable, or wide and deep enough to allow the passage of ships. The **Seine** (SAYN) **River** flows through **Paris,** the capital of France. Most of the country's major roads, railways, and canals meet in Paris. From the Paris area, barges transport goods by river and canal to Marseille, France's busiest Mediterranean port, and from there to other countries. Other important French rivers are the Loire (LWAHR), Rhône, Garonne, and the Rhine. The **Rhine River,** a major inland waterway of Europe, forms part of France's eastern border.

Climate Regions Most of France has a marine west coast climate with cool summers, mild winters, and plenty of rainfall. This climate is ideal for agriculture. The

Views of France

Tourists in Nice (NEES) enjoy the Mediterranean Sea (left). Wine grapes are harvested in one of France's grape-growing valleys (right).

Place What details in the photos suggest that France has a mild climate?

NATIONAL GEOGRAPHIC **On Location**

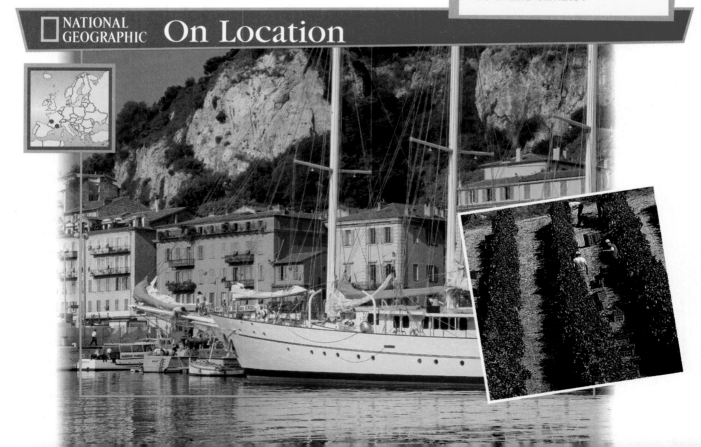

mountain areas have a highland climate. Along the Mediterranean coast, summers are hot and winters are mild. Blue skies, rocky cliffs, and lovely beaches draw many tourists to this region.

> **Reading Check** What bodies of water border France?

France's Economy

France's well-developed economy relies on agriculture and manufacturing. Most people, however, work in service industries such as banking, commerce, communications, and tourism. Tourists from all over the world flock to visit France's historic and cultural sites, such as palaces and museums.

Agriculture France produces more food than any other nation in Europe. In fact, it ranks as the second-largest food exporter in the world, after the United States. Yet only 5 percent of French workers labor on farms. Their success is a tribute to France's fertile soil and good climate.

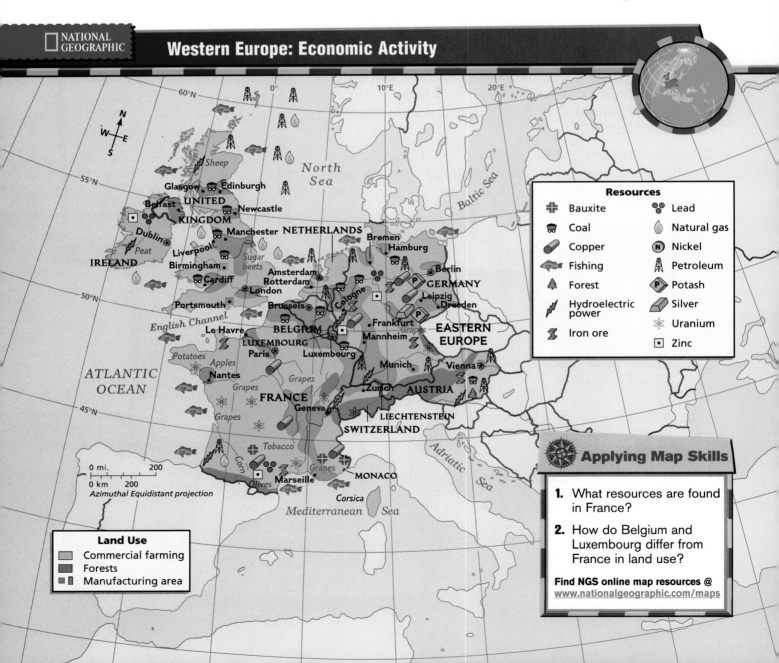

NATIONAL GEOGRAPHIC

Western Europe: Economic Activity

Resources

- ✚ Bauxite
- 🛢 Coal
- Copper
- 🐟 Fishing
- 🌲 Forest
- ⚡ Hydroelectric power
- Iron ore
- Lead
- 💧 Natural gas
- (N) Nickel
- Petroleum
- (P) Potash
- Silver
- ✳ Uranium
- ◉ Zinc

Land Use
- Commercial farming
- Forests
- Manufacturing area

0 mi. 200
0 km 200
Azimuthal Equidistant projection

Applying Map Skills

1. What resources are found in France?

2. How do Belgium and Luxembourg differ from France in land use?

Find NGS online map resources @ www.nationalgeographic.com/maps

French farmers grow grains, sugar beets, fruits, and vegetables. They also raise beef and dairy cattle. In addition, vineyards are a common sight. The grapes are used to make famous French wines. Olives are grown along the warm, dry Mediterranean coast. Most French food exports go to other members of the European Union. The United States imports mainly wine and cheese.

Mining and Manufacturing France's natural resources include bauxite, iron ore, and coal. France has small petroleum reserves and little hydroelectric power. How does the nation power its industries? About 80 percent of France's energy comes from nuclear power plants.

Workers produce a variety of manufactured goods, including steel, chemicals, textiles, airplanes, cars, and computers. France is also a leading center of commerce, with an international reputation in fashion.

☑ Reading Check From where does most of France's energy come?

The French People

"Liberté . . . Egalité . . . Fraternité" (Liberty, Equality, Fraternity)—France's national motto—describes the spirit of the French people. Although they have regional differences, the French share a strong national loyalty. Most French trace their ancestry to the Celts, Romans, and Franks of early Europe. They speak French, and about 90 percent of them are Roman Catholic.

History and Government During the 50s B.C., the area that is now France came under Roman rule. Then in the A.D. 400s, a group called the Franks arrived from central Europe and gave France its name. By A.D. 800, the Frankish ruler Charlemagne (SHAR•luh•MAYN) had created an empire that stretched across much of Western Europe. After Charlemagne's death, France emerged as a separate kingdom.

Kings ruled France for nearly 800 years and made the nation a major European power. Royalty and nobles built elegant country houses along the Loire River. King Louis XIV, the "Sun King," erected a magnificent palace at Versailles (vehr•SY) in the 1600s, when France was at a peak of its cultural and political influence.

In 1789 the French people overthrew their king in the bloody French Revolution. They set up a **republic,** a strong national government headed by elected leaders. In 1799 General Napoleon Bonaparte seized power. He conquered much of Europe before armies from a group of European countries defeated him.

In the 1900s, France was a battleground during the two world wars. The country survived and today plays an important role in world affairs. France has been a strong supporter of the European Union.

France's government is known as the Fifth Republic. A powerful president, elected for a seven-year term, leads the nation. The French president manages the country's foreign affairs. He or she appoints a prime minister to run the day-to-day affairs of government. France also has a legislature with two parts: the Senate and the more powerful National Assembly. Under certain conditions, the French president can

Reims Cathedral

Majestic Gothic cathedrals like this one at Reims draw tourists to France. Several features characterize Gothic architecture. One is the pointed arch. Another is the use of ribbed vaults, or stone "ribs" that hold the vaulted roof. A third is the use of flying buttresses, or stone archways that extend from the main part of the church to an outer part.

The cathedral at Reims was begun in 1211 and took 80 years to complete, although the decorations continued for centuries. It is almost 500 feet long, making it one and a half times the length of a football field. Twenty-five kings of France received their crowns here.

disband the National Assembly and order new elections. This gives him or her more power over France than the president of the United States holds over America.

An Urban Population About three-fourths of France's 59.1 million people live in cities and towns. Paris, the capital, is the largest city and the main transportation and communications center. Other large cities include Lyon, Marseille, Bourdeaux, and Toulouse. A superb railroad system connects these and other cities. French passenger trains are among the fastest in the world. *Trains à Grande Vitesse* (TGVs) or "trains of great speed" can go as fast as 168 miles (270 km) per hour.

The city of Paris was founded more than 2,000 years ago on a small island in the Seine River. Gradually the city spread onto both banks of the river. Today Paris and its suburbs have a population of more than 9 million people. The city is home to many universities, museums, and other cultural sites. Outstanding cultural figures who lived in Paris include the writer Victor Hugo and the painters Claude Monet and Pierre-Auguste Renoir. Each year, millions of tourists flock to the City of Light, as Paris is called. They visit such sites as the Eiffel Tower, the cathedral of Notre Dame, and the Louvre (LOOV), one of the world's most famous art museums.

✓ Reading Check **What is the main religion in France?**

Assessment

Defining Terms
1. **Define** navigable, republic.

Recalling Facts
2. **Movement** What structural innovation makes it possible for goods and people to move quickly between France and England?
3. **Economics** Name five of France's agricultural products.
4. **Culture** What are three famous landmarks found in Paris?

Critical Thinking
5. **Analyzing Information** What characteristics have created a strong sense of national loyalty in France?
6. **Drawing Conclusions** France is the second-largest food exporter in the world. Why is that remarkable?

Graphic Organizer
7. **Organizing Information** Draw a diagram like the one below. At the end of each arrow, list a fact that indicates a power of the French president.

Applying Geography Skills

8. **Analyzing Maps** Turn to the physical map on page 291. What river in France empties into the English Channel?

Geography Skill

Reading a Vegetation Map

Vegetation maps show the kinds of plants that naturally grow in a given area. Climate largely determines the vegetation of an area. For example, evergreen trees (also called *coniferous*) such as firs and spruces grow in cool climates. In the year-round warmth of the tropics, evergreens with broad leaves, such as palm trees and rubber trees, can grow. Between these two extremes, broad-leaved *deciduous* trees are common. Deciduous trees have broad leaves, but they shed them in autumn. In dry or Mediterranean climates, grasses and shrubs are found because there is not enough water to support tree growth. Highland climates may have *alpine* vegetation—small shrubs and wildflowers. Extremely cold or dry climates may have little or no vegetation.

France: Vegetation

Natural Vegetation
- Deciduous forest
- Coniferous forest
- Mixed forest (coniferous and deciduous)
- Mediterranean vegetation
- Alpine vegetation

Lambert Azimuthal Equal-Area projection

Learning the Skill

To read a vegetation map, follow these steps:

- Read the title of the map.
- Study the map key.
- Find examples of each vegetation zone on the map.
- Look at other aspects of the area's geography, such as rivers, oceans, and landforms to explain the vegetation patterns.

Practicing the Skill

Look at the map above to answer the following questions.

1. What vegetation covers most of France?
2. What type of vegetation is found along France's Mediterranean coast?
3. From the map, what conclusions can you draw about the amount of rain the regions of France receive?

Applying the Skill

Find a vegetation map of your state. What types of vegetation are common in your part of the country?

Section 4

Germany, Switzerland, and Austria

Guide to Reading

Main Idea

Germany, Switzerland, and Austria are known for their mountain scenery and prosperous economies.

Terms to Know

- autobahn
- Holocaust
- communist state
- federal republic
- reunification
- infrastructure
- neutrality

Places to Locate

- Germany
- Switzerland
- Austria
- Danube River
- Liechtenstein

Reading Strategy

Create a diagram like this one. Under the headings fill each oval with facts about each country. Put statements that are true of all three countries where the ovals overlap.

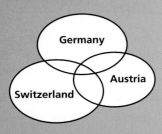

NATIONAL GEOGRAPHIC Exploring Our World

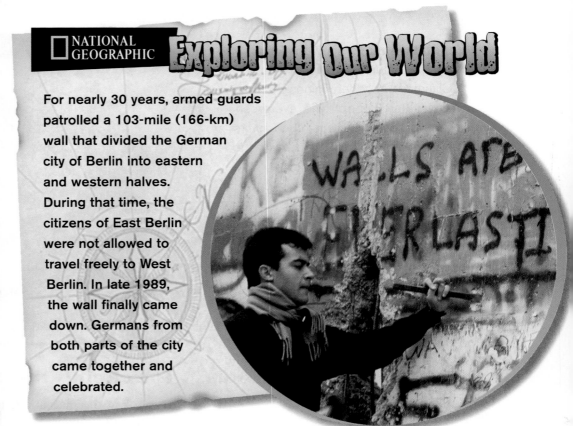

For nearly 30 years, armed guards patrolled a 103-mile (166-km) wall that divided the German city of Berlin into eastern and western halves. During that time, the citizens of East Berlin were not allowed to travel freely to West Berlin. In late 1989, the wall finally came down. Germans from both parts of the city came together and celebrated.

From the 1940s to the late 1980s, the countries of **Germany, Switzerland,** and **Austria** lay on—or close to—the "iron curtain." This was the name given after World War II to the imaginary barrier that separated Communist Eastern Europe from democratic Western Europe. Today all three countries are adjusting to the changes sweeping Europe since the fall of communism's iron curtain.

Germany

About the size of Montana, Germany lies in the heart of Europe. Mountains in the south and plains in the north form the physical landscape of Germany. The Alps rise in the southern German state of Bavaria. The lower slopes of these mountains—a favorite destination for skiers—are covered with forests. Mountain peaks, many capped with snow year-round, tower above the trees. Below the mountains and their foothills are dairy farms, bogs, and sparkling lakes.

CHAPTER 10

Have you ever heard a cuckoo clock announce the hour? That clock may have come from Germany's Black Forest. Lying just north of the Alps, this region is famous for its beautiful scenery and for its wood products. Is the forest really black? No, but in places the trees grow so close together that the forest appears to be black. In recent decades the Black Forest has suffered severe damage from acid rain. Smoke containing sulfur and other chemicals from factories and automobiles causes this problem. The Germans have not yet found a solution to the acid rain problem, because much of the pollution is formed in other countries.

From the Black Forest area flows one of Europe's most important waterways—the **Danube River.** It winds eastward across southern Germany. Rivers are also important in northern Germany, which is part of the vast North European Plain. The Rhine, Elbe, and Weser Rivers flow across this region. Linked by canals, these rivers transport raw materials and manufactured goods.

On Germany's northern plain, lakes and marshes dot a landscape scattered with deposits of earth and stones left by long-ago glaciers. In places the soil is suitable only for pasture, which supports an important dairy farming industry. The plain's southern edge, however, has fertile soil and small farms. Because of the rivers and productive land, the northern plain has many cities and towns. Berlin, Germany's largest city and capital, is the major center of the northeast. To the west lies Hamburg, Germany's largest port city, located on the Elbe River.

Germany's Climate   Westerly Atlantic Ocean winds crossing Europe help keep Germany mild in winter and cool in summer. As you can see from the climate map on page 297, most of Germany has a marine west coast climate. The mountainous southern areas have a highland climate with colder winters.

An Economic Power   Germany is a global economic power and a leader in the European Union. In fact, an area in western Germany called the Ruhr ranks as one of the world's most important industrial centers. The Ruhr developed around rich deposits of coal and iron ore. Battles have been fought among Europe's leaders for control of this productive area. Factories here produce high-quality steel, ships, cars, machinery, chemicals, and electrical equipment. Cities such as Cologne (kuh•LAWN), Essen, and Dortmund cover the Ruhr.

In the eastern part of Germany you will find another large industrial urban area. Metals, chemicals, machines, and textiles are produced in cities such as Leipzig (LYP•sik) and Dresden. The explosive growth of factories, service industries, and high technology has used up the supply of workers. Thus, a growing number have come from Turkey, Italy, Greece, and the former Yugoslav republics.

Germany imports about one-third of its food, although it is a leading producer of beer, wine, and cheese. Farmers raise livestock and grow grains, vegetables, and fruits. Superhighways called autobahns, along with railroads, rivers, and canals, link Germany's cities.

Bavaria, Germany

The beauty of the Bavarian Alps attracts many tourists.

Location In what part of Germany are mountains located?

People and Culture Most of Germany's 82 million people trace their ancestry to groups who settled in Europe from about the A.D.100s to 400s. The people speak German, a language that is related to English. Roman Catholics and Protestants make up most of the population and are fairly evenly represented.

Throughout history, Germans have made important contributions to Europe's cultural heritage. Johann Sebastian Bach (BAHK) and Ludwig van Beethoven (BAY•TOH•vuhn) composed some of the world's greatest classical music. Munich (MYOO•nikh), the largest city in southern Germany, is known for its theaters, museums, and concert halls. Berlin has also emerged as a cultural center.

Influences of History For hundreds of years, Germany was a collection of small territories ruled by princes. During the 1500s, a German priest named Martin Luther tried to reform some of the practices of the Catholic Church. When that failed, he began a new form of Christianity known as Protestantism. Luther's followers, called Protestants, followed forms of worship that they felt were closer to the Bible's teachings. Years of warfare between Protestants and Catholics further divided the German territories.

During the late 1800s, a German leader named Otto von Bismarck finally united the territories into a single nation—Germany. The country's efforts to become a world power helped lead to the European tensions that caused World War I. Germany's defeat in this war caused great anger and distress among the German people.

In 1933 dictator Adolf Hitler and his Nazi Party gained control of the country. Hitler increased German military power and invaded neighboring countries, setting off World War II. One of the horrors of World War II was the **Holocaust,** the systematic murder of more than 6 million Jews by Nazis. Another 6 million people also were murdered. When the Allies—led by the United States, the United Kingdom, and the Soviet Union—defeated Germany, the Holocaust finally ended.

After World War II, the Allies divided Germany. One part—West Germany—became a democracy with close ties to the United States. It built a prosperous free market economy. The other part—East Germany—was a communist state with ties to the Soviet Union. A **communist state** is a country whose government has strong control over the economy and society as a whole.

In 1989 protests calling for democracy swept through the Communist lands of Eastern Europe. The economies of these countries were a disaster, and people were tired of low standards of living. East Germany's government fell, and the two parts of Germany were reunited a year later.

Germany's Government   Like the United States, Germany is a federal republic in which a national government and state governments share powers. An elected president serves as Germany's head of state, but he or she carries out only ceremonial duties. Another official, the chancellor, is the real head of the government. The chancellor is chosen by one of the two houses of parliament. Usually, the majority party in the parliament selects the chancellor.

Berlin   For many years, Berlin was Germany's capital. When Germany was split after World War II, Berlin was divided, too. East Germany made East Berlin its capital. West Berlin was linked to West Germany, but it was entirely surrounded by East German land. West Berlin became a prosperous and exciting place. East Berlin, however, experienced slower growth and became run-down. Meanwhile, West Germany set up a separate capital in the city of Bonn on the Rhine River.

Built by the East Germans in 1961, the Berlin Wall finally came down in 1989. Ten years later, Berlin became the capital of Germany again. The old parliament building, known as the Reichstag, has been rebuilt to hold the government of the new Germany. Companies are putting up new buildings across the city—especially in the eastern part.

Germany Reunited   Germans were overjoyed when their country was reunited. Families that had been split by living in the two separate countries could now visit each other. Reunification—bringing the two parts together under one government—has been difficult, however. Workers in East Germany had less experience and training in modern technology than workers in West Germany. After reunification, many old and inefficient factories in the east could not compete with the more advanced industries in the west. Having no Communist government to support them, the factories were forced to close. As a result, the number of people without jobs has risen in the eastern part of Germany.

One of the challenges for the united Germany has been to close the economic gap between the two parts of the country. The German government has spent billions of dollars to provide needed training, education, and modernization. Huge amounts of money have also been spent in the east to improve the infrastructure, or the transportation and communications networks on which an economy depends.

✓ Reading Check **What was the Holocaust?**

Switzerland

The Alps form most of the landscape in Switzerland, Austria, and **Liechtenstein.** That is why they are called the Alpine countries. Switzerland and Austria together cover an area about the size of

Alabama. Sandwiched between them is tiny Liechtenstein, covering only 60 square miles (155 sq. km).

The rugged Swiss Alps have always created a natural barrier to travel between northern and southern Europe. For centuries, landlocked Switzerland guarded the few routes that cut through this barrier.

Because of its location, Switzerland has practiced neutrality— refusing to take sides in disagreements and wars between countries. In fact, the Swiss have not been involved in a foreign war since the early 1500s. As a result of Switzerland's peaceful heritage, the Swiss city of Geneva is today the center of many international organizations.

NATIONAL GEOGRAPHIC On Location

Swiss Chocolate

Switzerland's factories produce some of the best chocolate in the world.

Place What other products are made in Switzerland?

Alpine Land and Climate The Alps make Switzerland the continental divide of central Europe. A continental divide is a high place from which rivers flow in different directions. Several rivers, including the Rhine and the Rhône, begin in the Swiss Alps. Dams built on many of Switzerland's rivers produce great amounts of hydro-electric power.

Another mountain range, the Jura Mountains, runs across northwest Switzerland. Between the Jura Mountains and the Alps lies a plateau known as the *Mittelland,* or "Middle Land." Most of Switzerland's industries and its richest farmlands are found here. Bern, Switzerland's capital, and Zurich, its largest city, are also located on this plateau.

The Alps shape Switzerland's climate. Though temperatures differ with elevation, in most parts of the country winters are cold, and summers are warm.

Manufacturing and Service Industries Although it has few natural resources, Switzerland is a thriving industrial nation. Using imported materials, Swiss workers make high-quality goods such as machinery, pharmaceuticals, high-tech electronic equipment, and clocks and watches. They also produce chemicals and gourmet foods such as chocolate and cheese. Tourism is an important industry, as are banking and insurance. Zurich and Geneva are important international financial centers.

Culture and Government As you might expect given its location, Switzerland has many different ethnic groups and religions. Did you know that the country has four national languages? They are German, French, Italian, and Romansch. Most Swiss speak German, and many speak more than one language. Only a tiny percentage speak Romansch, an old language based on Latin.

Just under half of the Swiss people are Catholic. A slightly smaller number are Protestants. The Protestant Reformed faith developed in Switzerland during the 1500s. It later spread to many countries, including the United States.

The Swiss have enjoyed a stable democratic government for more than 700 years. Like Germany, Switzerland is a federal republic. In Switzerland, however, the units of local government—which are called cantons—have a great deal of power. Heading the national government is a seven-member panel called the Federal Council. The council chooses one of its members to be Switzerland's president.

✓ Reading Check **In what part of Switzerland are most of its industries and farmlands found?**

Austria

The political map on page 290 shows you that Austria is a land-locked country lying in the heart of Europe. Small when compared to some other European countries, Austria is slightly larger than the state of Maryland.

Yet, believe it or not, Austria once was one of Europe's largest countries. Under the powerful Hapsburg royal family, the Austrian Empire ruled much of central Europe from the late 1200s to the early 1900s. World War I destroyed that empire. By 1918 Austria had become a small republic, with the borders it has today.

A Mountainous Country The Alps cover three-fourths of Austria. In fact, Austria is one of the most mountainous countries in the world. Have you ever seen the movie *The Sound of Music*? It took place in Austria's spectacular mountains.

Vienna, Austria

People enjoy relaxing in one of Vienna's many coffeehouses.

Place What famous musicians lived or performed in Vienna?

NATIONAL GEOGRAPHIC **On Location**

The Danube River flows from west to east across the country's northern region. Austria's climate is similar to Switzerland's. In winter, lowland areas receive rain, and mountainous regions have snow. Summers are cooler here than they are in Switzerland.

Austria's Economy   Austria's economy is strong and varied. The mountains provide valuable timber and hydroelectric power. They yield iron ore and coal as well. The mountains also attract millions of tourists who enjoy the fresh air and fine hiking and skiing.

Austrian factories produce machinery, chemicals, metals, and vehicles. The country also has some petroleum and natural gas. Austrian farmers raise dairy cattle and other livestock, sugar beets, grains, potatoes, and fruits.

Austria's People   Austria is home to about 8.1 million people. Most Austrians live in cities and towns and work in manufacturing or service jobs. The majority of people speak German. About 90 percent of the people are Roman Catholic.

Vienna, on the Danube River, is the capital and largest city. It has a rich history as a center of culture and learning. Some of the world's greatest composers, including Mozart, Schubert, and Haydn, lived or performed in Vienna. The city's concert halls, historic palaces and churches, and grand architecture continue to draw musicians today.

✔ **Reading Check** What economic benefits do Austria's mountains provide?

Assessment

Defining Terms
1. **Define** autobahn, Holocaust, communist state, federal republic, reunification, infrastructure, neutrality.

Recalling Facts
2. **Human/Environment Interaction** What has damaged the Black Forest?

3. **History** Why have battles been fought over the Ruhr?

4. **Culture** What are Switzerland's four languages?

Critical Thinking
5. **Understanding Cause and Effect** What problems have emerged as a result of German reunification?

6. **Analyzing Information** How have the Alps helped Switzerland maintain its neutrality?

Graphic Organizer
7. **Organizing Information** Draw a diagram like the one below. On the lines list two facts about Austria's physical features, two facts about Austria's people, and four facts about Austria's economy.

Physical Features — Austria — Economy
People

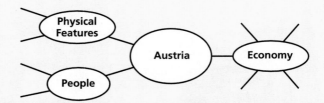

Applying Geography Skills

8. **Analyzing Maps** Look at the political map on page 290. The city of Frankfurt is located at what degree of latitude?

5 The Benelux Countries

Guide to Reading

Main Idea

Belgium, the Netherlands, and Luxembourg are small countries with long histories of international trade.

Terms to Know

- polder
- multinational company
- multilingual

Places to Locate

- Belgium
- Netherlands
- Luxembourg
- Brussels
- North Sea
- Rotterdam
- Amsterdam

Reading Strategy

Create a diagram like this one. Fill in the names of the three Benelux countries and write at least one key fact about the people from each country.

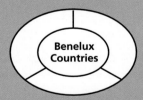

Benelux Countries

NATIONAL GEOGRAPHIC Exploring Our World

Tulips create colorful carpets in the Netherlands every spring. Originally from Asia, these vivid blooms are more than a beautiful sight—they are an important cash crop. Dutch growers export flowers and bulbs around the world. The Dutch have another reason to feel proud of this beautiful display. These flowers grow on land the Dutch reclaimed from the sea.

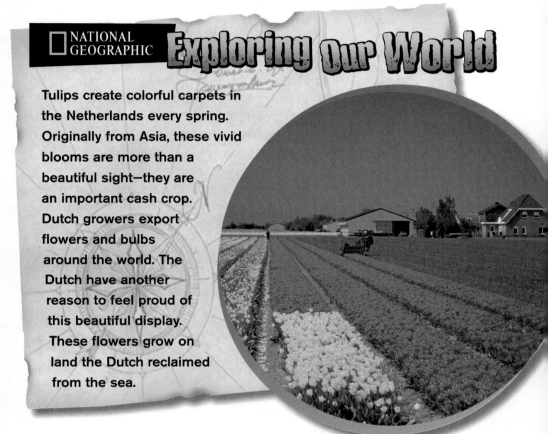

The name *Benelux* comes from combining the first letters of three countries' names: **Belgium,** the **Netherlands,** and **Luxembourg.** These small countries have much in common. Their lands are low, flat, and densely populated. Most people live in cities, work in businesses or factories, and enjoy a high standard of living. All three nations are members of the European Union. They are also parliamentary democracies with constitutional monarchs. In a parliamentary democracy, voters elect members of a lawmaking body who then choose a prime minister to head the government. In a constitutional monarchy, a king or queen is head of state, but elected officials actually run the government.

Belgium

About the size of Maryland, Belgium touches France, Luxembourg, Germany, and the Netherlands. From a wide coastal plain, the land rolls gently upward to a hilly, forested region in the interior known as

313

Ardennes. Because Belgium is centrally located in Western Europe, it has been a battleground over the centuries. Most recently, British and American troops fought the Germans there in World War II. Lying near Western Europe's major industrial regions, Belgium has long been a trade and manufacturing center.

Belgium's Economy   Belgian lace, Belgian chocolate, and Belgian diamond-cutting all enjoy a worldwide reputation for excellence. With few natural resources of their own, the Belgian people import metals, fuels, and raw materials. They use these materials to make vehicles, chemicals, and textiles, which are then exported. Belgium's excellent road system also helps its economic success.

Belgium's People   Most of the people are Roman Catholic. The country has two main cultural and language groups. The Flemings in the north speak Flemish, a language based on Dutch. The south is home to the French-speaking Walloons. Tensions sometimes arise between the two groups, especially because there is more wealth and industry in the north than in the south.

Most Belgians live in crowded urban areas. **Brussels**—the capital and largest city—is an international center for trade. Many world

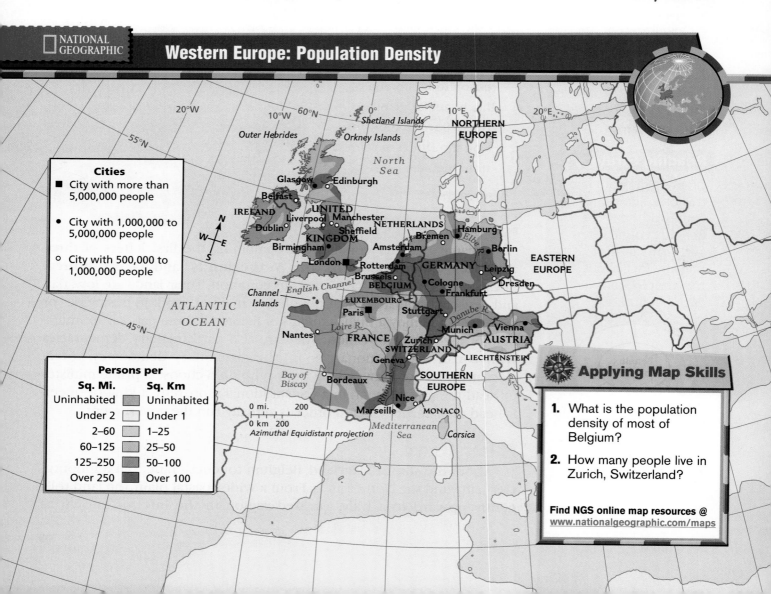

Western Europe: Population Density

NATIONAL GEOGRAPHIC

Cities
- ■ City with more than 5,000,000 people
- ● City with 1,000,000 to 5,000,000 people
- ○ City with 500,000 to 1,000,000 people

Persons per

Sq. Mi.	Sq. Km
Uninhabited	Uninhabited
Under 2	Under 1
2–60	1–25
60–125	25–50
125–250	50–100
Over 250	Over 100

0 mi.   200
0 km   200
Azimuthal Equidistant projection

Applying Map Skills

1. What is the population density of most of Belgium?

2. How many people live in Zurich, Switzerland?

Find NGS online map resources @
www.nationalgeographic.com/maps

organizations have located their headquarters here as well. By law, Brussels recognizes both Flemish and French speakers. Its street signs and official documents are printed in both languages.

√ Reading Check **Why has Belgium's location been an advantage and a disadvantage?**

The Netherlands

The Netherlands—about half the size of Maine—is one of the most densely populated countries in the world. Sometimes called Holland, it is located on the **North Sea** to the northeast of Belgium.

Netherlands means "lowlands." True to its name, nearly half of this small, flat country lies below sea level. Over the centuries, the people of the Netherlands, known as the Dutch, have learned how to protect their land and how to reclaim more from the sea. The Dutch method of reclaiming land is simple, but it takes hard work. First they strengthen sand dunes along the coasts and build dikes, or dams, to keep the sea out. Then they cut ditches and canals for drainage and pump the wetlands dry. Once run by windmills, pumps are now driven by steam or electricity. These drained lands, called polders, have rich farming soil. The Dutch also build factories, airports, and even towns on them.

Without defenses against the sea, high tides would flood much of the country twice a day. An engineering feat known as the Delta Plan Project was completed in 1986. It consists of four huge barriers that keep the North Sea from overflowing the countryside during storms.

The Dutch Economy and People On polders, Dutch farmers raise dairy cattle, grow food, and produce tulips and other flowers. They use extremely efficient farming methods. High technology makes small farms so productive that the Dutch can export cheese, vegetables, and flowers. In fact, the Netherlands ranks third in the world—after the United States and France—in the value of its agricultural exports.

Though farming is important, most people work in service industries, manufacturing, and trade. During the 1600s, the Netherlands was a major sea power, and world trade brought the country much wealth. Still active in trading, the Netherlands has one of the world's busiest ports at **Rotterdam,** located near the Rhine River.

About 90 percent of the Dutch live in cities and towns. **Amsterdam** is the capital and largest city. Canals crisscross Amsterdam's historic districts, which preserve the art and architecture of the Dutch "golden age" of the 1600s. Living in a densely populated country, the Dutch make good use of their space. Houses are narrow but tall, and apartments are often built over highways. One of Amsterdam's most famous

NATIONAL GEOGRAPHIC **On Location**

Rotterdam

Rotterdam's architects are challenged to find ways of cramming more buildings into an already overcrowded city.

Location Near what river is Rotterdam located?

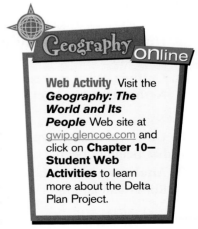

Geography Online

Web Activity Visit the *Geography: The World and Its People* Web site at gwip.glencoe.com and click on **Chapter 10— Student Web Activities** to learn more about the Delta Plan Project.

people was Anne Frank. Her diary tells about her life as she and her family hid from the German Nazis during World War II.

About one-third of the Dutch people are Roman Catholic. Others are Protestants, and a small number of immigrants from other lands are Muslims. The people of the Netherlands speak Dutch.

✓ Reading Check How much of the Netherlands lies below sea level?

Luxembourg

Southeast of Belgium lies Luxembourg, one of Europe's smallest countries. Its capital and largest city is also called Luxembourg. The entire country is only about 55 miles (89 km) long and about 35 miles (56 km) wide, making it slightly smaller than the state of Rhode Island.

Despite its size, Luxembourg is prosperous. Many **multinational companies,** or firms that do business in several countries, have their headquarters here. It is home to the second-largest steel-producing company in Europe and is a major banking center as well. By the late 1990s, well over 200 banks were operating in Luxembourg.

Why is Luxembourg so attractive to foreign companies? First, the country is centrally located. Second, most people in this tiny land are **multilingual,** or able to speak several languages. They speak Luxembourgian, a blend of old German and French; French, the official language of the law; and German, used in most newspapers. Almost all the people of Luxembourg are Roman Catholic.

✓ Reading Check What industries are important in Luxembourg?

Assessment

Defining Terms

1. **Define** polder, multinational company, multilingual.

Recalling Facts

2. **Culture** What are the two major cultures and languages of Belgium?

3. **Human/Environment Interaction** What is the Delta Plan Project?

4. **Culture** What percentage of the people in the Netherlands live in cities and towns?

Critical Thinking

5. **Understanding Cause and Effect** Why has Belgium been a battleground?

6. **Analyzing Information** Why is Rotterdam an important city?

Graphic Organizer

7. **Organizing Information** Draw a diagram like this one. In the center circle list five characteristics that Belgium, the Netherlands, and Luxembourg share.

Belgium → ⟨ ⟩ ← Netherlands
↑
Luxembourg

Applying Geography Skills

8. **Analyzing Maps** Turn to the physical map on page 291. What body of water borders Belgium and the Netherlands?

Reading Review

Section 1 | The United Kingdom

Terms to Know

moor   currency
loch   devolution
parliamentary
   democracy
constitutional
   monarchy

Main Idea

The United Kingdom is a major industrial country that once ruled a vast global empire.

✓ **Economics** The United Kingdom has a strong economy and is rich in energy resources.

✓ **Government** The government of the United Kingdom recently began to give more power to regional governments.

Section 2 | The Republic of Ireland

Terms to Know

peat
bog

Main Idea

Once a war-torn, agricultural country, the Republic of Ireland is now enjoying economic growth.

✓ **Economics** Ireland has attracted foreign investors and improved its economy.

✓ **History** After years of conflict, a peace plan was adopted in Northern Ireland.

Section 3 | France

Terms to Know

navigable
republic

Main Idea

A center of European culture, France is also a major agricultural and manufacturing country.

✓ **Economics** France balances agriculture, manufacturing, and service industries.

✓ **Culture** Paris is a world center of art, learning, and culture.

Section 4 | Germany, Switzerland, and Austria

Terms to Know

autobahn
Holocaust
communist state
federal republic
reunification
infrastructure
neutrality

Main Idea

Germany, Switzerland, and Austria are known for their mountain scenery and prosperous economies.

✓ **Economics** The German economy is very strong, but the eastern region—reunited with the west in 1990—lags behind the western region.

✓ **Economics** Switzerland produces high-quality manufactured goods.

✓ **Human/Environment Interaction** Austria's strong economy makes good use of its mountainous terrain.

Section 5 | The Benelux Countries

Terms to Know

polder
multinational
   company
multilingual

Main Idea

Belgium, the Netherlands, and Luxembourg are small countries with long histories of international trade.

✓ **Location** Belgium's location has made it an international center for trade.

✓ **Economics** The Dutch grow valuable crops on reclaimed land.

✓ **Economics** Luxembourg is home to many multinational companies.

Using Key Terms

Match the terms in Part A with their definitions in Part B.

A.

1. moor
2. multilingual
3. neutrality
4. loch
5. polder
6. constitutional monarchy
7. multinational company
8. Holocaust
9. infrastructure
10. currency

B.

a. type of money
b. transportation and communication networks on which an economy depends
c. a king or queen shares power with elected officials
d. land reclaimed from the sea
e. treeless, windy highland area
f. refusing to take sides
g. slaughter of Jews by the Nazis
h. narrow bay cut into highland coasts
i. company that has offices in several countries
j. able to speak several languages

Reviewing the Main Ideas

Section 1 The United Kingdom

11. **Region** What regions make up the United Kingdom?
12. **History** How have the British influenced other parts of the world?

Section 2 The Republic of Ireland

13. **Culture** What is the major language(s) and religion of the Republic of Ireland?
14. **History** What has been the major source of conflict in Northern Ireland?

Section 3 France

15. **Place** Why is Paris an important city?
16. **Government** What is the Fifth Republic?

Section 4 Germany, Switzerland, and Austria

17. **History** Why was Germany divided after World War II?
18. **Location** Why is Geneva the center of many international organizations?

Section 5 The Benelux Countries

19. **Place** What is the capital of Belgium?
20. **Economics** Why do the Dutch reclaim land from the sea?

 **Western Europe**

Place Location Activity

On a separate sheet of paper, match the letters on the map with the numbered places listed below.

1. United Kingdom
2. English Channel
3. France
4. Germany
5. Alps
6. Rhine River
7. Ireland
8. Paris
9. Berlin
10. London

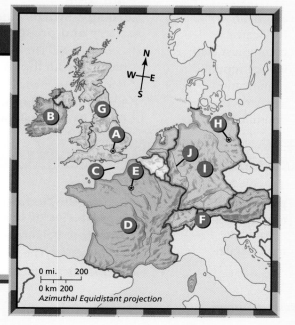

0 mi. 200
0 km 200
Azimuthal Equidistant projection

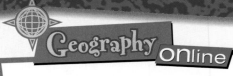

Critical Thinking

21. **Making Generalizations** How do you think the Industrial Revolution affected the lives of people who lived during that time? How do you think it has shaped your life today?

22. **Sequencing Information** Draw a time line like this one, then list five events with dates from Germany's history.

GeoJournal Activity

23. **Writing a News Broadcast** Research an important event that took place in one of the countries studied in this chapter. Write a script for a news broadcast about the event. Present the broadcast as a news anchor.

Mental Mapping Activity

24. **Focusing on the Region** Draw a simple outline map of Western Europe, then label the following:

- United Kingdom
- Republic of Ireland
- Rhine River
- English Channel
- Switzerland

- France
- Germany
- Austria
- Alps
- Paris

Technology Skills Activity

25. **Using a Spreadsheet** List the names of the nine Western European countries in a spreadsheet, beginning with cell A2 and continuing down the column. Find each country's population and record the figures in column B. In column C list each country's area in square miles. Title column D "Population Density." Then divide column B by column C to determine the population density. Print out and share your spreadsheet with the class.

Standardized Test Practice

Directions: Study the map, then answer the question that follows.

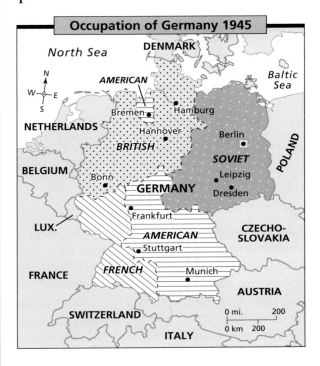

Occupation of Germany 1945

1. **In 1945 what country controlled the land surrounding Berlin, Germany's capital?**

   F   the United Kingdom

   G   the Soviet Union

   H   the United States

   J   France

Test-Taking Tip: This question asks you to synthesize information on the map with prior knowledge. Notice that the map does not specifically state that the United Kingdom, for example, controlled a portion of Germany. Instead, it refers to this area as the "British" zone.

319

1961—the Wall goes up

GERMANY:
Together Again

It was one of the biggest parties in German history. On November 9, 1989, thousands of people poured into the streets—cheering, singing, and hugging each other. The Berlin Wall could no longer keep the German people apart.

The Fall of the Wall

Today there is only one Germany—but after World War II, there were two. After surrendering in 1945, Germany was carved into four parts. The United States, the United Kingdom, and France merged their three parts into one country—West Germany. The Soviet Union's portion became East Germany.

Deep within Soviet territory lay the city of Berlin. It also was divided into four parts. The Soviet section became East Berlin. The remaining sections combined into West Berlin, which became part of West Germany.

In West Germany, the United States and its allies encouraged a democratic government that allowed people to choose their leaders from among competing political parties. The Soviets instituted a one-party, Communist government that gave East Germans no political choices and little control over their own lives.

East Germans soon grew unhappy with their government. During the early 1960s, refugees poured into West Germany through Berlin. Desperate to stop the flow, East Germany's government erected the Berlin Wall in 1961. Built of concrete and topped with barbed wire, the Wall separated West Berlin from East Berlin, blocking the exit route for East Germans. Those caught trying to escape to West Berlin were often shot by guards who had orders to kill.

The Wall held Germans apart for more than 20 years. By 1989, however, people across Eastern Europe were demanding change. East Germans wanted freedom. They forced their leaders to open the Wall on November 9, and Germany reunited within a year.

Growing Pains

However, after 40 years of living under separate political systems, many differences had developed between East Germans and West Germans. Former West Germans support the competition of a market economy. They live a more prosperous lifestyle than do former East Germans.

Former East Germans regret the loss of free health care and the guaranteed jobs that came with communism. Although East Germany was an industrial giant, it lacked modern necessities such as adequate transportation systems and reliable telephone service. Today Germany is investing heavily to improve services for all of its citizens.

QUESTIONS

1 Why was Germany split into two countries?

2 What are some difficulties Germans face in becoming one country?

1989—the Wall comes down ▶

The Two Germanys: 1945

UNITED
KINGDOM

North
Sea

DENMARK

West
Berlin

East
Berlin

NETHERLANDS

POLAND

WEST
GERMANY

EAST
GERMANY

BELGIUM

LUX.

CZECHOSLOVAKIA

FRANCE

AUSTRIA

HUNGARY

SWITZ.

0 mi.   200
0 km   200

N
W   E
S

Chapter 11

Southern Europe

The World and Its People — NATIONAL GEOGRAPHIC

To learn more about the people and places of Southern Europe, view *The World and Its People* **Chapter 11** video.

Geography Online

Chapter Overview Visit the *Geography: The World and Its People* Web site at gwip.glencoe.com and click on **Chapter 11—Chapter Overviews** to preview information about Southern Europe.

Spain and Portugal

Guide to Reading

Main Idea

Spain and Portugal are working toward building prosperous economies.

Terms to Know

- plateau
- dry farming
- navigable
- colony
- parliamentary republic
- dialect

Places to Locate

- Spain
- Portugal
- Andorra
- Iberian Peninsula
- Pyrenees
- Tagus River
- Guadalquivir River
- Madrid
- Barcelona
- Lisbon

Reading Strategy

Draw a diagram like this one. List information about Spain and Portugal under their names in the outer parts of the ovals. Where the ovals overlap, write statements that are true of both countries.

Spain ⟋ ⟍ Portugal

NATIONAL GEOGRAPHIC Exploring Our World

The "running of the bulls" is an annual and controversial event in Pamplona, a city in northern Spain. Although animal rights groups object to it, each morning during the weeklong Festival of San Fermín, a half dozen bulls are released to run along the city's narrow streets. Young men risk their lives running ahead of the bulls. Their goal is to stay in the race as long as possible.

Spain and its neighbors, **Portugal** and **Andorra,** make up the **Iberian Peninsula.** The physical map on page 325 shows you that the Iberian Peninsula is separated from the rest of Europe by the **Pyrenees** mountain range. Off the peninsula's southern tip lies the Strait of Gibraltar, a narrow passageway between the Atlantic Ocean and the Mediterranean Sea. The strait divides the Iberian Peninsula—and Europe—from Africa. Only 9 miles (14 km) separate the two continents at this point.

Spain—about twice the size of Oregon—takes up about 80 percent of the Iberian Peninsula. Portugal—slightly smaller than Indiana—sits on the peninsula's western edge and occupies most of the remaining space. Tiny Andorra, with only 175 square miles (453 sq. km), perches high in the Pyrenees.

The Land and Climate

As you can tell from the physical map, most of the Iberian Peninsula has high elevation. The Pyrenees forms the border between Spain and France. For a long time, these very steep mountains isolated the Iberian Peninsula from the rest of Europe. Today, with advances in technology and communication, the peninsula's people are closely linked to European events.

A huge central plateau covers about two-thirds of the peninsula. A plateau, you may recall, is flat land with high elevation. Because this plateau lies in the rain shadow of the mountains, it is the driest part of the peninsula. In many areas the reddish-yellow soil is poor, and the land is dry-farmed to grow crops such as wheat and vegetables. Dry farming is a method in which the land is left unplanted every few years so that it can store moisture. Some farmers also herd sheep, goats, and cattle.

Many of the Iberian Peninsula's major rivers begin in the central plateau. The **Tagus** (TAY•guhs) **River**—the peninsula's longest river—flows from central Spain to the Atlantic Ocean. Most of these rivers are not navigable, or wide and deep enough to allow the passage of ships. The **Guadalquivir** (GWAH•thahl•kee•VEER) **River** is one of the few navigable rivers.

On the climate map on page 331, you see that Portugal and most of Spain have a Mediterranean climate. People in these areas generally enjoy mild winters and hot summers. Yet elevation and closeness to

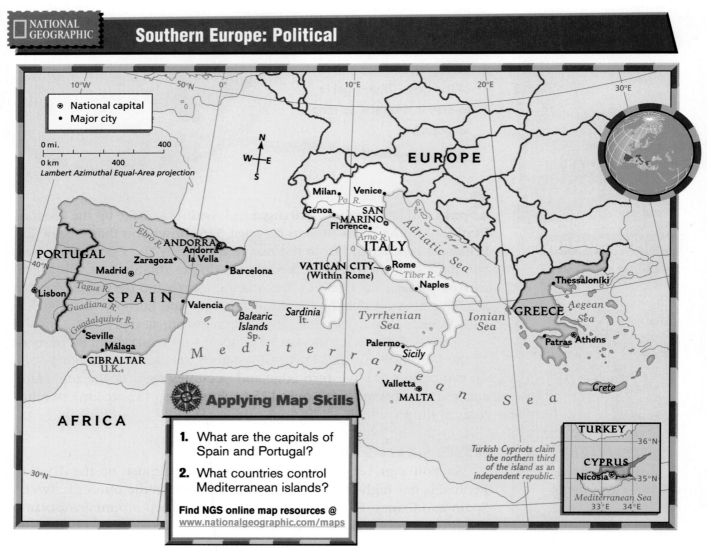

Southern Europe: Political

NATIONAL GEOGRAPHIC

⊗ National capital
• Major city

0 mi. 400
0 km 400
Lambert Azimuthal Equal-Area projection

EUROPE

Milan• Venice•
Genoa• SAN
MARINO
Florence•
Po R.
Arno R.
ITALY
VATICAN CITY— ⊗ Rome
(Within Rome)
Tiber R.
Naples•
Adriatic Sea

Ebro R.
ANDORRA
Andorra la Vella
PORTUGAL
Zaragoza•
Madrid ⊗
⊗Lisbon
SPAIN
Tagus R.
Guadiana R.
Guadalquivir R.
•Valencia
Barcelona•
Balearic Islands Sp.
Sardinia It.
Tyrrhenian Sea
Ionian Sea
GREECE
Thessaloníki•
Aegean Sea
Patras• Athens⊗

Seville•
•Málaga
•GIBRALTAR U.K.
Mediterranean
Palermo• Sicily
Valletta⊗
• MALTA
Crete
Sea

AFRICA

Turkish Cypriots claim the northern third of the island as an independent republic.

TURKEY
36°N
CYPRUS
Nicosia⊗
35°N
Mediterranean Sea
33°E 34°E

Applying Map Skills

1. What are the capitals of Spain and Portugal?

2. What countries control Mediterranean islands?

Find NGS online map resources @
www.nationalgeographic.com/maps

the sea cause differences in climate. In northwestern Spain, warm winds from the Atlantic Ocean drop much rain, and winters are chilly. You find a dry steppe climate in southeastern Spain because of hot winds from North Africa.

√Reading Check **What type of landform covers central Spain?**

Growing Economies

Spain and Portugal both belong to the European Union. Once slow in developing manufacturing, the two countries in recent years have worked hard to catch up economically with other European Union nations. Agriculture remains important to Spain and Portugal, but most of their people now work in manufacturing and service industries. Andorra draws millions of tourists each year who flock to its duty-free shops.

Although much of Spain is dry and rugged, Spanish farmers grow a wide variety of grains. On eastern coastal plains, they use irrigation to grow citrus fruits and olive trees. Acres and acres of olive groves make Spain one of the world's leading producers of olive oil. Vegetables and

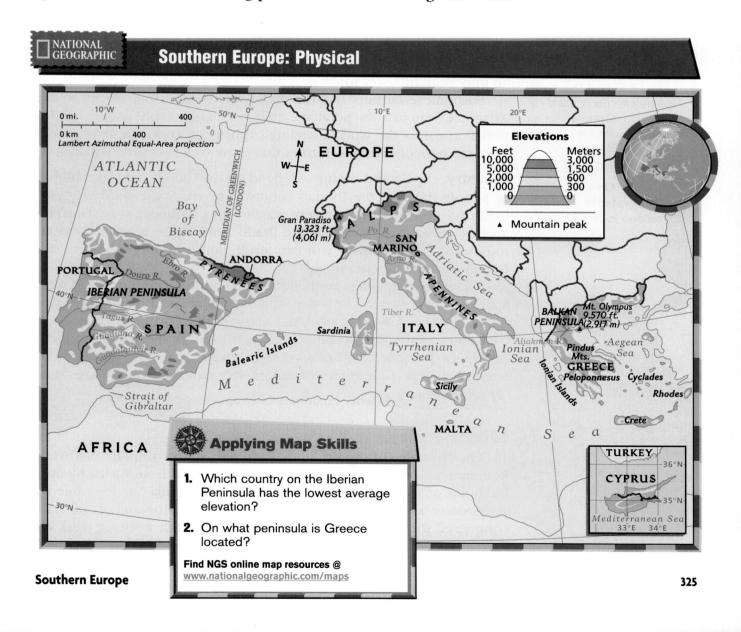

NATIONAL GEOGRAPHIC

Southern Europe: Physical

Applying Map Skills

1. Which country on the Iberian Peninsula has the lowest average elevation?

2. On what peninsula is Greece located?

Find NGS online map resources @ www.nationalgeographic.com/maps

NATIONAL GEOGRAPHIC On Location

Spain and Portugal

Barcelona (above) is the main city of the Spanish region called Catalonia. Cork is stripped from a tree in central Portugal (right).

Location On what body of water is Barcelona located?

grapes are other important Spanish crops. Grass grows well in the north, which has more rain. People there raise beef and dairy cattle. Fishing is an important industry, too.

Portuguese farmers use the well-watered land that slopes to the Atlantic Ocean to grow potatoes, grains, fruits, olives, and grapes. In addition, Portugal is the world's leading exporter of cork. The cork comes from the bark of certain oak trees that grow well in central Portugal.

Industry Spain and Portugal depend on the tourist industry. Millions of people travel to the Iberian Peninsula to enjoy the sunny climate, beautiful beaches, and ancient castles and cathedrals. In **Madrid,** Spain's capital, you can visit the Prado, one of the world's great art museums. In **Barcelona** you can stroll along the city's flower-lined boulevard called the Ramblas.

Manufacturing industries benefit both countries' economies as well. Spanish workers mine rich deposits of iron ore and make processed foods, clothing, footwear, steel, and automobiles. Portugal's economy is based on traditional products such as textiles, cork and wood products, beverages, porcelain, and glassware. In recent years, the country also strengthened its position in the automobile and electronics industries.

✓ Reading Check In what export does Portugal lead the world?

History and Government

The people of Spain and Portugal share similar histories. Remember that the Iberian Peninsula is separated from Africa by only the Strait of Gibraltar. Because of this, the peninsula has long been a crossroads as well as a target for invaders. Romans from Italy conquered the peninsula in the 100s B.C. They left a strong mark on Iberian life and culture. Most languages spoken in Spain and Portugal are based on Latin, the language of Rome. Roman law forms the basis for law in both countries. In addition, most people in Spain and Portugal are Roman Catholic.

Beginning in the A.D. 700s, Muslims—followers of the religion of Islam—crossed the Strait of Gibraltar from North Africa and conquered much of the Iberian Peninsula. Spanish and Portuguese Christians fought for centuries to drive out the Muslims. The Spanish won their final victory over the Muslims in 1492.

That same year, the Spanish king and queen gave money to an explorer named Christopher Columbus. Columbus sailed west hoping to find a new sea route to Asia. Instead, he touched ground in the Americas. Soon Spain had a vast empire in South America and North America.

Portugal's Prince Henry "the Navigator" likewise encouraged exploration. In the 1500s, the country had become a trading nation and built its own empire. Portuguese merchants and soldiers founded colonies in Africa, South America, India, and Southeast Asia.

These overseas empires brought Spain and Portugal wealth from gold, silver, and trade goods. By the late 1800s, however, Spain had lost most of its colonies, or overseas territories. Portugal's power also declined, but it ruled parts of Africa and Asia until the late 1900s.

The Government   As their empires crumbled, Spain and Portugal suffered a series of wars and rebellions. From 1936 to 1939, Spain had a violent civil war. General Francisco Franco's forces won, and he ruled as dictator of Spain until his death in 1975. During this same period, Portugal also was under a dictatorship. Today both countries are modern democracies. Spain is a constitutional monarchy, in which a king or queen is head of state, but elected officials run the government. Portugal is a parliamentary republic, with a president as head of state. A prime minister, chosen by the legislature, is the head of government. Andorra is a semi-independent principality—a region governed by both Spain and France.

✓ Reading Check   **What two things happened in 1492 in this region?**

The People

Despite similar histories, the people of Spain and Portugal have their cultural differences. Portugal developed a unified culture based on the Portuguese language, while Spain remained a "country of different countries." The Spanish people do not all speak the same language or even have a single culture.

The region of Castile, in central Spain, dominated the culture for centuries. The dialect of the Castilian people became the form of Spanish spoken by most Spaniards. A dialect is a local form of a language. People in other regions of Spain, however, regard themselves as separate groups. In the Mediterranean region of Catalonia, the people speak Catalan, a language related to French. Galicia in northwestern Spain has its own language—Galician. Andalusia in the south has its unique dialects and cultural traditions.

Southern Europe

The Alhambra

The Muslim rule of Spain brought economic success to the Iberian Peninsula. The Muslims brought scientific knowledge, methods of irrigation, and new crops. They also brought literature, music, and art to Spain.

The Alhambra Palace in southern Spain is a spectacular example of Muslim art. Islam forbids art that includes human forms. As a result, Muslim artists created complex patterns and elaborate designs. This is the Court of the Lions. In the center is a fountain where running water cooled the people of the palace in Spain's hot midday sun.

The Basque people in the Pyrenees see themselves as completely separate from Spain. They speak Basque, a language unlike any other in the world. Having lived in Spain longer than any other group, many Basques want independence in order to preserve their way of life. Some Basque groups have even used violence against the Spanish government.

Rural and City Life   Almost two-thirds of the people in Portugal live in rural areas. **Lisbon,** with a population of just about 2 million, is Portugal's capital and largest city. Porto is its second-largest city.

In contrast, more than three-fourths of Spain's 39.4 million people live in cities and towns. Madrid has more than 3 million people and ranks as one of Europe's leading cultural centers. Madrid faces the usual urban challenges of heavy traffic and air pollution. Fast-paced Barcelona is Spain's leading seaport and industrial center.

You find some centuries-old traditions even in the modern cities. For example, most Spanish families usually do not eat dinner until 9 or 10 o'clock at night. On special occasions, Spaniards enjoy paella, a traditional dish of shrimp, lobster, chicken, ham, and vegetables mixed with seasoned rice.

Rock and jazz music are popular with young Spaniards and Portuguese, but the people of each region have their own traditional songs and dances as well. Spanish musicians often accompany singers and dancers on guitars, castanets, and tambourines. Spanish dances such as the bolero and flamenco and soulful Portuguese music known as fado have spread throughout the world.

√ Reading Check   **What are the five culture regions of Spain?**

Section 1 Assessment

Defining Terms
1. Define plateau, dry farming, navigable, colony, parliamentary republic, dialect.

Recalling Facts
2. Location What three countries are located on the Iberian Peninsula?
3. Economics Name four types of goods manufactured in Portugal.
4. History What group controlled Spain from the A.D. 700s until 1492?

Critical Thinking
5. Drawing Conclusions Why do you think Spain is more urban than Portugal?
6. Understanding Cause and Effect Why do the Basque people feel separate from the rest of Spain?

Graphic Organizer
7. Organizing Information In a chart like this one, list facts about Spain and Portugal for each category.

	Spain	Portugal
Land		
Economy		
Cities		
People		

Applying Geography Skills

8. Analyzing Maps Turn to the physical map on page 325. What rivers flow through both Spain and Portugal?

Geography Skill

Reading a Population Map

Population density is the number of people living in a square mile or square kilometer. A **population density map** shows you where people live in a given region. Mapmakers use different colors to represent different population densities. The darker the color, the more dense, or crowded, the population is in that particular area. Cities that are shown by dots or squares also represent different population sizes.

Learning the Skill

To read a population density map, follow these steps:

- Read the title of the map.
- Study the map key to determine what the colors mean.
- On the map, find the areas that have the lowest and highest population density.
- Identify what symbols are used to show how heavily populated the cities are.

Practicing the Skill

Look at the map below to answer the following questions.

1. What color stands for 125–250 people per square mile (50–100 per sq. km)?
2. Which cities have more than 1 million people?
3. Which areas have the lowest population density? Why?

Applying the Skill

Obtain a population density map of your state. What is the population density of your area? What is the nearest city with 1 million people?

GO TO

Practice key skills with **Glencoe Skillbuilder Interactive Workbook, Level 1.**

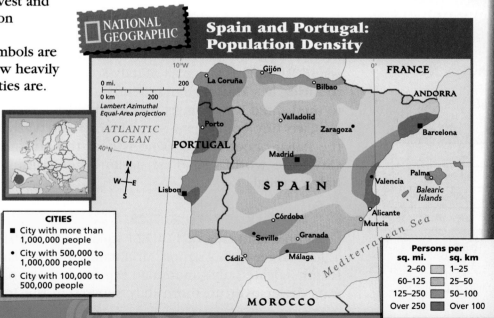

NATIONAL GEOGRAPHIC

Spain and Portugal: Population Density

CITIES
- ■ City with more than 1,000,000 people
- • City with 500,000 to 1,000,000 people
- ○ City with 100,000 to 500,000 people

Persons per	
sq. mi.	sq. km
2–60	1–25
60–125	25–50
125–250	50–100
Over 250	Over 100

Guide to Reading

Main Idea

Once the center of a mighty empire, Italy has influenced Europe's religions, cultures, and governments.

Terms to Know

- sirocco
- city-state
- Renaissance
- coalition government
- pope

Places to Locate

- Italy
- Sicily
- Malta
- San Marino
- Vatican City
- Rome
- Apennines
- Po River
- Venice

Reading Strategy

Create a chart like this one and fill in at least one key fact about Italy under each category heading.

Italy	
Land	History
Climate	Government
Economy	People

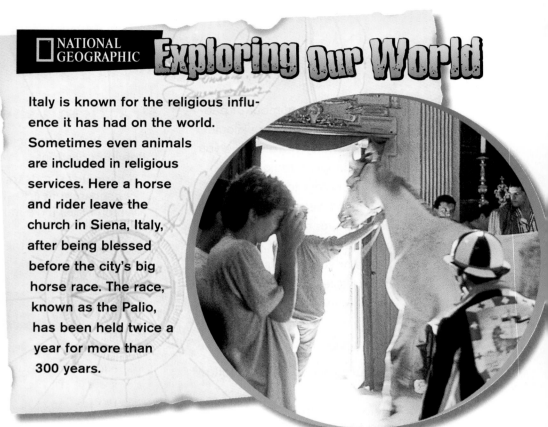

NATIONAL GEOGRAPHIC Exploring Our World

Italy is known for the religious influence it has had on the world. Sometimes even animals are included in religious services. Here a horse and rider leave the church in Siena, Italy, after being blessed before the city's big horse race. The race, known as the Palio, has been held twice a year for more than 300 years.

The country of **Italy** is a land of seacoasts. Look at the map on page 324. You see that the Italian peninsula sticks out from Southern Europe into the center of the Mediterranean Sea. The peninsula looks like a boot about to kick a triangle-shaped football. The "football" is **Sicily,** an island that belongs to Italy. Another large Mediterranean island, Sardinia, is also part of Italy. South of Sicily, just before you come to Africa, lies **Malta,** an independent island country that has close ties to the United Kingdom as well as to Italy.

Two tiny countries—**San Marino** and **Vatican City**—lie within the Italian "boot." San Marino is near the northeastern coast. Vatican City—the world's smallest nation—lies completely within **Rome,** Italy's capital.

Italy's Land and Climate

With a land area of 116,324 square miles (301,277 sq. km), Italy is a little larger than Arizona. Mountains and highlands run down Italy's length. Locate Italy's mountain regions on the map on page 325. The Alps tower over northern Italy, separating the Italian peninsula from

France, Switzerland, Austria, and Slovenia. Another mountain range, the **Apennines** (A•puh•NYNZ), runs down the center of Italy to the toe of the boot. Low hills flank these mountains on the east and west sides.

The rumbling of volcanic mountains echoes through the southern part of the peninsula and the island of Sicily. Mount Etna, Europe's highest active volcano, rises on Sicily's eastern coast. Throughout history, the southern part of Italy has experienced volcanic eruptions and earthquakes.

Although mountains and hills dominate most of Italy, you can also find lowland regions. In the northeast, the largest lowland is the **Po River** valley. The Po, Italy's longest river, flows 400 miles (644 km) from the French border to the Adriatic Sea.

The map below shows you that most of Italy has a mild Mediterranean climate of sunny summers and rainy winters. In spring and summer, hot dry winds called siroccos blow across Italy from North Africa. In fall and winter, the land receives cool, moist air from the Atlantic Ocean. Most of Italy receives enough rain to grow crops.

Italy has two other climate regions. The Po River valley has a humid subtropical climate, with warm, humid summers and cool, rainy winters. The Alps in the far north of Italy give this area a highland climate. Temperatures there tend to be cool or cold.

✓ Reading Check Where are Italy's two main mountain regions located?

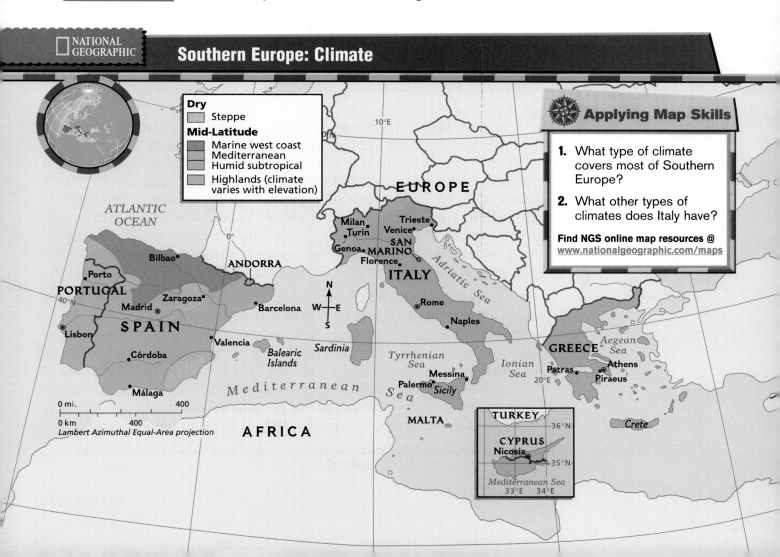

Southern Europe: Climate

Dry
- Steppe

Mid-Latitude
- Marine west coast
- Mediterranean
- Humid subtropical
- Highlands (climate varies with elevation)

Applying Map Skills

1. What type of climate covers most of Southern Europe?

2. What other types of climates does Italy have?

Find NGS online map resources @ www.nationalgeographic.com/maps

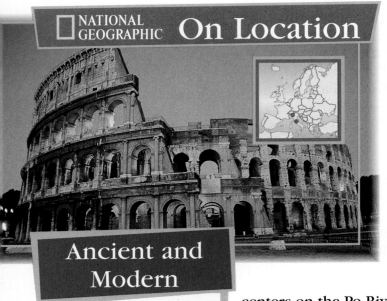

Ancient and Modern

Italy has both the old—Rome's Colosseum—and the new—the stylish models of Milan's fashion industry.

Place How has Italy's economy changed in the past 50 years?

Italy's Economy

In the past 50 years, Italy has changed from a mainly agricultural country into one of the world's leading industrial economies. A member of the European Union, Italy has a large market for its goods, which include cars and home appliances. Many of those goods are produced by small, family-owned businesses rather than by large corporations. Italian businesses are known for creating new designs and methods for making products.

Prosperous North and Center Fertile soil and steady, year-round rainfall give northern Italy many benefits. Most of Italy's agriculture centers on the Po River valley. Farmers grow sugar beets, corn, wheat, vegetables, and fruits. In hilly areas of northern and central Italy, farmers grow grapes for wine. Italy is one of the world's major wine-producing countries.

The Po River valley is even more important as a manufacturing center. Factories here produce most of Italy's manufactured goods. Skilled workers in the cities of Milan and Turin make cars, machinery, chemicals, clothing, and leather goods. Genoa, a thriving port city, also has factories.

Italy has few natural resources, so most of the raw materials used in these factories must be imported. Italy also imports most of its energy. Fast-running rivers in the Alps do provide some hydroelectric power, though.

Tourism is also important in northern and central Italy. Resorts in the Alps attract skiers. **Venice,** to the northeast, is built on 117 islands. You find no cars in this city, which is crisscrossed by canals and relies on boats for transportation. Other cities also have their unique features. The city of Florence is known for its art treasures. Pisa takes pride in its famous leaning tower, and Milan in its opera house. In central Italy lies the star attraction—Rome. The seat of the ancient Roman Empire, it is Italy's capital and the center of Christianity. Here you can see ancient Roman ruins and magnificent churches and palaces.

The Developing South Southern Italy is poorer and less developed than northern and central Italy. It lacks hydroelectric power and the rich soil found in the Po River valley, although the soils are good for growing citrus fruits, olives, and grapes. Still, farming methods are often outdated, and much of the region's dry land is used for pasturing animals. While southern Italy attracts tourists, especially to the Roman ruins at Pompeii, unemployment and poverty are common. Many southern Italians have moved to northern Italy or to other parts of Europe.

✔ **Reading Check** Why are northern and central Italy more prosperous than southern Italy?

Italy's History and Culture

For hundreds of years, Italy was the heart of Western civilization. The city of Rome became a powerful force in the Mediterranean in the 200s B.C. Roman armies conquered a huge empire that stretched from the eastern Mediterranean to Spain and from North Africa to the British Isles. Rome influenced the government, arts, and architecture of much of Europe.

After the fall of the Roman Empire in the A.D. 400s, Italy broke into many small territories and city-states. Each **city-state** consisted of a city and its surrounding countryside. In the 1300s, wealthy merchant families arose in many city-states. Their money helped artists and scholars explore new ways of thinking. The result was the **Renaissance**—a French word meaning "rebirth." From the 1300s to 1600, this time of profound cultural achievements in the arts and learning spread throughout Europe.

European powers such as Spain and Austria took control of parts of Italy, however. In addition, rivalry grew among some of Italy's city-states. As a result, Italy did not become a unified nation until the mid-1800s. From the 1920s to the early 1940s, dictator Benito Mussolini ruled the country. He joined with Germany's Adolf Hitler and pulled Italy into World War II. Italy was defeated in the war, and Mussolini was removed from power.

Italy's Government After World War II, Italy became a democratic republic. Yet democracy did not bring a stable government. Since the late 1940s, political power has constantly changed hands. Rivalry between the wealthy north and the poorer south has caused political tensions. In addition, many political parties exist, and no single

Architecture

For 800 years the Leaning Tower of Pisa has stood as a monument to construction mistakes. Begun in 1173, the tower began to tilt even before it was finished. Over time the tower moved even more, until by 1990, it leaned 12 feet (3.7 m) to the south. Fearing the tower would fall over, experts closed it. They added 800 tons (726 t) of lead weights to its base. They also removed 30 tons (27 t) of subsoil from underneath the north side of the tower in hopes that it would sink the opposite way. Visitors have once more returned to the tower.

Looking Closer Why do you think experts fixed the tower's problem but still left it leaning?

Traveling Water Boulevards

Carlo lives in Venice, Italy—a city made up of 117 islands. He gets around by jumping on a *vaporetto* (water bus) or gondola (also called a water taxi). You see, most of the city's streets are canals. For fun, Carlo and his friends cruise the "streets" of Venice by boat.

party has been strong enough to gain control. Instead, Italy has seen many coalition governments, where two or more political parties work together to run a country.

Italy's People   If you look at the population density map on page 337, you will see that parts of Italy are very crowded. About two-thirds of Italy's 57.7 million people live on only one-fourth of the land. Why? Mountains take up much of the country. The Po River valley is the most densely populated region. Many Italians have moved to northern cities, such as Milan and Turin, from the poorer southern region in order to find jobs.

About 70 percent of Italy's people live in towns and cities, and more than 90 percent of Italians work in manufacturing and service industries. Most Italians—more than 95 percent—are Roman Catholic. Celebrating the church's religious festivals is a widely shared part of Italian life. Vatican City in Rome is the headquarters of the Roman Catholic Church. The pope, who is the head of the church, lives and works here. Vatican City has many art treasures as well as the world's largest church, St. Peter's Basilica.

The people of Italy speak Italian, which developed from Latin, the language of ancient Rome. Italian is closely related to French and Spanish. Do you know any Italian words? Many have been adopted into the English language. Pasta, made from flour and water, is the basic dish in Italy. Some pasta dishes are spaghetti, lasagna, and ravioli.

✓ Reading Check   **Why have coalition governments been necessary in Italy?**

Section 2 Assessment

Defining Terms

1. **Define** sirocco, city-state, Renaissance, coalition government, pope.

Recalling Facts

2. **Place** What is the capital of Italy?
3. **Location** Where is Italy's major agricultural region? Why?
4. **Culture** What is Vatican City?

Critical Thinking

5. **Summarizing Information** What are at least five things tourists can see in Italy?
6. **Understanding Cause and Effect** Why are there no cars in Venice?

Graphic Organizer

7. **Organizing Information** Draw a time line like this one and label it with four important events and dates in Italy's history.

├─────────┼─────────┼─────────┤

Applying Geography Skills

8. **Analyzing Maps** Look at the climate map on page 331. In which climate zone is the city of Venice located?

Leonardo da Vinci

The Italian Leonardo da Vinci is considered one of the greatest artists of the Renaissance. He painted the *Mona Lisa* and the *Last Supper*, two of the world's best-known paintings. He was also a talented architect, engineer, and inventor.

The Artist

Leonardo da Vinci was born in 1452 in a small town near Florence, Italy. As the son of a wealthy man, he received the best education that Florence could offer. Leonardo became known for his ability to create sculptures and paintings that looked almost lifelike. Much of his success in this area came from his keen interest in nature. He also studied human anatomy and used this knowledge to make his figures realistic.

The Inventor

As a child, Leonardo was fascinated with machines and began to draw his own inventions. The first successful parachute jump was made from the top of a French tower in 1783—but Leonardo had sketched a parachute in 1485. He designed flying machines, armored tanks, and aircraft landing gear. He even drew a diver's suit that used tubes and air chambers to allow a swimmer to remain underwater for long periods of time.

Leonardo's Notebooks

Much of what we know about Leonardo comes from the thousands of pages of notes and sketches he kept in his notebooks. He used mirror, or reverse, writing, starting at the right side of the page and moving across to the left. No one is sure why Leonardo wrote this way. Some think he was trying to keep people from reading and stealing his ideas. He may also have been trying

to hide his thoughts from the Roman Catholic Church, whose teachings sometimes conflicted with his ideas. From a practical standpoint, writing in reverse probably helped him avoid smearing wet ink, since he was left-handed.

▲ The *Mona Lisa*

▲ Leonardo da Vinci

Making the Connection

1. What are two of Leonardo's best-known works?

2. Why might Leonardo have written his notebooks in mirror writing?

3. **Understanding Cause and Effect** In what way did Leonardo's interest in the world around him influence his work?

Greece

Guide to Reading

Main Idea

With a rugged landscape, Greece relies on sea trade and tourism to build its economy.

Terms to Know

- mainland
- elevation
- suburb

Places to Locate

- Athens
- Greece
- Balkan Peninsula
- Pindus Mountains
- Peloponnesus
- Crete
- Aegean Sea
- Cyprus

Reading Strategy

Create a diagram like this one. In each box give an example of something that would attract tourists to Greece.

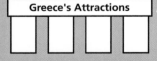

Greece's Attractions

NATIONAL GEOGRAPHIC **Exploring Our World**

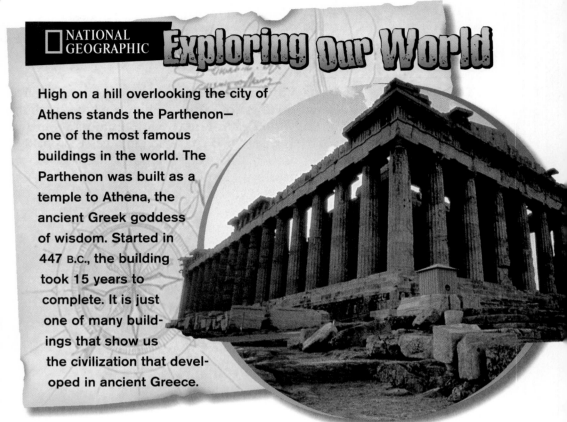

High on a hill overlooking the city of Athens stands the Parthenon—one of the most famous buildings in the world. The Parthenon was built as a temple to Athena, the ancient Greek goddess of wisdom. Started in 447 B.C., the building took 15 years to complete. It is just one of many buildings that show us the civilization that developed in ancient Greece.

In ancient times, **Athens** was a small town. Today the city is a modern urban center that is growing rapidly in size and population. The ancient Greeks would be amazed to see the changes that have taken place in Athens over the centuries. Yet they would find many other places in **Greece** that have changed very little. Outside the large cities, much of Greek life still follows traditional ways.

Rugged Land, Mild Climate

Much of Greece sits on the southern tip of the **Balkan Peninsula** that juts out from Europe into the Mediterranean Sea. This is the Greek mainland—the major part of a country connected to a large landmass. Greece also includes 2,000 islands around the mainland. Greece—both mainland and islands—is about the size of Alabama.

Greece has a Mediterranean climate with hot, dry summers and mild, rainy winters. As you learned in Chapter 2, however, climate may vary, depending on elevation, or height above sea level. High-elevation areas in Greece are often cooler and wetter than lowlands.

The physical map on page 325 shows you that most of Greece is rocky and mountainous. Rugged ranges separate valleys and plains. Long arms of the sea reach into the coast, forming many smaller peninsulas. Like other Mediterranean areas, Greece is often shaken by earthquakes.

The **Pindus Mountains** run through the center of the Greek mainland. The Pindus and other smaller ranges divide Greece into many separate regions. Historically, this has kept people in one region isolated from people in other regions. Because of the poor, stony soil, most people living in the highlands graze sheep and goats. Constant grazing over the centuries has destroyed natural forests. Scrubby plants continue to grow, but they do little to halt soil erosion.

To the east of the Pindus Mountains lie two fertile lowlands—the Plain of Thessaly and the Macedonia-Thrace Plain. These two plains are Greece's major farming areas. At the southern end of the Pindus range is still another lowland, the Plain of Attica. About one-third of Greece's people live here. Athens, the Greek capital, lies on the Plain of Attica.

Southwest of the Plain of Attica lies the **Peloponnesus** (PEH•luh•puh•NEE•suhs). This large peninsula has rugged mountains and deep valleys. If you had traveled through this region about

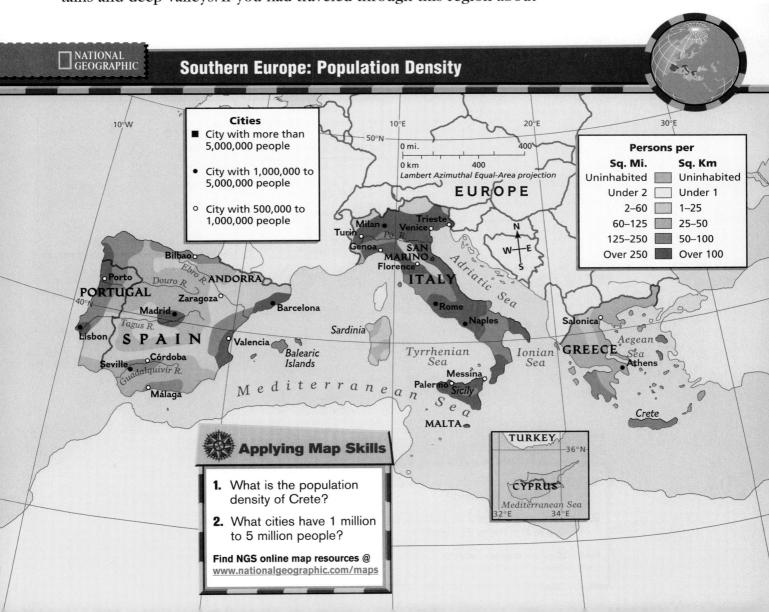

NATIONAL GEOGRAPHIC

Southern Europe: Population Density

Cities
- ■ City with more than 5,000,000 people
- ● City with 1,000,000 to 5,000,000 people
- ○ City with 500,000 to 1,000,000 people

0 mi. 400
0 km 400
Lambert Azimuthal Equal-Area projection

Persons per

Sq. Mi.	Sq. Km
Uninhabited	Uninhabited
Under 2	Under 1
2–60	1–25
60–125	25–50
125–250	50–100
Over 250	Over 100

Applying Map Skills

1. What is the population density of Crete?

2. What cities have 1 million to 5 million people?

Find NGS online map resources @
www.nationalgeographic.com/maps

2,600 years ago, you might have heard crowds cheering on athletes competing in the first Olympic games.

The Islands   Of the 2,000 Greek islands, only about 170 have people living on them. The largest Greek island, covering more than 3,000 square miles (7,770 sq. km), is **Crete.** Some of the islands are grouped together and named after the body of water in which they are located. The Ionian Islands lie in the Ionian Sea. The Aegean Islands lie in the **Aegean Sea.**

Farther east in the Mediterranean is the island country of **Cyprus.** Once under Turkish and then British rule, Cyprus became independent in 1960. For centuries Greeks and Turks have lived on Cyprus, but fighting between the two groups has resulted in a divided country. A United Nations peacekeeping force patrols a strip of land 112 miles (180 km) long. This strip separates the south (ruled by Greek Cypriots) from the north (controlled by Turkish Cypriots).

✓Reading Check **Where are Greece's major farming areas?**

NATIONAL GEOGRAPHIC

Southern Europe: Economic Activity

Land Use
- Commercial farming
- Forests
- Manufacturing area

Resources
- ✚ Bauxite
- 🚃 Coal
- Copper
- Fishing
- 🌲 Forest
- ⚡ Hydroelectric power
- Iron ore
- Lead
- ◈ Natural gas
- Ⓟ Potash
- Petroleum
- ✳ Uranium
- ▣ Zinc

Applying Map Skills

1. What kinds of animals are raised on the Greek mainland?

2. Where is the largest manufacturing region in Southern Europe?

Find NGS online map resources @ www.nationalgeographic.com/map

Greece's Economy

Greece belongs to the European Union but has one of the least developed economies in Europe. It must import food, fuels, and many manufactured goods. Because of Greece's rocky soil, only about one-third of the land can be used for farming. Yet almost 20 percent of Greek workers make their living from the land. Most Greek farms are small and lie in valleys, where the land is flat and more fertile. Farmers cultivate sugar beets, grains, citrus fruits, and tobacco. Like the other countries of Southern Europe, Greece has important crops of olives, used for olive oil, and grapes, used for wine.

Manufacturing has grown rapidly in recent years. Greek workers produce food and beverages, tobacco products, cement, textiles, and chemicals. Almost 60 percent of workers have jobs in service industries.

Shipping and Tourism   No part of Greece is more than 85 miles (137 km) from the sea. Have you ever read the famous epic poem, the *Odyssey*? Composed by an early poet named Homer, the *Odyssey* tells of a hero who took a long sea voyage. Greece still has one of the largest shipping fleets in the world, including oil tankers, cargo ships, fishing boats, and passenger vessels. Shipping is vital to the economy.

Tourism is another key industry. Each year millions of visitors come to Greece to visit historic sites such as the Parthenon in Athens and the temple of Apollo at Delphi. Others come to relax on beaches and to enjoy the beautiful island scenery. Cruise ships ferry passengers from island to island, while hotels and restaurants cater to the needs of tourists.

Although Greece and other Mediterranean countries rely on the sea for income, their growing populations threaten the sea's environment. Untreated sewage, garbage, and industrial chemicals often are dumped into the sea. These wastes pollute Mediterranean waters.

√ Reading Check   How has Greece's growing population affected the sea?

NATIONAL GEOGRAPHIC **On Location**

Greek Islands

Sparkling blue waters and brilliant white architecture attract tourists to the Greek islands.

Movement In what ways does tourism help the Greek economy?

Greece's History and Culture

Many "firsts" mark the history of Greece. The early Greeks developed theories of geometry, medical science, astronomy, physics, and government. In fact, much of Western civilization grew out of the achievements of ancient Greece.

The early Greeks reached the height of their cultural development during the mid-400s B.C.—the Golden Age of Greece. During this time, the city-state of Athens gave birth to the ideals of Western democracy. Our word "democracy" comes from the Greek language and means "power of the people." The ancient Greeks prized freedom and valued the importance of the individual. Greek thinkers such as Plato and Aristotle laid the foundations of Western philosophy and science. Greek writers such as Sophocles (SAH•fuh•KLEEZ) created dramas that explored human thoughts and feelings.

Lying near the crossroads of Europe and Asia, Greece was attacked and conquered by neighboring peoples. Not until 1829 did Greece become an independent country. During World War II, the country was brutally occupied by German forces. After this conflict, a bitter civil war was fought between Greek conservatives and Communists. In 1968 the Greeks set up their present form of government—a parliamentary republic. A president performs ceremonial duties, while a prime minister and cabinet hold actual political power.

Daily Life   About 65 percent of Greece's 10.5 million people live in urban areas. Athens, the capital and largest city, is home to more than 750,000 people. Another 2.5 million live in its suburbs—the smaller communities that surround a city. The Greeks of today have much in common with their ancestors. They debate political issues with great enthusiasm, and they value the art of storytelling.

More than 95 percent of Greeks are Greek Orthodox Christians. Religion influences much of Greek life, especially in rural areas. Easter is the most important Greek holiday. Traditional holiday foods include lamb, fish, and feta cheese—made from sheep's or goat's milk.

√ Reading Check **How are the Greeks of today like their ancestors?**

Web Activity Visit the *Geography: The World and Its People* Web site at gwip.glencoe.com and click on **Chapter 11—Student Web Activities** to learn more about Greek culture.

Section 3 Assessment

Defining Terms
1. **Define** mainland, elevation, suburb.

Recalling Facts
2. **Place** What is the Peloponnesus?
3. **Government** Why does a United Nations peacekeeping force patrol the island of Cyprus?
4. **Economics** Why could one say that Greece's economy is dependent on the sea?

Critical Thinking
5. **Analyzing Information** The mid-400s B.C. is known as the Golden Age of Greece. List facts that support that title.
6. **Understanding Cause and Effect** Why has much of Greece's land suffered from land erosion?

Graphic Organizer
7. **Organizing Information** On a diagram like this one, list facts about Greece under each of the four topic headings.

History — Land — Greece — People — Economy

Applying Geography Skills

8. **Analyzing Maps** Study the physical map on page 325. What body of water separates Italy and Greece?

Reading Review

Section 1 — Spain and Portugal

Terms to Know
plateau
dry farming
navigable
colony
parliamentary republic
dialect

Main Idea
Spain and Portugal are working toward building prosperous economies.

✓ Location Spain and Portugal occupy the Iberian Peninsula, which is blocked from the rest of Europe by the Pyrenees and dominated by a huge central plateau.

✓ Economics Agriculture is still important to both countries, but the majority of people now work in manufacturing and service industries.

✓ Culture Spain has several distinct cultural regions, each with its own language or dialect.

Section 2 — Italy

Terms to Know
sirocco
city-state
Renaissance
coalition government
pope

Main Idea
Once the center of a mighty empire, Italy has influenced Europe's religions, cultures, and governments.

✓ Economics The Po River valley, in the north, is Italy's main agricultural region and its principal industrial center.

✓ History The ancient Romans made important contributions to Western civilization in language, government, and architecture.

✓ History The Renaissance developed in Italy and spread renewed learning throughout Europe.

✓ Government Rome, Italy's capital, surrounds Vatican City, the world's smallest nation and headquarters of the Roman Catholic Church.

Section 3 — Greece

Terms to Know
mainland
elevation
suburb

Main Idea
With a rugged landscape, Greece relies on sea trade and tourism to build its economy.

✓ Place Greece consists of a mountainous mainland and 2,000 islands.

✓ Economics Shipping and tourism are vital to Greece's economy.

✓ History Ancient Greeks laid the foundations for Western civilization— including science, art, philosophy, government, and drama.

▲ Mount Athos, Greece

Assessment and Activities

Using Key Terms

Match the terms in Part A with their definitions in Part B.

A.

1. mainland
2. Renaissance
3. coalition government
4. sirocco
5. dry farming
6. city-state
7. plateau
8. dialect
9. suburb
10. pope

B.

a. rebirth of learning and the arts
b. land is left unplanted every few years
c. head of the Roman Catholic Church
d. the main landmass of a country
e. high, flat land
f. two or more political parties share power to run a country
g. smaller community surrounding a central city
h. local form of a language
i. hot, dry wind from Africa
j. independent city and the lands around it

Reviewing the Main Ideas

Section 1 Spain and Portugal

11. **Location** What narrow body of water separates Africa and the Iberian Peninsula?
12. **Human/Environment Interaction** How do Spain's farmers grow crops on the central plateau?
13. **Culture** Why did the Castilian dialect become the major language of Spain?

Section 2 Italy

14. **Economics** Why is the Po River valley important to Italy?
15. **Place** Why is Rome an important city?
16. **Culture** What does daily life center on in Italy?

Section 3 Greece

17. **Location** Which mountains are located in the center of the Greek mainland?
18. **Human/Environment Interaction** How does the rocky landscape influence Greece's economic activities?
19. **History** Identify three important people of ancient Greece. Why were they important?

 **Southern Europe**

Place Location Activity

On a separate sheet of paper, match the letters on the map with the numbered places listed below.

1. Mediterranean Sea
2. Strait of Gibraltar
3. Portugal
4. Spain
5. Rome
6. Cyprus
7. Adriatic Sea
8. Greece
9. Aegean Sea
10. Pyrenees

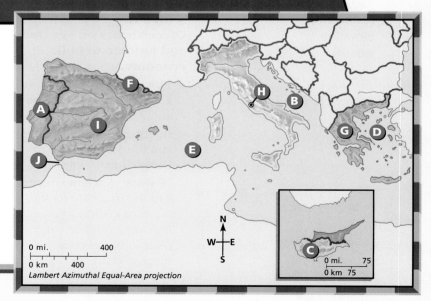

0 mi. 400
0 km 400
Lambert Azimuthal Equal-Area projection

0 mi. 75
0 km 75

Critical Thinking

20. **Drawing Conclusions** Why is the *Odyssey* an appropriate story for a Greek hero?

21. **Making Comparisons** On a diagram like this one, compare the ways in which ancient Greece and Rome influenced Western civilization by writing information under the "Greece" and "Rome" heads.

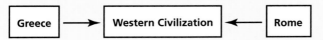

Greece → Western Civilization ← Rome

GeoJournal Activity

22. **Writing and Drawing** Andorra, Vatican City, Malta, and San Marino are four of Europe's smallest countries. The others are Liechtenstein and Monaco. Choose one of these countries and research it. Among other things, these countries are known for their beautiful postage stamps. As part of your research, design a stamp that displays one of the country's important features.

Mental Mapping Activity

23. **Focusing on the Region** Draw a simple outline map of Europe, then label the following:

- Portugal
- Spain
- Pyrenees
- Italy
- Po River
- Greece
- Rome
- Crete
- Sicily
- Mediterranean Sea

Technology Skills Activity

24. **Developing a Multimedia Presentation** Southern Europe is full of places that attract tourists. Imagine that you have taken a two-week trip to this region. Create a multimedia presentation that showcases the places you have visited and what you have learned about the region.

Standardized Test Practice

Directions: Read the paragraphs below, then answer the question that follows.

The ancient Greeks held the Olympic Games in Olympia every four years. The games were a religious festival in honor of Zeus, the Greeks' chief god. Trading and wars stopped while the games took place. The first Greek calendar began with the supposed date of the first Olympic Games in 776 B.C.

Athletes came from all over the Greek-speaking world to compete. Only male athletes, however, were allowed to take part, and women were not permitted even as spectators. Olympic events at first consisted only of a footrace. Later the broad jump, the discus throw, boxing, and wrestling were added. The Greeks crowned Olympic winners with wreaths of olive leaves and held parades in their honor.

1. **Based on the paragraphs, which of the following statements about Greek culture is NOT correct?**

   F The Greeks valued individual achievements.

   G The Greeks believed in many gods.

   H The Greeks believed in being healthy.

   J The Greeks discouraged individual glory.

Test-Taking Tip: Read all the choices carefully before choosing the one that does NOT describe Greek culture. Eliminate answers that you know are incorrect. For example, all the Olympic events were performed by individuals, not by teams. Therefore, answer F *does* describe Greek culture. The question, however, is asking for the statement that does NOT describe Greek culture.

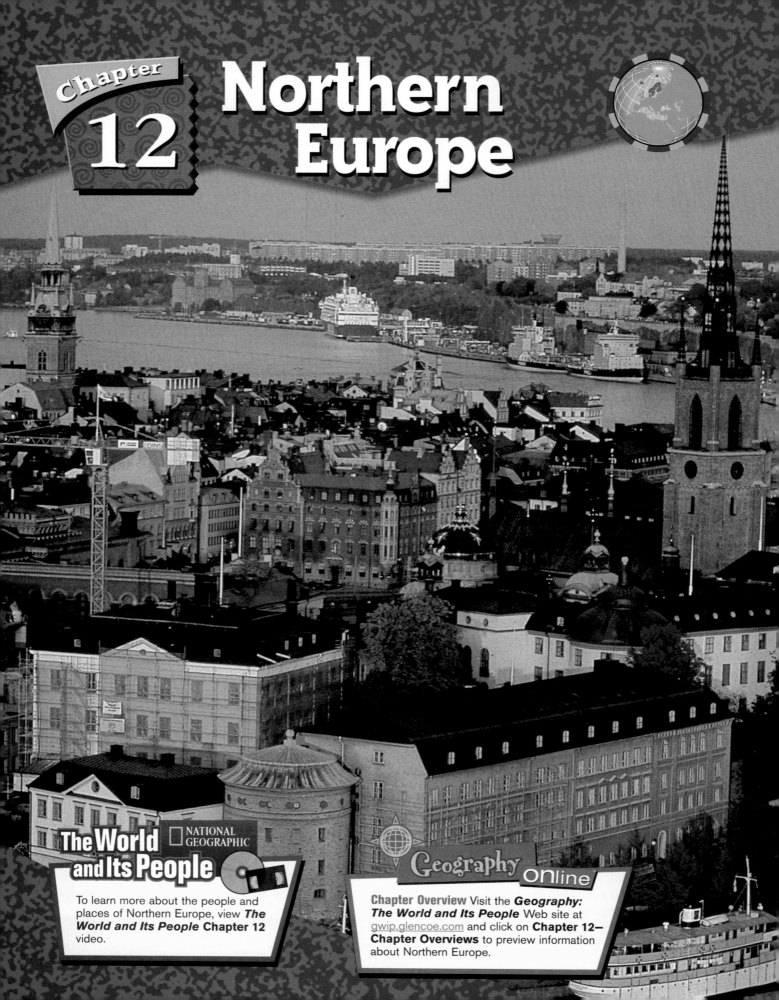

Chapter 12
Northern Europe

The World and Its People — NATIONAL GEOGRAPHIC

To learn more about the people and places of Northern Europe, view *The World and Its People* Chapter 12 video.

Geography Online

Chapter Overview Visit the *Geography: The World and Its People* Web site at gwip.glencoe.com and click on **Chapter 12— Chapter Overviews** to preview information about Northern Europe.

Norway, Sweden, and Finland

◀ Stockholm, Sweden

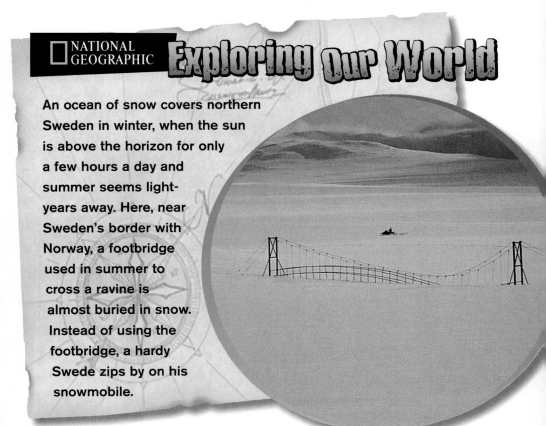

NATIONAL GEOGRAPHIC **Exploring Our World**

An ocean of snow covers northern Sweden in winter, when the sun is above the horizon for only a few hours a day and summer seems light-years away. Here, near Sweden's border with Norway, a footbridge used in summer to cross a ravine is almost buried in snow. Instead of using the footbridge, a hardy Swede zips by on his snowmobile.

The region of Northern Europe—also known as **Scandinavia**—is made up of five countries: **Norway, Sweden, Finland, Denmark,** and **Iceland.** People in these countries have standards of living that are among the highest in the world.

Norway

Jutting out to sea on the Scandinavian peninsula lie Norway and Sweden. Along the peninsula's western edge runs the kingdom of Norway. Its long, jagged coastline on the Atlantic Ocean includes many **fjords** (fee•AWRDS), or steep-sided valleys that are inlets of the sea. Thousands of years ago glaciers slowly moved across the mountainous land. On the seacoast they carved deep, narrow valleys. When the glacial ice eventually melted, the sea level rose. Water then flooded the valleys, producing the fjords. Today the fjords provide Norway with sheltered harbors and beautiful scenery popular with tourists.

The glacier-covered **Kjølen** (CHOO•luhn) **Mountains** tower over northern Norway. Rivers rushing down from mountains provide hydroelectricity to farms, factories, and homes. Only 3 percent of Norway is suitable for agriculture, much of it in the southeast.

Forests cover about 25 percent of Norway. Acid rain is slowly destroying many forested areas, however. Turn to page 362 to learn more about acid rain and its effects on Northern Europe.

About one-third of Norway lies north of the Arctic Circle. This rugged area is often called Land of the Midnight Sun. Here the sun never sets in the midsummer months. In the midwinter months, the sun never rises. Turn to page 352 to find out more about the Midnight Sun.

Norway's far northern location results in a mostly cold climate. However, a mild climate is found along Norway's southern and western coasts, even though this area lies at the same latitude as Alaska. Winds blowing over the North Atlantic Current raise temperatures on the land. Most of Norway's 4.5 million people live in the south within 10 miles (16 km) of the coast, chiefly in urban areas. **Oslo,** the capital and largest city, lies at the end of a fjord on the southern coast.

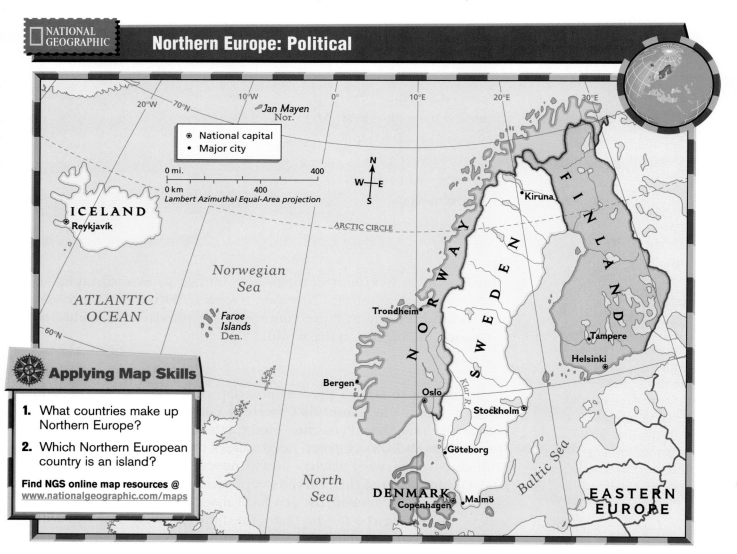

NATIONAL GEOGRAPHIC

Northern Europe: Political

⊛ National capital
• Major city

0 mi.   400
0 km   400
Lambert Azimuthal Equal-Area projection

N
W—E
S

Jan Mayen
Nor.

Kiruna

ICELAND
⊛ Reykjavík

ARCTIC CIRCLE

Norwegian Sea

FINLAND

N
O
R
W
A
Y

S
W
E
D
E
N

ATLANTIC OCEAN

Faroe
Islands
Den.

60°N

Tampere

Helsinki

Trondheim

Bergen•

Oslo

Klar R.

Stockholm ⊛

Applying Map Skills

1. What countries make up Northern Europe?

2. Which Northern European country is an island?

Find NGS online map resources @
www.nationalgeographic.com/maps

Göteborg

Baltic Sea

North Sea

DENMARK
Copenhagen

• Malmö

EASTERN EUROPE

Norway's Economy Norway is a wealthy country, partly because of the seas that lap its coast. Norway began to extract oil and natural gas from beneath the North Sea in the 1970s. Today it is the world's second-largest oil exporter, after Saudi Arabia. The seas themselves provide an important export—fish. The city of Bergen is a major port and fish market. Warm ocean currents keep Bergen and most of Norway's other harbors ice-free all year. Norway's large fleet of commercial ships and cruise ships carries cargo and people around the world.

Norway's History and People Norway's first settlers arrived about 10,000 years ago. They followed herds of reindeer that migrated north as the glaciers retreated. During the A.D. 700s and 800s, Norway's Vikings, seeking land and adventure, raided and traded throughout Europe. They often founded settlements along the way. You can still see unique stave churches that reflect traditions of the Vikings as well as early Christian traditions. These churches are among the world's oldest wooden buildings. About A.D. 1000, a Viking named Leif Eriksson became possibly the first European to explore North America's coast.

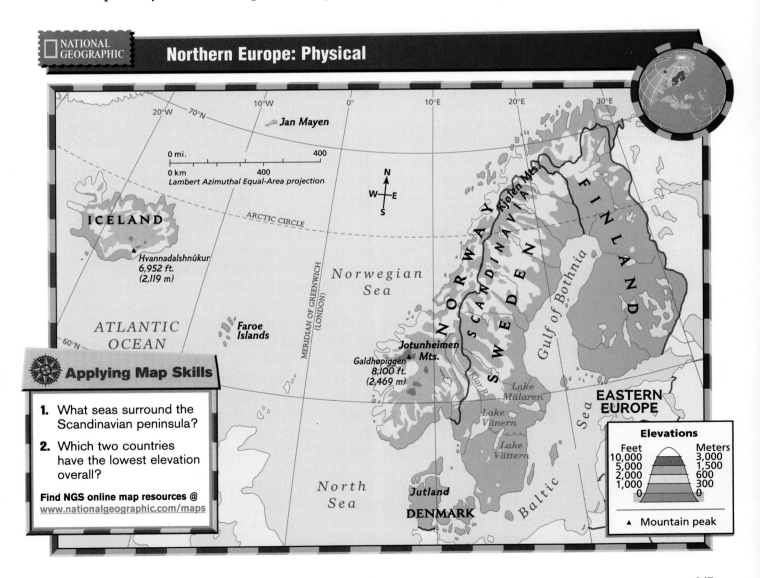

NATIONAL GEOGRAPHIC

Northern Europe: Physical

0 mi. 400
0 km 400
Lambert Azimuthal Equal-Area projection

Jan Mayen

ARCTIC CIRCLE

ICELAND

Hvannadalshnúkur
6,952 ft.
(2,119 m)

ATLANTIC
OCEAN

Faroe
Islands

Norwegian
Sea

MERIDIAN OF GREENWICH (LONDON)

NORWAY

SCANDINAVIA

Kjølen Mts.

SWEDEN

FINLAND

Gulf of Bothnia

Jotunheimen
Mts.
Galdhøpiggen
8,100 ft.
(2,469 m)

Klar R.

Lake
Mälaren

Lake
Vänern

Lake
Vättern

EASTERN
EUROPE

North
Sea

Jutland

DENMARK

Baltic Sea

Applying Map Skills

1. What seas surround the Scandinavian peninsula?

2. Which two countries have the lowest elevation overall?

Find NGS online map resources @
www.nationalgeographic.com/maps

Elevations

Feet		Meters
10,000		3,000
5,000		1,500
2,000		600
1,000		300
0		0

▲ Mountain peak

Norway's Economy

A shopper goes from boat to boat looking for bargains in Bergen's water market (above). Europe's richest oil and natural gas fields are found in the North Sea (right).

Movement What keeps Norway's harbors ice-free all year?

In 1387 Norway came under the rule of tiny but much more populous Denmark to the south—a union that lasted more than 400 years. Then in 1814 the people of Norway came under the rule of Sweden to the east. In 1905 Norway finally became independent. The country is a constitutional monarchy and a parliamentary democracy, like the United Kingdom. A king or queen is head of state, but a prime minister and other elected officials actually run the government.

Norway—with its profitable farming, fishing, and oil industries—is one of Europe's most prosperous nations. Wanting to keep control of its economy, Norway voted not to join the European Union (EU) in 1994. EU membership is still hotly debated, however.

The people of Norway share many cultural characteristics with their Scandinavian neighbors in Sweden, Denmark, and Iceland. The Norwegian language is closely related to their languages. Most Norwegians follow the Protestant Lutheran faith. This form of Christianity came from Germany during the 1500s.

The people of Norway hold on to cultural traditions. You might see them wearing elaborate folk dress at weddings and village festivals. Norwegians are a very modern people, though. Three-fourths of the population lives in cities, and more than one-third owns computers. When they are not typing on keyboards, they may be skiing or riding snowmobiles.

The Sami are an ethnic group that lives north of the Arctic Circle in Finland, Sweden, and Norway. In the past, the Sami herded reindeer and constantly moved. Today many work in mining and forestry.

✓**Reading Check** What type of government does Norway have?

Sweden

Sweden is almost the size of California. Inland snow-covered mountains adjoin forested highlands, fertile lowlands, and then coastal skerries, or rocky islands. Sweden's long coastline touches the **Baltic Sea,** the **Gulf of Bothnia,** and a narrow arm of the North Sea. The country has about 100,000 lakes, most carved by glaciers.

Sweden is colder than Norway. Why is that? Sweden's mountains block the warm winds of the North Atlantic Current. This causes northern Sweden to have cool summers and cold winters. Many coastal ports are frozen for at least a couple of months during the winter. The North Atlantic winds provide the far south with a milder climate.

Sweden's Economy Sweden is a wealthy, industrial country. Its prosperity comes from abundant natural resources. Sweden's powerful northern rivers produce hydroelectricity. Iron ore deposits in the Arctic region supply steel to factories that manufacture cars, machinery, and ships. Timber from Sweden's forests provides lumber for furniture and wood pulp for newsprint. Exports include machinery, motor vehicles, paper products, wood, and electronic products. Only about 8 percent of Sweden's land can be used for farming. Swedish farmers have developed efficient ways to grow crops, and their farms supply most of the nation's food.

Roads and railroads crisscross the southern region. In 2000 a bridge and tunnel system was opened, joining Sweden and Denmark for the first time. This system is 10 miles (16 km) long and connects Malmö, Sweden, with Copenhagen, Denmark's capital.

Sweden's History and People The Vikings had an important role in Sweden's early history. In 1523 Sweden became a separate kingdom apart from Denmark and Norway. King Gustav Vasa turned Sweden from a Roman Catholic to a Protestant country. During the 1600s, Swedish armies conquered much of the area around the Baltic Sea.

Sweden's agricultural economy suffered during the 1800s. Many Swedes emigrated, or moved to other countries. About 1 million Swedes settled in the United States. A turnaround began during the late 1800s. Cities and factories grew, and a new middle class arose.

Sweden's economic wealth enabled it to become a welfare state. A welfare state is a country that uses tax money to help people who are sick, needy, jobless, or retired. Since the 1970s, economic slowdowns and high taxes have limited government spending for welfare. To help its economic growth, Sweden joined the European Union in 1995. The country is a parliamentary democracy.

Most of Sweden's almost 9 million people live in cities in the southern lowlands. **Stockholm** is the country's capital and largest city. Most of the people are Swedes and speak Swedish. Sweden's high standard of living has attracted more than 1 million immigrants from nearby Norway and Denmark and distant Turkey and Vietnam.

✓ Reading Check What three natural resources have helped make Sweden wealthy?

The Sami

The Sami live in northern Finland, Sweden, and Norway. Experts believe the original Sami came to Finland after the last Ice Age, following herds of reindeer. Many Sami still follow a traditional way of life based on herding reindeer, although many also fish, hunt, and work in manufacturing. How important are reindeer to the Sami culture? The various Sami languages have about 400 names for reindeer according to gender, age, color, shape, and so on. Today about 30,000 Sami live in Norway, 20,000 in Sweden, 6,000 in Finland, and 2,000 in Russia.

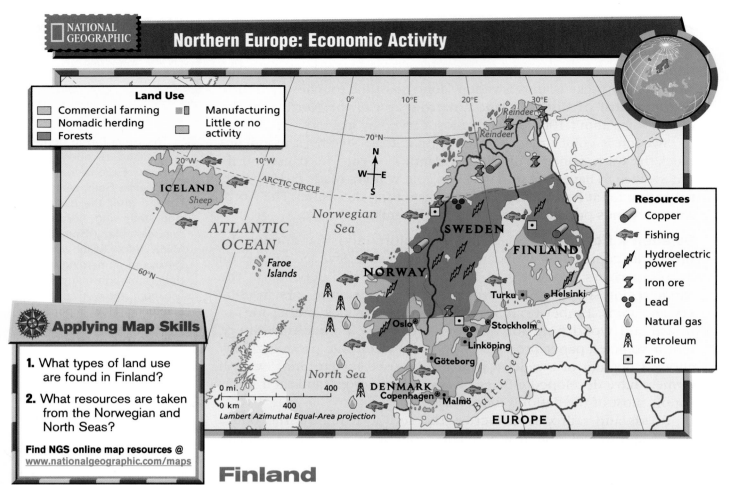

Land Use
- Commercial farming
- Nomadic herding
- Forests
- Manufacturing
- Little or no activity

Resources
- Copper
- Fishing
- Hydroelectric power
- Iron ore
- Lead
- Natural gas
- Petroleum
- Zinc

Applying Map Skills

1. What types of land use are found in Finland?

2. What resources are taken from the Norwegian and North Seas?

Find NGS online map resources @
www.nationalgeographic.com/maps

Finland

Finland lies on a flat plateau broken by small hills and valleys. Its inland areas hold some of the largest unspoiled wilderness in Europe. Thick forests cover two-thirds of the country. Thousands of glacier-formed lakes dot the countryside. If you include marshes and bogs, water covers about 10 percent of Finland's land area.

Finland can get extremely cold in winter. Like Sweden, it lies far from the warm North Atlantic Current. As a result, the country has humid continental and subarctic climates.

Finland's Economy Most of Finland's wealth comes from its huge forests of spruce, pine, and birch. Paper and wood production are important exports. As in Ireland, peat bogs provide fuel. Rivers, flowing from Finland's abundant lakes, yield hydroelectric power.

The Finns have long traded with neighboring Russia. Now they are expanding their markets in the west. In 1995 Finland joined the European Union. In recent years, heavy industry—or manufactured goods such as machinery—has driven Finland's economy. The Finns are also leaders in the electronic communications industry. In fact, Finns as young as 10 carry mobile phones to school.

Finland's best farmland lies in the southwestern part of the country. Farmers raise livestock for dairy products and meat, meeting all of the country's needs. They also grow potatoes and grains. Because of the short growing season, however, Finland must import fruits and vegetables.

Finland's History and People The ancestors of the Finns settled in the region thousands of years ago. These people probably came from what is now Siberia in eastern Russia. As a result, Finnish language and culture differ from those of Finland's Scandinavian neighbors.

By the A.D. 1000s, Swedish Vikings controlled Finland. For almost 700 years, Finland was part of Sweden. Some Swedish customs remain in the culture. Along with Finnish, Swedish is an official language.

In 1809 the Finns came under the control of Russia. During the 1800s, nationalism, or the desire for an independent country, swept through Finland. With the fall of the Russian Empire in 1917, Finland declared its independence as a republic. Finland then became, and remains, a parliamentary democracy. A president serves as head of state, and a prime minister runs the government.

Most of Finland's more than 5 million people live in towns and cities on the southern coast. **Helsinki,** the capital, has over 900,000 people, but the city has still kept a small-town atmosphere. Helsinki, for example, has no high-rise buildings.

Most Finns belong to the Finnish ethnic group. Their language, Finnish, is a Uralic language. Because of centuries under Swedish rule, most Finns practice the Protestant Lutheran faith. With snow on the ground for about half of the year, Finns enjoy cross-country skiing. After outdoor activities, many Finns enjoy relaxing in saunas, or wooden rooms heated by water sizzling on hot stones.

✓ Reading Check **Why is Finnish culture different from the cultures in the rest of Scandinavia?**

Assessment

Defining Terms
1. Define fjord, skerry, emigrate, welfare state, heavy industry, nationalism, sauna.

Recalling Facts
2. Location What countries make up Scandinavia?

3. Movement How was the landscape of Norway created?

4. Economics What resource produces most of Norway's wealth?

Critical Thinking
5. Summarizing Information Why is Sweden called a welfare state?

6. Understanding Cause and Effect Why must Finland import most of its fruits and vegetables?

Graphic Organizer
7. Organizing Information Draw a diagram like the one below. At the end of each of the three arrows, list one way that the sea affects life in Norway.

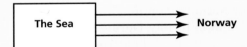

The Sea → Norway

Applying Geography Skills
8. Analyzing Maps Study the physical map on page 347. Which country in this region extends the farthest north? Which country in the region has the lowest overall elevation?

Midnight Sun, Shine On!

At the Equator, the number of hours of daylight is nearly constant year-round. Elsewhere, daylight varies with the latitude and the changing seasons of the year. North of the Arctic Circle, however, the sun shines both night and day for part of the summer. This period is known as the Midnight Sun.

▲ A worker herds the "lawnmowers" for Stockholm's city parks at 3:00 A.M., lit by the Midnight Sun.

Endless Day

The Midnight Sun in Scandinavia is a period of uncommon beauty. During the Midnight Sun, the sun always stays above the horizon. Sunlight shines around the clock, and people can take part in outdoor activities regardless of the time. For many people, the Midnight Sun brings a feeling of celebration. Tourists from around the world flock to the area. The length of the Midnight Sun can vary from a few days to several months, depending on latitude.

Endless Night

In winter, the pattern is reversed. The sun dips below the horizon and stays there. Days or months go by without daylight. Most people continue to carry out their regular routines. Floodlights are turned on to make outdoor activities possible. However, without daylight, the body's own sense of time can be affected. Research has shown that the human body tends to operate on a 24- or 25-hour cycle. Your body's cycle helps keep you alert during the day. It also helps you relax at night. Light and darkness can influence this cycle.

Sunlight and Mood

Although most people are not severely affected by the continuous period of darkness, others develop more serious problems. Some people report trouble falling asleep or difficulty in staying awake during the daytime. Others gain weight. For some, the dark period is a time when they feel continuously depressed. This condition, known as seasonal affective disorder (SAD), also affects people in other parts of the world during winter.

Medical professionals sometimes use lights to help people affected by SAD. Through exposure to special bright lights, patients may be able to adjust their body cycles.

➤ Making the Connection

1. What is the Midnight Sun?
2. What happens to the sun in winter north of the Arctic Circle?
3. **Understanding Cause and Effect** How can one's body be affected by lack of sunlight?

Denmark and Iceland

Guide to Reading

Main Idea

Denmark and Iceland have related histories and rely on the sea for their economies.

Terms to Know

- archipelago
- moor
- geyser
- geothermal energy
- fault line
- saga

Places to Locate

- Jutland
- Copenhagen
- Reykjavík

Reading Strategy

Create a chart like this one, filling in at least two key facts for (1) the land, (2) the economy, and (3) the people of each country.

Denmark	(1)
	(2)
	(3)
Iceland	(1)
	(2)
	(3)

NATIONAL GEOGRAPHIC Exploring Our World

The island of Iceland is called the Land of Fire and Ice because it has both huge glaciers and live volcanoes. Underground hot springs are used to heat homes, buildings, and even swimming pools. Here, Icelanders swim in a pool heated by a geothermal power station. The electric power is produced by steam from the hot springs.

Iceland lies out in the Atlantic Ocean far from mainland Northern Europe. Despite this remote location, the people of Iceland consider themselves part of Northern Europe. They have close cultural ties with the other Scandinavian countries. Denmark is a small country that extends north from Germany toward the Scandinavian peninsula.

Denmark

South of Norway and Sweden, Denmark guards the channel that connects the North Sea and the Baltic Sea. Only about the size of Maryland, Denmark rules the large island of Greenland off the coast of Canada. It also governs the Faroe **archipelago,** or group of islands, in the North Atlantic between Norway and Iceland.

The map on page 347 shows you that most of Denmark is made up of a peninsula known as **Jutland.** The southern border of Jutland touches Germany, Denmark's only land connection to the European mainland. Denmark also includes nearly 500 islands, only 100 of which have people living on them. **Copenhagen,** Denmark's capital, lies on Zealand, the largest island. Throughout history, Denmark's location has made it a link for people and goods between Scandinavia and the rest of Europe. Ferries and bridges connect Jutland and the islands. A bridge and tunnel now join Denmark's Zealand to Sweden.

Most of Denmark is low, rolling grasslands, green hills, woods, and **moors,** or windy treeless land that is often wet. Unlike Norway and Sweden, Denmark has a relatively flat landscape. Its highest elevation, in Jutland's lake district, is only 568 feet (173 m) above sea level.

The North Atlantic Current sweeps northward along Denmark's western coast. Warm winds from this current give Denmark a mild, damp climate. Winter months are cold, and daylight hours are short. Spring and summer have more sunshine and warmer temperatures.

Denmark's Economy   A fairly fertile landscape and a moist, mild climate allow the Danes to farm more than 70 percent of their country. Denmark has some of the richest farmland in Northern Europe. Danish farm products include butter, cheese, bacon, and ham. Food that is exported helps pay for the machinery and raw materials that must be imported. The Danish also export ships, diesel engines, and beautifully designed furniture, silver, and porcelain. Royal Copenhagen porcelain is among the finest in the world. The Danes also invented and export the world-famous Lego toy building blocks.

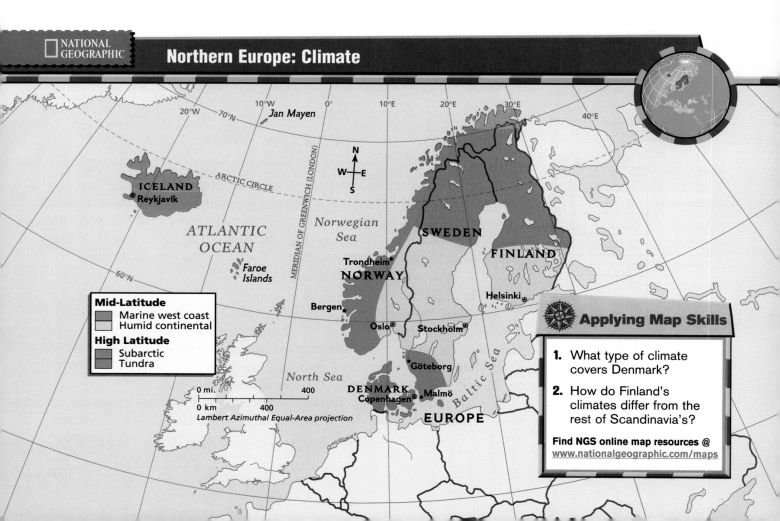

NATIONAL GEOGRAPHIC

Northern Europe: Climate

Mid-Latitude
Marine west coast
Humid continental

High Latitude
Subarctic
Tundra

0 mi.   400
0 km   400
Lambert Azimuthal Equal-Area projection

ICELAND
Reykjavík

ATLANTIC OCEAN

Faroe Islands

Norwegian Sea

NORWAY
Trondheim
Bergen
Oslo

SWEDEN

FINLAND
Helsinki

Stockholm

North Sea

Göteborg

DENMARK
Copenhagen   Malmö

Baltic Sea

EUROPE

Jan Mayen

ARCTIC CIRCLE

MERIDIAN OF GREENWICH (LONDON)

Applying Map Skills

1. What type of climate covers Denmark?

2. How do Finland's climates differ from the rest of Scandinavia's?

Find NGS online map resources @ www.nationalgeographic.com/maps

History and Government Historians believe that the Danes came from Sweden and settled the area that is now Denmark around A.D. 500. About 350 years later, Norwegian Viking warriors conquered Jutland. By the 1000s, their descendants had established a kingdom in Denmark and had converted the country to Christianity.

In the late 1300s, Denmark, Norway, and Sweden formed one kingdom under the Danish Queen Margarethe (mahr•GRAY•tuh). Opposed to Danish controls, Sweden and Norway eventually broke away. Today Denmark is a parliamentary democracy, with a king or queen as head of state and elected officials running the government.

Like Norway, Denmark tried to be neutral in the war-torn 1900s. The German Nazis, however, invaded and ran the country's affairs during World War II. After the war, Denmark prospered and set up a welfare state similar to that in Sweden. In 1993 it joined the European Union.

NATIONAL GEOGRAPHIC On Location

Copenhagen, Denmark

This attractive city is known for its old church spires and red-tiled roofs as well as its modern buildings and parks.

Place What sites attract tourists to Copenhagen?

The Danes The more than 5 million Danes enjoy a high standard of living. About 85 percent of them live in cities or towns. Copenhagen, Denmark's capital, is the largest city in Northern Europe. In the center of Copenhagen sits Tivoli Gardens—one of Europe's oldest amusement parks. In Copenhagen's harbor is another famous attraction: a statue of the Little Mermaid. She is a character from a story by the Danish author Hans Christian Andersen. Andersen, who lived and wrote during the 1800s, is Denmark's most famous writer.

The Danish language is similar to Swedish and Norwegian. Like their Scandinavian neighbors, the Danes mostly are Protestant Lutheran in religion. Although traditional customs remain, Danes pride themselves on being thoroughly modern. People here are less inclined to wear historical clothes and celebrate traditional festivals as people often do in other parts of Europe. Instead of noisy, elaborate occasions, Danes prefer quiet, relaxing evenings in their homes or cozy get-togethers with friends in small cafés.

✓ **Reading Check** What large Atlantic island does Denmark rule?

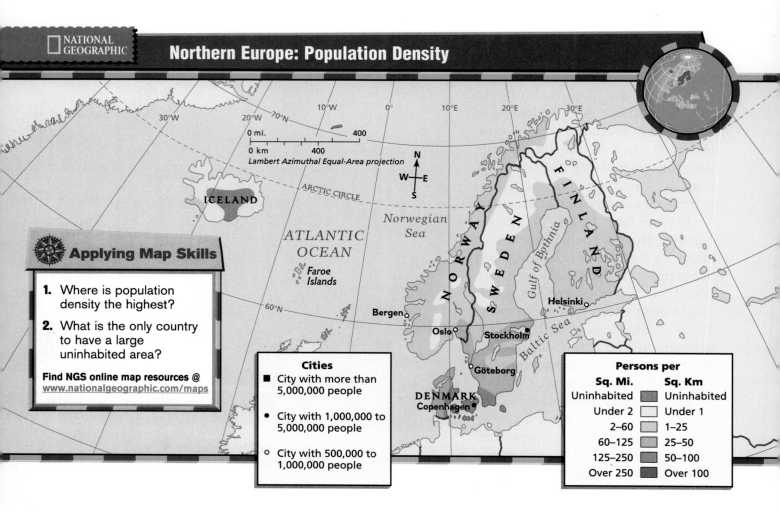

NATIONAL GEOGRAPHIC

Northern Europe: Population Density

Applying Map Skills

1. Where is population density the highest?

2. What is the only country to have a large uninhabited area?

Find NGS online map resources @ www.nationalgeographic.com/maps

Lambert Azimuthal Equal-Area projection

Cities

■ City with more than 5,000,000 people

● City with 1,000,000 to 5,000,000 people

○ City with 500,000 to 1,000,000 people

Persons per	
Sq. Mi.	**Sq. Km**
Uninhabited	Uninhabited
Under 2	Under 1
2–60	1–25
60–125	25–50
125–250	50–100
Over 250	Over 100

Iceland

Iceland was given its chilly name because of its far northern location. Yet in certain places the country's groundwater actually boils. Iceland is a land of hot springs and geysers—springs that shoot hot water and steam into the air. The people of Iceland make the most of this unusual environment. They use geothermal energy, or heat produced by natural underground sources, to heat most of their homes, buildings, and swimming pools.

What makes such natural wonders possible? In the North Atlantic, an undersea range of mountains runs along a fault line, or break in the earth's crust along which movement occurs. Sitting on top of the fault line, Iceland is at the mercy of constant volcanic activity. Every few years, one of the country's 200 volcanoes erupts. The volcanoes heat the springs that appear across the length of Iceland.

Almost 80 percent of Iceland's total land area is made up of glaciers, lakes, and treeless wilderness areas where few people live. Fast-flowing rivers provide a good source for hydroelectric power. The North Atlantic Current warms most of Iceland's coast and keeps temperatures from getting too cold. Iceland has more than its fair share of clouds and rainfall, though. In January, **Reykjavík** (RAY•kyah•veek), the capital, enjoys an average of only three sunny days.

Geography Online

Web Activity Visit the *Geography: The World and Its People* Web site at gwip.glencoe.com and click on **Chapter 12— Student Web Activities** to learn more about the culture of Iceland.

Iceland's Economy   The country's economy depends heavily on fishing. Fish exports provide the money for Iceland to buy food and consumer goods from other countries. For this reason, Iceland is concerned that overfishing will reduce the amount of fish available. In the 1970s, Iceland decided to enlarge the ocean area open only to Icelandic fishing boats. British fishing fleets did not agree with this decision. For a few years, British and Icelandic gunboats exchanged rounds of fire during what became known as the Cod Wars. To reduce its dependence on fishing, Iceland has introduced new manufacturing and service industries.

Iceland's People   Most Icelanders trace their heritage to Vikings who came from mainland Scandinavia during the A.D. 800s and 900s. Sagas, or long tales, written between A.D. 1180 and 1300, celebrate the achievements of Viking heroes and early Icelandic settlers. The country became a parliamentary republic in 1944. A president serves as head of state, and a prime minister is in charge of the government.

About 99 percent of the 300,000 Icelanders live in urban areas. More than half the people live in Reykjavík. The people have a passion for books, magazines, and newspapers. In fact, the literacy rate in Iceland is 100 percent—every adult can read and write.

✓ Reading Check   **How do the people of Iceland take advantage of the country's geysers?**

Cozy Ballet?
Helle Oelkers (far right) is a member of Northern Europe's finest ballet— the Royal Danish Ballet. Helle likes to think that her performance encourages audience members to feel *hygge. Hygge* means feeling cozy and snug. She explains, "The greatest compliment a Dane can give is to thank someone for a cozy evening."

Section 2 Assessment

Defining Terms
1. Define archipelago, moor, geyser, geothermal energy, fault line, saga.

Recalling Facts
2. **Location** With what other European country does Denmark share a border?
3. **Economics** What are five products made in Denmark?
4. **History** Who are the ancestors of today's Icelanders?

Critical Thinking
5. **Analyzing Information** How has Denmark's location affected its relationship with the rest of Europe?
6. **Understanding Cause and Effect** What events led to the Cod Wars?

Graphic Organizer
7. **Organizing Information** Create a diagram like the one below. In the second box list three effects on Iceland from its location on a fault line.

| Fault | → → → | Effects on Iceland |

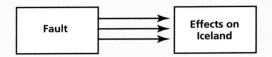

Applying Geography Skills
8. **Analyzing Maps** Study the physical map on page 347. What is Denmark's elevation? At what elevation is central Iceland?

Study and Writing Skill

Using Library Resources

Your teacher has assigned a major research report, so you go to the library. As you wander the aisles surrounded by books, you wonder: Where do I start my research? Which reference tools should I use?

Learning the Skill

Libraries contain many resources. Here are brief descriptions of important ones:

- **Encyclopedia:** set of books containing short articles on many subjects arranged alphabetically

- **Biographical Dictionary:** brief biographies listed alphabetically by last names

- **Atlas:** collection of maps and charts

- **Almanac:** reference updated yearly that provides current statistics and historical information on a wide range of subjects

- **Card Catalog:** listing of every book in the library, either on cards or computerized; search for books by author, subject, or title

- **Periodical Guide:** set of books listing topics covered in magazines and newspaper articles

- **Computer Database:** collections of information organized for rapid search and retrieval

- **World Wide Web:** collection of information on the Internet accessed with a Web browser *(Caution: Some information may not be reliable.)*

Practicing the Skill

Suppose you are assigned a research report dealing with Denmark. Read the questions below, then decide which of the resources on the left you would use to answer each question and why.

1. During which years did Queen Margarethe rule Denmark?
2. What is the current population of Denmark?
3. Besides "The Little Mermaid," what stories did Danish author Hans Christian Andersen write?

Applying the Skill

Using library resources, research the origins and main stories of Icelandic sagas. Find out if the sagas say anything about the land or environment of Iceland. Present the information you find to the class.

◀ The Little Mermaid statue in Copenhagen

Section 1   Norway, Sweden, and Finland

Terms to Know

fjord
skerry
emigrate
welfare state
heavy industry
nationalism
sauna

Main Idea

The economies of Norway, Sweden, and Finland rely on water, forests, and mineral resources.

✓ Region   Northern Europe—also known as Scandinavia—includes Norway, Sweden, Finland, Denmark, and Iceland.

✓ Economics   North Sea oil and gas have made Norway a wealthy nation.

✓ Human/Environment Interaction   Sweden's prosperity comes from vast forests, rich deposits of iron ore, and waterpower.

✓ Economics   Sweden is a welfare state, although government spending for welfare has been limited.

✓ Economics   With its thick forests, Finland is a major producer of wood and paper products.

Urnes Stave Church near Sognefjord, Norway ▶

Section 2   Denmark and Iceland

Terms to Know

archipelago
moor
geyser
geothermal energy
fault line
saga

Main Idea

Denmark and Iceland have related histories and rely on the sea for their economies.

✓ Movement   Throughout history, Denmark's location has made it a link for people and goods between Scandinavia and the rest of Europe.

✓ Economics   With rich farmland, Denmark exports butter, cheese, bacon, and ham.

✓ Government   Denmark and most other Scandinavian countries are parliamentary democracies, with a king or queen as head of state and elected officials running the government.

✓ Place   Most of Iceland has glaciers, lakes, and treeless wilderness areas where few people live.

✓ History   Most Icelanders trace their heritage to Vikings who came from mainland Scandinavia during the A.D. 800s and 900s.

Assessment and Activities

Using Key Terms

Match the terms in Part A with their definitions in Part B.

A.

1. fjord
2. saga
3. emigrate
4. skerry
5. geothermal energy
6. heavy industry
7. geyser
8. fault line
9. nationalism
10. sauna

B.

a. long tale or story
b. heat produced by underground steam
c. rocky island
d. production of industrial goods
e. to move to another country
f. steep-sided inlet of the sea
g. wooden rooms heated by water sizzling on hot stones
h. hot spring that spouts hot water
i. loyalty or pride in one's country
j. cracks in the earth's crust

Reviewing the Main Idea

Section 1 Norway, Sweden, and Finland

11. **Location** Why is part of Norway called the Land of the Midnight Sun?
12. **Human/Environment Interaction** Which of Norway's industries depend on the sea?
13. **Culture** Who are the Sami? Where do they live?
14. **History** Why have so many people from other parts of the world immigrated to Sweden?
15. **Economics** What resource produces most of Finland's wealth?
16. **Location** Why is Finland so cold?

Section 2 Denmark and Iceland

17. **Government** Which North Atlantic islands does Denmark control?
18. **Location** On which peninsula is Denmark located?
19. **Economics** How do Icelanders obtain the money to buy consumer goods and food from other countries?
20. **Culture** What is Iceland's literacy rate?

NATIONAL GEOGRAPHIC **Northern Europe**

Place Location Activity

On a separate sheet of paper, match the letters on the map with the numbered places listed below.

1. North Sea
2. Baltic Sea
3. Iceland
4. Finland
5. Sweden
6. Kjølen Mountains
7. Denmark
8. Norway
9. Helsinki
10. Copenhagen

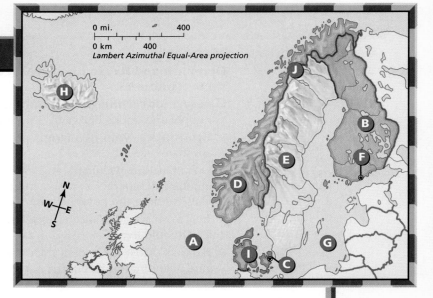

0 mi. 400
0 km 400
Lambert Azimuthal Equal-Area projection

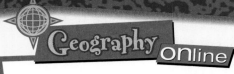

Critical Thinking

21. **Analyzing Information** Why is the name Land of Fire and Ice appropriate for Iceland?
22. **Organizing Information** Create an outline of each country in this section. Use the following guide as your base outline.

    I. Name of Country
      A. Land
        1. Physical features
        2. Climate

      B. Economy
        1. Agriculture
        2. Manufacturing

      C. People

GeoJournal Activity

23. **Writing a News Article** From 1963 to 1967, the island of Surtsey was born in a series of volcanic eruptions off the southern coast of Iceland. Research this event, and write a newspaper article describing it.

Mental Mapping Activity

24. **Focusing on the Region** Draw a simple outline map of Northern Europe, then label the following:

    • Arctic Ocean
    • Baltic Sea
    • Finland
    • Sweden
    • Norway
    • Denmark

    • Iceland
    • Copenhagen
    • Oslo
    • Kjølen Mountains

Technology Skills Activity

25. **Building a Database** Search the Internet to find important facts about one of the countries in this chapter. Use the information to create a database for visitors to the country. Include climate information, currency, foods, and holidays.

Standardized Test Practice

Directions: Study the graph below, then answer the question that follows.

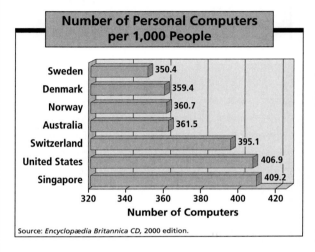

Number of Personal Computers per 1,000 People

Country	Number
Sweden	350.4
Denmark	359.4
Norway	360.7
Australia	361.5
Switzerland	395.1
United States	406.9
Singapore	409.2

Number of Computers

Source: *Encyclopædia Britannica CD*, 2000 edition.

1. **Which Scandinavian country has the highest number of personal computers per 1,000 people?**

   A Singapore
   B Switzerland
   C Denmark
   D Norway

Test-Taking Tip: Use the information on the graph to help you answer this question. Look carefully at the information on the bottom and the side of a bar graph to understand what the bars represent. The important word in the question is *Scandinavian*. Other countries may have more personal computers, but which Scandinavian country listed on the graph has the most personal computers per 1,000 people?

RAIN, RAIN
Go Away

Acid Rain Have you ever sucked on a lemon slice? Yow! Lemons make you pucker up because they are high in acid. Rainwater can be acidic, too. Any form of precipitation that contains high amounts of acid is known as acid rain. In some parts of the world, rain or snow falls that is as acidic as lemon juice.

Why does this happen? When cars and trucks burn gasoline, or when factories and power plants burn coal, sulfur and nitrogen compounds are produced. High in the atmosphere, these gases mix with moisture to form sulfuric acid and nitric acid. These acids make rainwater much more acidic than normal. Acid rain is a problem because it

- harms fish and other animals in lakes and streams;
- damages trees and crops;
- washes nutrients out of soils.

Taking Action Europeans are very concerned about acid rain and its effects. Half of the trees in Germany's Black Forest are sick or dying. Forests in Norway, Austria, Poland, France, and the Czech Republic have also been damaged. In Sweden, 20 percent of the lakes contain few or no fish. The same is true of most lakes in southern Norway.

Many European countries are trying to reduce acid rain by

- installing filters on factory smokestacks;
- putting special exhaust systems on motor vehicles;
- building new factories that do not burn coal.

Acid rain eats away at a statue in Rome.

A German factory spews chemicals that cause acid rain.

High acid rain
Medium acid rain
Low acid rain

ICELAND

NORWAY SWEDEN FINLAND

ESTONIA

IRELAND UNITED KINGDOM DENMARK LATVIA LITH.

NETH. BELG. LUX. GERMANY POLAND BELARUS

CZECH REP. SLOVAK. UKRAINE

FRANCE SWITZ. AUST. SLOV. HUNGARY MOLDOVA

PORTUGAL ANDORRA SPAIN CROATIA BOSN.& HERZG. ROMANIA

ITALY YUG. BULGARIA MACED. ALBANIA

MALTA GREECE CYPRUS

Making a Difference

Acid Rain 2000 A project called Acid Rain 2000 is giving students across Europe a chance to study acid rain and its effects. From 2000 to 2005, participating students will be collecting four kinds of environmental data at study sites in Europe.

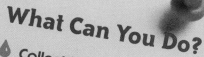
Student collects weather data

- **WEATHER** — Each day, students record the wind direction and the acidity of precipitation.
- **PLANTS** — Once a month, students check the condition of trees and other plants at their sites.
- **SOIL** — Once a month, students test the soil at their sites for acid and plant nutrient levels.
- **LICHENS** — Twice a year, students record the condition of plants called lichens. Since lichens die if the air is too polluted, they are good indicators of a site's air quality.

Acid Rain 2000 participants e-mail the data they collect to Northamptonshire Grammar School, near Northampton, England. There, students and staff process the data and publish the project's findings on the Internet. Acid Rain 2000 hopes to show which areas in Europe are most sensitive to acid rain.

What Can You Do?

Collect Data
Although Acid Rain 2000 is a European project, you can collect similar kinds of data at a study site in your community. For more information about how to set up a site and collect data, contact Acid Rain 2000 at www.brixworth.demon.co.uk/acidrain2000

Investigate
Does acid rain affect your community? If so, what impact has acid rain had on the environment? What are local industries doing to combat the problem? Motor vehicle exhaust contributes to acid rain. What can you do to limit vehicle use on a daily basis?

Use the Internet
Learn more about the international problem of acid rain. Good sites include www.brixworth.demon.co.uk/acidrain2000/sites.htm and the acid rain home page of the U.S. Environmental Protection Agency at www.epa.gov/docs/airmarkets/acidrain

Lichens

The World and Its People NATIONAL GEOGRAPHIC

To learn more about the people and places of Eastern Europe, view **The World and Its People Chapter 13** video.

Geography ONLINE

Chapter Overview Visit the **Geography: The World and Its People** Web site at gwip.glencoe.com and click on **Chapter 13— Chapter Overviews** to preview information about Eastern Europe.

Poland

Guide to Reading

Main Idea

Poland, a large country with a rich history, is undergoing many changes.

Terms to Know

- bog
- communist state
- acid rain
- republic
- pope

Places to Locate

- Poland
- North European Plain
- Carpathian Mountains
- Vistula River
- Oder River
- Warta River
- Warsaw

Reading Strategy

Make a chart like this one, writing three facts in the right column for each category in the left column.

Poland	
Land	
Economy	
People	

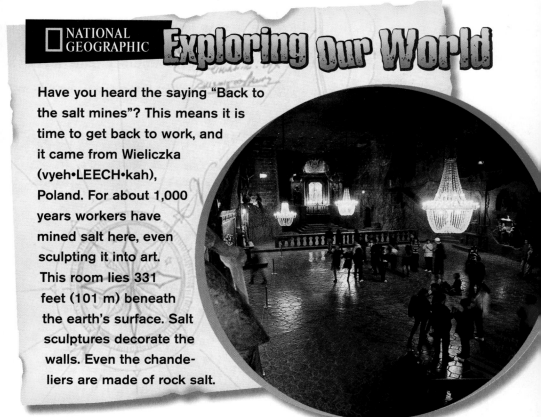

NATIONAL GEOGRAPHIC

Exploring Our World

Have you heard the saying "Back to the salt mines"? This means it is time to get back to work, and it came from Wieliczka (vyeh•LEECH•kah), Poland. For about 1,000 years workers have mined salt here, even sculpting it into art. This room lies 331 feet (101 m) beneath the earth's surface. Salt sculptures decorate the walls. Even the chandeliers are made of rock salt.

Poland is a large country in Eastern Europe. This was not the case in the late 1700s when Poland disappeared from world maps—gobbled up by larger neighbors. In the late 1900s, Poland became a democracy with a free market economy. Since then, many changes have occurred.

Poland's Land and Climate

Find Poland on the map on page 367. About the size of New Mexico, Poland lies on the huge **North European Plain** that stretches from France to Russia. Thick forests once blanketed this flat landscape, but most of the trees were cut down long ago to create farmland. Many Polish people live in this fertile central region.

North toward the Baltic Sea, you find gently rolling land that cradles several thousand glacier-carved lakes. Forests and bogs, or low swampy lands, also dot this northern part of the country. Sandy beaches wind along the Baltic coastline. In the south, the low Sudeten (zoo•DAYT•uhn) Mountains stretch along Poland's border with the Czech Republic. The higher **Carpathian** (kahr•PAY•thee•uhn) **Mountains** form Poland's border with Slovakia.

Poland's rivers begin in the mountains, then twist and turn northward to eventually drain into the Baltic Sea. The **Vistula** (VISH•chuh•luh) **River** begins in the Carpathians and flows more than 680 miles (1,094 km) on an S-shaped course. The **Oder River** and **Warta River** flow through western Poland.

Western Poland has a marine west coast climate. Warm winds blowing across Europe from the Atlantic Ocean bring mild weather year-round. If you like cooler weather, visit eastern Poland. It experiences a humid continental climate of cool summers and cold winters.

√ **Reading Check** What three major rivers flow through Poland?

Eastern Europe: Political

Applying Map Skills

1. What is the capital of Poland?

2. What Eastern European countries border the Adriatic Sea?

Find NGS online map resources @ www.nationalgeographic.com/maps

Legend:
⊛ National capital
• Major city

0 mi.   200
0 km   200
Azimuthal Equidistant projection

A Changing Economy

In the past, Poland was a communist state, or a country in which the government has strong control over the economy and society as a whole. The Polish government decided what, how, and how many goods would be produced. In 1989 Poland started moving to a free market economy. The change has been difficult. In their communist state, workers had jobs for life, even if business was slow. Today businesses lay off workers if they cannot afford to keep a large staff. Poland has met many economic challenges, though. Many people have started businesses, and Poles no longer suffer from shortages of goods.

NATIONAL GEOGRAPHIC

Eastern Europe: Physical

Applying Map Skills

1. What plain covers most of Poland?

2. What four mountain ranges are found in this region?

Find NGS online map resources @ www.nationalgeographic.com/maps

ESTONIA

LATVIA

LITHUANIA

Baltic Sea

Western Dvina R.

NORTH EUROPEAN PLAIN

BELARUS

Warta R.

Vistula R.

Oder R.

Neman R.

POLAND

Pinsk Marshes

RUSSIA

Elbe R.

CZECH REPUBLIC

Danube R.

SLOVAKIA

CARPATHIAN MTS.

Dniester R.

U K R A I N E

Dnieper R.

Triglav 9,393 ft. (2,863 m)

HUNGARY

Tisza R.

Siret R.

MOLDOVA

Lake Balaton

Great Hungarian Plain

SLOVENIA

Sava R.

Drava R.

ROMANIA

Sea of Azov

DINARIC ALPS

CROATIA

Mures R.

Transylvanian Alps

Crimean Peninsula

Adriatic Sea

BOSNIA & HERZEGOVINA

Morava R.

Danube R.

Black Sea

YUGOSLAVIA

BALKAN MTS.

ITALY

MACEDONIA

BULGARIA

▲ Musala 9,596 ft. (2,925 m)

ALBANIA

Aegean Sea

0 mi. 200

0 km 200

Azimuthal Equidistant projection

Elevations

Feet		Meters
10,000		3,000
5,000		1,500
2,000		600
1,000		300
0		0

▲ Mountain peak

Poland's Economy

Coal mining is one of Poland's most important industries and is concentrated near the Czech Republic (above). Polish shoppers do not face most of the shortages that occurred in the past (right).

Human/Environment Interaction What are some products manufactured in Poland?

Poland is dotted with almost 2 million small farms, on which about 28 percent of Poles work. Farmers here grow more potatoes and rye than any other country in Europe. Other crops include wheat, sugar beets, fruits, and vegetables. Some farmers raise cattle, pigs, and chickens.

Most mining and manufacturing takes place in Poland's central and southern regions. The mountains hold copper, zinc, and lead. Poland also has petroleum and natural gas, and produces hydroelectric power. Today—as throughout much of Poland's history—coal mining is one of the most important industries. Factories process foods and make machines, transportation equipment, and chemicals. The city of Gdańsk (guh•DAHNSK), a Baltic Sea port, is an important shipbuilding center.

Environmental Challenges Under Communist rule, Polish factories caused some of the worst water and air pollution in Europe. Since 1989 the government has moved to clean up the environment. Still, problems continue because Polish factories rely on burning coal. Factory smoke causes acid rain, or rain containing chemical pollutants.

✓Reading Check What is one of Poland's most important industries?

Poland's People and History

About 38.7 million people live in Poland. Almost all of them are ethnic Poles. Poles belong to the large ethnic group called Slavs and speak Polish, a Slavic language. Look at the chart on page 388 to see other Slavic languages.

Poland's location and lack of mountains on its eastern and western borders have made the country an easy target for invading armies. Founded in the A.D. 900s, Poland was a powerful kingdom during its early years. By the 1800s, it had fallen victim to powerful neighbors—Germany, Russia, and Austria. In 1939 German troops overran western

Poland, beginning World War II. Poles suffered greatly during the conflict. **Warsaw,** the capital, was bombed to ashes. Some 6 million European Jews and 6 million others were murdered in brutal prison camps set up by the Germans in Poland and elsewhere.

After the war, the Soviet Union swallowed up lands in eastern Poland. In exchange, the Poles gained western areas belonging to defeated Germany. In 1947 a Communist government came to power in Poland. Resisting its rule, workers and farmers in 1980 formed Solidarity, a labor group that struggled peacefully for democratic change. The Communist government finally allowed free elections in 1989, and a new democratic government was formed. A year later, Solidarity leader Lech Walesa (LEHK vah•LEHN•suh) was elected Poland's first democratic president. Today Poland is a democratic republic, a government headed by elected leaders. The people vote for a president who serves for five years. They also elect all the lawmakers.

Daily Life Poland is more rural than nations in other parts of Europe. About one-third of the people live in the countryside. As Poland's economy changes, more people are moving to cities such as Warsaw.

Poles feel a deep loyalty to their country. Religion unites Poles as well. Most are Roman Catholic, and religion has a strong influence on daily life. Poles were very proud in 1978 when Karol Wojtyla (voy•TEE•wah) was named pope, or head of the Roman Catholic Church. Taking the name John Paul II, he was the first Pole ever to become pope.

✓ Reading Check **What two beliefs or attitudes unite the Polish people?**

Section 1 Assessment

Defining Terms
1. Define bog, communist state, acid rain, republic, pope.

Recalling Facts
2. Human/Environment Interaction What happened to most of Poland's forests?
3. Economics Why is Gdańsk important?
4. Culture What is Poland's language and major religion?

Graphic Organizer
5. Organizing Information On a time line like this one, label five important events and their dates in Poland's history.

|—————|—————|—————|—————|

Critical Thinking
6. Understanding Cause and Effect Which of Poland's physical features has made it an easy target for invading armies? Why?
7. Making Comparisons What is the difference in job security under a communist state and the new free market economy?

Applying Geography Skills

8. Analyzing Maps Refer to the physical map on page 367. The Vistula River empties into what body of water? Now turn to the population density map on page 387. What is the population density of the area surrounding Kraków?

Study and Writing Skill

Taking Notes

Effective note taking involves more than just writing facts in short phrases. It involves breaking up information into meaningful parts so that it can be remembered.

Learning the Skill

To take good notes, follow these steps:

- Write key points and important facts and figures quickly and neatly. Use abbreviations and phrases.
- Copy words, statements, or diagrams drawn on the board.
- Ask the teacher to repeat important points you do not understand.
- When studying textbook material, organize your notes into an outline (see page 758).
- For a research report, take notes on cards. Note cards should include the title, author, and page number of sources.

Practicing the Skill

Suppose you are writing a research report on Eastern Europe. First, identify main idea questions about this topic, such as "Who has ruled Poland?" or "What economic activities are found in the Czech Republic?" Then find material about each question.

Using this textbook as a source, read the material on pages 368 and 378 and prepare notes like this:

Main Idea: Who has ruled Poland?
1. Powerful kingdom founded in A.D. 900s
2. 1800s–taken over by Germany, Russia, Austria
3. 1939–Germans invade
4. After WWII, Soviet Union takes over
Main Idea: What economic activities are found in the Czech Republic?
1.
2.
3.

Applying the Skill

In an encyclopedia or on the Internet, find information about Poland's coal industry and the environmental consequences of burning coal. Take notes by writing the main idea and supporting facts. Then rewrite the article using only your notes.

◀ A Czech teenager displays Soviet souvenirs for tourists who flock to Prague.

The Baltic Republics

Guide to Reading

Main Idea

People in Estonia, Latvia, and Lithuania are trying to change their economies, yet keep their cultures.

Terms to Know

- oil shale
- peat

Places to Locate

- Baltic Sea
- Estonia
- Latvia
- Lithuania
- Tallinn
- Rīga
- Vilnius

Reading Strategy

Make a chart like this one, listing ways in which Estonia, Latvia, and Lithuania are similar and different. Write the facts under the correct heading.

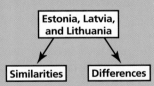

NATIONAL GEOGRAPHIC **Exploring Our World**

The Estonians have a rich tradition of dance, music, and story-telling. Every four years, Estonians stage the Song and Dance Festival to celebrate their people's music. Up to 500,000 people of Estonian descent may take part, many coming from as far away as the United States, Canada, and Australia.

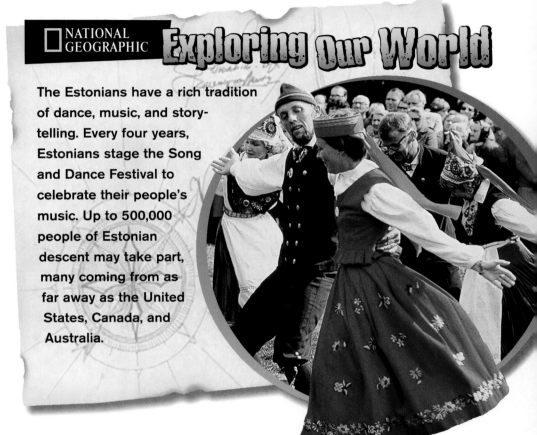

The **Baltic Sea** washes the western shores of three small countries. These lands—**Estonia, Latvia,** and **Lithuania**—have several things in common. In 1940 the Soviet Union took control of all three. Hoping for independence, the countries finally won that freedom in 1991. Today they are all democratic republics with elected leaders. They also have free market economies.

Estonia

The smallest Baltic republic, Estonia is mostly a flat lowland of lakes and rivers. Forests cover about one-third of the land, and about 800 islands dot Baltic coastal areas. Winds from the Baltic Sea give the western part of Estonia cool summers and mild winters. Farther inland, winters can be fairly severe.

Much of Estonia is a wetland and poor for agriculture. On soil that *is* farmable, farmers work hard to grow potatoes and grains. They also raise beef and dairy cattle. Food processing, Estonia's largest manufacturing industry, is followed by textiles, chemicals, cement, and wood products.

Estonia can produce most of the energy it needs. It has large deposits of oil shale, a rock that contains oil. Another resource is peat, wet ground with decaying plants that can be dried and used for fuel. Some peat bogs are as much as 20 feet (6 m) thick.

The Estonians   About two-thirds of Estonia's people are ethnic Estonians. The Estonian language, very different from those of the other Baltic republics, is related to Finnish and Hungarian. Most ethnic Estonians speak Russian as well as Estonian. In fact, about 30 percent of the people are Russian.

Most of Estonia's 1.4 million people live in towns and cities. **Tallinn,** the capital, is on the Baltic coast. Estonia's much older German heritage appears in Tallinn's beautiful churches and castles built hundreds of years ago. Because of historical ties with Germany, about 80 percent of Estonians are Protestant Lutheran.

✓ Reading Check   **What languages do Estonians speak?**

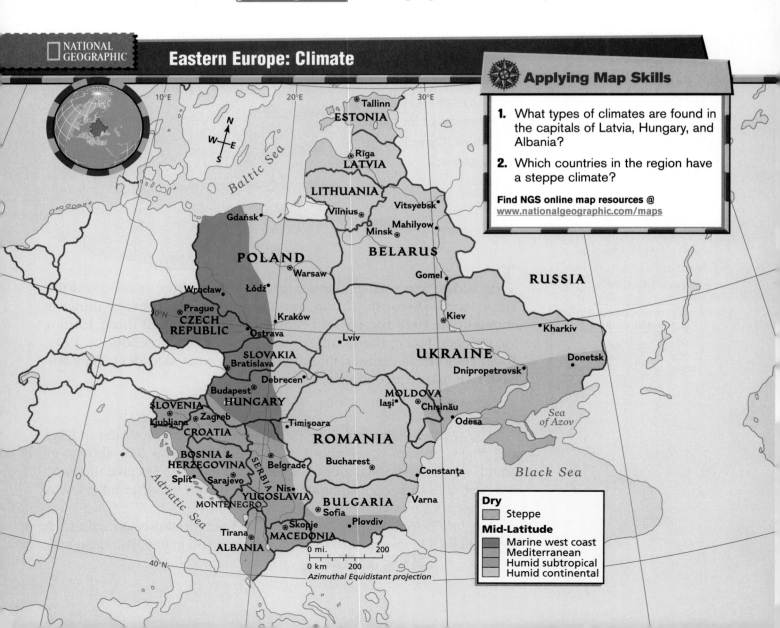

NATIONAL GEOGRAPHIC

Eastern Europe: Climate

Applying Map Skills

1. What types of climates are found in the capitals of Latvia, Hungary, and Albania?

2. Which countries in the region have a steppe climate?

Find NGS online map resources @ www.nationalgeographic.com/maps

Dry
Steppe

Mid-Latitude
Marine west coast
Mediterranean
Humid subtropical
Humid continental

0 mi.   200
0 km   200
Azimuthal Equidistant projection

Latvia

South of Estonia lies Latvia, about the size of West Virginia. Crossing Latvia, you see a landscape of coastal plains, low hills, and forests. The Gulf of Rīga pokes into Latvia's northern coast. This Baltic location has helped make Latvia a trading center. It also has given the country a moderate climate of cool summers and mild winters.

After years of Communist rule, Latvia is working hard to create a free market economy. Dairy and livestock farming is at the heart of agriculture. Latvian farmers also grow potatoes, grains, sugar beets, fruits, and vegetables. Workers produce chemicals, vehicles, wood products, electrical machinery, and textiles. **Rīga,** the capital and largest city, is an important shipping and industrial center.

Latvia's industry is more developed than industry in the other two Baltic republics. The rapid building of factories during Communist days led to widespread pollution, though. Latvia faces the challenge of cleaning up its environment.

The Latvians   Like Estonia, Latvia has been invaded many times throughout its history. The most recent conquerors were the Soviets in the 1940s. At that time, many Latvians fled to the West. More than 200,000 people of Latvian descent live in the United States, the United Kingdom, Canada, and Australia.

Ethnic Latvians now make up only about 56 percent of Latvia's 2.4 million people. Russians—who make up 30 percent—actually outnumber Latvians in the country's largest cities. The government has recently passed a law requiring that all citizens know the Latvian language, which has upset some Russians who do not speak Latvian fluently.

The Latvian language is similar to that of Latvia's southern neighbors, the Lithuanians. Most Latvians are Lutheran like their northern neighbors, the Estonians. In all three countries, large midsummer celebrations begin in late June, when people flock to the countryside to enjoy the lakes and forests.

✔ Reading Check **What is the capital of Latvia?**

Lithuania

About the size of Ireland, Lithuania is the largest of the Baltic republics. The country's mostly flat lowlands are crossed by rivers, dotted with lakes, and only lightly forested. Small hills formed by glacial

NATIONAL GEOGRAPHIC **On Location**

Latvia's Economy

Fish caught in the nearby Baltic Sea are sold in Rīga's Central Market.

Place What products are manufactured in Latvia?

rock deposits rise in the south and east of the country. Thanks to Baltic winds, mild winter and summer temperatures prevail.

Lithuania is the most rural of the Baltic republics. With fertile soil and a mild climate, the country has productive farms that grow potatoes, grains, and sugar beets. Farmers also raise dairy cattle and livestock.

Under Soviet rule, Lithuania developed factories fueled by nuclear power. Workers process foods and make textiles, chemicals, and wood products. Some quarry limestone, a stone often used in making cement. Lithuania is also a major source of amber, the fossilized sap of pine trees, used to make jewelry.

The Lithuanians With 3.7 million people, Lithuania is the most populous of the Baltic republics. Ethnic Lithuanians make up a greater part of their population than do ethnic Estonians or Latvians in their own countries. More than 80 percent of the people are ethnic Lithuanian, while less than 10 percent are Russian. Unlike the other Baltic peoples, Lithuania's people are mostly Roman Catholic. You will find beautiful old Catholic churches in the city of **Vilnius,** the capital.

During its early history, Lithuania had close ties with Poland to the west. In the 1700s, part of Poland and most of Lithuania were absorbed into the Russian Empire. After World War I, the Lithuanians enjoyed a brief period of freedom until the Soviet takeover in the 1940s. With the fall of the Soviet Union in 1991, Lithuania finally became independent.

√ Reading Check **What percentage of Lithuania's people are ethnic Lithuanians?**

Assessment

Defining Terms
1. Define oil shale, peat.

Recalling Facts
2. Human/Environment Interaction What resources enable Estonia to meet most of its own energy needs?

3. History Why are only 56 percent of Latvia's people of Latvian descent?

4. Culture What is the difference between the religion of Lithuania and that of the other Baltic Republics?

Critical Thinking
5. Understanding Cause and Effect What results of Estonia's years of rule by other countries are still visible today?

6. Analyzing Information How does Latvia's location help make it a trading center?

Graphic Organizer
7. Organizing Information Create a diagram like this one. Then list three resources or products for each country at the end of each line.

Estonia	Latvia	Lithuania

Applying Geography Skills

8. Analyzing Maps Refer to the political map on page 366. Which city is farther west, Tallinn, Estonia, or Vilnius, Lithuania?

Hungary, the Czech Republic, and Slovakia

Guide to Reading

Main Idea

Hungary, the Czech Republic, and Slovakia are changing to free market economies.

Terms to Know

- landlocked
- bauxite
- nomad
- spa
- nature preserve
- privatize

Places to Locate

- Hungary
- Czech Republic
- Slovakia
- Danube River
- Great Hungarian Plain
- Budapest
- Prague
- Bratislava

Reading Strategy

Fill in three charts like this one with facts about the past and present of each of the following countries: Hungary, the Czech Republic, and Slovakia.

Country	
Past	
Present	

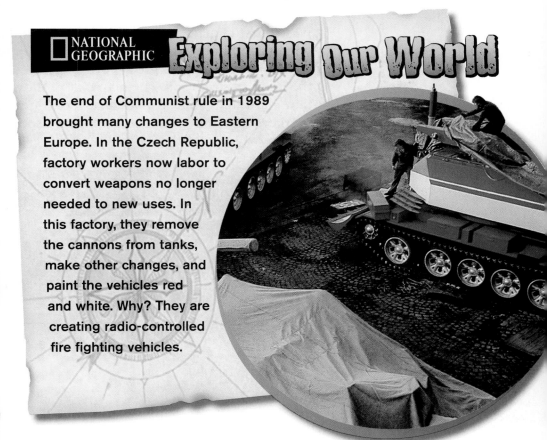

NATIONAL GEOGRAPHIC *Exploring Our World*

The end of Communist rule in 1989 brought many changes to Eastern Europe. In the Czech Republic, factory workers now labor to convert weapons no longer needed to new uses. In this factory, they remove the cannons from tanks, make other changes, and paint the vehicles red and white. Why? They are creating radio-controlled fire fighting vehicles.

In the center of Eastern Europe, you find **Hungary,** the **Czech (CHEHK) Republic,** and **Slovakia** (sloh•VAH•kee•uh). The Czech Republic and Slovakia once were partners in a larger country known as Czechoslovakia.

Hungary

Hungary, almost the size of Indiana, is landlocked, meaning it has no land bordering a sea or an ocean. Hungary depends on the **Danube River** for trade and transportation. This mighty river twists and turns through Hungary and several other countries. Its waters flow 1,776 miles (2,858 km) before emptying into the Black Sea.

The **Great Hungarian Plain** runs through eastern Hungary. This vast lowland area, dotted with farms, has excellent soil for farming and for grazing animals. Many Hungarians here also raise horses. The Danube River separates the Great Hungarian Plain from a very different region to the west. This region is called Transdanubia because it lies "across the Danube." Rolling hills, wide valleys, and forests cover the landscape. Lake Balaton, one of Europe's largest lakes, also lies in Transdanubia. Many Hungarians spend their vacations in this part of the country.

The Carpathian Mountains rise in northern Hungary. In this scenic area, you can wander through thick forests, find strange rock formations, and explore underground caves.

The map on page 372 shows that parts of Hungary have a marine west coast climate and others a humid continental climate with cold winters and hot summers. Western Hungary receives the most rainfall.

Hungary's farmers grow corn, sugar beets, wheat, and potatoes in the country's rich soil. The map below shows that Hungary has important natural resources, such as coal, petroleum, and natural gas. Workers also mine **bauxite,** a mineral used to make aluminum. Foods,

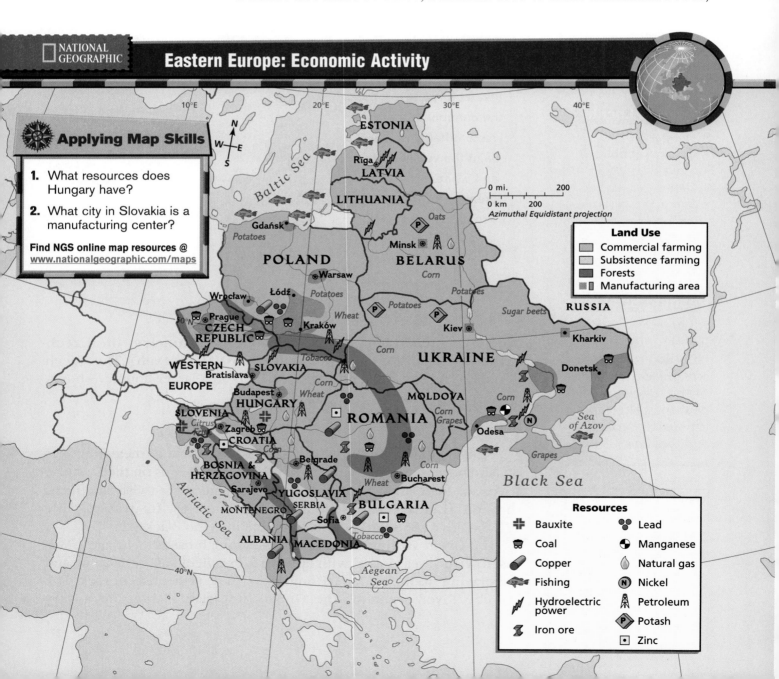

NATIONAL GEOGRAPHIC

Eastern Europe: Economic Activity

Applying Map Skills

1. What resources does Hungary have?
2. What city in Slovakia is a manufacturing center?

Find NGS online map resources @ www.nationalgeographic.com/maps

0 mi. 200
0 km 200
Azimuthal Equidistant projection

Land Use
- Commercial farming
- Subsistence farming
- Forests
- Manufacturing area

Resources
- ✚ Bauxite
- Coal
- Copper
- Fishing
- Hydroelectric power
- Iron ore
- Lead
- Manganese
- Natural gas
- (N) Nickel
- Petroleum
- (P) Potash
- Zinc

Budapest, Hungary

Hungary's capital extends along both banks of the Danube River.

Place What two physical regions does the Danube River separate?

beverages, and tobacco products are manufactured, as well as machines, chemicals, and metals.

In the late 1940s, Hungary's economy came under Communist control. Industry was rapidly increased to meet military needs. When communism collapsed in the 1980s, a free market economy returned. Service industries, such as financial services and tourism, now thrive.

The Hungarians Magyars came to the Danube River valley from central Asia about 1,000 years ago. They were nomads, or people who move from place to place with herds of animals. Skilled horse riders, the Magyars used the grassy plains to feed their animals. They set up a large kingdom in Eastern Europe and adopted Catholicism.

Hungarians value their unique Magyar culture. They struggled to protect it even when they lost their freedom to other peoples. Beginning in the 1500s, the Ottoman Turks and later the Austrians ruled most or all of Hungary.

In 1867 Hungary and Austria became partners in a large empire. When this empire was defeated in World War I, Hungary lost much of its territory and became the small, landlocked nation it is today. Hungary became Communist under the control of the Soviet Union after World War II. In 1956 Hungarians revolted, but Soviet tanks crushed the revolt, killing many people. In 1989 Hungarians finally ended Communist rule and set up a democracy.

Eastern Europe

About 90 percent of Hungary's 10 million people are descended from the Magyars. Almost all speak the Hungarian language. About two-thirds are Roman Catholic, while another one-fourth are Protestant. Two-thirds of Hungarians live in towns and cities. **Budapest** (BOO•duh•PEHST), the capital and largest city, is called "the Paris of Eastern Europe." It is actually two different cities divided by the Danube River. On the western bank lies the old city of Buda, full of beautiful churches and palaces. Bridges link this older settlement to the newer city of Pest, which has factories and tall, modern buildings.

✓ Reading Check **What river is important to Hungary, and why?**

The Czech Republic

The Czech Republic is a landlocked country. About the size of South Carolina, it includes two historic political regions—Bohemia, in the west, and Moravia, in the east. Many areas are known for their natural beauty. In the mountains to the north and south, you can visit spas, or resorts with hot mineral springs that people bathe in to regain their health. You may explore nature preserves, or protected areas for plants and animals. The mountains also show scars caused by Czech industries. You can see the bare trunks of trees killed by pollution and acid rain.

The Czechs enjoy a high standard of living compared to other Eastern European countries. Large fertile areas make the Czech Republic a major agricultural producer. Farmers grow grains, sugar beets, potatoes, and other foods. Manufacturing forms the backbone of the country's economy, however. Factories make machinery, vehicles, metals, and textiles. Food processing is an important industry as well. **Prague** (PRAHG), the capital, is a center of service industries, tourism, and high-technology manufacturing.

The country has some petroleum and natural gas. Minerals include limestone, coal, and kaolin, a fine clay used for pottery. The brown coal that powers factories creates heavy smoke, though, so Czech leaders are trying to move toward nuclear energy.

Before 1989 the Czech Republic had a government-controlled economy based on communism. Since then, it has moved toward a free market. Yet the Communist past continues to haunt the present. Many factories are old, inefficient, and harmful to the environment. The Czechs are trying to modernize them to continue their prosperity.

The Czechs The ancestors of the Czechs were Slavic groups that settled much of Eastern Europe in the A.D. 400s and 500s. By 900, the Czechs had accepted Christianity. They also had formed a kingdom called Bohemia that lasted 600 years. Bohemia became part of the Austrian Empire in the 1500s.

Austrian rule lasted until 1918, when Czechoslovakia was created. Czechoslovakia was under Communist rule from 1948 until 1989, when a new democratic government peacefully voted the Communists out of office. In 1993 the Czechs and Slovaks agreed to split into the Czech Republic and the Republic of Slovakia. Today the Czech Republic is a parliamentary democracy, with a powerful president assisted by a prime minister.

Most of the Czech Republic's 10.3 million people belong to a Slavic ethnic group called Czechs and speak the Czech language. About two-thirds live in cities, many in crowded high-rise apartment buildings. Prague, with about 1.2 million people, is often called "the city of a hundred spires" because of its many church steeples. The entire country is famous for the architectural splendor of its historic buildings and monuments. Musical contributions range from classical to punk. More recently, the country has been a leading European center of jazz.

The Czechs also have produced great literature. Even government leaders are known for their writing skills. The first president of the Czech Republic, Vaclav Havel (VAHT•slahf HAH•vehl), is a noted author of plays.

✓ Reading Check What country ruled Czechoslovakia from the early 1500s until 1918?

Slovakia

About twice the size of New Hampshire, Slovakia is largely a mountainous land. The Carpathian Mountains tower over the northern region. Rugged peaks, thick forests, and blue lakes make this area a popular vacation spot. Farther south, vineyards and farms spread across fertile lowlands that stretch to the Danube River. Farmers grow barley, corn, potatoes, sugar beets, and wine grapes. Under Communist rule, factories were built for heavy industry. Although communism is

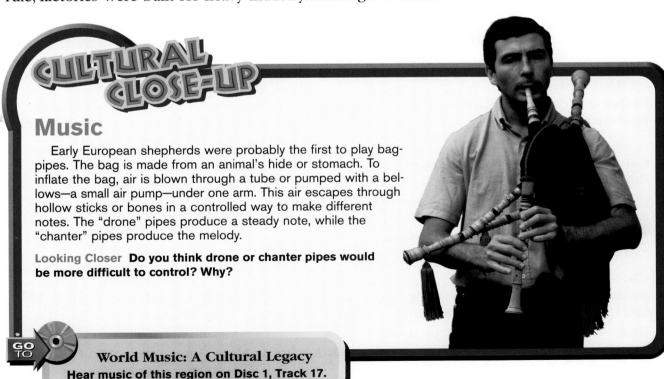

CULTURAL CLOSE-UP

Music

Early European shepherds were probably the first to play bagpipes. The bag is made from an animal's hide or stomach. To inflate the bag, air is blown through a tube or pumped with a bellows—a small air pump—under one arm. This air escapes through hollow sticks or bones in a controlled way to make different notes. The "drone" pipes produce a steady note, while the "chanter" pipes produce the melody.

Looking Closer Do you think drone or chanter pipes would be more difficult to control? Why?

GO TO

World Music: A Cultural Legacy
Hear music of this region on Disc 1, Track 17.

now gone, the push to develop industries continues. The Carpathian Mountains are rich in iron ore, lead, zinc, and copper. Factories use these minerals to produce iron and steel products. Workers also make cement, plastics, textiles, and processed foods.

Slovakia has had difficulty moving to a free market economy. Leaders set out to privatize businesses, or transfer the ownership of factories from the government to individual citizens. Some government officials acted corruptly, giving advantages to themselves or to their friends. This made few foreign companies willing to start new businesses here. Slovak factories also suffer from outdated technology that contributes to air and water pollution.

The Slovaks From the 900s to the early 1900s, Hungary ruled Slovakia. After Hungary's defeat in World War I, the Slovaks joined with the Czechs to form Czechoslovakia. In 1993 rising tensions between the two peoples led them to set up separate countries.

Slovaks make up most of the population. They have a language and culture different from the Czechs. Most Slovaks are Roman Catholic. Nearly 60 percent of Slovakia's 5.4 million people live in modern towns and cities. **Bratislava** (BRAH•tih•SLAH•vuh), a port on the Danube, is Slovakia's capital and largest city. Tourists visit villages to see people dress in traditional clothes for festivals. You might even see musicians playing folk music on shepherds' flutes and bagpipes.

✓ Reading Check What mineral resources does Slovakia have?

Assessment

Defining Terms
1. **Define** landlocked, bauxite, nomad, spa, nature preserve, privatize.

Recalling Facts
2. **Culture** To what ethnic group do most Hungarians belong?
3. **Place** What are three of the Czech Republic's natural resources?
4. **Economics** Why has Slovakia had difficulty in moving to a free market economy?

Critical Thinking
5. **Analyzing Information** Why do the Czechs have a high standard of living?
6. **Understanding Cause and Effect** How do years of Communist control still affect the Czech Republic and Slovakia today?

Graphic Organizer
7. **Organizing Information** Draw a diagram like the one below. Then add at least two facts under the headings in each outer oval.

People — Hungary — Land
History — Hungary — Economy

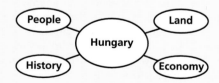

Applying Geography Skills

8. **Analyzing Maps** Turn to the political map on page 366. What countries border Hungary to the north? To the east?

Section 4

The Balkan Countries

Guide to Reading

Main Idea

The Balkan countries have greatly suffered from ethnic conflicts and economic setbacks.

Terms to Know

- consumer goods
- ethnic cleansing
- refugee
- mosque

Places to Locate

- Romania
- Bulgaria
- Albania
- Slovenia
- Croatia
- Bosnia and Herzegovina
- Yugoslavia (Serbia and Montenegro)
- Macedonia

Reading Strategy

Make a chart like this one. For each Balkan country, write a fact or cause in the left box. Then write an effect that results from that fact in the right box.

Cause ➝ Effect

NATIONAL GEOGRAPHIC Exploring Our World

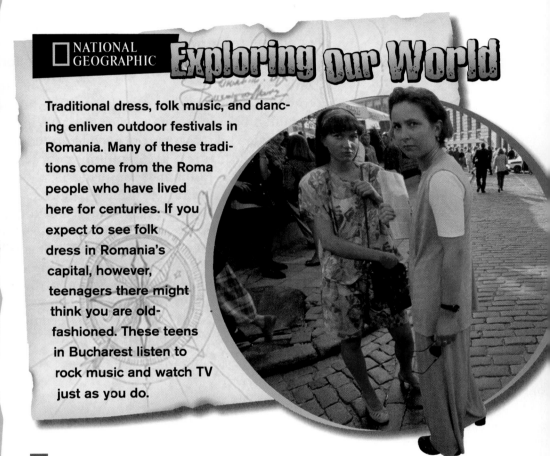

Traditional dress, folk music, and dancing enliven outdoor festivals in Romania. Many of these traditions come from the Roma people who have lived here for centuries. If you expect to see folk dress in Romania's capital, however, teenagers there might think you are old-fashioned. These teens in Bucharest listen to rock music and watch TV just as you do.

Europe's Balkan Peninsula lies between the Adriatic Sea and the Black Sea. The political map on page 366 shows you that several countries make up this Balkan region. They are **Romania, Bulgaria,** the former Yugoslav republics, and **Albania.**

Romania

Romania sits on the northeastern edge of the Balkan Peninsula. The Carpathian Mountains—home to bears, wolves, and other wildlife—take up about one-third of the country's land area. A vast plateau covers central Romania. A coastal region along the Black Sea includes the mouth of the Danube River. Many different birds and fish live in lakes and marshes here. Winters can be very cold and foggy, with much snow. Summers are hot and sunny, but rainfall is abundant.

Romania's economic activities include farming, manufacturing, and mining. The forested mountains and central plateau contain deposits of coal, petroleum, and natural gas. Oil wells stand in the south. Orchards and vineyards stretch along Romania's western, eastern, and southern borders. Farmers also grow grains, vegetables, and herbs here.

Despite abundant resources, Romania's economy has been held back by the Communist policies of the past. Under communism, Romania's factories produced steel, chemicals, and machinery. Few **consumer goods**—clothing, shoes, and other products made for people—were manufactured. Romania now has a free market economy to supply these goods, but aging factories must be updated for Romania's economy to grow. In addition, the country needs to heal an environment widely damaged by air and water pollution.

The Romanians   About 56 percent of Romania's people live in towns and cities. Bucharest, the capital and largest city, has more than 2 million people. What does Romania's name tell you about its history? If you guessed that the Romans once ruled this region, you are correct. Most of Romania's 22.5 million people are descended from the Romans. The Romanian language is closer to French, Italian, and Spanish—which are all Romance languages—than it is to other Eastern European languages. In other ways, the Romanians are more like their Slavic neighbors. Many Romanians are Eastern Orthodox Christians.

After centuries of rule by the Ottoman Turks and the Austrians, Romania finally emerged as an independent kingdom in the 1800s. During the 1900s, Romanians suffered through two world wars and, later, Communist rule. In 1989 they violently overthrew their Communist dictator. Romania then became a democratic republic and moved toward a free market economy.

✔ Reading Check **To what other languages is Romanian related?**

Bulgaria

Mountainous Bulgaria lies south of Romania. Two ranges—the Balkan Mountains and the Rhodope (RAH•duh•pee) Mountains—span most of the country. Fertile valleys and plains are tucked among these mountains. Bulgaria's coast along the Black Sea has warmer year-round temperatures than the mountainous inland areas.

Bulgaria's economy rests on both agriculture and manufacturing. Wheat, corn, and sugar beets grow in the fertile valleys. Roses are grown in the central Valley of the Roses. Their sweet-smelling oil is used in perfumes.

When the Soviet Union collapsed, Bulgaria lost the main market for its goods. The government has been trying to rebuild the economy ever since. Manufacturing depends on the country's deposits of zinc and coal. Factories produce machinery, metals, textiles, and processed foods. Tourism is growing as visitors flock to Bulgaria's resorts on the Black Sea.

The Bulgarians   Most of Bulgaria's 8.2 million people trace their ancestry to the Slavs and other groups from central Asia. Most Slavic people use the Cyrillic (suh•RIH•lihk)

Transylvania

The region of central Romania known as Transylvania has long grabbed our imaginations. Transylvania's mountains and eerie cliff-hanging castles were the setting for English author Bram Stoker's novel *Dracula,* a story about a vampire. Recently, a Spanish doctor noticed that many myths about vampires matched the symptoms of rabies, including pain from bright lights. The doctor discovered that rabies had spread through the region at the same time that the vampire tales began.

alphabet, which was first created to write the Russian language. The Bulgarian language, similar to Russian, is written in this Cyrillic alphabet. Most Bulgarians practice the Eastern Orthodox Christian religion. About 9 percent of the people are Muslim Turks.

For centuries Turkish officials ruled Bulgaria. In the late 1800s, the Bulgarians won their freedom with Russian help. After World War II, Bulgaria's strong ties to Russia made it easy for Communists to take power. Communism in Bulgaria ended in 1989. High joblessness, crime, and rising prices have troubled the change to democracy.

Sofia, with over 1 million people, is the capital and largest city. During the summer, Bulgarians join vacationers from other countries at resorts on the Black Sea coast. Here, modern hotels line wide, sandy beaches.

✓Reading Check What alphabet is used in many Slavic languages?

Former Yugoslav Republics

The former Yugoslav republics used to be one country called Yugoslavia. For years, a Communist dictator named Joseph Broz Tito held the country together. He died in 1980, and communism itself collapsed about 10 years later. In the early 1990s, long-simmering disputes among ethnic groups boiled to the surface and tore the country apart. Five countries emerged: **Slovenia, Croatia, Bosnia and Herzegovina** (HEHRT•seh•GAW•vee•nah), **Yugoslavia** (made up of **Serbia and Montenegro**), and **Macedonia,** also known as the Former Yugoslav Republic of Macedonia (or F.Y.R.O.M.).

After the breakup, Serbia, the strongest country, kept the name of Yugoslavia. It wished to regain control of the other former Yugoslav republics. Serbia also wanted to make sure that Serbs living in the other republics would not lose their rights. As a result, wars erupted throughout the 1990s. Some countries forced people from other ethnic groups to leave their homes, a policy called ethnic cleansing. Hundreds of thousands of people died. About the same number became refugees, or people who flee to another country to escape danger or disaster. These wars left the region badly scarred without promise of long-term peace.

Find the former Yugoslav republics on the physical map on page 367. You see that mountains form the backbone of the landscape. Plains lie in the north where the Danube River flows. The southwest borders the Adriatic Sea. Inland areas are warm in summer and cold in winter. The long coastal strip has warm summers and cool winters.

Slovenia Slovenia, in the northwest of the Balkans region, has rugged mountains and fertile, densely populated valleys. Of all the countries of the old Yugoslavia, Slovenia is the most peaceful and prosperous. With many factories and service industries, it also has the region's highest standard of living. About 52 percent of the 2 million Slovenians live in towns and cities. Most are Roman Catholic.

Croatia Croatia spreads along the island-studded coast of the Adriatic Sea. Then it suddenly swings inland, encompassing rugged mountains and a fertile plain. Zagreb, the capital and largest city, lies in

Albania

Rugged mountains have isolated Albania from neighboring countries.

Movement Why have many Albanians fled the country in recent years?

this inland area. An industrialized republic, Croatia supports agriculture as well. Tourists once flocked to Croatia's beautiful Adriatic beaches, but war has damaged many places.

The Croats, a Slavic group, make up 78 percent of Croatia's 4.6 million people. Another 12 percent are Serbs. Both Croats and Serbs speak the same Serbo-Croatian language, but they use different alphabets. The Croats use the Latin alphabet, the same one that you use for English. The Serbs write with the Cyrillic alphabet. Religion also divides Croats and Serbs. Croats are mainly Roman Catholic, while Serbs are Eastern Orthodox Christians.

Bosnia and Herzegovina Mountainous and poor, Bosnia and Herzegovina has an economy based mainly on crops and livestock. Sarajevo (SAR•uh•YAY•voh), the capital, has the look of an Asian city, with its marketplaces and mosques, or Muslim houses of worship. Many of the Bosnian people are Muslims, followers of the religion of Islam. Others are Eastern Orthodox Serbs or Roman Catholic Croats. Serbs began a bitter war after Bosnia's independence in 1992. The Dayton Peace Accords divided Bosnia into two regions under one government in 1995. American and other troops came as peacekeepers.

Yugoslavia (Serbia and Montenegro) All that is left of Yugoslavia is Serbia and its reluctant partner, Montenegro. Inland plains and mountains cover the area. The economies of these two republics are based on agriculture and industry. The region's largest city is Belgrade. The 10.6 million Serbs and Montenegrins practice the Eastern Orthodox faith.

Serbia has faced growing unrest in some of its local provinces. Muslim Albanians living in the province of Kosovo want independence from Serbia. Also living in Kosovo is a smaller group of Eastern Orthodox Serbs. For centuries, Albanians and Serbs here have felt a deep anger toward each other. In 1999 Serb forces tried to push the Albanians out of Kosovo. The United States and other nations bombed Serbia to force it to withdraw its troops. Peace in Kosovo remains shaky.

Macedonia (F.Y.R.O.M.) Macedonia's 2 million people, mostly farmers, are a mix of different ethnic groups from the Balkans. In Skopje (SKAW•pyeh), Macedonia's capital, there is an amazing mix of ancient Christian churches, timeworn Turkish markets, and modern shopping centers. Close to Kosovo, Macedonia handled a huge wave of ethnic Albanian refugees from Kosovo, who fled Serb forces in 1999.

Reading Check What nations were formed from the former Yugoslavia?

Albania

Bordering the Adriatic Sea, Albania is slightly larger than the state of Maryland. Mountains cover most of the country, contributing to Albania's isolation from neighboring countries. A small coastal plain runs along the Adriatic Sea. Most of Albania has a Mediterranean climate.

Albania is a very poor country. Although the country has valuable mineral resources, it lacks the money to mine them. Most Albanians farm, growing corn, grapes, olives, potatoes, sugar beets, and wheat in mountain valleys.

The Albanians   Almost two-thirds of Albanians live in the countryside. The capital and largest city, Tirana, and its suburbs have a population of more than 500,000. Although 3.5 million people live in Albania, another 3.2 million Albanians live in nearby countries. These refugees fled Albania to escape the violence that swept the country after communism's fall in the early 1990s.

About 70 percent of Albanians are Muslims. The rest are Christian—either Eastern Orthodox or Roman Catholic. While the Communists opposed religion, Albania's democratic government has allowed people to practice their faith. As a result, many mosques and churches have opened across the country. The most famous Albanian in recent times was the Catholic nun Mother Teresa, who served the poor in Calcutta, India.

✓ Reading Check **What is the main religion in Albania?**

Assessment

Defining Terms
1. Define consumer goods, ethnic cleansing, refugee, mosque.

Recalling Facts
2. Place What is the capital of Romania?

3. Economics How are roses used in Bulgaria?

4. History Which of the former Yugoslav republics is most prosperous?

Critical Thinking
5. Drawing Conclusions How do you think people in the Balkans feel about the recent changes in their countries?

6. Understanding Cause and Effect What factors have contributed to the recent conflicts in the Balkans?

Graphic Organizer
7. Organizing Information Draw a chart like the one below and complete it by filling in two facts under each country name.

Romania	Bulgaria	Slovenia	Croatia
Bosnia and Herzegovina	Yugoslavia	Macedonia	Albania

Applying Geography Skills

8. Analyzing Maps Study the physical map on page 367. What mountain range is found on the east coast of the Adriatic Sea?

Ukraine, Belarus, and Moldova

Section

5

NATIONAL GEOGRAPHIC **Exploring Our World**

On April 26, 1986, Reactor 4 of the Chernobyl (chuhr•NOH•buhl) Nuclear Power Plant in Ukraine exploded. An estimated 5,000 people died, and about 30,000 were disabled by exposure to radiation. Some 30,000 square miles of good farmland were poisoned. These vehicles have been permanently scrapped after being used to clean up after the explosion.

Ukraine, **Belarus** (BEE•luh•ROOS), and **Moldova** (mawl•DAW•vuh) once belonged to the Soviet Union. When the Soviet Union broke apart in late 1991, Ukraine, Belarus, and Moldova became independent. Since then, they have struggled to build new economies.

Ukraine

Slightly smaller than Texas, Ukraine is by far the largest Eastern European country. The Carpathian Mountains rise along its southwestern border. Farther east, a vast steppe, or gently rolling, partly wooded plains, makes up the country. Nearly 23,000 rivers twist across the steppe. The most important waterway, the **Dnieper** (NEE•puhr) **River,** has been made navigable so ships can carry goods to distant markets. The **Crimean Peninsula** juts into the Black Sea. Most of Ukraine has a humid continental climate with cold winters and warm summers.

Rich dark soil covers nearly two-thirds of Ukraine. Farms are very productive, earning the country the name "breadbasket of Europe." Farmers grow sugar beets, potatoes, and grains and raise cattle and sheep. Factories make machinery, processed foods, and chemicals.

The Ukrainians   Early Slavic groups settled and traded along the rivers of the region. During the A.D. 800s, warriors from Scandinavia united these groups into a large state centered on the city of **Kiev** (KEE•ihf). A century later, the people of Kiev accepted the Eastern Orthodox faith and built one of Europe's most prosperous civilizations. After 300 years of freedom, the people of Kiev were conquered by Mongols, then Lithuanians and Poles, and finally the Russians.

In the 1930s, Soviet dictator Joseph Stalin brought Ukraine's farms under government control. This action caused a famine in which millions of Ukrainians starved. Millions more died when Germans invaded

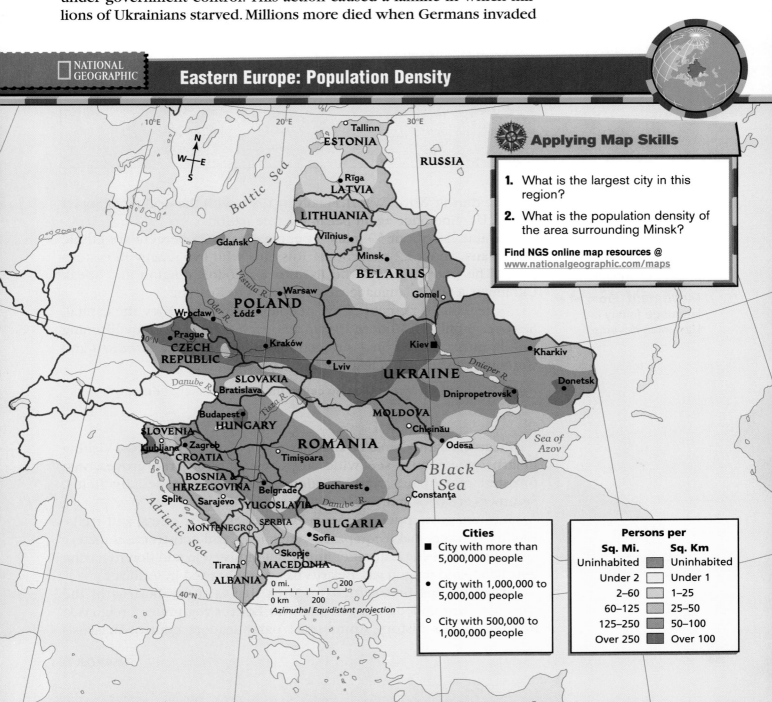

NATIONAL GEOGRAPHIC

Eastern Europe: Population Density

Applying Map Skills

1. What is the largest city in this region?

2. What is the population density of the area surrounding Minsk?

Find NGS online map resources @ www.nationalgeographic.com/maps

Cities

■ City with more than 5,000,000 people

● City with 1,000,000 to 5,000,000 people

○ City with 500,000 to 1,000,000 people

Persons per	
Sq. Mi.	**Sq. Km**
Uninhabited	Uninhabited
Under 2	Under 1
2–60	1–25
60–125	25–50
125–250	50–100
Over 250	Over 100

0 mi.   200
0 km   200
Azimuthal Equidistant projection

Language Families of Europe

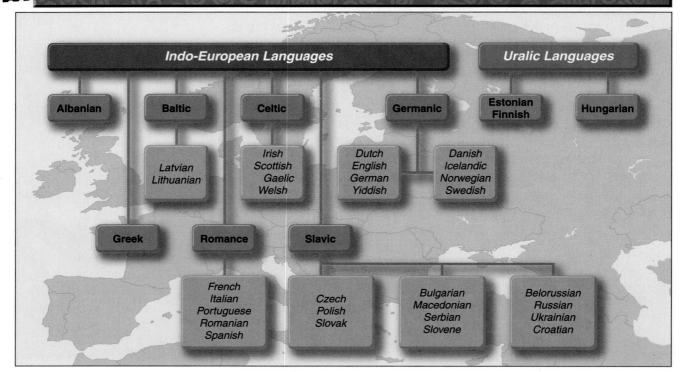

Indo-European Languages				Uralic Languages	

Albanian | **Baltic** | **Celtic** | **Germanic** | **Estonian Finnish** | **Hungarian**

Baltic: Latvian, Lithuanian

Celtic: Irish, Scottish, Gaelic, Welsh

Germanic: Dutch, English, German, Yiddish / Danish, Icelandic, Norwegian, Swedish

Greek | **Romance** | **Slavic**

Romance: French, Italian, Portuguese, Romanian, Spanish

Slavic: Czech, Polish, Slovak / Bulgarian, Macedonian, Serbian, Slovene / Belorussian, Russian, Ukrainian, Croatian

Analyzing the Chart

Seven main language families stem from Indo-European origins.

Movement From what language family did Ukrainian develop?

Ukraine during World War II. Finally in 1991, with the decline of Soviet power, Ukraine once again became a free nation.

Ukraine has about 50 million people. Nearly 75 percent are ethnic Ukrainians. About 22 percent are Russians who live mainly in eastern areas. The people follow the Eastern Orthodox religion and speak Ukrainian, a Slavic language closely related to Russian.

More than 70 percent of the people live in cities. Kiev, the capital, has over 5 million. Modern Ukrainians, even teenagers, enjoy listening to folk music played on a stringed instrument called a *bandura* and watching the acrobatic leaps of the *hopak* dance.

✓ Reading Check **Why is Ukraine called the "breadbasket of Europe"?**

Belarus and Moldova

Belarus, slightly smaller than Kansas, is largely lowlands. Visiting Belarus, you would see wide stretches of birch tree groves, vast forested marshlands, and wooden villages surrounded by fields. Summers are cool and wet, and winters are cold.

Farmers grow potatoes, grains, vegetables, sugar beets, and fruits. Factory workers make equipment, chemicals, and construction materials. Food processing is another important industry. In addition to having petroleum and natural gas, Belarus has potash, used in fertilizer.

Slavic groups first settled the area that is today Belarus in the A.D. 600s. Surrounded by larger countries, Belarus was under foreign rule for most of its history. Communist party leaders control Belarus's government and have maintained close ties with Russia. Foreign companies have been unwilling to do business here. In addition, Belarus is still linked to neighboring Russia's weak economy. For these reasons, the Belarussians have faced many hardships in recent years.

The 10.2 million people of Belarus are mostly Eastern Orthodox Slavs. Their Belarussian language is closely related to Russian and Ukrainian and is written in Cyrillic. Two-thirds of Belarus's people live in cities. **Minsk,** the largest city, is the capital.

Moldova Moldova is mostly a rolling hilly plain sliced by rivers. These waterways form valleys that hold rich fertile soil. This soil, along with mild winters and warm summers, provides productive farmland. Farmers grow sugar beets, grains, potatoes, apples, and tobacco. Some grow grapes used to make wine. Factories turn out processed foods, machinery, metals, construction materials, and textiles.

Moldova's flag looks similar to Romania's flag. Why? Moldova once was part of Romania. About two-thirds of the people trace their language and culture to that country. Moldova's eastern region, home to many Russians, Ukrainians, and Turks, recently declared independence. This is an important issue to Moldova, because that region also produces about 80 percent of the country's electricity.

Moldova has 4.3 million people. About half live in cities, but much of Moldova's culture is still based on a rural way of life. Villagers celebrate special occasions with lamb, cornmeal pudding, and goat's milk cheese. The chief city is the capital, **Chişinău** (KEE•shee•NOW).

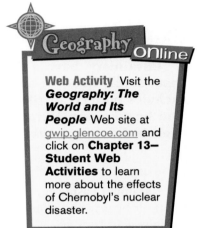

Geography Online

Web Activity Visit the **Geography: The World and Its People** Web site at gwip.glencoe.com and click on **Chapter 13— Student Web Activities** to learn more about the effects of Chernobyl's nuclear disaster.

✓**Reading Check** With what nation does Belarus have close ties?

Section 5 Assessment

Defining Terms
1. Define steppe, potash.

Recalling Facts
2. **Location** Where is the Crimean Peninsula located?
3. **Government** What type of government does Belarus have?
4. **Economics** Name three of Moldova's agricultural products.

Critical Thinking
5. **Categorizing Information** List four agricultural products and three manufactured products of Ukraine.

6. **Understanding Cause and Effect** Why is the culture of Moldova similar to that of Romania?

Graphic Organizer
7. **Organizing Information** Draw a time line like this one. Then label five important periods or events and dates in Ukraine's history.

├─────────┼─────────┼─────────┼─────────┤

Applying Geography Skills

8. **Analyzing Maps** Compare the physical and population maps on pages 367 and 387. What is the population density around Ukraine's Dniester River?

Ukrainian Easter Eggs

Ukrainians have a rich folk art tradition that dates back thousands of years. It includes pottery, textiles, and woodworking. The best-known Ukrainian art form, however, is that of *pysanky,* or decorated eggs.

History

Ukrainian Easter eggs are known worldwide for their beauty and skillful designs. Many of the designs date back to a time when people in the region worshiped a sun god. According to legend, the sun god preferred birds over all other creatures. Birds' eggs became a symbol of birth and new life, and people believed the eggs could ward off evil and bring good luck. Eggs were decorated with sun symbols and used in ceremonies that marked the beginning of spring.

When Christianity took hold in Ukraine in A.D. 988, the tradition of decorative eggs continued. The egg came to represent religious rebirth and new life. People decorated eggs in the days before Easter, then gave them as gifts on Easter morning.

Technique

The word *pysanky* comes from Ukrainian words meaning "things that are written upon." This phrase helps explain the wax process used to decorate the eggs. An artist uses a pin or a tool called a *kistka* to "write" a design in hot wax onto the egg. The egg is then dipped into yellow dye, leaving the wax-covered portion of the eggshell white. After removing the egg from the dye, the artist writes with hot wax over another section of the egg. This portion stays yellow as the egg is dipped into a second dye color. The process continues, with the artist adding wax and dipping the egg into a darker and darker color. At the end, the artist removes the wax layers to reveal the multicolored design.

▶ Making the Connection

1. What is *pysanky* and when did it originate?
2. What role does placing wax onto the eggshell play in creating a decorative egg?
3. **Drawing Conclusions** What purposes, other than entertainment, might folk art accomplish?

◄ Ukrainian Easter eggs

Reading Review

Poland

Terms to Know
bog
communist state
acid rain
republic
pope

Main Idea
Poland, a large country with a rich history, is undergoing many changes.
✓ Place Poland is a large country with southern mountains and northern plains.
✓ Economics The change to a free market economy has brought challenges.
✓ Culture The Poles feel deep loyalty to their country and the Catholic Church.

The Baltic Republics

Terms to Know
oil shale
peat

Main Idea
People in Estonia, Latvia, and Lithuania are trying to change their economies yet keep their cultures.
✓ Place The Baltic countries of Estonia, Latvia, and Lithuania are working to increase manufacturing and service industries.
✓ History These countries are trying to strengthen their ethnic heritage.

Hungary, the Czech Republic, and Slovakia

Terms to Know
nomad   spa
landlocked
bauxite
nature preserve
privatize

Main Idea
Hungary, the Czech Republic, and Slovakia are changing to free market economies.
✓ Economics The Czech Republic is prosperous but must modernize its factories.
✓ Economics Because of corruption, Slovakia has had difficulty moving to a free market economy.

The Balkan Countries

Terms to Know
consumer goods
ethnic cleansing
refugee
mosque

Main Idea
The Balkan countries have greatly suffered from ethnic conflicts and economic setbacks.
✓ Culture The people of Romania are not related to the Slavic peoples who form the populations of most Eastern European countries.
✓ History Ethnic conflicts have torn apart the former Yugoslav republics.
✓ Economics Albania is rich in minerals but is too poor to develop them.

Ukraine, Belarus, and Moldova

Terms to Know
steppe
potash

Main Idea
Past ties to Russia have had different effects on the economies and societies of Ukraine, Belarus, and Moldova.
✓ Geography Ukraine's rich soil allows it to grow large amounts of food.
✓ Culture The people of Belarus want to maintain ties with Russia.
✓ Culture Moldova suffers from disagreements among different ethnic groups.

Chapter 13 Assessment and Activities

Using Key Terms

Match the terms in Part A with their definitions in Part B.

A.

1. spa
2. ethnic cleansing
3. nomad
4. pope
5. steppe
6. privatize
7. mosque
8. bog
9. refugee
10. oil shale

B.

a. head of the Roman Catholic Church
b. resort with hot mineral springs
c. layered rock that contains oil
d. government transfers ownership of businesses to individuals
e. low-lying marshy land
f. Muslim house of worship
g. people who move from place to place
h. forcing people from other groups to leave their homes
i. person who must flee to another country
j. dry, treeless grasslands

Reviewing the Main Ideas

Section 1 Poland

11. **Location** Poland borders what large body of water?
12. **Economics** What is Poland's most important industry?

Section 2 The Baltic Republics

13. **History** Who controlled the Baltic Republics from 1940 to 1991?
14. **Region** Which Baltic Republic is most industrialized?

Section 3 Hungary, the Czech Republic, and Slovakia

15. **Place** What river divides Hungary?
16. **Place** What is the Czech Republic's capital?

Section 4 The Balkan Countries

17. **Government** What problems have accompanied Bulgaria's transition to democracy?
18. **History** What caused Yugoslavia to fall apart?

Section 5 Ukraine, Belarus, and Moldova

19. **Place** What is the capital of Ukraine?
20. **Government** What is unusual about Belarus's government?

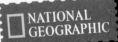 **NATIONAL GEOGRAPHIC** **Eastern Europe**

Place Location Activity

On a separate sheet of paper, match the letters on the map with the numbered places listed below.

1. Danube River
2. Black Sea
3. Croatia
4. Albania
5. Latvia
6. Hungary
7. Warsaw
8. Carpathian Mountains
9. Baltic Sea
10. Ukraine

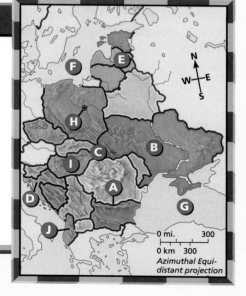

0 mi. 300
0 km 300
Azimuthal Equidistant projection

Self-Check Quiz Visit the *Geography: The World and Its People* Web site at gwip.glencoe.com and click on **Chapter 13— Self-Check Quizzes** to prepare for the Chapter Test.

Critical Thinking

21. **Understanding Cause and Effect** How have the Balkan Mountains contributed to the ethnic diversity of that region?

22. **Categorizing Information** In a chart like the one below, identify Eastern European countries that are succeeding and ones that continue to struggle economically. Include one fact that explains each situation.

Countries That Are Succeeding	Countries That Are Struggling

GeoJournal Activity

23. **Writing a Poem** A cinquain is a poem with five lines. The first line is a one-word title. Line two has two words that describe the title. Line three has three action words that describe the title. Line four is a four-word phrase that expresses a feeling about the subject. Line five is one word that is a synonym or restatement of the title. Write a cinquain about one Eastern European country.

Mental Mapping Activity

24. **Focusing on the Region** Draw a map of Eastern Europe, then label the following:

- Poland
- Czech Republic
- Hungary
- Black Sea
- Ukraine
- Albania
- Serbia
- Lithuania
- Danube River
- Adriatic Sea

Technology Skills Activity

25. **Building a Database** Create a database of Eastern European countries. Include fields for capital, size, population, government, and products. After analyzing your database, predict which countries have a better chance of improving their standards of living.

Standardized Test Practice

Directions: Study the map below, then answer the question that follows.

European Union 2000

Existing members

To be considered for membership

1. **Which of these Eastern European countries is NOT being considered for membership in the European Union?**

   A Ukraine

   B Poland

   C Estonia

   D Czech Republic

Test-Taking Tip: This is a tricky question because all the choices are Eastern European countries. You need to study the map to see which answer choice is not listed on the map. Use the process of elimination to narrow your answer choices.

Workers on the statue *Motherland Calls*, Volgograd

Russians in front of St. Basil's Cathedral, Moscow

Russia

If you had to describe Russia in one word, that word would be "BIG"! Russia is the largest country in the world in area. Its almost 6.6 million square miles (17 million sq. km) are spread across two continents—Europe and Asia. As you can imagine, such a large country faces equally large challenges. Since 1991, when Russia emerged as an independent country, it has been struggling to unite its many ethnic groups, set up a democratic government, and build a stable economy.

◀ Siberian tiger in a forest in eastern Russia

NGS ONLINE
www.nationalgeographic.com/education

395

Focus on:

Russia

RUSSIA IS THE WORLD'S LARGEST COUNTRY. Spanning 11 time zones, it is almost twice the size of the United States. Russia's far northern location affects its climate and the lifestyle of its people. Being somewhat isolated from the rest of the world has played a major role in the country's history, politics, and economic development.

The Land

Stretching nearly halfway around the globe, Russia sprawls from the Bering Sea in the east to the Baltic Sea in the west. Its northernmost lands lie above the Arctic Circle; its southern border winds through the middle of Asia.

Russia is a land of sweeping plains and plateaus interrupted by mountain ranges. The Ural Mountains run north to south, dividing the country into a European region and a much larger Asian region. West of the Urals is the fertile North European Plain—home to three-fourths of the country's population. East of the Urals lies Siberia, which means "sleeping land." Immense and sparsely populated, Siberia is an area of harsh, forbidding landscapes. Although rich in natural resources, much of it remains a wilderness.

The Caspian Sea—actually a saltwater lake—lies at the base of the Caucasus Mountains in Russia's southwest. It is the largest inland body of water in the world. Farther east is Lake Baikal, the world's deepest lake. Many rivers wind through Russian landscapes. The Volga flows southward to empty into the Caspian Sea. The Lena, Yenisey, and Ob Rivers all flow north to the Arctic Ocean.

The Climate

Most of Russia has a cold climate due to its northern location. In Siberia's far north, the landscape is dominated by tundra, a treeless plain. Winters on the tundra are long, dark, and fiercely cold. During the brief summers, only the top few inches of soil thaw out. Deeper down is permafrost—permanently frozen ground.

South of the tundra are immense evergreen forests. This vast woodland area, known as the taiga, is the largest continuous expanse of forest on the earth. Snow blankets the taiga for as much as eight months of the year.

Farther south, the taiga gives way to flat, grass-covered plains, or steppes. Here the climate is less harsh, and the soil quite rich. The steppes make up Russia's most productive agricultural area.

Wildflowers and wooden churches on the North European Plain, in north-western Russia ▼

◀ Reindeer pulling sled across the tundra, Siberia

The Economy

For nearly 70 years in the twentieth century, Russia and the other republics of the Soviet Union shared a single economy that was controlled by Communist authorities. Wheat and other crops were grown on huge government-owned farms. The top economic priority was heavy industry—the manufacturing of goods such as machinery and military equipment. Russia's rich deposits of minerals, coal, and oil supplied the raw materials and energy for many of its industries. The push to industrialize, however, led to widespread pollution of the air, soil, and water. Industrial growth also was more important than the needs of the people. Shortages of consumer goods—clothing and household products, for example—were common.

In the 1990s, when Russia and the other republics of the Soviet Union became independent countries, each took charge of its own economy. Today Russia is struggling to make the transition to a free enterprise system, in which people run their own businesses and farms.

The People

Roughly 147 million people live in Russia. Most live west of the Ural Mountains, where the climate is mildest and the land most fertile. Moscow and St. Petersburg are the region's largest cities.

Although people of many ethnic groups can be found within Russia's borders, most Russians are descendants of Slavic peoples, or Slavs. Centuries ago, Slavs from northeastern Europe settled in what is now western Russia. Their settlements grew into city-states ruled by princes. By the 1400s, these states were united under the rule of a czar. For more than 400 years, Russia was governed by a series of powerful, and often ruthless, czars. Their armies conquered surrounding lands to gradually create a huge Russian Empire.

In 1917 the last czar was overthrown, and a Communist dictatorship emerged. The Russian Empire became the Union of Soviet Socialist Republics. The Communists made the Soviet Union an industrial power but denied the people basic freedoms. In 1991 the Communists fell from power, and the Soviet Union disintegrated.

Exploring the Region

1. **Why might Russia's north-flowing rivers be difficult to travel in winter?**
2. **Why would it be hard to grow crops on the tundra?**
3. **What was the top economic priority of Communist authorities?**
4. **To what ethnic group do most Russians belong?**

◀ **Russian worker inspecting tractors in a factory**

▲ Young people strolling and singing in St. Petersburg

REGIONAL ATLAS

Russia

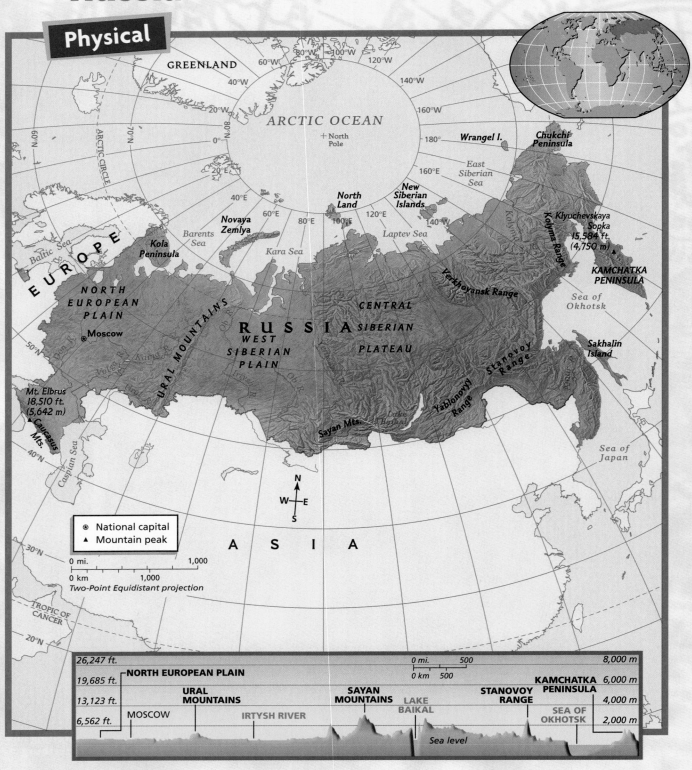

Physical

GREENLAND

ARCTIC OCEAN

+ North Pole

Wrangel I.

Chukchi Peninsula

East Siberian Sea

North Land

New Siberian Islands

Laptev Sea

Novaya Zemlya

Barents Sea

Kara Sea

Klyuchevskaya Sopka 15,584 ft. (4,750 m)

KAMCHATKA PENINSULA

Kola Peninsula

E U R O P E

Sea of Okhotsk

Verkhoyansk Range

Kolyma Range

N O R T H E U R O P E A N P L A I N

U R A L   M O U N T A I N S

R U S S I A

CENTRAL SIBERIAN PLATEAU

Sakhalin Island

Moscow

WEST SIBERIAN PLAIN

Stanovoy Range

Volga R.

Kama R.

Ob R.

Irtysh R.

Yablonovyy Range

Mt. Elbrus 18,510 ft. (5,642 m)

Caucasus Mts.

Sayan Mts.

Lake Baikal

Sea of Japan

Caspian Sea

Baltic Sea

N
W E
S

⊕ National capital
▲ Mountain peak

A S I A

0 mi. 1,000
0 km 1,000
Two-Point Equidistant projection

TROPIC OF CANCER

ARCTIC CIRCLE

26,247 ft.				0 mi. 500	8,000 m
19,685 ft.	NORTH EUROPEAN PLAIN			0 km 500	6,000 m
13,123 ft.	URAL MOUNTAINS	SAYAN MOUNTAINS		KAMCHATKA PENINSULA	4,000 m
6,562 ft.	MOSCOW	IRTYSH RIVER	LAKE BAIKAL	STANOVOY RANGE	SEA OF OKHOTSK	2,000 m

Sea level

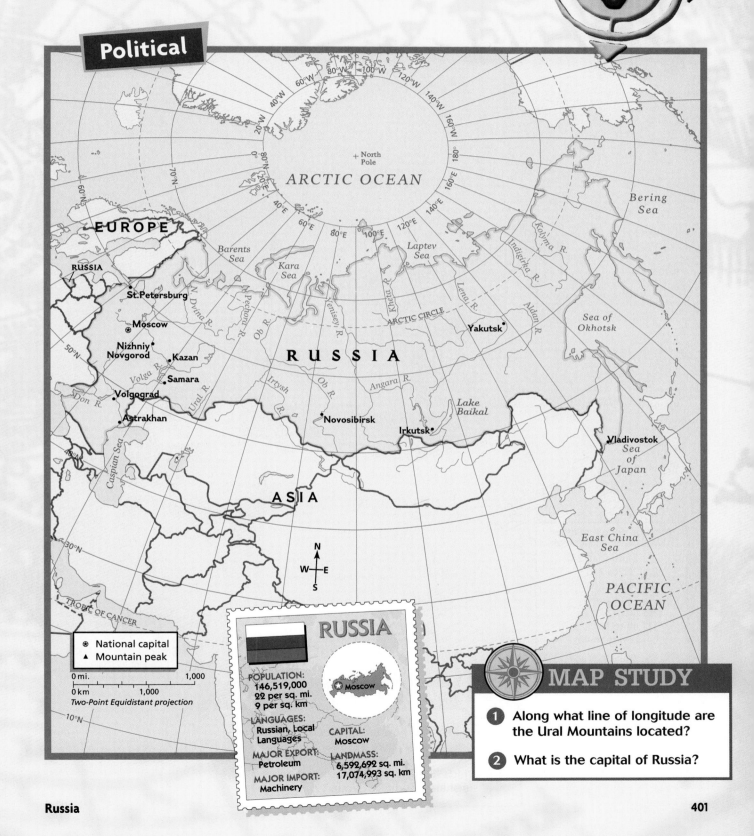

Political

North Pole

ARCTIC OCEAN

EUROPE

RUSSIA

St. Petersburg

⊗ Moscow

Nizhniy
Novgorod

Kazan

Samara

Volgograd

Astrakhan

Caspian Sea

Barents
Sea

Kara
Sea

N. Dvina R.

Pechora R.

Ob R.

Volga R.

Don R.

Ural R.

Irtysh R.

Ob R.

Yenisey R.

ARCTIC CIRCLE

R U S S I A

Novosibirsk

Irkutsk

Lake
Baikal

Angara R.

Laptev
Sea

Khatanga R.

Lena R.

Yakutsk

Aldan R.

Indigirka R.

Kolyma R.

Bering
Sea

Sea of
Okhotsk

Vladivostok
Sea
of
Japan

East China
Sea

ASIA

N
W E
S

PACIFIC
OCEAN

TROPIC OF CANCER

⊗ National capital
▲ Mountain peak

0 mi. 1,000
0 km 1,000
Two-Point Equidistant projection

RUSSIA

⊕ Moscow

POPULATION:
146,519,000
22 per sq. mi.
9 per sq. km

LANGUAGES:
Russian, Local
Languages

MAJOR EXPORT:
Petroleum

MAJOR IMPORT:
Machinery

CAPITAL:
Moscow

LANDMASS:
6,592,692 sq. mi.
17,074,993 sq. km

MAP STUDY

1 Along what line of longitude are the Ural Mountains located?

2 What is the capital of Russia?

Russia

The Russian Winter

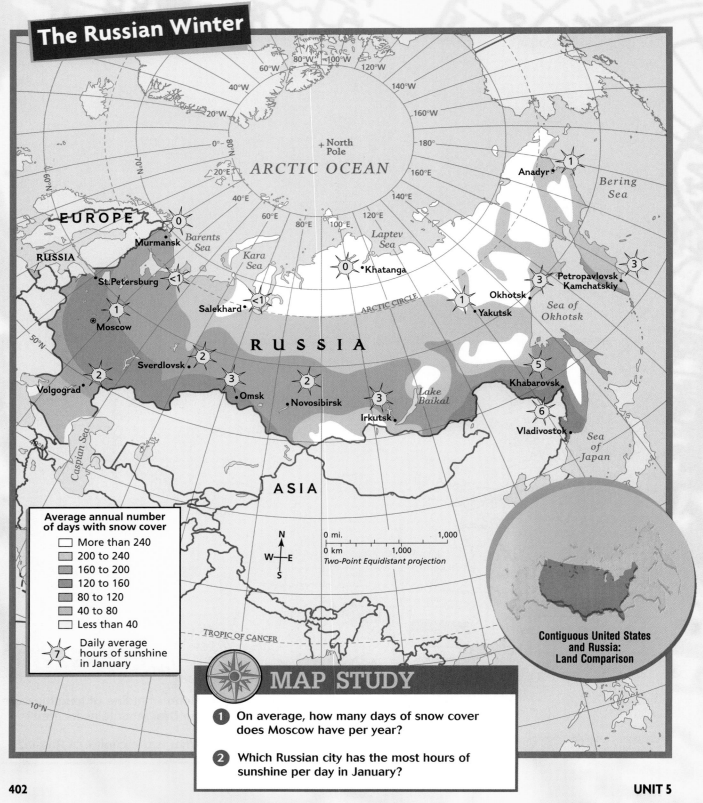

+ North Pole

ARCTIC OCEAN

EUROPE

RUSSIA

Murmansk

St. Petersburg

Moscow

Volgograd

Sverdlovsk

Omsk

Novosibirsk

Irkutsk

R U S S I A

ASIA

Salekhard

Khatanga

Laptev Sea

Kara Sea

Barents Sea

ARCTIC CIRCLE

Lake Baikal

Yakutsk

Okhotsk

Petropavlovsk Kamchatskiy

Sea of Okhotsk

Anadyr

Bering Sea

Khabarovsk

Vladivostok

Sea of Japan

Caspian Sea

TROPIC OF CANCER

Average annual number of days with snow cover

- More than 240
- 200 to 240
- 160 to 200
- 120 to 160
- 80 to 120
- 40 to 80
- Less than 40

7 Daily average hours of sunshine in January

N
W — E
S

0 mi.   1,000
0 km   1,000
Two-Point Equidistant projection

Contiguous United States and Russia: Land Comparison

MAP STUDY

1 On average, how many days of snow cover does Moscow have per year?

2 Which Russian city has the most hours of sunshine per day in January?

Geo Extremes

① **HIGHEST POINT**
Mount Elbrus
18,510 ft.
(5,642 m) high

② **LOWEST POINT**
Caspian Sea
92 ft. (28 m)
below sea level

③ **LONGEST RIVER**
Ob-Irtysh
3,362 mi.
(5,411 km) long

④ **LARGEST LAKE**
Caspian Sea
143,244 sq. mi.
(371,000 sq. km)

⑤ **DEEPEST LAKE**
Lake Baikal
5,315 ft.
(1,620 m) deep

⑥ **LARGEST ISLAND**
Sakhalin
29,500 sq.mi.
(76,405 sq. km)

COMPARING POPULATION:
United States and Russia

UNITED STATES

RUSSIA

🧍 = 25,000,000

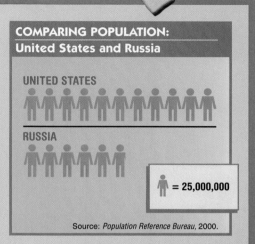

Source: *Population Reference Bureau*, 2000.

COMPARING AREA AND POPULATION:
Russia East and West of the Ural Mountains

AREA
West of Urals 24.4%
East of Urals 75.6%

POPULATION
East of Urals 26%
West of Urals 74%

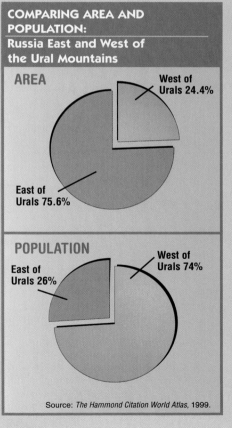

Source: *The Hammond Citation World Atlas*, 1999.

GRAPHIC STUDY

1 What two "extremes" does the Caspian Sea lay claim to?

2 What percentage of Russia's people live west of the Ural Mountains?

Russia—A Eurasian Country

The World and Its People NATIONAL GEOGRAPHIC

To learn more about Russia's land and economy, view *The World and Its People* Chapter 14 video.

Geography ONLINE

Chapter Overview Visit the *Geography: The World and Its People* Web site at <u>gwip.glencoe.com</u> and click on **Chapter 14— Chapter Overviews** to preview information about Russia.

The Russian Land

Guide to Reading

Main Idea

Russia is a huge country with a cold climate due to its far northern location.

Terms to Know

- steppe
- peninsula
- taiga
- tundra
- permafrost

Places to Locate

- Ural Mountains
- North European Plain
- Moscow
- St. Petersburg
- Volgograd
- West Siberian Plain
- Siberia
- Caucasus Mountains

Reading Strategy

Draw a chart like this one. List two specific names of each type of physical feature given.

Russia	
Plains	
Mountains	
Plateaus	
Rivers	

NATIONAL GEOGRAPHIC

Exploring Our World

Siberian tigers hunt in the eastern forests of Russia—sometimes even climbing trees to find food. Only a few hundred now live in the wild, though. The animals they hunt—elk, deer, and wild boar—are dwindling, and the tigers themselves are hunted by people. Poachers who kill the tigers illegally can sell a skin for $15,000. Russia is enforcing laws more strictly to save these animals.

Russia is the world's largest country. Nearly twice as big as the United States, Russia is called a *Eurasian* country because its lands lie on two continents—Europe and Asia. The western part of Russia borders Eastern European countries such as Belarus and Ukraine. As you move east, you run into the **Ural Mountains**—the dividing line between Europe and Asia. The rest of Russia stretches across Asia to the Pacific Ocean. Along the way, it shares borders with 14 other countries. Russia is so wide that it includes 11 time zones from east to west. When it is 12:00 noon in eastern Russia, it is 1:00 A.M. in western Russia.

Russia is also a northern country. As you can see from the map on page 406, its southern border is in the middle latitudes but the north reaches past the Arctic Circle. Russia has a long coastline along the Arctic Ocean. Ice makes shipping difficult or impossible most of the year. Even many of Russia's ports on the Baltic Sea and Pacific Ocean are closed by ice part of the year.

Russia's gigantic size and harsh climates make transportation difficult within the country as well. If you visited Russia, you would

◀ **Red Square in Moscow, Russia**

discover that, unlike in the United States, railroads still are an important means of getting around. With 93,771 miles (151,000 km) of track, railroads are the leading movers of people and goods in Russia.

Plains Areas

Two large lowland plains cover the western half of Russia. The Ural Mountains divide these plains. Find the Ural Mountains on the map on page 407. To the west of the Urals, you find the **North European Plain.** You may recall that this fertile plain begins in France in Western Europe. This plain has Russia's mildest climate, and about 75 percent of the population live here. This region holds Russia's capital, **Moscow,** and other important cities, such as **St. Petersburg** and **Volgograd.** Much of Russia's industry is also found here. Good farmland lies to the south, along the Don and Volga Rivers. This area is part of the steppe, the nearly treeless grassy plain that stretches through Ukraine.

East of the Urals lies the **West Siberian Plain.** This huge lowland—the world's largest area of flat land—is part of **Siberia.** Because

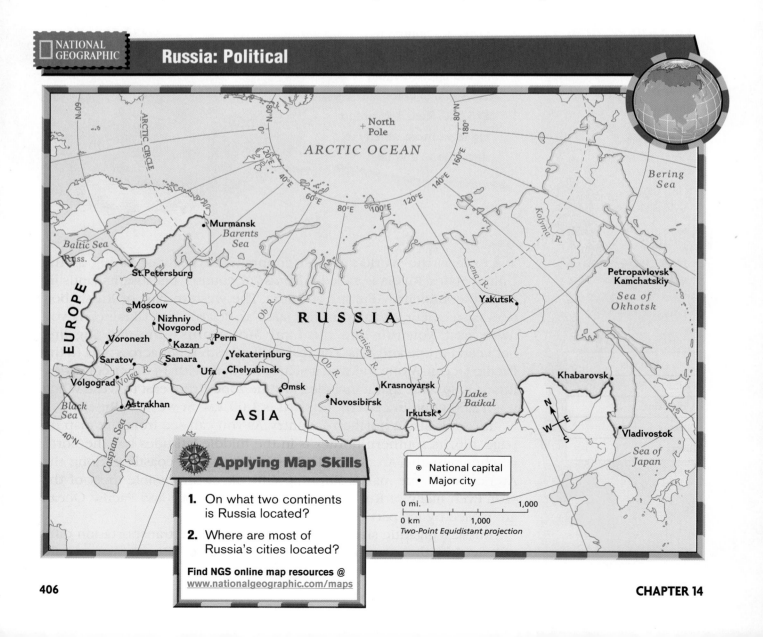

NATIONAL GEOGRAPHIC

Russia: Political

Applying Map Skills

1. On what two continents is Russia located?

2. Where are most of Russia's cities located?

Find NGS online map resources @ www.nationalgeographic.com/maps

⊛ National capital
• Major city

0 mi.   1,000
0 km   1,000
Two-Point Equidistant projection

of its many rivers—and the fact that these rivers often flow over ground that is frozen—much of the West Siberian Plain is marshy. Many people live in the southern part of this plain, which is for the most part drier and warmer than the plain to the north.

✓ Reading Check Which part of Russia is home to most of its cities and industries? Why?

Mountains and Plateaus

Two mountain ranges rise in western Russia—the Urals and the Caucasus (KAW•kuh•suhs). The Ural Mountains, very old and worn by erosion, do not reach very high. Their length is extensive, though, running about 2,500 miles (4,023 km) from the Arctic Ocean to Russia's southern border. As you read earlier, these mountains form the boundary between the continents of Europe and Asia.

The **Caucasus Mountains** separate southwestern Russia from Southwest Asia. Thickly covered with pines and other trees, the Caucasus are much taller than the Urals. They include Mt. Elbrus,

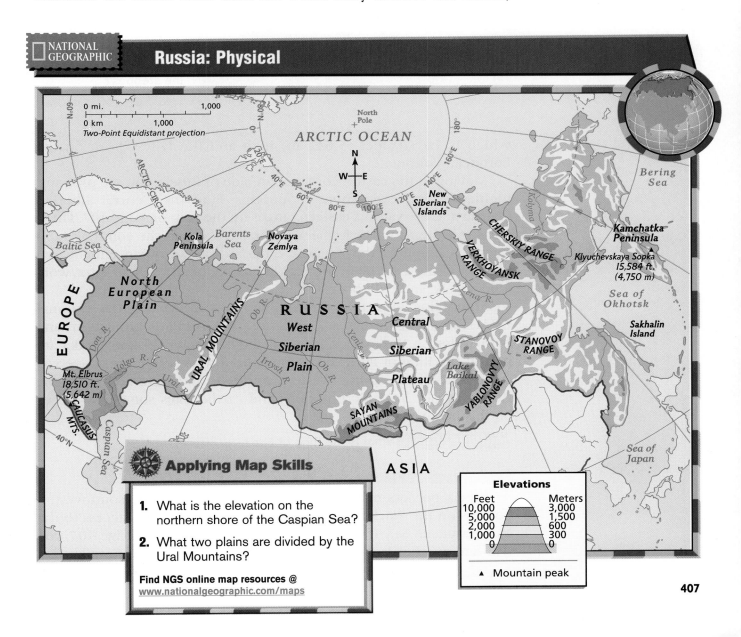

NATIONAL GEOGRAPHIC

Russia: Physical

0 mi. 1,000
0 km 1,000
Two-Point Equidistant projection

North Pole
ARCTIC OCEAN

New Siberian Islands

Bering Sea

Kola Peninsula
Barents Sea
Novaya Zemlya

Kolyma

Kamchatka Peninsula
Klyuchevskaya Sopka 15,584 ft. (4,750 m)

Baltic Sea

CHERSKIY RANGE

VERKHOYANSK RANGE

Sea of Okhotsk

EUROPE

North European Plain

RUSSIA

Lena R.

Sakhalin Island

Don R.

Ob R.

Irtysh R.

URAL MOUNTAINS

West Siberian Plain

Yenisey R.

Ob R.

Central Siberian Plateau

Lake Baikal

STANOVOY RANGE

YABLONOVYY RANGE

Volga R.

Ural R.

Mt. Elbrus 18,510 ft. (5,642 m)

CAUCASUS MTS.

Caspian Sea

SAYAN MOUNTAINS

ASIA

Sea of Japan

Applying Map Skills

1. What is the elevation on the northern shore of the Caspian Sea?

2. What two plains are divided by the Ural Mountains?

Find NGS online map resources @
www.nationalgeographic.com/maps

Elevations

Feet	Meters
10,000	3,000
5,000	1,500
2,000	600
1,000	300
0	0

▲ Mountain peak

The Steppes

Many Russians earn their living by farming and ranching on the plains near the Caucasus Mountains.

Place What type of vegetation grows on the steppes?

which is 18,510 feet (5,642 m) high and the tallest peak on the European continent.

Moving to the eastern, or Siberian, side of Russia, you see plateaus, highlands, and even more mountains. Find the Central Siberian Plateau and the various eastern ranges on the map on page 407. These areas stair-step to ever higher mountain ranges in the south. Eastern Russia is home to the majestic Siberian tiger, now an endangered species. Other wildlife in the region include bear, reindeer, lynx, wolf, wildcat, elk, and wild boar.

Mountains also rise on the far eastern Kamchatka (kuhm•CHAHT•kuh) Peninsula. A peninsula, you recall, is a piece of land with water on three sides. Many of the mountains in Kamchatka are part of the Ring of Fire. This name is used to describe the active volcanic zone that forms the western, northern, and eastern edges of the Pacific Ocean. Volcanic eruptions and earthquakes sometimes occur on this peninsula.

Reading Check What is the Ring of Fire?

Inland Water Areas

Russia borders many inland bodies of water. In the southwest, it borders on the Black Sea. Through this sea, Russia gains access to the Mediterranean Sea. If you look at the physical map on page 407, you will find another large sea in southwestern Russia—the Caspian Sea.

About the size of California, the Caspian Sea is actually the largest inland body of water in the world. Like the Great Salt Lake in Utah, the Caspian Sea has salt water, not freshwater. Russia shares this sea with four other countries.

High in the Central Siberian Plateau, you find Lake Baikal. This is the world's deepest freshwater lake. In fact, Lake Baikal holds almost 20 percent of the world's supply of unfrozen freshwater. Tourists travel by train to see the lake's shimmering blue waters.

Russia has several important rivers. The Volga—the longest river in Europe—is an important transportation route. It and other rivers of European Russia are connected by canals. Many rivers also flow through the Asian side of Russia. Most of these rivers begin in the mountains of southern Siberia and flow north across the lowlands to empty into the frigid Arctic Ocean. The Lena (LEE•nuh), the Yenisey (YIH•nih•SAY), and the Ob (AHB) are among the longest rivers in the world.

✓ Reading Check How do the waters of the Caspian Sea and Lake Baikal differ?

NATIONAL GEOGRAPHIC

Russia: Climate

Dry
- Steppe

Mid-Latitude
- Humid continental

High Latitude
- Subarctic
- Tundra

0 mi.   1,000
0 km   1,000
Two-Point Equidistant projection

🧭 Applying Map Skills

1. What high latitude climate zones cover much of Russia?

2. What type of climate does Moscow have?

Find NGS online map resources @
www.nationalgeographic.com/maps

Russia's Climate

European Russia generally enjoys warmer temperatures than Asian Russia. Most of the European region has a humid continental climate. Summers are warm and rainy, while winters are cold and snowy. Moscow can have snow five months out of the year.

Areas near the Black Sea in southern Russia enjoy mild temperatures with a steppe climate throughout the year. Less rain falls—the amount varies from year to year. In years of plentiful rain, farmers can produce abundant crops. In dry years, though, the wheat crop fails.

In contrast, Siberia has high latitude climates. In the south, summers are short and cool, and winters are long and very snowy. Siberia may receive snow as much as eight months of the year. Huge forests of evergreen trees grow in this cool climate. These forests—called the taiga (TY•guh)—cover southern Siberia.

Northern Siberia has one of the coldest climates in the world. Not even hardy evergreens can grow there. Instead, you find tundra, a vast and rolling treeless plain in which only the top few inches of ground thaw out in summer. The permanently frozen lower layers of soil are called permafrost and cover 40 percent of Russia.

✓Reading Check **What are the evergreen forests of Siberia called?**

Section 1

Assessment

Defining Terms

1. **Define** steppe, peninsula, taiga, tundra, permafrost.

Recalling Facts

2. **Place** What is the largest area of flat land in the world?

3. **Location** What mountain range separates Europe and Asia?

4. **Region** How many countries does Russia border?

Critical Thinking

5. **Analyzing Information** Why do you think trains are more important than other kinds of vehicles for moving people and goods across Russia?

6. **Drawing Conclusions** How do you think permafrost contributes to the marshy lands of the West Siberian Plain?

Graphic Organizer

7. **Organizing Information** Create a chart like this one. Then place each of the following items into the column in which it is located: Moscow, Lake Baikal, Yakutsk, Kamchatka Peninsula, St. Petersburg, Vladivostok, Yenisey River, and Ob River.

North European Plain	West Siberian Plain	Eastern Russia

Applying Geography Skills

8. **Analyzing Maps** Turn to the political map on page 406. Name a Russian city located north of the Arctic Circle. Now look at the climate map on page 409. What type of climate does the city north of the Arctic Circle experience?

Russian Poetry

Boris Pasternak is one of Russia's best-known writers. He wrote both poetry and fiction, and in 1958 he won the Nobel Prize for Literature for his novel *Doctor Zhivago*. Officials in what was then the Soviet Union did not approve of the work. Today, however, the Russian government recognizes Pasternak's great contribution to Russian literature.

Is It Not Time for the Birds to Sing
by Boris Pasternak (1890–1960)

The wind pokes about
With a branch of wet lilac
Like a tiny wet sparrow:
Is it not time for the birds to sing?

Raindrops heavy as cuff links
And the garden all shiny
Like a pond dotted
With a million blue tears.

Nursed by grief
And still prickly with pain
The garden revives
Filled with whispers and scents.

All through the night
It had knocked at the window.
Suddenly clothes
Smell musty and wet.

Awakened by the magic
Roll call of other days and names,
Today looks out
With eyes like anemones [flowers].

"Is It Not Time for the Birds to Sing" from *My Sister, Life and Other Poems* by Boris Pasternak. Copyright © 1976 by Reich Verlag. Reprinted by permission of Harcourt, Inc.

▲ Springtime arrives in a Russian garden.

▶ Making the Connection

1. What lines in the poem hint that the garden wants to bloom?

2. Given the Russian climate, why is the coming of spring significant?

3. **Making Inferences** Is the poet saying it is or is not time for the birds to sing? Explain.

Russia's Economic Regions

Guide to Reading

Main Idea

With many resources, Russia could have a large and diverse economy.

Terms to Know

- heavy industry
- light industry
- consumer goods
- hydroelectric power
- bauxite

Places to Locate

- Kaliningrad
- St. Petersburg
- Baltic Sea
- Murmansk
- Vladivostok
- Astrakhan
- Caspian Sea
- Volga River

Reading Strategy

Create a chart like this one, then list at least two facts about the economy of each region.

Region	Facts
Moscow	
Port Cities	
Volga & Urals	
Siberia	

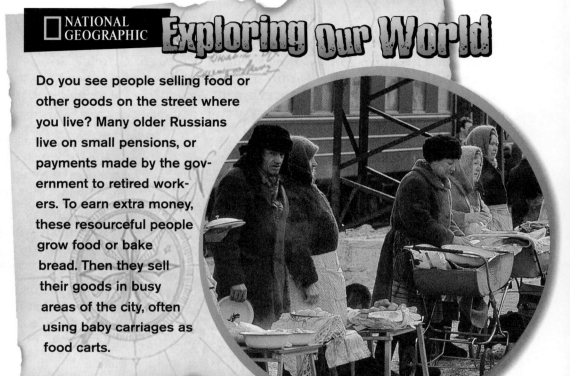

NATIONAL GEOGRAPHIC

Exploring Our World

Do you see people selling food or other goods on the street where you live? Many older Russians live on small pensions, or payments made by the government to retired workers. To earn extra money, these resourceful people grow food or bake bread. Then they sell their goods in busy areas of the city, often using baby carriages as food carts.

With its many resources, Russia could have a growing economy. It has large deposits of coal, oil, and gas. It also has many minerals, including nickel, iron ore, tin, and gold. The southwestern area can produce rich yields of grains. Russia's fishing industry is among the largest in the world. The vast forests of Siberia provide plenty of timber.

Despite all these advantages, Russia's economy has struggled in recent decades. You will learn why in Chapter 15. For now, you will see what Russia's economy produces. Geographers divide Russia into four different economic regions: the Moscow Region, Port Cities, the Volga and Urals Region, and Siberia.

The Moscow Region

About 800 years old, Moscow is the political and cultural center of Russia. The largest city in Russia, Moscow is the country's economic center and largest transportation hub as well. Look at the economic activity map below. Many of Russia's manufacturing centers are located in the western part of Russia. In the past, many of the country's factories focused on **heavy industry,** or making goods such as machinery, mining equipment, and steel. In recent years, more factories have shifted to **light industry,** or the making of such goods as clothing, shoes, furniture, and household products.

These economic changes are reflected in Moscow. Its older industries include textiles, electrical equipment, and automobiles. Today factories make more **consumer goods,** or household and electronics products. High-technology services have also emerged in the city. Many people in Moscow travel to their jobs by taking the Metro, the name for Moscow's subway system.

Some farming also takes place in the Moscow region. Farmers raise dairy cattle and such grains as barley and oats. They also grow potatoes, corn, and sugar beets. Other crops include flax, which is used to make textiles.

✓ **Reading Check** What is the difference between heavy industry and light industry?

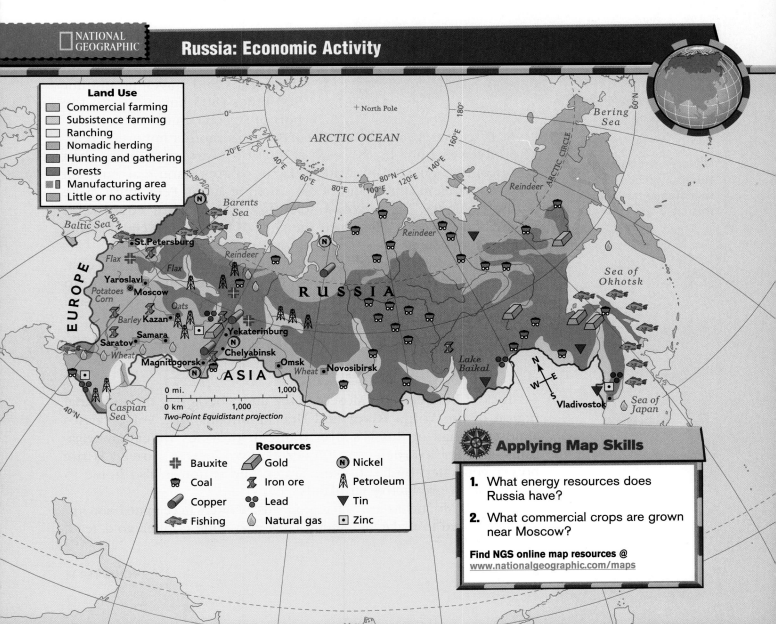

NATIONAL GEOGRAPHIC

Russia: Economic Activity

Land Use
- Commercial farming
- Subsistence farming
- Ranching
- Nomadic herding
- Hunting and gathering
- Forests
- Manufacturing area
- Little or no activity

Resources
- ✚ Bauxite
- Coal
- Copper
- Fishing
- Gold
- Iron ore
- Lead
- Natural gas
- Ⓝ Nickel
- Petroleum
- ▼ Tin
- ⊡ Zinc

Applying Map Skills

1. What energy resources does Russia have?

2. What commercial crops are grown near Moscow?

Find NGS online map resources @
www.nationalgeographic.com/maps

Kaliningrad

Tall cranes dot the harbor of Kaliningrad.

Movement How are Russian officials trying to increase trade in this city?

Port Cities

Russia has two important northwestern ports—**Kaliningrad** and **St. Petersburg.** Look at the map on page 416. Do you see that Russia owns a small piece of land on the **Baltic Sea** separated from the rest of the country? The port of Kaliningrad is located on this land. This city is Russia's only Baltic port that remains free of ice year-round. Russian officials, hoping to increase trade here, have eliminated all taxes on foreign goods brought to this city. Still, companies that deliver goods to this port face an obstacle. They must transport their goods another 200 miles (322 km) through other countries to reach the nearest inland part of Russia. In summer, they have the choice of keeping the goods on their ships and traveling another 500 miles (805 km) north to the port of St. Petersburg.

St. Petersburg, once the capital of Russia, is another important port. The city was built in the early 1700s on a group of more than 100 islands connected by bridges. Large palaces stand gracefully on public squares in this beautiful city. Factories in St. Petersburg make light machinery, textiles, and scientific and medical equipment. Located on the Neva River near the Gulf of Finland, the city is also a shipbuilding center.

Murmansk, in Russia's far north, and **Vladivostok,** in the east, are other important port cities. So is **Astrakhan,** on the **Caspian Sea.** Trade in these port cities brings needed goods to the Russian people.

✓ Reading Check What are the two main ports in northwestern Russia?

The Volga and Urals Region

Tucked between the Moscow area and Siberia lies the industrial region of the **Volga River** and Ural Mountains. The Volga carries almost one-half of Russia's river traffic. It provides water for irrigation—and for hydroelectric power. Hydroelectric power is the power generated by fast-flowing water. The region is also home to Russia's most productive farmlands.

The Ural Mountains are rich in minerals. Workers here mine bauxite, a mineral used to make aluminum. They also mine copper, gold, lead, and nickel. The mountains have such energy resources as coal, oil, and natural gas as well. Find these resources on the economic activity map on page 413.

✓ Reading Check **What energy resources are found in the Volga and Urals region?**

Siberia

East of the Urals lies Siberia. This region has the largest supply of minerals in Russia, including iron ore, uranium, gold, diamonds, and coal. Huge deposits of oil and natural gas lie beneath the permafrost of northern Siberia. About two-thirds of Siberia is covered with trees that could support a timber industry.

Tapping all these resources is very difficult, however. Siberia is mostly undeveloped because of its harsh climate. Another problem is isolation—it can take eight or more days to travel across all of Russia by train. Finding a way to obtain these resources is important to Russia's future economic success, though. Many of the minerals and fuels of western Russia have been used up. The industrial centers there need the resources from Siberia.

✓ Reading Check **Why are Siberia's mineral resources important?**

Geography online

Web Activity Visit the **Geography: The World and Its People** Web site at gwip.glencoe.com and click on **Chapter 14— Student Web Activities** to learn more about Siberia.

Assessment

Defining Terms
1. **Define** heavy industry, light industry, consumer goods, hydroelectric power, bauxite.

Recalling Facts
2. **Location** Why is Moscow the economic center of Russia?
3. **Location** Why is Kaliningrad a very important city?
4. **Economics** What are five mineral resources found in the Ural Mountains?

Critical Thinking
5. **Drawing Conclusions** Why would consumers want the Russian economy to change from relying on heavy industry to a greater emphasis on light industry?
6. **Understanding Cause and Effect** How would eliminating taxes on foreign goods brought through Kaliningrad increase trade with other countries?

Graphic Organizer
7. **Organizing Information** Draw a chart like the one below. On your chart, fill in at least three examples of heavy industry and three examples of light industry.

Heavy Industry	Light Industry
1.	1.
2.	2.
3.	3.

Applying Geography Skills

8. **Analyzing Maps** Look at the transportation map on page 416. What three major ports are located on Russia's eastern borders? What two cities signify the start and end stations of the Trans-Siberian Railroad?

Geography Skill

Reading a Transportation Map

Transportation maps show how people and goods move throughout a region. Lines, colors, and symbols represent different types of transportation. A route normally follows the least difficult path between cities. Because of the physical landscape, though, the least difficult route may not be the shortest route.

Learning the Skill

When reading a transportation map, follow these steps:

- Read the map title and the map key.
- Locate major population centers.
- Identify the routes on the map and points of connection between the routes.
- Draw conclusions about how the physical landscape affects transportation and trade.

Practicing the Skill

Study the map below, then answer the following questions.

1. What kinds of transportation are shown?
2. Which major railroad extends to the far eastern part of Russia? Why does this railroad not follow a straight line?
3. Which general region of Russia has the most trade, industry, and people? How can you tell?

Applying the Skill

On a road map of your state, select a city in the north and another city in the south. Find the quickest route you would take to get from one city to the other.

Russia: Transportation Routes

NATIONAL GEOGRAPHIC

Legend:
- National boundary
- Sea route
- Trans-Siberian Railroad
- Other railroad
- ⊛ National capital
- • Other city
- ⊛ Seaport

Chapter 14 Reading Review

Section 1 — The Russian Land

Terms to Know
steppe
peninsula
taiga
tundra
permafrost

Main Idea

Russia is a huge country with a cold climate due to its far northern location.

✓ Location Spanning two continents—Europe and Asia—Russia is the world's largest country.

✓ Region The western half of Russia is mostly lowland. The eastern half is covered with mountains and plateaus.

✓ Movement Inland waterways are important for moving goods through Russia, but many long rivers drain north into the frigid Arctic Ocean and freeze in winter.

✓ Region European Russia has the mildest climate, while most of Siberia, or Asian Russia, has cold high-latitude climate zones.

Section 2 — Russia's Economic Regions

Terms to Know
heavy industry
light industry
consumer goods
hydroelectric power
bauxite

Main Idea

With many resources, Russia could have a large and diverse economy.

✓ History Russia has many resources, but its economy has struggled in recent years.

✓ Economics Moscow, with many industries, is the economic center of Russia.

✓ Movement Ports in the northwest, southwest, and east carry on trade between Russia and other countries.

✓ Location Siberia has many resources, but the area is so cold and remote that it is difficult to tap these resources.

◄ A train of the Trans-Siberian Railroad zips along Lake Baikal.

Assessment and Activities

✵ Using Key Terms

Match the terms in Part A with their definitions in Part B.

A.

1. permafrost
2. light industry
3. bauxite
4. steppe
5. consumer goods
6. taiga
7. hydroelectric power
8. peninsula
9. heavy industry
10. tundra

B.

a. huge, subarctic evergreen forests
b. dry, treeless plains in the high latitudes
c. dry, treeless grasslands
d. permanently frozen lower layers of soil
e. land with water on three sides
f. production of consumer goods
g. electricity generated by water
h. production of industrial goods
i. mineral from which aluminum is made
j. products for personal use, such as clothing

✵ Reviewing the Main Ideas

Section 1 The Russian Land

11. **Human/Environment Interaction** Why is Russia unable to use ports along its Arctic coast for most of the year?
12. **Location** Which area of Russia has the mildest climate?
13. **Place** In what way is Mt. Elbrus significant?
14. **Region** Why is the Kamchatka Peninsula considered part of the Ring of Fire?
15. **Place** What is the longest river in Europe?

Section 2 Russia's Economic Regions

16. **Place** What city is the political and cultural center of Russia?
17. **Economics** Name four agricultural products from the Moscow region.
18. **Economics** Name four mineral resources found in Siberia.
19. **Movement** What is an important means of transportation for people in Russia?
20. **Economics** Which of Russia's economic regions is the greatest potential source of minerals for the country?

NATIONAL GEOGRAPHIC **Russia—A Eurasian Country**

Place Location Activity

On a separate sheet of paper, match the letters on the map with the numbered places listed below.

1. Ural Mountains
2. Kamchatka Peninsula
3. Lake Baikal
4. Volga River
5. Moscow
6. Lena River
7. West Siberian Plain
8. Caspian Sea
9. Caucasus Mountains
10. Sakhalin Island

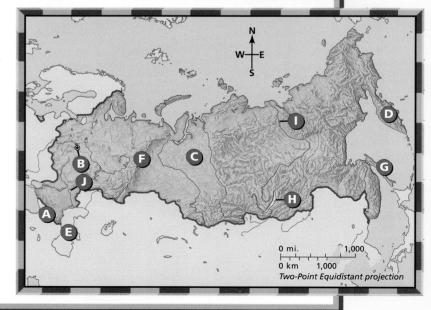

Two-Point Equidistant projection

Critical Thinking

21. **Making Generalizations** How have recent changes in Russia affected its economy?

22. **Understanding Cause and Effect** On a diagram like the one below, label factors that contribute to Siberia's climate.

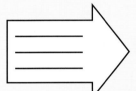

Siberia's Climate

GeoJournal Activity

23. **Writing a Paragraph** Choose one of the photographs from this chapter or the Unit Regional Atlas on pages 394–399 and imagine yourself in the scene. Write a description using vivid words to portray the sights, sounds, and smells you would encounter. What is the temperature? What things can you touch? How do they feel? Describe what is happening.

Mental Mapping Activity

24. **Focusing on the Region** Draw a simple outline map of Russia, then label the following:

- Arctic Ocean
- Pacific Ocean
- Caspian Sea
- Ural Mountains
- Lake Baikal
- Caucasus Mountains
- Volga River
- Barents Sea
- Yenisey River
- Kamchatka Peninsula

Technology Skills Activity

25. **Using the Internet** Search the Internet for information on Siberian tigers. Create a display that includes a map showing the habitats, pictures, and facts about these endangered animals.

The Princeton Review

Standardized Test Practice

Directions: Read the paragraph below, then answer the following question.

For 70 years, the Communist government of the Soviet Union stopped at nothing to industrialize the country. Soviet leaders gave little thought to the health of the people or to the land they were ruining. From St. Petersburg to Vladivostok, across more than 8 million square miles, the Russian environment shows decay. Today Russia's great rivers are sewers of chemicals and human waste. Air in more than 100 cities is at least five times more polluted than standards allow, putting millions at risk of lung diseases. Tons of nuclear waste lie under Arctic waters, and the use of toxic fertilizers has poisoned the soil.

Adapted from "The U.S.S.R.'s Lethal Legacy" by Mike Edwards, *National Geographic*, August 1994.

1. **Which of the following best explains why the Russian environment is so polluted today?**

   F Soviet leaders did not care about the health of the people.

   G Soviet leaders wanted to industrialize the country at all costs.

   H Farmers were careless about the chemicals they spread on their fields.

   J Eastern Russia is not as polluted as western Russia.

Test-Taking Tip: Notice that the question asks for the *best* explanation. Read all the choices carefully before choosing the best one. Although all of the answer statements may be true, you need to find the best answer to the question.

EYE on the Environment

RUSSIA'S
Lethal Legacy

Polluted Empire When you turn on a faucet, you expect clean, safe drinking water to come out. Unfortunately, many Russians do not enjoy this simple luxury. People in Russia and the former Soviet Republics are struggling with pollution—a tragic reminder of the days of Communist rule.

For decades, Soviet leaders were determined to turn the USSR into a mighty industrial power. They did not care what happened to the environment. The results of their actions are an environmental nightmare.

Kola Peninsula

St. Petersburg

R U S S I A

☐ Severe pollution
▨ Acid rain
— Polluted river

- Coal-burning factories puffed chemicals into the air. Russian city dwellers endure some of the worst air pollution in the world.

- Toxic wastes and untreated sewage were dumped into rivers, resulting in polluted drinking water.

- Pesticides and heavy metals poisoned the soil in many areas.

- Radioactive wastes were emptied into seas and rivers.

- Large oil spills were common and were not cleaned up.

Cleaning Up Since the fall of communism, people in Russia have been trying to clean up their polluted air, land, and water. Yet little money is available, and progress is slow.

A steel plant in Siberia pollutes the air.

Worker cleans up an oil spill in Russian waters.

Making a Difference

Tough Trees Many young Russians are growing up in environmental disaster zones created by past generations. Nevertheless, some students are working to find solutions.

Maria Pestova and Andrey Ignashov live on Russia's Kola Peninsula, near a human-made wasteland called the Kuzomenskikh Desert. They spent two years determining that planting a local pine tree could best turn the desert back into a forest. Then they devised a new planting method to increase the young pines' chances of survival.

Maria Pestova

Detergents for a Clean Environment Ulya Loginova and Natalia Skorohodova live in St. Petersburg, Russia. The Neva River, which flows through the city, is heavily polluted. Household detergents contribute to the problem. Ulya and Natalia studied the effects of several detergents and shampoos and identified which ones caused the least environmental damage.

Ulya and Natalia, along with Maria and Andrey, each won an environmental award for their research.

Ulya Loginova and
Natalia Skorohodova

What Can You Do?

Check Out Your Community
Does your community have any pollution problems? What laws or restrictions does your community have concerning pollution? Create an information pamphlet about these laws.

Save Lake Baikal
Russia's Lake Baikal is the world's oldest and deepest lake. A group called Baikal Watch is working to protect the environment of the lake. Find out more about Baikal Watch at www.earthisland.org/baikal

Use the Internet
Learn more about pollution around the world by reading *Pollution Magazine* online at www.gtonline.net/private/pmag. Also read about a project closer to home called Keep America Beautiful at www.kab.org. Suggest a way for your class to become involved.

Russia–Past and Present

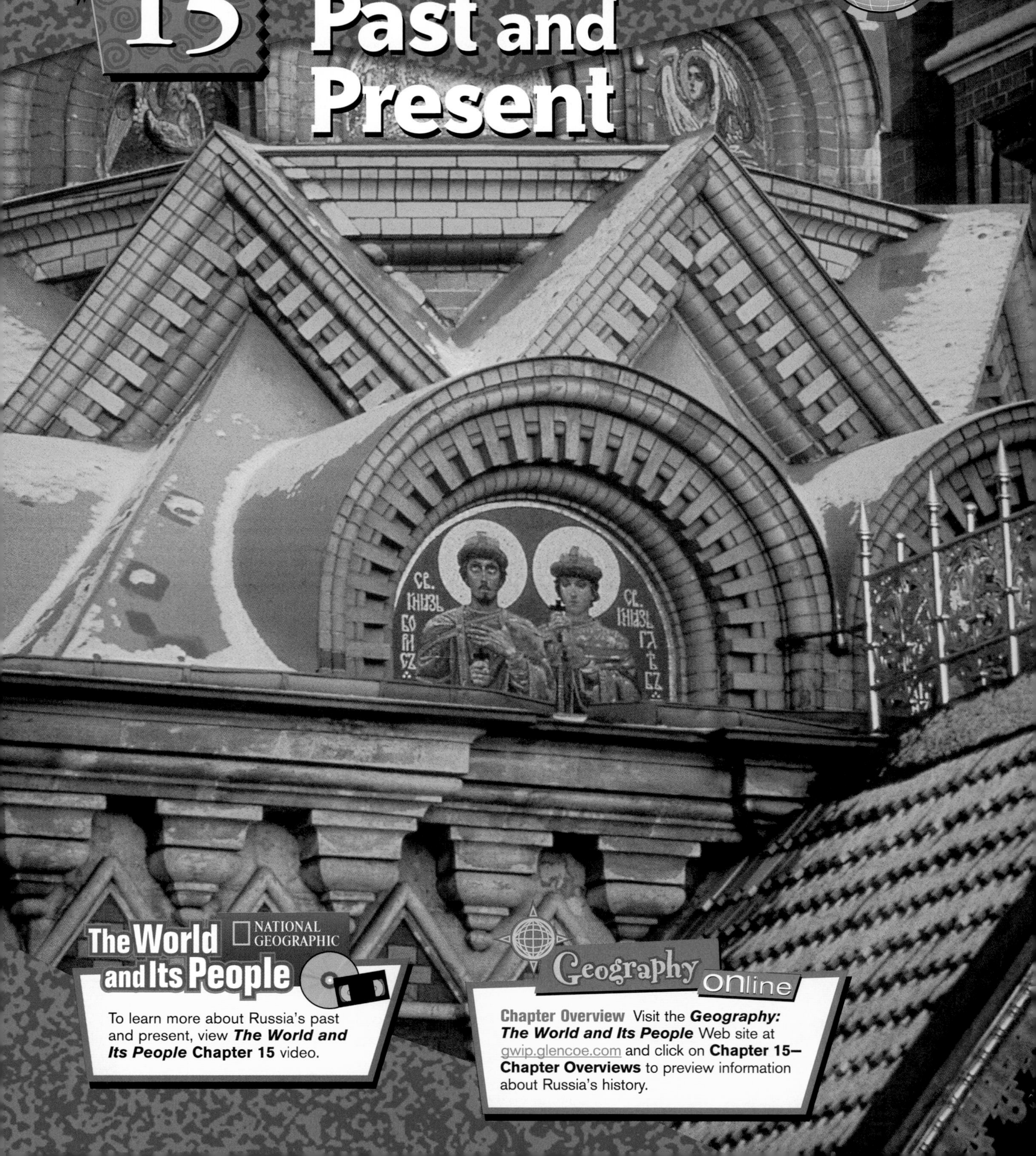

The World and Its People NATIONAL GEOGRAPHIC

To learn more about Russia's past and present, view *The World and Its People* **Chapter 15** video.

Geography online

Chapter Overview Visit the *Geography: The World and Its People* Web site at gwip.glencoe.com and click on **Chapter 15— Chapter Overviews** to preview information about Russia's history.

A Troubled History

Guide to Reading

Main Idea

The harsh rule of powerful leaders has often sparked violent uprisings in Russia.

Terms to Know

- czar
- serf
- industrialize
- communist state
- cold war
- ethnic group

Places to Locate

- Moscow
- St. Petersburg
- Vladivostok
- Siberia

Reading Strategy

Make a time line like this one. Then list five important years in Russian history and describe what happened in those years.

NATIONAL GEOGRAPHIC Exploring Our World

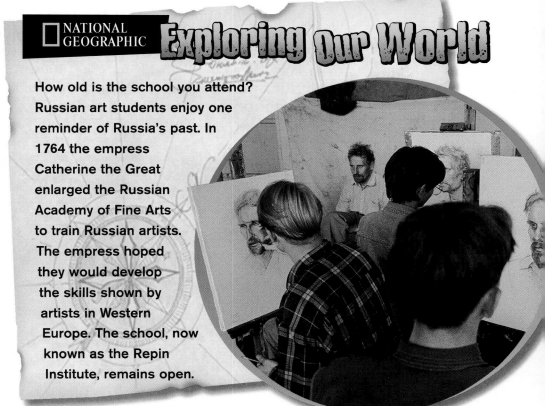

How old is the school you attend? Russian art students enjoy one reminder of Russia's past. In 1764 the empress Catherine the Great enlarged the Russian Academy of Fine Arts to train Russian artists. The empress hoped they would develop the skills shown by artists in Western Europe. The school, now known as the Repin Institute, remains open.

As you read in Chapter 14, Russia is the largest country in the world. Early in its history, however, it was a small territory on the edge of Europe. Strong rulers gradually expanded Russian territory. Their harsh rule led to unrest, eventually leading to two major upheavals—one in 1917, the other in 1991.

The History of Russia

To understand the challenges facing Russia today, let us go back through Russia's history. Modern Russians descend from early groups of Slavs who settled along the rivers of what is today Ukraine and Russia. During the A.D. 800s, these early Slavs built a civilization around the city of Kiev, today the capital of Ukraine. This civilization was called Kievan Rus (kee•AY•vuhn ROOS). By the A.D. 1000s, the ruler and people of Kievan Rus had accepted Eastern Orthodox Christianity. They prospered from trade with the Mediterranean world and Western Europe.

Then Mongols swept in from central Asia in the 1200s. They took control of Kiev and its surrounding lands. Mongol rule, which lasted about 200 years, greatly reduced Kiev's wealth and power. Meanwhile,

Moscow—a town to the north—became the center of a new Slavic territory called Muscovy (muh•SKOH•vee). In 1480 Ivan III, a prince of Muscovy, drove out the Mongols and made the territory independent.

Rise of the Czars   Muscovy slowly developed into the country we know today as Russia. Russian rulers expanded their power, built up their armies, and seized land and other resources. They called themselves czars, or emperors. They had complete and total control over the government. As a citizen of Muscovy, you would have feared Czar Ivan IV, who ruled during the 1500s. Known as "the Terrible" or "the Awesome," Ivan used a secret police force to tighten his iron grip on the people and control their lives.

Expansion of Russia

ALASKA
Russian
until 1867

ARCTIC OCEAN

North Pole

Bering Sea

FINLAND

EUROPE

LATVIA   ESTONIA
LITHUANIA

BELARUS

St. Petersburg (Leningrad)

Muscovy of Ivan III, 1462

Moscow

UKRAINE

Kiev

ARCTIC CIRCLE

Sea of Okhotsk

R U S S I A

MOLDOVA

Omsk

GEORGIA

KAZAKHSTAN

Vladivostok

ARMENIA

AZERBAIJAN

TURKMENISTAN   UZBEKISTAN

A S I A

KYRGYZSTAN

TAJIKISTAN

Barents Sea

0 mi. 1,000
0 km 1,000
Two-Point Equidistant projection

Russian Expansion

▨ Kievan Rus	■ 1917–1945
▢ 1462–1505	— Boundary of Soviet
▨ 1505–1689	Union in 1945
▨ 1689–1917	— Present boundaries

Applying Map Skills

1. In what present-day country is Kiev located?

2. What parts of the Soviet Union are now separate countries?

Find NGS online map resources @ www.nationalgeographic.com/maps

As the map on page 424 shows, the czars gradually conquered surrounding territories. As a result, many non-Russian peoples became part of the growing Russian Empire. Czars, such as Peter I and Catherine II, pushed the empire's borders southward and westward. They also tried to make Russia modern and more like Europe. Peter built a new capital—**St. Petersburg**—in the early 1700s. Built close to Europe near the Baltic coast, St. Petersburg was designed like a European city with its elegant palaces, public squares, and canals. If you had been a Russian noble at this time, you would have spoken French as well as Russian. You also would have put aside traditional Russian dress, worn European clothes, and attended fancy balls and parties.

The czar and the nobles enjoyed rich, comfortable lives. At the bottom of society, however, were the great masses of people. Most were serfs, or farm laborers who could be bought and sold along with the land. These people lived hard lives, working on the nobles' country estates or in city palaces. Few could read and write. They remained fiercely loyal to traditional Russian ways and practiced the Eastern Orthodox religion.

Dramatic Changes In 1812 a French army led by Napoleon Bonaparte invaded Russia. Brave Russian soldiers and the fierce winter weather finally forced the French to retreat. The year 1812 became a symbol of Russian patriotism. Have you ever heard the *1812 Overture,* with its dramatic ending that includes the ringing of bells and the bursts of cannon fire? Written by the Russian composer Peter Tchaikovsky (chy•KAWF•skee), this musical masterpiece celebrates the Russian victory over Napoleon.

In the late 1800s, Russia entered a period of economic and social change. In 1861 Czar Alexander II, known as the Czar-Liberator, freed the serfs from being tied to the land. His new law did little to lift them out of poverty, though. Russia began to industrialize, or change its economy to rely more on manufacturing and less on farming. Railroads, including the famous Trans-Siberian Railroad, spread across the country. It linked Moscow in the west with **Vladivostok** on Russia's Pacific coast.

Yet Russia did not progress politically. Educated Russians in the cities were stirred by democratic ideas from Western Europe. They pushed for freedom, equality, and a share in governing the country. The czars ignored these demands and continued to rule harshly.

√ Reading Check **What was the name of the civilization that early Slavs built in the area that is today Ukraine?**

The Soviet Era

In 1914 World War I broke out in Europe. Russian and German armies met and fought bloody battles in Eastern Europe. Unprepared for war, Russia suffered many defeats and few victories. As the fighting dragged on, shortages of food in Russian cities caused much starvation. Discontent grew among the Russian people.

Peter the Great
In 1698 Peter I, known as Peter the Great, came to the Russian throne. Nearly 7 feet (2 m) tall, Peter had boundless energy. After becoming czar, he took an 18-month tour of England and the Netherlands. He visited shipyards, factories, and laboratories. He learned carpentry and enough skill in surgery and dentistry to want to practice on others. When he returned home, Peter forced the Russian nobility to adopt the ways of Western Europe. In fact, those who refused to study math and geometry were not allowed to get married.

Web Activity Visit the **Geography: The World and Its People** Web site at gwip.glencoe.com and click on **Chapter 15— Student Web Activities** to learn more about the Russian Revolution.

The Russian Revolution In 1917 political leaders, soldiers, and factory workers forced Czar Nicholas II to give up the throne. Later that year, a political rebel named Vladimir Lenin led the Russian Revolution and seized control. He and his followers set up a communist state, a country whose government has strong control over the economy and society as a whole. Threatened at first by outside invasion, the Communists moved Russia's capital from coastal St. Petersburg to Moscow in the heart of the country.

Growth of Soviet Power By 1922 Russia's Communist leaders were securely in power. In that year, they formed the Union of Soviet Socialist Republics (U.S.S.R.), or the Soviet Union. This vast territory included Russia and most of the conquered territories of the old Russian Empire. After Lenin died in 1924, Communist officials disagreed over who was to lead the country.

Within a few years, Joseph Stalin had won out over the others and became the Soviet Union's leader. Under Stalin's orders, the government took control of all farming. Stalin also set up five-year plans to industrialize the country. These were programs that set economic goals for a five-year period. They led to an amazing growth of cities. Factory managers were told what to make and how to make it. The government took complete control over the economy. Those who opposed Stalin's actions were killed or sent to remote prison camps deep in the vast forests of icy **Siberia.** Millions of people suffered under Stalin's rule.

In 1941 Nazi Germany invaded the Soviet Union, bringing the country into World War II. During the conflict, the Soviets joined with Great Britain and the United States to defeat the Germans. Millions of Russians—soldiers and civilians—died in what Russians call the Great Patriotic War.

CULTURAL CLOSE-UP

Art

Peter Carl Fabergé was no ordinary Russian jeweler. His successful workshop designed extravagant jeweled flowers, figures, and animals. He is most famous for crafting priceless gold Easter eggs for the czar of Russia and other royalty in Europe and Asia. Each egg was unique and took nearly a year to create. Lifting the lid of the egg revealed a tiny surprise. One egg Fabergé created (shown here) held an intricate ship inside.

Looking Closer Why do you think Fabergé's workshop closed after the Russian Revolution of 1917?

Fabergé egg ▶

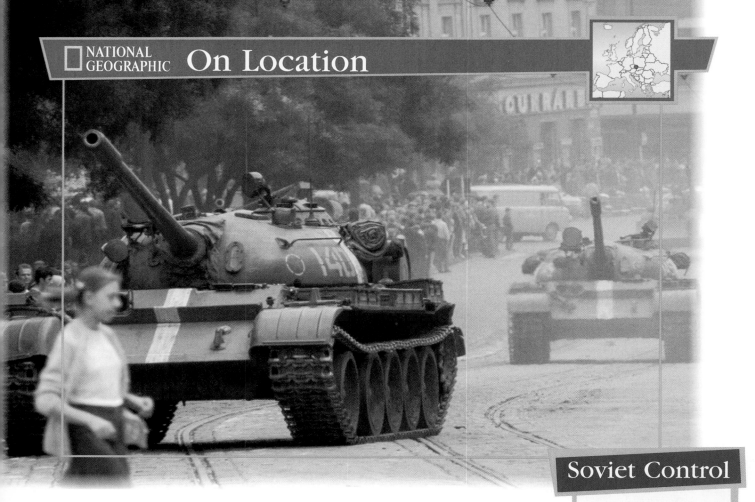

NATIONAL GEOGRAPHIC On Location

Soviet Control

Czechoslovakia tried to throw off Soviet control in 1968. Soviet tanks and troops poured into Bratislava to crush the revolt.

Movement Why did Stalin set up Communist governments in Eastern European countries?

A Superpower   When World War II ended, Stalin wanted to protect the Soviet Union from any more invasions. He sent troops to set up Communist governments in neighboring Eastern European countries. Stalin and the leaders who followed him strengthened the military and built powerful nuclear weapons. The Soviet Union became one of the two most powerful nations in the world. The other superpower—the United States—opposed Soviet actions. From the late 1940s to the late 1980s, these two nations waged a cold war. They competed for world influence without actually fighting each other. They even competed in areas *outside* the world. Both the Soviet Union and the United States launched rockets in a bid to be first in outer space.

The Cold War Years   During the Cold War years, the Soviet economy faced many problems. Under government control and with no competition, factories became inefficient and produced poor-quality goods. The government focused on making tanks and airplanes, not cars and refrigerators. As a result, people had few consumer goods to buy. Food often became scarce, and people had to wait in long lines to buy bread, milk, and other basics. Once again, discontent spread among the Russian people and those living in Soviet-controlled areas.

　　The Soviet Union had another challenge. This vast empire included not only Russians but also people from many other ethnic groups, or people who share a common culture, language, or history. Instead of being scattered throughout the country, people in each of these other groups generally lived together in the same area. They resented the control of the government in Moscow, which they believed favored ethnic Russians. They wanted to leave the Soviet Union and form their own countries.

Soviet Collapse In 1985 Mikhail Gorbachev (GAWR•buh•CHAWF) became the leader of the Soviet Union. Without totally abandoning communism, Gorbachev hoped to lessen the government's control of the economy and society. He tried to allow farmers and factory managers to make their own decisions. He allowed people to speak freely about the government and important issues, a policy called *glasnost,* or "openness." Instead of strengthening the country, however, his policies only made people doubt the communist system even more. People's demands for more and more changes eventually led to the collapse of both communism and the Soviet Union.

In late 1991, each of the 15 republics that made up the Soviet Union declared its independence. The Soviet Union no longer existed. Russia emerged as the largest and most powerful of those republics. Although a rough road lay ahead, many Russians were thrilled by the end of communism and the chance to enjoy freedom.

Reading Check **What two superpowers struggled in the Cold War?**

Assessment

Defining Terms
1. **Define** czar, serf, industrialize, communist state, cold war, ethnic group.

Recalling Facts
2. **History** Why is Czar Alexander II known as the Czar-Liberator?
3. **History** Who led the 1917 revolution in Russia?
4. **History** What happened to the Soviet Union in 1991?

Critical Thinking
5. **Understanding Cause and Effect** How did the enormous size of the Soviet Union create problems for that nation?
6. **Analyzing Information** How did a lack of competition affect the economy of the Soviet Union?

Graphic Organizer
7. **Organizing Information** In a chart like this one, write facts that show the contrast between the nobles and the serfs of Russia.

Nobles	Serfs

Applying Geography Skills

8. **Analyzing Maps** Study the historical map on page 424. When did Russia give up control of Alaska? In what present-day countries was Kievan Rus located?

Living in Space

In 1986 the Russian space station *Mir* began to orbit Earth. Since that time *Mir* has been home to more than 60 cosmonauts— Russian astronauts—and astronauts from more than a dozen countries.

History

The Soviet Union built the space station *Mir,* the Russian word for "peace," to establish a permanently staffed laboratory in space. Each mission to *Mir* lasted an average of six months to a year. Other spacecraft delivered fresh food, new equipment, and mail from home.

Mir cosmonauts and astronauts spent most of their day in the core module, which contained the living quarters and work compartment. Other modules provided equipment for producing oxygen, recycling water, and performing science experiments. The experiments focused on the long-term effects of weightlessness on plants and animals, including humans. Crew members also checked ozone levels in Earth's atmosphere, recorded changes in the oceans, and observed distant galaxies.

In 1993 the United States and Russia decided to build an International Space Station. The Russian government decided to scrap *Mir* and let it burn up in the atmosphere in 2001.

▶ Making the Connection

1. What was *Mir* and when was it first launched?
2. What kind of scientific research did the *Mir* crew conduct?
3. **Drawing Conclusions** What value do you think a permanent space station can offer?

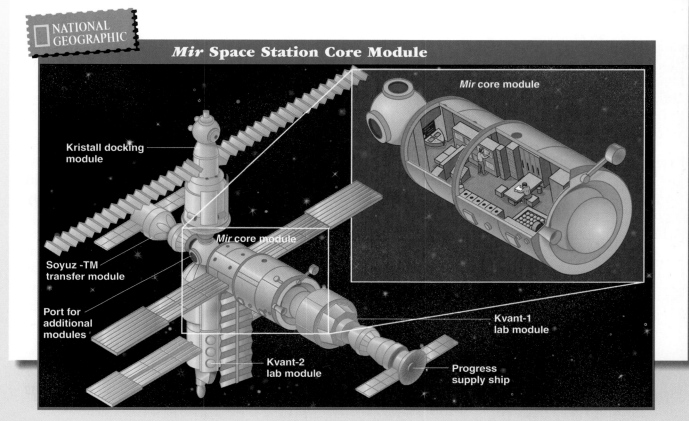

NATIONAL GEOGRAPHIC

Mir Space Station Core Module

Kristall docking module

Mir core module

Mir core module

Soyuz -TM transfer module

Port for additional modules

Kvant-1 lab module

Kvant-2 lab module

Progress supply ship

A New Russia

Guide to Reading

Main Idea

Russia has a rich cultural heritage, but faces challenges in adopting a new economic system and government.

Terms to Know

- free enterprise system
- nuclear energy
- life expectancy
- democracy
- federal republic
- urban
- suburb
- rural

Places to Locate

- Chechnya
- Caspian Sea
- Moscow
- St. Petersburg

Reading Strategy

Draw a diagram like this one. Then write four challenges facing Russia.

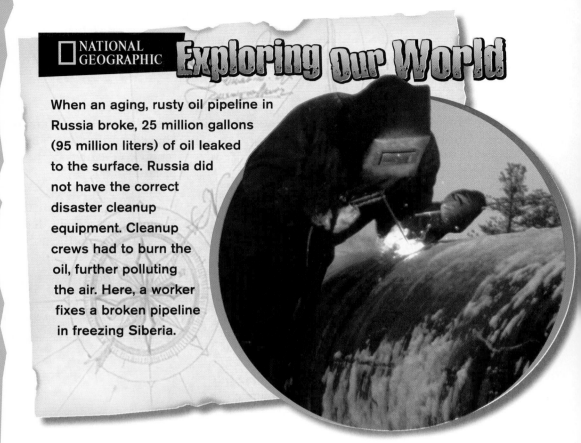

NATIONAL GEOGRAPHIC Exploring Our World

When an aging, rusty oil pipeline in Russia broke, 25 million gallons (95 million liters) of oil leaked to the surface. Russia did not have the correct disaster cleanup equipment. Cleanup crews had to burn the oil, further polluting the air. Here, a worker fixes a broken pipeline in freezing Siberia.

The Russian people have been working to move from the strict, tightly controlled rule of the past toward an open and free economy and government. They have found that these changes do not come easily.

From Communism to Free Enterprise

Under communism, the government controlled Russia's economy. Government officials decided what—and how much—farmers should grow. They told factory managers what products to make, how many to make, and to whom they could sell them.

The fall of communism turned Russia's economy upside down. The Russian government adopted a market economy, the economic system followed in the United States. Under a market economy, or **free enterprise system,** people start and run businesses with little government intervention. Today Russian factory managers can decide what products to make. People can open businesses—restaurants, stores, or computer companies—and choose their own careers.

The Russian people gained freedom, but that has not always helped them. People now can make their own decisions, but those decisions do not always lead to success. They may become unemployed. Under communism, everybody had jobs. Workers today can lose their jobs when a business declines.

In addition, the government no longer sets prices for food and other goods. When prices were set low, the Russian people could afford the goods but often faced shortages. Why? Producers would not supply goods if they could not make a profit. Without government controls, prices have risen. Higher prices make it harder to buy necessities like food and clothing. It is hoped that, in the future, factories will start producing more goods and services. Now that manufacturers know they can receive higher prices and more profits for their goods, supplies should increase.

Environmental Issues The old Communist government did little to protect the environment. Factories belched dirty smoke into the air and dumped poisons into rivers and lakes. Pollution fouled the air, water, and land. The government also built plants to make **nuclear energy,** or power made by creating a controlled atomic reaction. Although nuclear power plants do not belch dirty smoke, they do leave dangerous by-products. In addition, if uncontrolled like the nuclear explosion at Chernobyl, the consequences are disastrous. Turn to page 386 to find out more about Chernobyl.

Pollution has harmed the health of the Russian people, too. Air pollution has caused many people to suffer from diseases of the lungs. Rising numbers of people have cancer, and life expectancy in Russia has fallen. **Life expectancy** is the number of years that an average person is expected to live. Falling life expectancy reveals that the population is growing less healthy.

NATIONAL GEOGRAPHIC On Location

Food Shortages

Under communism, prices were low, but people faced shortages and had to wait in line to buy goods.

Place Why can many Russians today not afford goods and services?

√ Reading Check Who made economic decisions under Communist rule?

Political Challenges

Under communism, members of the Communist Party controlled Russia's government and told people how to vote. Today Russia is a **democracy,** a government in which people freely elect their leaders. Russia also is a **federal republic.** This means that power is divided between national and state governments with a president who leads the nation.

A Russian president has stronger powers than an American president. For example, the Russian president can issue orders that become laws even if they are not passed by the legislature. Russia's first two presidents—Boris Yeltsin (BOH•rehs YEHL•tzehn) and Vladimir Putin (VLAH•deh•meehr POO•tihn)—used their powers to push for economic changes in Russia.

In adjusting to a new form of government, Russians face important political challenges. They have to learn how to act in a democracy. Democracy is built on the idea of the rule of law. This means that laws govern not just ordinary people but also government officials. In the past, Russian leaders did what they wanted to do. In the new system, they must learn to follow the law. Also, past governments punished people who criticized their decisions. Now officials have to learn to accept disagreements over government policies.

Another challenge results from the fact that Russia is home to many different ethnic groups. Some of these groups want to form their own countries. Among these groups are the Chechens (CHEH•chehnz), who live in **Chechnya** (CHEHCH•nee•uh) near the **Caspian Sea** in southern Russia. Find Chechnya on the map on page 439. Russian troops have fought Chechen forces to keep Chechnya a part of Russia.

✓ Reading Check **Why is being home to different ethnic groups a challenge for Russia?**

The Russian People

Russia is one of the most populous countries in the world, with 146.5 million people. Most of them are ethnic Russians, but the country includes many other groups as well. About 90 percent of the people speak the Russian language. Look at the map on page 433. You will find western Russia, especially around **Moscow,** to be the most densely populated area of the country. About 75 percent of Russia's people live in cities.

Living in the City Russia's urban, or city, areas are large and modern with stone or concrete buildings and wide streets. Tall buildings hold apartments for hundreds of families. Many of these apartments are small and cramped, however. When people in cities relax, they spend time with their families and friends, take walks through parks, or attend concerts, movies, and the circus.

Russian cities have changed in recent years. Some people have benefited from the economic changes sweeping the country. Many of these prosperous people have clustered in Moscow. Their new cars speed down the city's streets. They are building large houses outside the city limits, where few people lived before. As a result, Russia is developing its first suburbs, or smaller communities that surround a city.

Yet a large number of Russia's city dwellers remain poor. These people lack the money to buy the consumer goods that are now more and more available. Many survive only by standing in long lines to receive food given away by government agencies. Some of the poor resent the success of the newly wealthy people.

Living in the Country In Russia's **rural** areas, or countryside, most people live in houses built of wood. The quality of health care and education is lower in rural areas than in the cities. Over the years, many people have left rural areas to find work in Russia's cities.

✓ **Reading Check** What percentage of Russia's people live in cities?

Culture in Russia

Russia has a rich tradition of art, music, and literature. Russians view these cultural achievements with pride. Some of their most beloved works of art are based on religious, historical, or folk themes.

Religion The Soviet Union's Communist rulers put severe limits on the practice of religion. They closed many houses of worship and persecuted outspoken religious leaders. Since the fall of communism, many Russians have returned to religious traditions, including Judaism, Islam,

NATIONAL GEOGRAPHIC

Russia: Population Density

Persons per

Sq. Mi.		Sq. Km
Uninhabited		Uninhabited
Under 2		Under 1
2–60		1–25
60–125		25–50
125–250		50–100
Over 250		Over 100

Cities

■ City with more than 5,000,000 people

● City with 1,000,000 to 5,000,000 people

○ City with 500,000 to 1,000,000 people

Applying Map Skills

1. What general area of Russia has the highest population density?

2. What is the population density of most of eastern Russia?

Find NGS online map resources @ www.nationalgeographic.com/maps

433

Urban vs. Rural

The GUM state department store (above) is very similar to modern malls in the United States. In some rural areas, however, people still live without heat, electricity, or plumbing (right).

Movement How have economic changes affected where Russia's people live?

and Christianity. The largest number practices a Russian form of the Eastern Orthodox Christian faith. Tourists enjoy seeing the onion-shaped domes and elaborate decorations of Russian Orthodox churches.

Music, Art, and Literature   If you wanted to see Russian culture at its best, where would you go? You might visit **St. Petersburg's** Mariinsky (MAH•ree•IHN•skee) Theater. One of Russia's top ballet companies dances here, and Russian ballet dancers are famous around the world. Composer Peter Tchaikovsky wrote some of the world's favorite ballets, including *Sleeping Beauty* and *The Nutcracker Suite*. You remember that his *1812 Overture* celebrated the Russian defeat of the French invasion of 1812. Nikolay Rimsky-Korsakov used Russian folktales and tunes in his operas and other works. Igor Stravinsky's *Firebird Suite* came from a Russian legend.

If you enjoy painting, you would like to stroll through St. Petersburg's Hermitage Museum. Originally built to hold the art collection of the czars, the museum now displays these works for the public. It has paintings by Russian and European painters and sculptors.

The great works of Russian literature also reflect Russian themes. Poet Alexander Pushkin protested the lack of freedom under the czars. Leo Tolstoy's novel *War and Peace* recounts how Russians rallied to defeat the French in 1812. Fyodor Dostoyevsky (DAHS•tuh•YEHF•skee) wrote many novels that explored Russian life during the late 1800s. In the 1970s, Alexander Solzhenitsyn (SOHL•zhuh•NEET•suhn) wrote novels that revealed the harsh conditions of Communist society.

Celebrations   Russians enjoy small family get-togethers as well as national holidays. New Year's Eve is the most festive nonreligious holiday for Russians. Children decorate a fir tree and exchange presents with others in their families. Russians also celebrate the first day of May with parades and speeches. May Day honors Russian workers.

Food   If you have dinner with a Russian family, you might begin with a big bowl of borscht, a soup made from beets, or *shchi,* a soup made from cabbage. Next, you might have meat turnovers called piroshki. For the main course, you are likely to eat meat, poultry, or fish with boiled potatoes. On special occasions, Russians like to eat caviar. This delicacy is eggs of the sturgeon, a fish from the Caspian Sea.

Sports and Recreation   Russians love soccer and tennis. They also enjoy hiking, camping, and mountain climbing. With Russia's cold climate, winter sports are popular. Russians like to ski, skate, and play ice hockey. Have you ever watched the Olympics? If so, you probably have seen Russian hockey players, figure skaters, and gymnasts standing on the victory podium. They thrill spectators with their skill and grace.

✓ Reading Check   **Which Russian author wrote about the harsh conditions of Communist society?**

Assessment

Defining Terms
1. **Define** free enterprise system, nuclear energy, life expectancy, democracy, federal republic, urban, suburb, rural.

Recalling Facts
2. **Government** What is one power of the Russian president?
3. **Government** Why has there been fighting with Chechnya?
4. **Culture** What is the major religion of Russia?

Critical Thinking
5. **Synthesizing Information** Describe the problems Russians face living in a free society after years of Communist rule.
6. **Making Predictions** Art ideas are frequently drawn from life. What themes do you think you will see in future Russian arts?

Graphic Organizer
7. **Organizing Information** Draw a diagram like this one, and list two facts for each topic in the four outer ovals.

Food Cities

Russian Life

Religion Sports

Applying Geography Skills

8. **Analyzing Maps** Look at the population density map on page 433. What city is located at about 38°E longitude and 55°N latitude? What is the estimated population of this city?

Critical Thinking Skill

Understanding Cause and Effect

Understanding cause and effect involves considering *why* an event occurred. A *cause* is the action or situation that produces an event. What happens as a result of a cause is an *effect*.

Learning the Skill

To identify cause-and-effect relationships, follow these steps:

- Identify two or more events or developments.
- Decide whether one event caused the other. Look for "clue words" such as *because, led to, brought about, produced, as a result of, so that, since,* and *therefore*.
- Look for logical relationships between events, such as "She overslept, and then she missed her bus."
- Identify the outcomes of events. Remember that some effects have more than one cause, and some causes lead to more than one effect. Also, an effect can become the cause of yet another effect.

Practicing the Skill

For each number below, identify which statement is the cause and which is the effect.

1. (A) The Communists moved Russia's capital from coastal St. Petersburg to Moscow in the heart of the country.
   (B) The capital of Russia was threatened by an outside invasion.

▲ Revolutionary leaders and philosophers Lenin, Engels, and Marx

2. (A) Revolutionary leaders seized control of the Russian government.
   (B) During World War I, shortages of food in Russian cities caused much starvation.
   (C) Discontent grew among the Russian people.
3. (A) The Soviet government kept prices for goods and services very low.
   (B) Many goods and services were in short supply in the Soviet Union.

Applying the Skill

In your local newspaper, read an article describing a current event. Determine at least one cause and one effect of that event. Show the cause-and-effect relationship in a diagram like the one here:

Cause ⟶ Effect

GO TO

Practice key skills with **Glencoe Skillbuilder Interactive Workbook, Level 1.**

Chapter 15 Reading Review

Section 1   A Troubled History

Terms to Know
czar
serf
industrialize
communist state
cold war
ethnic group

Main Idea

The harsh rule of powerful leaders has often sparked violent uprisings in Russia.

✓ History Emperors called czars ruled the Russian Empire from 1480 to 1917.

✓ Movement The czars expanded Russian territory to reach from Europe to the Pacific.

✓ Government Under the Communists, Russia became part of the Soviet Union.

✓ History In 1991 the Soviet Union broke apart, and Russia became an independent republic.

Section 2   A New Russia

Terms to Know
free enterprise system
nuclear energy
life expectancy
democracy
federal republic
urban
suburb
rural

Main Idea

Russia has a rich cultural heritage, but faces challenges in adopting a new economic system and government.

✓ Economics The change to a free enterprise economy has been a challenge for Russians as they face rising unemployment and rising prices.

✓ Government Russians have had to learn how to live in a democracy.

✓ Government Some non-Russian ethnic groups want to create independent nations out of Russia.

✓ Culture Russians practice many different religions, but more follow Russian Orthodox Christianity than any other faith.

✓ Culture Russian artists, composers, and writers often used themes based on Russian history or traditions.

◀ The Hermitage in St. Petersburg

Assessment and Activities

 Using Key Terms

Match the terms in Part A with their definitions in Part B.

A.

1. cold war
2. nuclear energy
3. rural
4. czar
5. life expectancy
6. industrialize
7. suburb
8. serf
9. urban
10. communist state

B.

a. smaller community surrounding a city
b. changing an economy to manufacturing
c. farm laborers bound to the land they worked
d. period of tension without actual fighting
e. relating to a city
f. former emperor of Russia
g. power from a controlled atomic reaction
h. relating to the countryside
i. number of years that the average person is expected to live
j. government with strong control over the economy and society as a whole

 Reviewing the Main Ideas

Section 1 A Troubled History

11. **History** Which czar used a secret police force to maintain strict control over the people?
12. **History** When was the Union of Soviet Socialist Republics formed?
13. **Government** Why did Stalin send people to Siberia?
14. **History** What took place in 1991?

Section 2 A New Russia

15. **Economics** What type of economic system has the new Russia adopted?
16. **Culture** What does falling life expectancy reveal?
17. **Government** Who were Russia's first two presidents?
18. **Culture** What do the *1812 Overture* and *War and Peace* have in common?
19. **Culture** Who wrote *The Nutcracker Suite*?
20. **Human/Environment Interaction** Give at least four examples of the environmental damage that occurred under the Soviet Communist government.

 **Russia—Past and Present**

Place Location Activity

On a separate sheet of paper, match the letters on the map with the numbered places listed below.

1. St. Petersburg
2. Caspian Sea
3. Moscow
4. Baltic Sea
5. Vladivostok
6. Irkutsk
7. Omsk
8. Sea of Japan

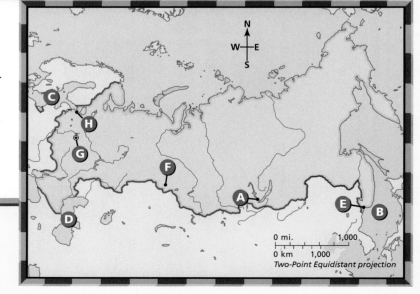

Self-Check Quiz Visit the *Geography: The World and Its People* Web site at gwip.glencoe.com and click on **Chapter 15— Self-Check Quizzes** to prepare for the Chapter Test.

Critical Thinking

21. **Understanding Cause and Effect** How did World War I help lead to the Russian Revolution?

22. **Organizing Information** Draw a diagram like this one. Complete it with four characteristics of the free enterprise system in Russia.

Free enterprise in Russia ⟹ _____

GeoJournal Activity

23. **Writing a News Article** Learn more about the nuclear disaster at Chernobyl. Find out what happened and its effects on the local people, animals, and land. Write a news article that might have appeared a few days after the explosion. Then write a follow-up article on the long-term effects of the nuclear explosion on neighboring countries.

Mental Mapping Activity

24. **Focusing on the Region** Draw a simple outline map of Russia, then label the following:

- Arctic Ocean
- Pacific Ocean
- Vladivostok
- Moscow
- Bering Sea
- St. Petersburg
- Siberia
- Baltic Sea

Technology Skills Activity

25. **Developing Multimedia Presentations** Choose one of the czars of Russia. Research your choice and create a multimedia presentation on this person. Include information on when he or she lived, what he or she accomplished (or did not accomplish), and other interesting facts. Use pictures, maps, and time lines to make your presentation more visual.

Standardized Test Practice

Directions: Study the map below, then answer the question that follows.

Chechnya

1. **Which of the following statements about this map is NOT true?**

   F Chechnya lies along Russia's southern border.

   G Chechnya is situated between the Black and Caspian Seas.

   H Chechnya's landscape is mostly flat, fertile farmland.

   J Chechnya's nearest neighbor is Georgia.

Test-Taking Tip: Be careful when you see the words NOT or EXCEPT in a question. Read all the answer choices and choose the one that *does not* fit with the question. Quickly eliminate answers that are true. Make sure that your answer choice is supported by information *on the map.*

Napoleon and troops retreat from Russia.

RUSSIA'S STRATEGY:
Freeze Your Foes

Winter weather can cancel school, bring flu outbreaks, and stop traffic. It can even change history. Such was the case when French ruler Napoleon thought he had conquered the Russian Empire.

In fact, Napoleon did not want to conquer Russia. His real enemy was Great Britain. Napoleon wanted Russia and other countries to stop trading with Great Britain. Yet Russia's czar, Alexander I, refused. By 1812, Napoleon was determined to change Alexander's mind. In June, leading an army of more than half a million soldiers, Napoleon invaded Russia. To reach Moscow and the czar, Napoleon had to fight his way across the Russian countryside.

By the time Napoleon's battle-weary forces reached Moscow, supplies were scarce. All along the route, Russians had burned villages as they retreated, leaving no food or shelter. Reaching Moscow, Napoleon found the city in flames and nearly empty of people. The czar had moved to St. Petersburg. Napoleon took Moscow without a fight, but most of the city was in ashes.

Winter Wins a War

With winter approaching, Napoleon waited in Moscow for Alexander I to offer peace. The czar remained silent, however. With dwindling supplies and many of his troops lacking winter clothes, Napoleon was forced to retreat. He tried to take a new way back, but the Russians made Napoleon use the same ruined route he had used before. Armed bands of Russians attacked at every turn. Starving and desperate to escape the bitter cold, several of Napoleon's soldiers threw themselves into burning buildings. Most of Napoleon's troops never made it out of Russia.

History Repeats

More than a century later, during World War II, Russia's winter was again a mighty foe. On June 22, 1941, Adolf Hitler's German army invaded Russia, then part of the Soviet Union. As the German army fought its way to Moscow, Soviet leader Joseph Stalin issued his own "scorched-earth policy." Soviet citizens burned anything of use to the invaders. By December, German troops were within sight of the Kremlin, Moscow's government center, when winter struck.

Snow buried the invaders. Temperatures fell below freezing. Grease in guns and oil in vehicles froze solid. German soldiers suffered frostbite and died. The Soviets were better clothed and had winterized their tanks and trucks. Stalin's troops pushed back the German army. Once again the Russians triumphed with help from "General Winter."

QUESTIONS

1 After Napoleon conquered Moscow in 1812, why did he retreat?

2 How did Russia's winter affect fighting in World War II?

German prisoners of Russia's winter ▶

NATIONAL
GEOGRAPHIC
SOCIETY

Average Winter Temperatures

EUROPE

⊕ Moscow

RUSSIA

ASIA

Napoleon's Advance,
June–October 1812

German Forces Front
Line, December 1941

< -40°F	
-40° to -31°F	0° to 10°F
-30° to -21°F	11° to 20°F
-20° to -11°F	21° to 30°F
-10° to 0°F	> 30°F

EXPLORING OTHER REGIONS

Use this handbook to explore other regions of the world.

So far, in *Geography: The World and Its People, Volume One,* you have studied several regions of the world. Read this special Exploring Other Regions handbook to learn about other parts of your world—the regions of North Africa, Southwest Asia, and Central Asia; Africa South of the Sahara; Asia; and Australia, Oceania, and Antarctica. As you read the information, remember what you have already learned about the physical and cultural geography of the United States and Canada, Latin America, Europe, and Russia, and use your knowledge to compare the regions of the world.

Contents

Nomad in
North Africa

Polynesia

Masai dance, Tanzania

Japanese architecture

Red kangaroo in Australia

Pyramids in Egypt

Photo Credits: CORBIS

Regions 1

Unit 6

Business district at dusk, Dubai, United Arab Emirates

Young woman in national dress, Turkmenistan ▶

North Africa, Southwest Asia, and Central Asia

Ancient Egyptian pyramids overlook industrial smokestacks. Three-thousand-year-old stone temples tower over sparkling new oil derricks. Remote mountain villages and endless desert seas of sand and gravel contrast with modern beaches overrun by tourists. All of these extremes can be found within the culture region of North Africa, Southwest Asia, and Central Asia.

Shepherd tending sheep, Atlas Mountains, Morocco

NGS ONLINE
www.nationalgeographic.com/education

Focus on:

North Africa, Southwest Asia, and Central Asia

LYING AT THE INTERSECTION of Europe, Asia, and Africa, this sprawling region has long been a meeting place for diverse peoples and cultures. Troubled by bitter conflicts and plagued by a scarcity of water, the region also is extremely rich in oil and other natural resources.

The Land

Glance at a physical map of this region and you will see a jumble of mountain chains. In the west, the Atlas Mountains—Africa's longest range—run through Morocco and Algeria. The Caucasus Mountains span the land between the Black and Caspian Seas and form the northern border of Georgia and Azerbaijan. Slanting southeast through Turkey and Iran are the Zagros Mountains, where earthquakes frequently occur. Farther east are the Hindu Kush in Afghanistan and the Tian Shan, or "heavenly mountains," in Kyrgyzstan—home to some of the world's largest glaciers.

Mountains block moist winds, helping to create vast deserts across much of the region. The Sahara, in North Africa, is the world's largest hot desert. The Rub' al Khali, or Empty Quarter, covers about one-fourth of the Arabian Peninsula. The Garagum and Qizilqum lie in Turkmenistan and Uzbekistan.

Through these arid landscapes flow great rivers that bring life-giving water. The world's longest river, the Nile, runs 4,241 miles (6,825 km) through Egypt to the Mediterranean Sea. The Tigris and Euphrates Rivers flow southeast through Turkey, Syria, and Iraq.

The Climate

Water is precious in much of this region. Most areas receive a meager 10 inches (25 cm) or less of rainfall each year. In such arid lands, agriculture is only possible along rivers and canals or in places where natural springs bubble to the surface to create lush but isolated oases.

In areas with a steppe climate, where enough rain falls to support grasses, people raise livestock such as sheep, camels, and goats. Steppes cover parts of many Southwest and Central Asian countries. A narrow band of steppe runs along the northern edge of the Sahara, too.

The areas that border the Mediterranean, Black, and Caspian Seas enjoy a milder Mediterranean climate. Although summers are hot and dry, winters bring enough precipitation to turn coastal lowlands into green landscapes.

Nile River flowing through Aswan, Egypt

◀ Man gazing out across the vast expanses of the Sahara

REGIONAL ATLAS

The Economy

Like water, natural resources are distributed unevenly across the region. This helps to create great differences in living standards. The region includes some of the world's wealthiest nations—and some of its poorest. Enormous reserves of oil and natural gas lie in certain areas, including lands in central North Africa, along the Persian Gulf, and around the Caspian Sea. Countries such as Saudi Arabia and Kuwait, which export petroleum products to fuel-hungry societies, generally enjoy high standards of living.

On the other hand, those countries with economies based on agriculture have much lower standards of living. Only a small percentage of the region's land is suitable for growing crops. In river valleys and along the coasts, where there is water and fertile soil, farmers raise citrus fruits, grapes, dates, grains, and cotton. Nomadic herding is common across the large expanses of this region that are too dry for crops.

The People

Great pyramids, built as tombs for Egyptian rulers, rise above desert sands. They are a reminder that some of the world's oldest civilizations developed in this region. Roughly 5,000 years ago, the ancient Egyptians built a kingdom along the life-giving Nile River. The Sumerian civilization, an even older society, flourished in the fertile valley between the Tigris and Euphrates Rivers.

Water still dictates where people settle in this region. Most cities lie along seacoasts or rivers, or near desert oases. Among the largest cities are Cairo, Egypt; Istanbul, Turkey; and Tehran, Iran.

In North Africa and Southwest Asia, most of the people are Arabs. Turkic ethnic groups are the majority in Central Asia. Many other ethnic groups also live in the region. Despite the ethnic diversity, many who live here are united by religion. Most people practice Islam, which developed in this region centuries ago. Two other major religions, Judaism and Christianity, also began here. Georgia and Armenia are two of the world's oldest Christian countries, and the country of Israel is the Jewish national homeland.

Exploring the Region

1. **How do mountains help create deserts across much of the region?**
2. **What parts of the region receive the most rainfall?**
3. **How do most people in the region make their living?**
4. **What religion do most people in the region practice?**

◀ **Construction of an oil pipeline across a desert, Yemen**

Desert dwellers sharing
a meal, Saudi Arabia ▶

North Africa, Southwest Asia, and Central Asia

Physical

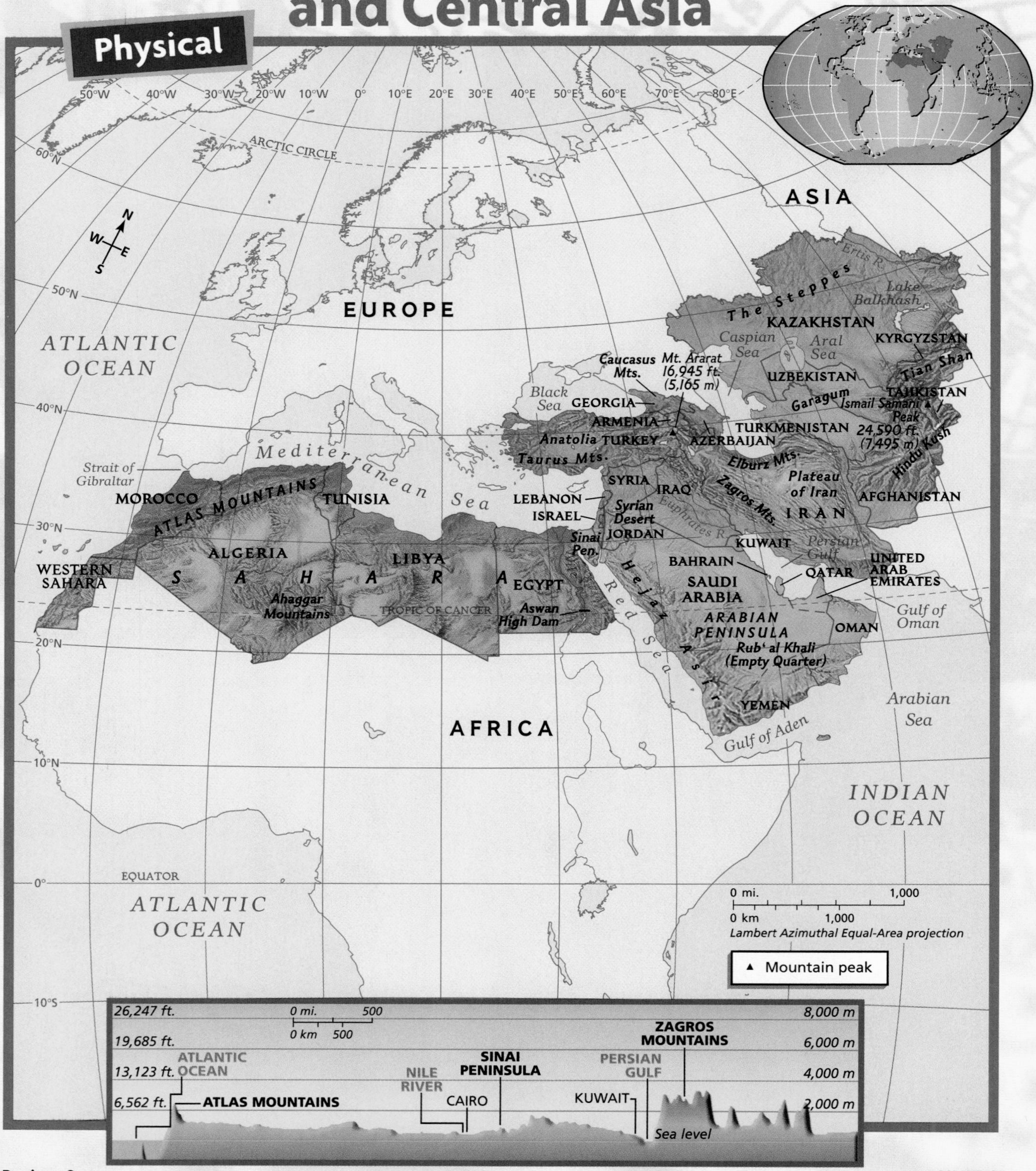

50°W 40°W 30°W 20°W 10°W 0° 10°E 20°E 30°E 40°E 50°E 60°E 70°E 80°E

ARCTIC CIRCLE

60°N

50°N

EUROPE

40°N

ATLANTIC OCEAN

Strait of Gibraltar

Mediterranean Sea

30°N

20°N

TROPIC OF CANCER

10°N

0° EQUATOR

ATLANTIC OCEAN

10°S

AFRICA

N
W — E
S

ASIA

The Steppes

Ertis R.

Lake Balkhash

KAZAKHSTAN

KYRGYZSTAN

Caspian Sea

Aral Sea

UZBEKISTAN

Tian Shan

Caucasus Mts.

Mt. Ararat 16,945 ft. (5,165 m)

Black Sea

GEORGIA

ARMENIA

Anatolia TURKEY

Taurus Mts.

AZERBAIJAN

Garagum

TURKMENISTAN

TAJIKISTAN

Ismail Samani Peak 24,590 ft. (7,495 m)

Hindu Kush

Elburz Mts.

Plateau of Iran

Zagros Mts.

AFGHANISTAN

SYRIA

IRAQ

IRAN

LEBANON

ISRAEL

Syrian Desert

JORDAN

Sinai Pen.

Euphrates R.

KUWAIT

Persian Gulf

BAHRAIN

QATAR

UNITED ARAB EMIRATES

Gulf of Oman

MOROCCO

TUNISIA

ATLAS MOUNTAINS

WESTERN SAHARA

ALGERIA

LIBYA

EGYPT

S A H A R A

Ahaggar Mountains

Aswan High Dam

Red Sea

Hejaz Mts.

SAUDI ARABIA

ARABIAN PENINSULA

Rub' al Khali (Empty Quarter)

OMAN

YEMEN

Gulf of Aden

Arabian Sea

INDIAN OCEAN

0 mi. 1,000
0 km 1,000
Lambert Azimuthal Equal-Area projection

▲ Mountain peak

26,247 ft. 0 mi. 500 8,000 m
0 km 500
19,685 ft. ZAGROS MOUNTAINS 6,000 m
ATLANTIC OCEAN
13,123 ft. NILE SINAI PENINSULA PERSIAN GULF 4,000 m
RIVER
6,562 ft. ATLAS MOUNTAINS CAIRO KUWAIT 2,000 m
Sea level

Political

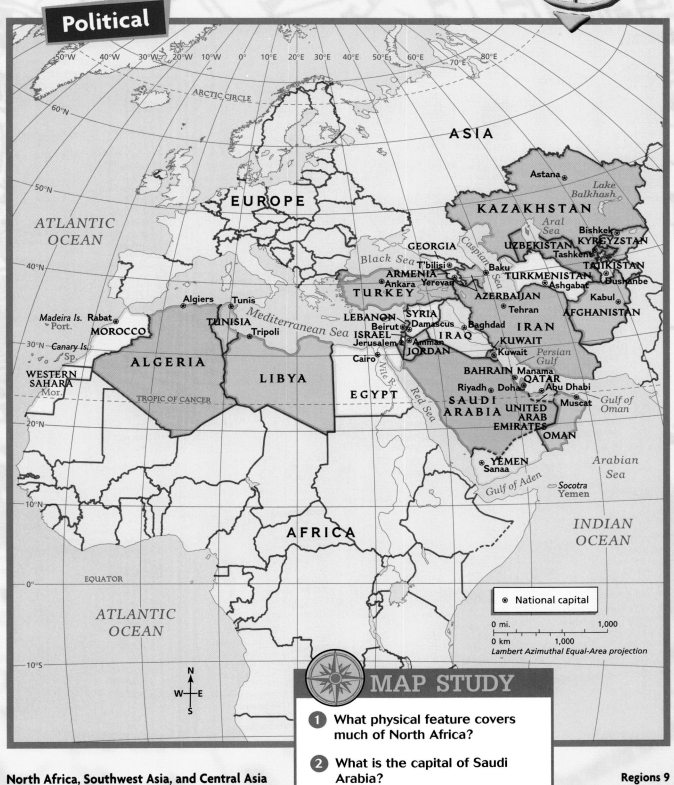

National capital

0 mi.   1,000

0 km   1,000

Lambert Azimuthal Equal-Area projection

MAP STUDY

1   What physical feature covers much of North Africa?

2   What is the capital of Saudi Arabia?

North Africa, Southwest Asia, and Central Asia   **Regions 9**

REGIONAL ATLAS

Country Profiles

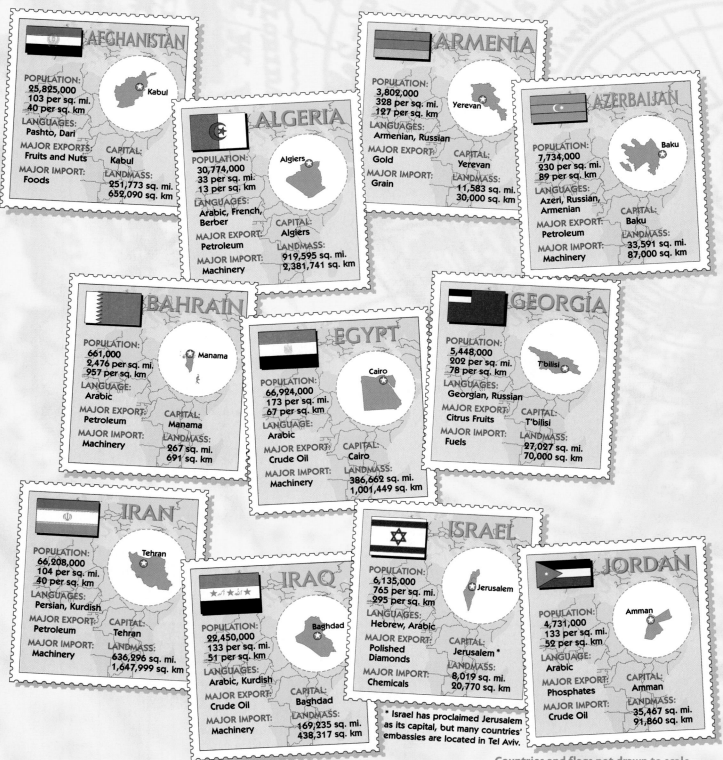

AFGHANISTAN

POPULATION:
25,825,000
103 per sq. mi.
40 per sq. km

LANGUAGES:
Pashto, Dari

MAJOR EXPORTS:
Fruits and Nuts

MAJOR IMPORT:
Foods

CAPITAL:
Kabul

LANDMASS:
251,773 sq. mi.
652,090 sq. km

ALGERIA

POPULATION:
30,774,000
33 per sq. mi.
13 per sq. km

LANGUAGES:
Arabic, French, Berber

MAJOR EXPORT:
Petroleum

MAJOR IMPORT:
Machinery

CAPITAL:
Algiers

LANDMASS:
919,595 sq. mi.
2,381,741 sq. km

ARMENIA

POPULATION:
3,802,000
328 per sq. mi.
127 per sq. km

LANGUAGES:
Armenian, Russian

MAJOR EXPORT:
Gold

MAJOR IMPORT:
Grain

CAPITAL:
Yerevan

LANDMASS:
11,583 sq. mi.
30,000 sq. km

AZERBAIJAN

POPULATION:
7,734,000
230 per sq. mi.
89 per sq. km

LANGUAGES:
Azeri, Russian, Armenian

MAJOR EXPORT:
Petroleum

MAJOR IMPORT:
Machinery

CAPITAL:
Baku

LANDMASS:
33,591 sq. mi.
87,000 sq. km

BAHRAIN

POPULATION:
661,000
2,476 per sq. mi.
957 per sq. km

LANGUAGE:
Arabic

MAJOR EXPORT:
Petroleum

MAJOR IMPORT:
Machinery

CAPITAL:
Manama

LANDMASS:
267 sq. mi.
691 sq. km

EGYPT

POPULATION:
66,924,000
173 per sq. mi.
67 per sq. km

LANGUAGE:
Arabic

MAJOR EXPORT:
Crude Oil

MAJOR IMPORT:
Machinery

CAPITAL:
Cairo

LANDMASS:
386,662 sq. mi.
1,001,449 sq. km

GEORGIA

POPULATION:
5,448,000
202 per sq. mi.
78 per sq. km

LANGUAGES:
Georgian, Russian

MAJOR EXPORT:
Citrus Fruits

MAJOR IMPORT:
Fuels

CAPITAL:
T'bilisi

LANDMASS:
27,027 sq. mi.
70,000 sq. km

IRAN

POPULATION:
66,208,000
104 per sq. mi.
40 per sq. km

LANGUAGES:
Persian, Kurdish

MAJOR EXPORT:
Petroleum

MAJOR IMPORT:
Machinery

CAPITAL:
Tehran

LANDMASS:
636,296 sq. mi.
1,647,999 sq. km

IRAQ

POPULATION:
22,450,000
133 per sq. mi.
51 per sq. km

LANGUAGES:
Arabic, Kurdish

MAJOR EXPORT:
Crude Oil

MAJOR IMPORT:
Machinery

CAPITAL:
Baghdad

LANDMASS:
169,235 sq. mi.
438,317 sq. km

ISRAEL

POPULATION:
6,135,000
765 per sq. mi.
295 per sq. km

LANGUAGES:
Hebrew, Arabic

MAJOR EXPORT:
Polished Diamonds

MAJOR IMPORT:
Chemicals

CAPITAL:
Jerusalem *

LANDMASS:
8,019 sq. mi.
20,770 sq. km

JORDAN

POPULATION:
4,731,000
133 per sq. mi.
52 per sq. km

LANGUAGE:
Arabic

MAJOR EXPORT:
Phosphates

MAJOR IMPORT:
Crude Oil

CAPITAL:
Amman

LANDMASS:
35,467 sq. mi.
91,860 sq. km

* Israel has proclaimed Jerusalem as its capital, but many countries' embassies are located in Tel Aviv.

Countries and flags not drawn to scale

KAZAKHSTAN

POPULATION:
15,417,000
15 per sq. mi.
6 per sq. km

LANGUAGES:
Kazakh, Russian

MAJOR EXPORT:
Petroleum

MAJOR IMPORT:
Machinery

CAPITAL:
Astana

LANDMASS:
1,049,039 sq. mi.
2,716,998 sq. km

KUWAIT

POPULATION:
2,076,000
302 per sq. mi.
117 per sq. km

LANGUAGE:
Arabic

MAJOR EXPORT:
Petroleum

MAJOR IMPORT:
Foods

CAPITAL:
Kuwait

LANDMASS:
6,880 sq. mi.
17,818 sq. km

KYRGYZSTAN

POPULATION:
4,728,000
62 per sq. mi.
24 per sq. km

LANGUAGES:
Kirghiz, Russian

MAJOR EXPORT:
Cotton

MAJOR IMPORT:
Grain

CAPITAL:
Bishkek

LANDMASS:
76,834 sq. mi.
198,999 sq. km

LEBANON

POPULATION:
4,070,000
1,014 per sq. mi.
391 per sq. km

LANGUAGES:
Arabic, French

MAJOR EXPORT:
Paper

MAJOR IMPORT:
Machinery

CAPITAL:
Beirut

LANDMASS:
4,015 sq. mi.
10,399 sq. km

LIBYA

POPULATION:
4,992,000
7 per sq. mi.
3 per sq. km

LANGUAGE:
Arabic

MAJOR EXPORT:
Crude Oil

MAJOR IMPORT:
Machinery

CAPITAL:
Tripoli

LANDMASS:
679,362 sq. mi.
1,759,540 sq. km

MOROCCO *

POPULATION:
28,248,000
103 per sq. mi.
40 per sq. km

LANGUAGES:
Arabic, French,
Berber

MAJOR EXPORT:
Foods

MAJOR IMPORT:
Manufactured
Goods

CAPITAL:
Rabat

LANDMASS:
275,117 sq. mi.
712,550 sq. km

OMAN

POPULATION:
2,460,000
30 per sq. mi.
12 per sq. km

LANGUAGE:
Arabic

MAJOR EXPORT:
Petroleum

MAJOR IMPORT:
Machinery

CAPITAL:
Muscat

LANDMASS:
82,030 sq. mi.
212,457 sq. km

QATAR

POPULATION:
541,000
127 per sq. mi.
49 per sq. km

LANGUAGE:
Arabic

MAJOR EXPORT:
Petroleum

MAJOR IMPORT:
Machinery

CAPITAL:
Doha

LANDMASS:
4,247 sq. mi.
11,000 sq. km

* Morocco claims the Western
Sahara area, but other countries
do not accept this claim.

SAUDI ARABIA

POPULATION:
20,899,000
25 per sq. mi.
10 per sq. km

LANGUAGE:
Arabic

MAJOR EXPORT:
Petroleum

MAJOR IMPORT:
Machinery

CAPITAL:
Riyadh

LANDMASS:
830,000 sq. mi.
2,149,690 sq. km

SYRIA

POPULATION:
16,033,000
226 per sq. mi.
87 per sq. km

LANGUAGES:
Arabic, Kurdish,
Armenian

MAJOR EXPORT:
Petroleum

MAJOR IMPORT:
Machinery

CAPITAL:
Damascus

LANDMASS:
71,044 sq. mi.
184,004 sq. km

TAJIKISTAN

POPULATION:
6,213,000
113 per sq. mi.
43 per sq. km

LANGUAGES:
Tajik, Russian

MAJOR EXPORT:
Aluminum

MAJOR IMPORT:
Fuels

CAPITAL:
Dushanbe

LANDMASS:
55,213 sq. mi.
143,001 sq. km

Country Profiles

TUNISIA

POPULATION:
9,498,000
150 per sq. mi.
58 per sq. km

LANGUAGES:
Arabic, French

MAJOR EXPORT:
Petroleum
Products

MAJOR IMPORT:
Machinery

CAPITAL:
Tunis

LANDMASS:
63,170 sq. mi.
163,610 sq. km

Tunis

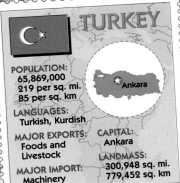

TURKEY

POPULATION:
65,869,000
219 per sq. mi.
85 per sq. km

LANGUAGES:
Turkish, Kurdish

MAJOR EXPORTS:
Foods and
Livestock

MAJOR IMPORT:
Machinery

CAPITAL:
Ankara

LANDMASS:
300,948 sq. mi.
779,452 sq. km

Ankara

TURKMENISTAN

POPULATION:
4,779,000
25 per sq. mi.
10 per sq. km

LANGUAGES:
Turkmen,
Russian, Uzbek

MAJOR EXPORT:
Natural Gas

MAJOR IMPORT:
Machinery

CAPITAL:
Ashgabat

LANDMASS:
188,418 sq. mi.
488,000 sq. km

Ashgabat

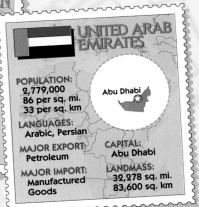

UNITED ARAB EMIRATES

POPULATION:
2,779,000
86 per sq. mi.
33 per sq. km

LANGUAGES:
Arabic, Persian

MAJOR EXPORT:
Petroleum

MAJOR IMPORT:
Manufactured
Goods

CAPITAL:
Abu Dhabi

LANDMASS:
32,278 sq. mi.
83,600 sq. km

Abu Dhabi

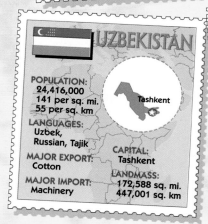

UZBEKISTAN

POPULATION:
24,416,000
141 per sq. mi.
55 per sq. km

LANGUAGES:
Uzbek,
Russian, Tajik

MAJOR EXPORT:
Cotton

MAJOR IMPORT:
Machinery

CAPITAL:
Tashkent

LANDMASS:
172,588 sq. mi.
447,001 sq. km

Tashkent

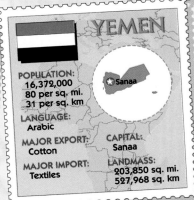

YEMEN

POPULATION:
16,372,000
80 per sq. mi.
31 per sq. km

LANGUAGE:
Arabic

MAJOR EXPORT:
Cotton

MAJOR IMPORT:
Textiles

CAPITAL:
Sanaa

LANDMASS:
203,850 sq. mi.
527,968 sq. km

Sanaa

GEO BEE: Questions From Buzz Bee!

The following questions are taken from National Geographic GeoBees. Use your textbook, the Internet, and other library resources to find the answers.

1. Which of the countries in North Africa has a king as head of state?

2. Which African capital city is only 150 miles (241 km) from Sicily?

3. The name of a group of people who have traditionally been nomadic herders comes from an Arabic word that means "desert dwellers." Name this group.

Anthias fish and coral reef,
Red Sea, Egypt

Waterfront of
Cape Town,
South Africa

Woman making
butter in Chad

Giraffe on a
plain in Kenya ▼

Africa
South of the
Sahara

The region of Africa south
of the Sahara is home to
more than 2,000 ethnic groups.
Its hot, humid forests and dry
grasslands support a variety of
wild animals. Both people and
animals face tough challenges
in this region. The people are
struggling to build stable governments and economies. Some
animals are threatened with
extinction as human activities
destroy natural habitats.

NGS ONLINE
www.nationalgeographic.com/education

FOCUS on:

Africa South of the Sahara

STRADDLING THE EQUATOR, Africa south of the Sahara lies almost entirely within the tropics. Famous for its remarkable wildlife, this region also has the world's fastest-growing human population. Settling ethnic rivalries and improving low standards of living are just two of the challenges facing the people in Africa south of the Sahara.

The Land

Africa south of the Sahara has the highest overall elevation of any world region. A narrow band of low plains hugs the coastline. Inland, the land rises from west to east in a series of steplike plateaus. Separating the plateaus are steep cliffs. The region has no long mountain ranges and few towering peaks, although Mt. Kenya and Kilimanjaro are exceptions. At 19,340 feet (5,895 m), Kilimanjaro's summit is the highest point on the African continent.

Great rivers arise in this region's interior highlands. As rivers spill from one plateau to the next, they create thundering waterfalls, such as the spectacular Victoria Falls (right). It is known locally as *Mosi oa Tunya*—"smoke that thunders." Although the Nile River is Africa's longest river, the Congo River is a giant in its own right, winding 2,715 miles (4,370 km) through Africa's heart, near the Equator.

The Great Rift Valley slices through eastern Africa like a steep-walled gash in the continent. Formed by movements of the earth's crust, the valley extends from Southwest Asia southward to the Zambezi River in Mozambique. It cradles a chain of deep lakes.

The Climate

If you started at the Equator and traveled north or south from there, you would pass through four major climate regions, one after the other. Tropical rain forests lie along the Equator and fill the great basin of the Congo River in central and western Africa. Head away from the Equator and you see rain forests give way to tropical savannas. These vast grasslands are home to some of the continent's most famous large mammals, including elephants, lions, rhinoceroses, and giraffes.

As you move farther from the Equator, rainfall becomes scarce, and savannas give way to tropical steppes. Finally you encounter very dry areas where deserts dominate the landscape. Deserts cover more of Africa than any other continent. The largest deserts south of the Sahara are the Namib and the Kalahari.

Victoria Falls, on the Zambezi River ▼

◄ Elephants browsing near
Kilimanjaro, Tanzania

The Economy

Africa south of the Sahara is rich in mineral resources, but these resources are not evenly distributed. Nigeria has huge reserves of oil. South Africa has fabulous deposits of gold and diamonds, making it the wealthiest country in the region. Overall, however, Africa south of the Sahara has the lowest standard of living of any world region.

Manufacturing plays only a small role in the region's economy. In the past, colonial rulers used Africa as a source of raw materials and left the continent largely undeveloped. Today the nations south of the Sahara are struggling to industrialize.

Most people in Africa south of the Sahara still depend on small-scale farming or livestock herding for their livelihoods. They usually are able to raise only enough food to feed their families. Some farmers work on plantations that grow crops for export to other countries. Such crops include coffee, cacao, cotton, peanuts, tea, bananas, and sisal (a fiber). Drought is a constant problem for the region's farmers.

The People

Thousands of years ago, great kingdoms and empires developed in Africa south of the Sahara. In the 1400s and 1500s, Europeans began trading with African societies, carrying away gold, spices, ivory, and enslaved people.

By the late 1800s, European nations had claimed almost all of Africa. For profit and political advantage, they carved up the continent into colonies. In the process, they ripped apart once-unified regions and threw together ethnic groups that did not get along.

Most African nations won their independence in the mid-1900s. Most emerged from colonial rule politically unstable and with crippled economies.

Today more than 625 million people inhabit Africa south of the Sahara. They represent some 2,000 ethnic groups and speak 800 different languages. Nearly three-fourths of the population lives in rural areas. Although Africa is the least urbanized continent, its cities are growing. Lured by the promise of better living conditions, people are flocking to African cities, which are among the fastest-growing urban areas in the world.

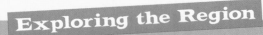
Exploring the Region

1. **What happens when Africa's rivers flow from one plateau to another?**
2. **What climate zone is centered on the Equator?**
3. **What makes South Africa the region's most prosperous country?**
4. **How did colonial rule affect Africa south of the Sahara?**

◄ **Woman fertilizing crops in Zimbabwe**

◀ **Crowded market in Lagos, Nigeria**

Africa South of the Sahara

Physical

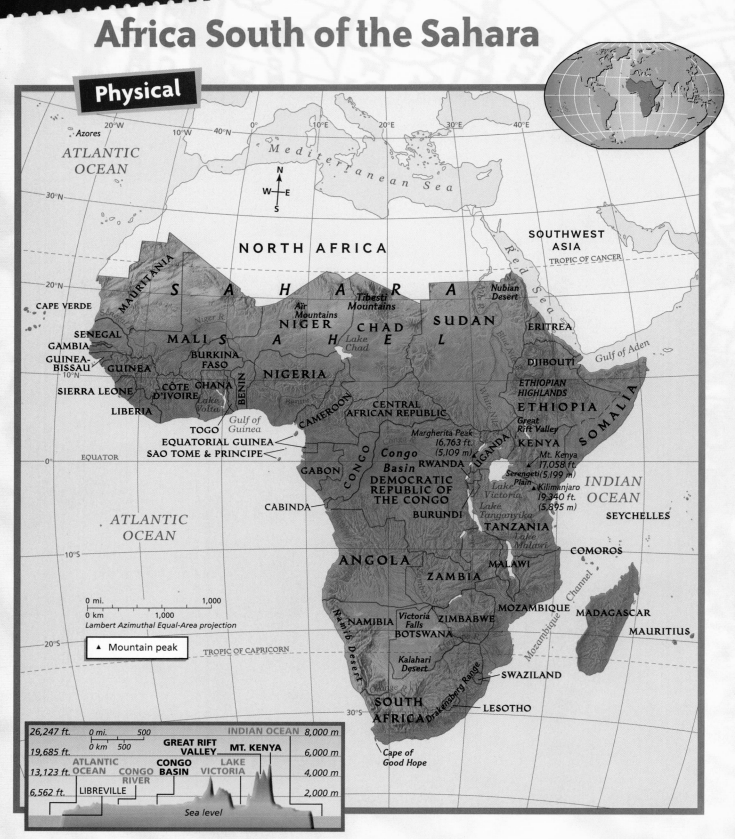

ATLANTIC OCEAN

20°W

10°W

0°

10°E

20°E

30°E

40°E

40°N

Mediterranean Sea

30°N

Azores

CAPE VERDE

NORTH AFRICA

SOUTHWEST ASIA

Red Sea

TROPIC OF CANCER

20°N

MAURITANIA

S A H A R A

A

Nubian Desert

Air Mountains

Tibesti Mountains

Niger R.

SENEGAL

MALI

NIGER

CHAD

SUDAN

ERITREA

GAMBIA

GUINEA-BISSAU

GUINEA

BURKINA FASO

S A H E L

Lake Chad

DJIBOUTI

Gulf of Aden

10°N

SIERRA LEONE

CÔTE D'IVOIRE

GHANA

BENIN

NIGERIA

Benue

ETHIOPIAN HIGHLANDS

ETHIOPIA

SOMALIA

LIBERIA

Lake Volta

CAMEROON

CENTRAL AFRICAN REPUBLIC

White Nile

Blue Nile

Great Rift Valley

TOGO

Gulf of Guinea

EQUATORIAL GUINEA

SAO TOME & PRINCIPE

Congo R.

Margherita Peak 16,763 ft. (5,109 m)

Congo Basin

RWANDA

UGANDA

KENYA

Mt. Kenya 17,058 ft. (5,199 m)

EQUATOR

0°

GABON

CONGO

DEMOCRATIC REPUBLIC OF THE CONGO

BURUNDI

Serengeti Plain

Lake Victoria

Kilimanjaro 19,340 ft. (5,895 m)

INDIAN OCEAN

SEYCHELLES

CABINDA

ATLANTIC OCEAN

TANZANIA

Lake Tanganyika

Lake Malawi

COMOROS

10°S

ANGOLA

ZAMBIA

MALAWI

MOZAMBIQUE

MADAGASCAR

Mozambique Channel

MAURITIUS

0 mi. 1,000

0 km 1,000

Lambert Azimuthal Equal-Area projection

▲ Mountain peak

Namib Desert

NAMIBIA

Victoria Falls

ZIMBABWE

BOTSWANA

Drakensberg Range

SWAZILAND

20°S

TROPIC OF CAPRICORN

Kalahari Desert

Orange R.

LESOTHO

30°S

SOUTH AFRICA

Cape of Good Hope

26,247 ft.

0 mi. 500

0 km 500

GREAT RIFT VALLEY

MT. KENYA

INDIAN OCEAN

8,000 m

19,685 ft.

6,000 m

ATLANTIC OCEAN

CONGO RIVER

CONGO BASIN

LAKE VICTORIA

13,123 ft.

4,000 m

6,562 ft.

LIBREVILLE

2,000 m

Sea level

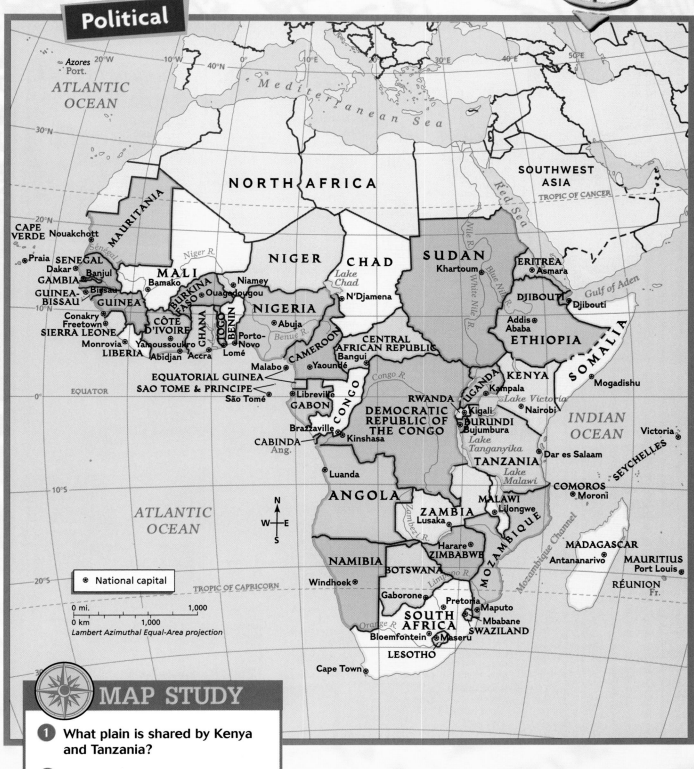

Political

Azores
Port.

ATLANTIC
OCEAN

Mediterranean Sea

NORTH AFRICA

SOUTHWEST
ASIA

TROPIC OF CANCER

Red Sea

MAURITANIA

CAPE
VERDE ⊛ Nouakchott

Praia ⊛ SENEGAL
Dakar ⊛
GAMBIA ⊛ Banjul
GUINEA- ⊛ Bissau
BISSAU GUINEA
Conakry ⊛
Freetown ⊛
SIERRA LEONE
Monrovia ⊛ Yamoussoukro ⊛
LIBERIA Abidjan

Sénégal R.
Niger R.

MALI
Bamako ⊛
BURKINA
FASO
⊛ Ouagadougou

CÔTE
D'IVOIRE
GHANA
TOGO
BENIN
Accra
Lomé

NIGER
Niamey ⊛

NIGERIA
⊛ Abuja
Porto-
Novo

Benue R.

CHAD
Lake Chad
⊛ N'Djamena

SUDAN
Khartoum ⊛

Nile R.
White Nile R.
Blue Nile R.

ERITREA
⊛ Asmara

DJIBOUTI
⊛ Djibouti

Gulf of Aden

CAMEROON
CENTRAL
AFRICAN REPUBLIC
Bangui ⊛
⊛ Yaoundé

Addis
Ababa ⊛
ETHIOPIA

SOMALIA

EQUATORIAL GUINEA
SAO TOME & PRINCIPE
São Tomé
⊛ Libreville
GABON
CONGO

Malabo ⊛

Congo R.

UGANDA
⊛ Kampala

KENYA
⊛ Nairobi

Mogadishu ⊛

EQUATOR

Lake Victoria

RWANDA
⊛ Kigali
BURUNDI
Bujumbura

INDIAN
OCEAN

Victoria ⊛

DEMOCRATIC
REPUBLIC OF
THE CONGO

Brazzaville ⊛
⊛ Kinshasa

CABINDA
Ang.

*Lake
Tanganyika*

TANZANIA
⊛ Dar es Salaam

SEYCHELLES

Luanda ⊛

*Lake
Malawi*

COMOROS
⊛ Moroni

ATLANTIC
OCEAN

ANGOLA

Zambezi R.

ZAMBIA
Lusaka ⊛

MALAWI
⊛ Lilongwe

MOZAMBIQUE

Mozambique Channel

MADAGASCAR

Antananarivo ⊛

MAURITIUS
Port Louis ⊛

NAMIBIA

Harare ⊛
ZIMBABWE

BOTSWANA

Limpopo R.

RÉUNION
Fr.

⊛ National capital

Windhoek ⊛

TROPIC OF CAPRICORN

Gaborone ⊛
Pretoria ⊛
⊛ Maputo

0 mi.   1,000
0 km   1,000
Lambert Azimuthal Equal-Area projection

Orange R.

SOUTH
AFRICA
Bloemfontein ⊛

Mbabane
⊛ SWAZILAND
⊛ Maseru
LESOTHO

Cape Town ⊛

N
W E
S

MAP STUDY

1 What plain is shared by Kenya
and Tanzania?

2 What is the capital of Nigeria?

Country Profiles

ANGOLA

POPULATION:
12,479,000
26 per sq. mi.
10 per sq. km

LANGUAGES:
Portuguese,
Local Languages

MAJOR EXPORT:
Crude Oil

CAPITAL:
Luanda

MAJOR IMPORT:
Machinery

LANDMASS:
481,354 sq. mi.
1,246,700 sq. km

BENIN

POPULATION:
6,186,000
142 per sq. mi.
55 per sq. km

LANGUAGES:
French, Fon, Yoruba

MAJOR EXPORT:
Cotton

CAPITAL:
Porto-Novo

MAJOR IMPORT:
Foods

LANDMASS:
43,484 sq. mi.
112,622 sq. km

BOTSWANA

POPULATION:
1,464,000
6 per sq. mi.
2 per sq. km

LANGUAGES:
English, Setswana

MAJOR EXPORT:
Diamonds

CAPITAL:
Gaborone

MAJOR IMPORT:
Foods

LANDMASS:
231,805 sq. mi.
600,372 sq. km

BURKINA FASO

POPULATION:
11,576,000
109 per sq. mi.
42 per sq. km

LANGUAGES:
French, Local
Languages

MAJOR EXPORT:
Cotton

CAPITAL:
Ouagadougou

MAJOR IMPORT:
Machinery

LANDMASS:
105,869 sq. mi.
274,200 sq. km

BURUNDI

POPULATION:
5,736,000
534 per sq. mi.
206 per sq. km

LANGUAGES:
Kirundi, French

MAJOR EXPORT:
Coffee

CAPITAL:
Bujumbura

MAJOR IMPORT:
Machinery

LANDMASS:
10,747 sq. mi.
27,834 sq. km

CAMEROON

POPULATION:
15,456,000
84 per sq. mi.
33 per sq. km

LANGUAGES:
French, English,
Local Languages

MAJOR EXPORT:
Crude Oil

CAPITAL:
Yaoundé

MAJOR IMPORT:
Machinery

LANDMASS:
183,569 sq. mi.
475,442 sq. km

CAPE VERDE

POPULATION:
406,000
261 per sq. mi.
101 per sq. km

LANGUAGES:
Portuguese, Crioulo

MAJOR EXPORT:
Shoes

CAPITAL:
Praia

MAJOR IMPORT:
Foods

LANDMASS:
1,557 sq. mi.
4,033 sq. km

CENTRAL AFRICAN REPUBLIC

POPULATION:
3,445,000
14 per sq. mi.
6 per sq. km

LANGUAGES:
French, Sango,
Arabic, Hunsa

MAJOR EXPORT:
Diamonds

CAPITAL:
Bangui

MAJOR IMPORT:
Foods

LANDMASS:
240,535 sq. mi.
622,984 sq. km

CHAD

POPULATION:
7,714,000
16 per sq. mi.
6 per sq. km

LANGUAGES:
French, Arabic,
Sara, Sango

MAJOR EXPORT:
Cotton

CAPITAL:
N'Djamena

MAJOR IMPORT:
Machinery

LANDMASS:
495,755 sq. mi.
1,284,000 sq. km

COMOROS

POPULATION:
563,000
783 per sq. mi.
302 per sq. km

LANGUAGES:
Arabic, French,
Comoran

MAJOR EXPORT:
Vanilla

CAPITAL:
Moroni

MAJOR IMPORT:
Rice

LANDMASS:
719 sq. mi.
1,862 sq. km

CONGO

POPULATION:
2,717,000
21 per sq. mi.
8 per sq. km

LANGUAGES:
French, Lingala,
Monokutuba

MAJOR EXPORT:
Crude Oil

CAPITAL:
Brazzaville

MAJOR IMPORT:
Machinery

LANDMASS:
132,047 sq. mi.
342,000 sq. km

Countries and flags not drawn to scale

CONGO
Democratic Republic of the

POPULATION:
50,481,000
56 per sq. mi.
22 per sq. km

LANGUAGES:
French, Lingala, Kingwana

MAJOR EXPORT:
Diamonds

MAJOR IMPORT:
Manufactured Goods

CAPITAL:
Kinshasa

LANDMASS:
905,568 sq. mi.
2,345,409 sq. km

CÔTE D'IVOIRE

POPULATION:
15,818,000
127 per sq. mi.
49 per sq. km

LANGUAGES:
French, Dioula

MAJOR EXPORT:
Cocoa

MAJOR IMPORT:
Foods

CAPITALS:
Yamoussoukro, Abidjan

LANDMASS:
124,504 sq. mi.
322,463 sq. km

DJIBOUTI

POPULATION:
629,000
70 per sq. mi.
27 per sq. km

LANGUAGES:
French, Arabic

MAJOR EXPORTS:
Hides and Skins

MAJOR IMPORT:
Foods

CAPITAL:
Djibouti

LANDMASS:
8,958 sq. mi.
23,200 sq. km

EQUATORIAL GUINEA

POPULATION:
442,000
41 per sq. mi.
16 per sq. km

LANGUAGES:
Spanish, French, Fang, Bubi, Ibo

MAJOR EXPORT:
Petroleum

MAJOR IMPORT:
Machinery

CAPITAL:
Malabo

LANDMASS:
10,831 sq. mi.
28,051 sq. km

ERITREA

POPULATION:
3,985,000
85 per sq. mi.
33 per sq. km

LANGUAGES:
Afar, Amharic, Arabic, Tigre

MAJOR EXPORT:
Livestock

MAJOR IMPORT:
Processed Foods

CAPITAL:
Asmara

LANDMASS:
46,842 sq. mi.
121,320 sq. km

ETHIOPIA

POPULATION:
59,680,000
140 per sq. mi.
54 per sq. km

LANGUAGES:
Amharic, Tigrinya, Orominga

MAJOR EXPORT:
Coffee

MAJOR IMPORTS:
Foods and Livestock

CAPITAL:
Addis Ababa

LANDMASS:
424,934 sq. mi.
1,100,574 sq. km

GABON

POPULATION:
1,197,000
12 per sq. mi.
4 per sq. km

LANGUAGES:
French, Local Languages

MAJOR EXPORT:
Crude Oil

MAJOR IMPORT:
Machinery

CAPITAL:
Libreville

LANDMASS:
103,347 sq. mi.
267,667 sq. km

GAMBIA

POPULATION:
1,268,000
291 per sq. mi.
112 per sq. km

LANGUAGES:
English, Mandinka, Fula, Wolof

MAJOR EXPORT:
Peanuts

MAJOR IMPORT:
Foods

CAPITAL:
Banjul

LANDMASS:
4,361 sq. mi.
11,295 sq. km

GHANA

POPULATION:
19,678,000
214 per sq. mi.
82 per sq. km

LANGUAGES:
English, Local Languages

MAJOR EXPORT:
Gold

MAJOR IMPORT:
Machinery

CAPITAL:
Accra

LANDMASS:
92,100 sq. mi.
238,537 sq. km

GUINEA

POPULATION:
7,539,000
79 per sq. mi.
31 per sq. km

LANGUAGES:
French, Local Languages

MAJOR EXPORT:
Bauxite

MAJOR IMPORT:
Petroleum Products

CAPITAL:
Conakry

LANDMASS:
94,926 sq. mi.
245,857 sq. km

GUINEA-BISSAU

POPULATION:
1,187,000
85 per sq. mi.
33 per sq. km

LANGUAGES:
Portuguese, Crioulo, Local Languages

MAJOR EXPORT:
Cashews

MAJOR IMPORT:
Foods

CAPITAL:
Bissau

LANDMASS:
13,948 sq. mi.
36,125 sq. km

Country Profiles

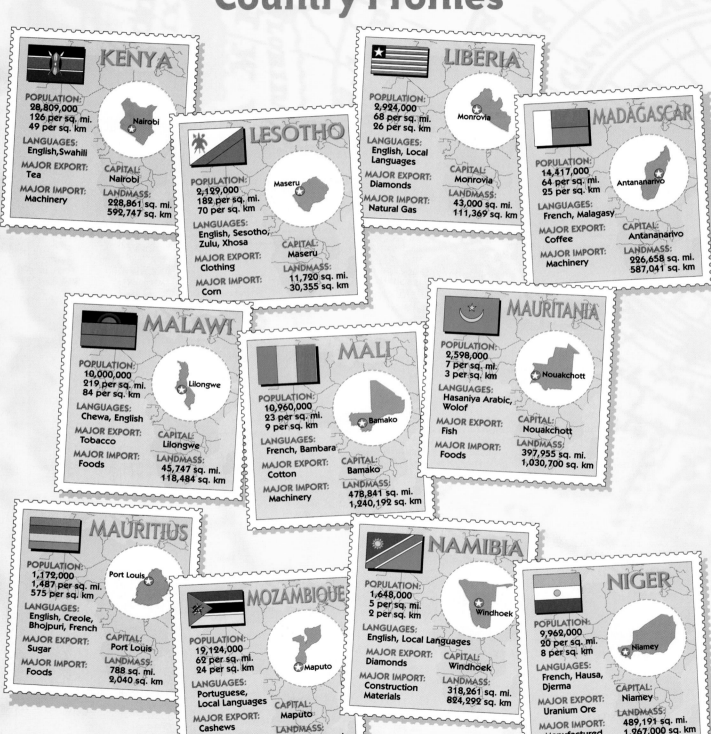

KENYA
POPULATION:
28,809,000
126 per sq. mi.
49 per sq. km

LANGUAGES:
English, Swahili

MAJOR EXPORT:
Tea

MAJOR IMPORT:
Machinery

CAPITAL:
Nairobi

LANDMASS:
228,861 sq. mi.
592,747 sq. km

LESOTHO
POPULATION:
2,129,000
182 per sq. mi.
70 per sq. km

LANGUAGES:
English, Sesotho,
Zulu, Xhosa

MAJOR EXPORT:
Clothing

MAJOR IMPORT:
Corn

CAPITAL:
Maseru

LANDMASS:
11,720 sq. mi.
30,355 sq. km

LIBERIA
POPULATION:
2,924,000
68 per sq. mi.
26 per sq. km

LANGUAGES:
English, Local
Languages

MAJOR EXPORT:
Diamonds

MAJOR IMPORT:
Natural Gas

CAPITAL:
Monrovia

LANDMASS:
43,000 sq. mi.
111,369 sq. km

MADAGASCAR
POPULATION:
14,417,000
64 per sq. mi.
25 per sq. km

LANGUAGES:
French, Malagasy

MAJOR EXPORT:
Coffee

MAJOR IMPORT:
Machinery

CAPITAL:
Antananarivo

LANDMASS:
226,658 sq. mi.
587,041 sq. km

MALAWI
POPULATION:
10,000,000
219 per sq. mi.
84 per sq. km

LANGUAGES:
Chewa, English

MAJOR EXPORT:
Tobacco

MAJOR IMPORT:
Foods

CAPITAL:
Lilongwe

LANDMASS:
45,747 sq. mi.
118,484 sq. km

MALI
POPULATION:
10,960,000
23 per sq. mi.
9 per sq. km

LANGUAGES:
French, Bambara

MAJOR EXPORT:
Cotton

MAJOR IMPORT:
Machinery

CAPITAL:
Bamako

LANDMASS:
478,841 sq. mi.
1,240,192 sq. km

MAURITANIA
POPULATION:
2,598,000
7 per sq. mi.
3 per sq. km

LANGUAGES:
Hasaniya Arabic,
Wolof

MAJOR EXPORT:
Fish

MAJOR IMPORT:
Foods

CAPITAL:
Nouakchott

LANDMASS:
397,955 sq. mi.
1,030,700 sq. km

MAURITIUS
POPULATION:
1,172,000
1,487 per sq. mi.
575 per sq. km

LANGUAGES:
English, Creole,
Bhojpuri, French

MAJOR EXPORT:
Sugar

MAJOR IMPORT:
Foods

CAPITAL:
Port Louis

LANDMASS:
788 sq. mi.
2,040 sq. km

MOZAMBIQUE
POPULATION:
19,124,000
62 per sq. mi.
24 per sq. km

LANGUAGES:
Portuguese,
Local Languages

MAJOR EXPORT:
Cashews

MAJOR IMPORT:
Foods

CAPITAL:
Maputo

LANDMASS:
308,642 sq. mi.
799,380 sq. km

NAMIBIA
POPULATION:
1,648,000
5 per sq. mi.
2 per sq. km

LANGUAGES:
English, Local Languages

MAJOR EXPORT:
Diamonds

MAJOR IMPORT:
Construction
Materials

CAPITAL:
Windhoek

LANDMASS:
318,261 sq. mi.
824,292 sq. km

NIGER
POPULATION:
9,962,000
20 per sq. mi.
8 per sq. km

LANGUAGES:
French, Hausa,
Djerma

MAJOR EXPORT:
Uranium Ore

MAJOR IMPORT:
Manufactured
Goods

CAPITAL:
Niamey

LANDMASS:
489,191 sq. mi.
1,267,000 sq. km

Countries and flags not drawn to scale

NIGERIA

POPULATION:
113,829,000
319 per sq. mi.
123 per sq. km

LANGUAGES:
English, Hausa,
Yoruba, Igbo

MAJOR EXPORT:
Petroleum

MAJOR IMPORT:
Machinery

CAPITAL:
Abuja

LANDMASS:
356,669 sq. mi.
923,768 sq. km

Abuja

RWANDA

POPULATION:
8,155,000
802 per sq. mi.
310 per sq. km

LANGUAGES:
Kinyarwanda,
French, English

MAJOR EXPORT:
Coffee

MAJOR IMPORT:
Foods

CAPITAL:
Kigali

LANDMASS:
10,169 sq. mi.
26,338 sq. km

Kigali

SAO TOME and PRINCIPE

POPULATION:
155,000
417 per sq. mi.
161 per sq. km

LANGUAGES:
Portuguese, Crioulo

MAJOR EXPORT:
Cocoa

MAJOR IMPORT:
Textiles

CAPITAL:
São Tomé

LANDMASS:
372 sq. mi.
964 sq. km

São Tomé

SENEGAL

POPULATION:
9,240,000
122 per sq. mi.
47 per sq. km

LANGUAGES:
French, Wolof,
Pulaar, Diola

MAJOR EXPORT:
Fish

MAJOR IMPORT:
Foods

CAPITAL:
Dakar

LANDMASS:
75,955 sq. mi.
196,722 sq. km

Dakar

SEYCHELLES

POPULATION:
80,000
457 per sq. mi.
177 per sq. km

LANGUAGES:
English, French,
Creole

MAJOR EXPORT:
Fish

MAJOR IMPORT:
Foods

CAPITAL:
Victoria

LANDMASS:
175 sq. mi.
453 sq. km

Victoria

SIERRA LEONE

POPULATION:
5,297,000
191 per sq. mi.
74 per sq. km

LANGUAGES:
English, Mende,
Temne, Krio

MAJOR EXPORT:
Diamonds

MAJOR IMPORT:
Foods

CAPITAL:
Freetown

LANDMASS:
27,699 sq. mi.
71,740 sq. km

Freetown

SOMALIA

POPULATION:
7,141,000
29 per sq. mi.
11 per sq. km

LANGUAGES:
Somali, Arabic

MAJOR EXPORT:
Livestock

MAJOR IMPORT:
Textiles

CAPITAL:
Mogadishu

LANDMASS:
246,201 sq. mi.
637,657 sq. km

Mogadishu

SOUTH AFRICA

POPULATION:
42,579,000
90 per sq. mi.
35 per sq. km

LANGUAGES:
Afrikaans,
English, Local
Languages

MAJOR EXPORT:
Gold

MAJOR IMPORT:
Transport Equip.

CAPITALS:
Pretoria, Cape Town,
Bloemfontein

LANDMASS:
471,445 sq. mi.
1,221,037 sq. km

Pretoria
Bloemfontein
Cape Town

SUDAN

POPULATION:
28,883,000
30 per sq. mi.
12 per sq. km

LANGUAGES:
Arabic, Nubian,
Ta Bedawie

MAJOR EXPORT:
Cotton

MAJOR IMPORT:
Petroleum
Products

CAPITAL:
Khartoum

LANDMASS:
963,600 sq. mi.
2,495,712 sq. km

Khartoum

SWAZILAND

POPULATION:
985,000
147 per sq. mi.
57 per sq. km

LANGUAGES:
English, Swazi

MAJOR EXPORT:
Soft Drink
Concentrates

MAJOR IMPORT:
Machinery

CAPITAL:
Mbabane

LANDMASS:
6,704 sq. mi.
17,364 sq. km

Mbabane

TANZANIA

POPULATION:
31,271,000
86 per sq. mi.
33 per sq. km

LANGUAGES:
Swahili, English

MAJOR EXPORT:
Coffee

MAJOR IMPORT:
Machinery

CAPITAL:
Dar es Salaam

LANDMASS:
364,900 sq. mi.
945,087 sq. km

Dar es Salaam

Country Profiles

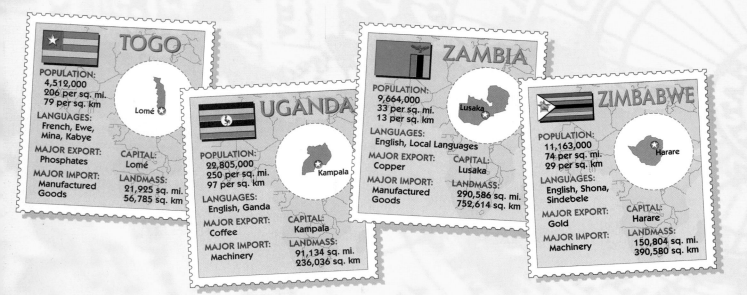

TOGO

POPULATION:
4,512,000
206 per sq. mi.
79 per sq. km

LANGUAGES:
French, Ewe, Mina, Kabye

MAJOR EXPORT:
Phosphates

MAJOR IMPORT:
Manufactured Goods

CAPITAL:
Lomé

LANDMASS:
21,925 sq. mi.
56,785 sq. km

Lomé ★

UGANDA

POPULATION:
22,805,000
250 per sq. mi.
97 per sq. km

LANGUAGES:
English, Ganda

MAJOR EXPORT:
Coffee

MAJOR IMPORT:
Machinery

CAPITAL:
Kampala

LANDMASS:
91,134 sq. mi.
236,036 sq. km

Kampala ★

ZAMBIA

POPULATION:
9,664,000
33 per sq. mi.
13 per sq. km

LANGUAGES:
English, Local Languages

MAJOR EXPORT:
Copper

MAJOR IMPORT:
Manufactured Goods

CAPITAL:
Lusaka

LANDMASS:
290,586 sq. mi.
752,614 sq. km

Lusaka ★

ZIMBABWE

POPULATION:
11,163,000
74 per sq. mi.
29 per sq. km

LANGUAGES:
English, Shona, Sindebele

MAJOR EXPORT:
Gold

MAJOR IMPORT:
Machinery

CAPITAL:
Harare

LANDMASS:
150,804 sq. mi.
390,580 sq. km

Harare ★

Countries and flags not drawn to scale

 **Questions From Buzz Bee!**

The following questions are taken from National Geographic GeoBees. Use your textbook, the Internet, and other library resources to find the answers.

1. The Dutch who settled in what is now South Africa became known by a Dutch word that means "farmer." By what name were these settlers known?

2. Volcanic activity created the highest peak along the Great Rift Valley. Name the peak.

3. The Hutu and the Tutsi are ethnic groups that have been fighting for control of two countries. Name these countries.

View over terraced fields
and small settlements,
Kabale, Uganda

8

Taj Mahal,
Agra, India

Macaques in a hot
spring, Japan

NATIONAL GEOGRAPHIC SOCIETY

Asia

For many people in the Western Hemisphere, the region of Asia—in the Eastern Hemisphere—brings to mind exotic images. Ancient temples stand in dense rain forests. Farmers work in flooded rice fields. Pandas nibble bamboo shoots. Yet bustling cities, gleaming skyscrapers, and high-technology industries also can be found here. Turn the page to learn more about this region and its 3 billion people.

Monks wrapping statue of Buddha in yellow cloth, Thailand

NGS ONLINE
www.nationalgeographic.com/education

Focus on:
Asia

THE REGION OF ASIA is made up of surprisingly diverse landscapes. It includes a large chunk of the Asian continent, together with island groups that fringe its southern and eastern shores. Some of the world's oldest civilizations and religions had their beginnings in Asia. Now more than 3 billion people call this region home.

The Land

Covering roughly 7.8 million square miles (20.2 sq. km), the Asian region stretches from the mountains of western Pakistan to the eastern shores of Japan. It reaches from the highlands of northeastern China to the tropical islands of Indonesia. The region's long, winding coastlines are washed by two major oceans—the Indian and the Pacific—as well as many seas.

Several mountain ranges slice through central Asia. Most famous are the towering Himalaya, the site of Mt. Everest—the earth's tallest peak. North of the Himalaya lies the vast Plateau of Tibet, so high it has been called the Roof of the World. Beyond the plateau are two immense deserts: the Taklimakan and the Gobi.

Other mountain ranges cut across northeastern China, slant down the Korean Peninsula, and sweep through the peninsulas of Southeast Asia. Offshore, Japan, Indonesia, and other mountainous islands lie along the Ring of Fire. This is an area where adjoining plates of the earth's crust slip and buckle, setting off earthquakes and volcanic eruptions.

Great rivers begin in Asia's lofty center. On their journey to the sea, they flow through fertile plains in several countries. The most important rivers include the Indus in Pakistan, the Ganges and Brahmaputra in India and Bangladesh, the Yangtze and Yellow in China, and the Mekong in Southeast Asia.

The Climate

A person traveling across Asia would need clothes to suit almost every imaginable climate. The snowcapped mountains and high, windswept plateaus of northern and central Asia can be bitterly cold. The deserts can shimmer with heat by day, yet be frosty at night. Lowlands and coastal plains enjoy milder climates. The peninsulas of Southeast Asia and the islands straddling the Equator have mostly tropical climates. They are cloaked in dense rain forests. Seasonal winds called monsoons blow across much of Asia, bringing dry weather in winter and drenching rains in summer.

Terraced rice fields,
Bali, Indonesia

◀ Street flooded by monsoon
rain, Tamil Nadu, India

The Economy

Agriculture is the major economic activity across most of Asia. The region's rugged mountains and vast deserts mean that only a small amount of the land is suitable for growing crops, however. For example, only about 10 percent of China's land can be used for agriculture. To feed the region's huge population, Asian farmers must make the most of every possible bit of farmland. Terraces allow farmers to grow rice even on steep hillsides. Rice, which grows well in places with warm temperatures and plenty of water, is the most important food crop in Asia. China, India, Indonesia, and Bangladesh are the leading rice producers in the world.

Most of Asia's manufacturing takes place in Japan, South Korea, Taiwan, China, and India. China and India both are rich in coal, iron ore, and other natural resources. Japan, however, has few mineral resources and must import fuel and nearly all the raw materials it needs. Still, Japan has become one of the world's leading manufacturers of cars, electronic products, and other goods. In some of the region's other countries, such as Laos, Vietnam, and Bhutan, industry is less well developed.

The People

Nestled in fertile river valleys, some of the world's oldest civilizations arose in Asia thousands of years ago. Ancient religions took root here as well. Both Hinduism and Buddhism, for example, originated in India. Hindus remain concentrated in India, but over time Buddhism spread throughout the region. The region's most widespread faith—Islam—began in Southwest Asia.

Europeans arrived in the region around 1500, bringing Christianity to some of the people. By the early 1800s, many Asian countries had become European colonies. Most of these countries became independent in the mid-1900s. In many cases, however, independence was followed by political turmoil and conflict between Communist and non-Communist forces. Today China, Vietnam, and North Korea have Communist governments; Nepal is ruled by a monarch; and military leaders control Myanmar. Japan, India, and the Philippines are democracies.

About 3.3 billion people live in Asia, but this huge population is very unevenly distributed. Most people live near the coasts and in the fertile river valleys. As a result, some parts of Asia are among the most crowded places in the world.

Exploring the Region

1. **Why is the Plateau of Tibet called the Roof of the World?**
2. **How do monsoons affect the region?**
3. **What is the most important food crop in Asia?**
4. **What are two religions that originated in the region?**

◀ **Robot welding car bodies in a factory, Japan**

先

士茂

門木

Three generations of a Chinese family

Asia

Physical

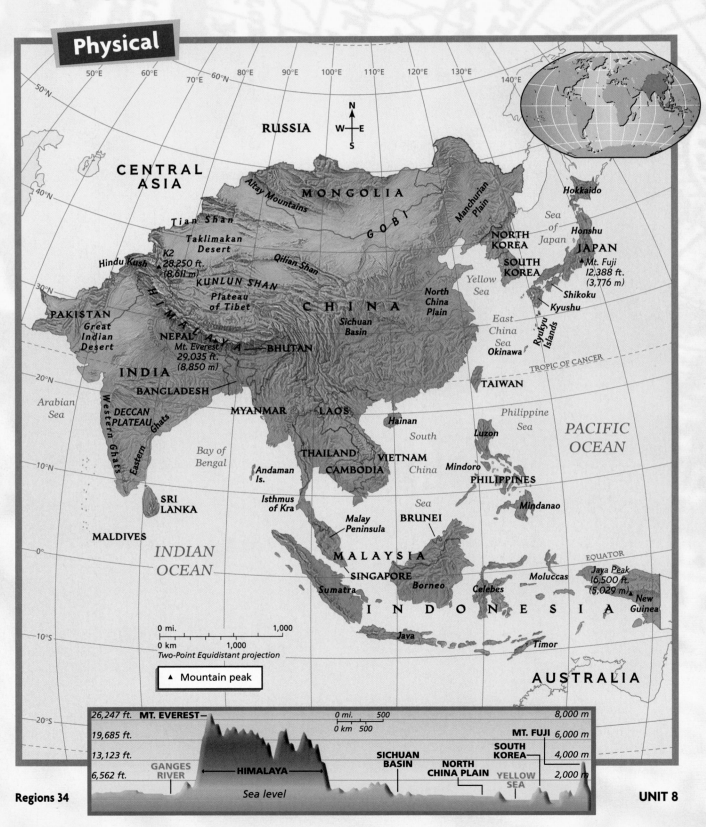

RUSSIA

N
W E
S

CENTRAL ASIA

Altay Mountains

MONGOLIA

GOBI

Manchurian Plain

Hokkaido

Sea of Japan

NORTH KOREA

Honshu

JAPAN

▲ Mt. Fuji 12,388 ft. (3,776 m)

SOUTH KOREA

Tian Shan

Taklimakan Desert

K2
▲ 28,250 ft.
(8,611 m)

Hindu Kush

Qilian Shan

KUNLUN SHAN

Plateau of Tibet

CHINA

Yellow R.

Yellow Sea

Shikoku

Kyushu

North China Plain

Sichuan Basin

East China Sea

Ryukyu Islands

Okinawa

PAKISTAN

Great Indian Desert

H I M A L A Y A

NEPAL

Mt. Everest 29,035 ft. (8,850 m)

BHUTAN

TROPIC OF CANCER

INDIA

BANGLADESH

TAIWAN

Arabian Sea

DECCAN PLATEAU

Western Ghats

Eastern Ghats

MYANMAR

LAOS

Hainan

South China

Philippine Sea

PACIFIC OCEAN

Luzon

Bay of Bengal

Andaman Is.

THAILAND

CAMBODIA

VIETNAM

Mindoro

PHILIPPINES

SRI LANKA

Isthmus of Kra

Mindanao

MALDIVES

Sea

Malay Peninsula

BRUNEI

INDIAN OCEAN

MALAYSIA

SINGAPORE

Sumatra

Borneo

Celebes

Moluccas

Jaya Peak 16,500 ft. (5,029 m) ▲

New Guinea

EQUATOR

Java

I N D O N E S I A

Timor

AUSTRALIA

0 mi. 1,000
0 km 1,000
Two-Point Equidistant projection

▲ Mountain peak

26,247 ft. **MT. EVEREST**

0 mi. 500
0 km 500

8,000 m

19,685 ft.

MT. FUJI 6,000 m

13,123 ft.

SICHUAN BASIN

NORTH CHINA PLAIN

SOUTH KOREA

4,000 m

6,562 ft.

GANGES RIVER

HIMALAYA

YELLOW SEA

2,000 m

Sea level

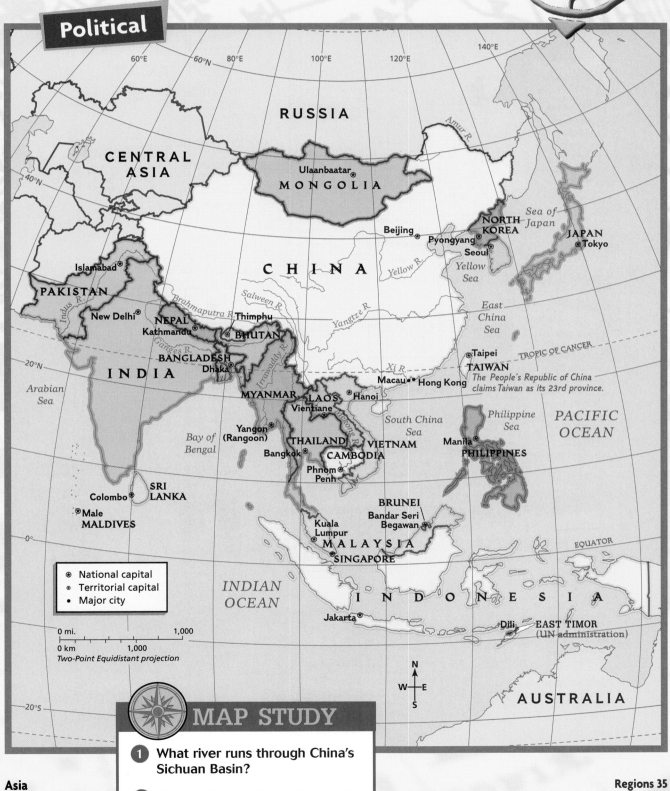

Political

RUSSIA

CENTRAL ASIA

Ulaanbaatar ⊛

MONGOLIA

NORTH KOREA

Sea of Japan

Beijing ⊛
Pyongyang ⊛

Seoul ⊛

JAPAN
⊛ Tokyo

C H I N A

Islamabad ⊛

PAKISTAN

Indus R.

Brahmaputra R.

Salween R.

Yangtze R.

Yellow R.

Yellow Sea

East China Sea

Thimphu ⊛

New Delhi ⊛

NEPAL
Kathmandu ⊛

BHUTAN

Ganges R.

BANGLADESH
Dhaka ⊛

TROPIC OF CANCER

Taipei ⊛

TAIWAN

Xi R.

Macau • • Hong Kong

The People's Republic of China claims Taiwan as its 23rd province.

I N D I A

MYANMAR

LAOS
Vientiane ⊛

Irrawaddy R.

• Hanoi

Arabian Sea

Yangon (Rangoon) ⊛

Bay of Bengal

THAILAND
Bangkok ⊛

Mekong R.

CAMBODIA

VIETNAM

South China Sea

Philippine Sea

PACIFIC OCEAN

Manila ⊛

PHILIPPINES

Phnom Penh ⊛

SRI LANKA
Colombo ⊛

⊛ Male
MALDIVES

BRUNEI
Bandar Seri Begawan ⊛

Kuala Lumpur ⊛

M A L A Y S I A

⊛ SINGAPORE

Legend

- ⊛ National capital
- ⊙ Territorial capital
- • Major city

INDIAN OCEAN

I N D O N E S I A

EQUATOR

0 mi.   1,000
0 km   1,000

Two-Point Equidistant projection

Jakarta ⊛

Dili • EAST TIMOR (UN administration)

N
W + E
S

AUSTRALIA

MAP STUDY

1 What river runs through China's Sichuan Basin?

2 What is the capital of Thailand?

REGIONAL ATLAS

Country Profiles

BANGLADESH

POPULATION:
125,721,000
2,261 per sq. mi.
873 per sq. km

LANGUAGE:
Bengali

MAJOR EXPORT:
Clothing

MAJOR IMPORT:
Machinery

CAPITAL:
Dhaka

LANDMASS:
55,598 sq. mi.
143,998 sq. km

BHUTAN

POPULATION:
800,000
44 per sq. mi.
17 per sq. km

LANGUAGES:
Dzonkha, Local
Languages

MAJOR EXPORT:
Cardamom

MAJOR IMPORT:
Fuels

CAPITAL:
Thimphu

LANDMASS:
18,147 sq. mi.
47,001 sq. km

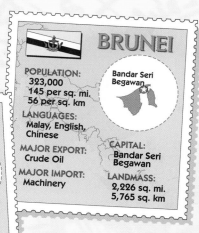

BRUNEI

POPULATION:
323,000
145 per sq. mi.
56 per sq. km

LANGUAGES:
Malay, English,
Chinese

MAJOR EXPORT:
Crude Oil

MAJOR IMPORT:
Machinery

CAPITAL:
Bandar Seri
Begawan

LANDMASS:
2,226 sq. mi.
5,765 sq. km

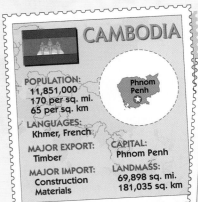

CAMBODIA

POPULATION:
11,851,000
170 per sq. mi.
65 per sq. km

LANGUAGES:
Khmer, French

MAJOR EXPORT:
Timber

MAJOR IMPORT:
Construction
Materials

CAPITAL:
Phnom Penh

LANDMASS:
69,898 sq. mi.
181,035 sq. km

CHINA

POPULATION:
1,254,062,000
339 per sq. mi.
131 per sq. km

LANGUAGE:
Mandarin Chinese

MAJOR EXPORT:
Machinery

MAJOR IMPORT:
Machinery

CAPITAL:
Beijing

LANDMASS:
3,705,820 sq. mi.
9,598,032 sq. km

INDIA

POPULATION:
1,000,849,000
788 per sq. mi.
304 per sq. km

LANGUAGES:
Hindi, English,
Local Languages

MAJOR EXPORTS:
Gems and
Jewelry

MAJOR IMPORT:
Crude Oil

CAPITAL:
New Delhi

LANDMASS:
1,269,346 sq. mi.
3,287,591 sq. km

INDONESIA

POPULATION:
211,806,000
286 per sq. mi.
110 per sq. km

LANGUAGES:
Bahasa Indonesia,
Local Languages

MAJOR EXPORT:
Textiles

MAJOR IMPORT:
Manufactured
Goods

CAPITAL:
Jakarta

LAND MASS:
741,101 sq. mi.
1,919,443 sq. km

JAPAN

POPULATION:
126,745,000
869 per sq. mi.
335 per sq. km

LANGUAGE:
Japanese

MAJOR EXPORT:
Machinery

MAJOR IMPORT:
Manufactured
Goods

CAPITAL:
Tokyo

LANDMASS:
145,875 sq. mi.
377,815 sq. km

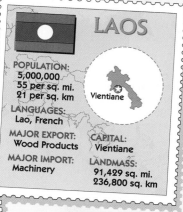

LAOS

POPULATION:
5,000,000
55 per sq. mi.
21 per sq. km

LANGUAGES:
Lao, French

MAJOR EXPORT:
Wood Products

MAJOR IMPORT:
Machinery

CAPITAL:
Vientiane

LANDMASS:
91,429 sq. mi.
236,800 sq. km

Countries and flags not drawn to scale

MALAYSIA

POPULATION:
22,710,000
178 per sq. mi.
69 per sq. km

LANGUAGES:
Malay, English,
Chinese

MAJOR EXPORT:
Electronic
Equipment

MAJOR IMPORT:
Machinery

CAPITAL:
Kuala Lumpur

LANDMASS:
127,317 sq. mi.
329,749 sq. km

Kuala
Lumpur

MALDIVES

POPULATION:
278,000
2,417 per sq. mi.
933 per sq. km

LANGUAGES:
Maldivian Divehi,
English

MAJOR EXPORT:
Fish

MAJOR IMPORT:
Machinery

CAPITAL:
Male

LANDMASS:
115 sq. mi.
298 sq. km

Male

MONGOLIA

POPULATION:
2,438,000
4 per sq. mi.
2 per sq. km

LANGUAGE:
Khalkha Mongol

MAJOR EXPORT:
Copper

MAJOR IMPORT:
Fuels

CAPITAL:
Ulaanbaatar

LANDMASS:
604,250 sq. mi.
1,565,000 sq. km

Ulaanbaatar

MYANMAR

POPULATION:
48,081,000
184 per sq. mi.
71 per sq. km

LANGUAGES:
Burmese,
Local Languages

MAJOR EXPORT:
Beans

MAJOR IMPORT:
Machinery

CAPITAL:
Yangon
(Rangoon)

LANDMASS:
261,218 sq. mi.
676,552 sq. km

Yangon
(Rangoon)

NEPAL

POPULATION:
24,303,000
447 per sq. mi.
173 per sq. km

LANGUAGE:
Nepali

MAJOR EXPORT:
Clothing

MAJOR IMPORT:
Petroleum
Products

CAPITAL:
Kathmandu

LANDMASS:
54,362 sq. mi.
140,797 sq. km

Kathmandu

NORTH KOREA

POPULATION:
21,386,000
460 per sq. mi.
177 per sq. km

LANGUAGE:
Korean

MAJOR EXPORT:
Minerals

MAJOR IMPORT:
Petroleum

CAPITAL:
Pyongyang

LANDMASS:
46,540 sq. mi.
120,538 sq. km

Pyongyang

PAKISTAN

POPULATION:
146,488,000
477 per sq. mi.
184 per sq. km

LANGUAGES:
Urdu, English,
Punjabi, Sindhi

MAJOR EXPORT:
Cotton

MAJOR IMPORT:
Petroleum

CAPITAL:
Islamabad

LANDMASS:
307,374 sq. mi.
796,095 sq. km

Islamabad

PHILIPPINES

POPULATION:
74,655,000
645 per sq. mi.
249 per sq. km

LANGUAGES:
Tagalog, English

MAJOR EXPORT:
Electronic
Equipment

MAJOR IMPORT:
Raw Materials

CAPITAL:
Manila

LANDMASS:
115,831 sq. mi.
300,001 sq. km

Manila

SINGAPORE

POPULATION:
3,999,000
16,732 per sq. mi.
6,471 per sq. km

LANGUAGES:
Chinese, Malay,
Tamil, English

MAJOR EXPORT:
Computer Equipment

MAJOR IMPORT:
Aircraft

CAPITAL:
Singapore

LANDMASS:
239 sq. mi.
618 sq. km

Singapore

Country Profiles

SOUTH KOREA
POPULATION:
46,873,000
1,226 per sq. mi.
473 per sq. km
LANGUAGE:
Korean
MAJOR EXPORT:
Electronic Equipment
MAJOR IMPORT:
Machinery
CAPITAL:
Seoul
LANDMASS:
38,230 sq. mi.
99,016 sq. km
Seoul

SRI LANKA
POPULATION:
19,003,000
750 per sq. mi.
290 per sq. km
LANGUAGES:
Sinhalese, Tamil, English
MAJOR EXPORT:
Textiles
MAJOR IMPORT:
Machinery
CAPITAL:
Colombo
LANDMASS:
25,332 sq. mi.
65,610 sq. km
Colombo

TAIWAN*
POPULATION:
22,000,000
1,583 per sq. mi.
611 per sq. km
LANGUAGE:
Mandarin Chinese
MAJOR EXPORT:
Textiles
MAJOR IMPORT:
Machinery
CAPITAL:
Taipei
LANDMASS:
13,900 sq. mi.
36,000 sq. km
Taipei

* The People's Republic of China claims Taiwan as its 23rd province.

THAILAND
POPULATION:
61,818,000
311 per sq. mi.
120 per sq. km
LANGUAGES:
Thai, Local Languages
MAJOR EXPORT:
Manufactured Goods
MAJOR IMPORT:
Machinery
CAPITAL:
Bangkok
LANDMASS:
198,457 sq. mi.
514,001 sq. km
Bangkok

VIETNAM
POPULATION:
79,490,000
625 per sq. mi.
241 per sq. km
LANGUAGES:
Vietnamese, Chinese
MAJOR EXPORT:
Crude Oil
MAJOR IMPORT:
Machinery
CAPITAL:
Hanoi
LANDMASS:
127,242 sq. mi.
329,556 sq. km
Hanoi

GEO BEE Questions From Buzz Bee!

The following questions are taken from National Geographic GeoBees. Use your textbook, the Internet, and other library resources to find the answers.

1 The Silk Road and modern highways skirt the largest desert in western China, which is characterized by shifting sand dunes. Name this desert.

2 Fortified monasteries called *dzongs* are found throughout the smallest country between China and India. Name this mountainous country.

View over city to Central Plaza at dusk,
Wanchai, Hong Kong

9

Fur seal on the beach, Antarctica

Boy selling fish, Samoa

NATIONAL
GEOGRAPHIC
SOCIETY

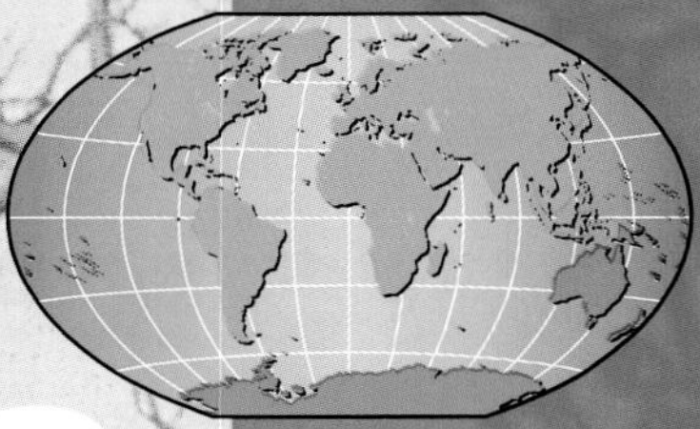

Australia, Oceania, and Antarctica

Australia, Oceania, and Antarctica are grouped together more because of their nearness to one another than because of any similarities among their peoples. These lands lie mostly in the Southern Hemisphere. Australia is a dry continent that is home to unusual wildlife. Oceania's 25,000 tropical islands spread out across the Pacific Ocean. Frozen Antarctica encompasses the South Pole.

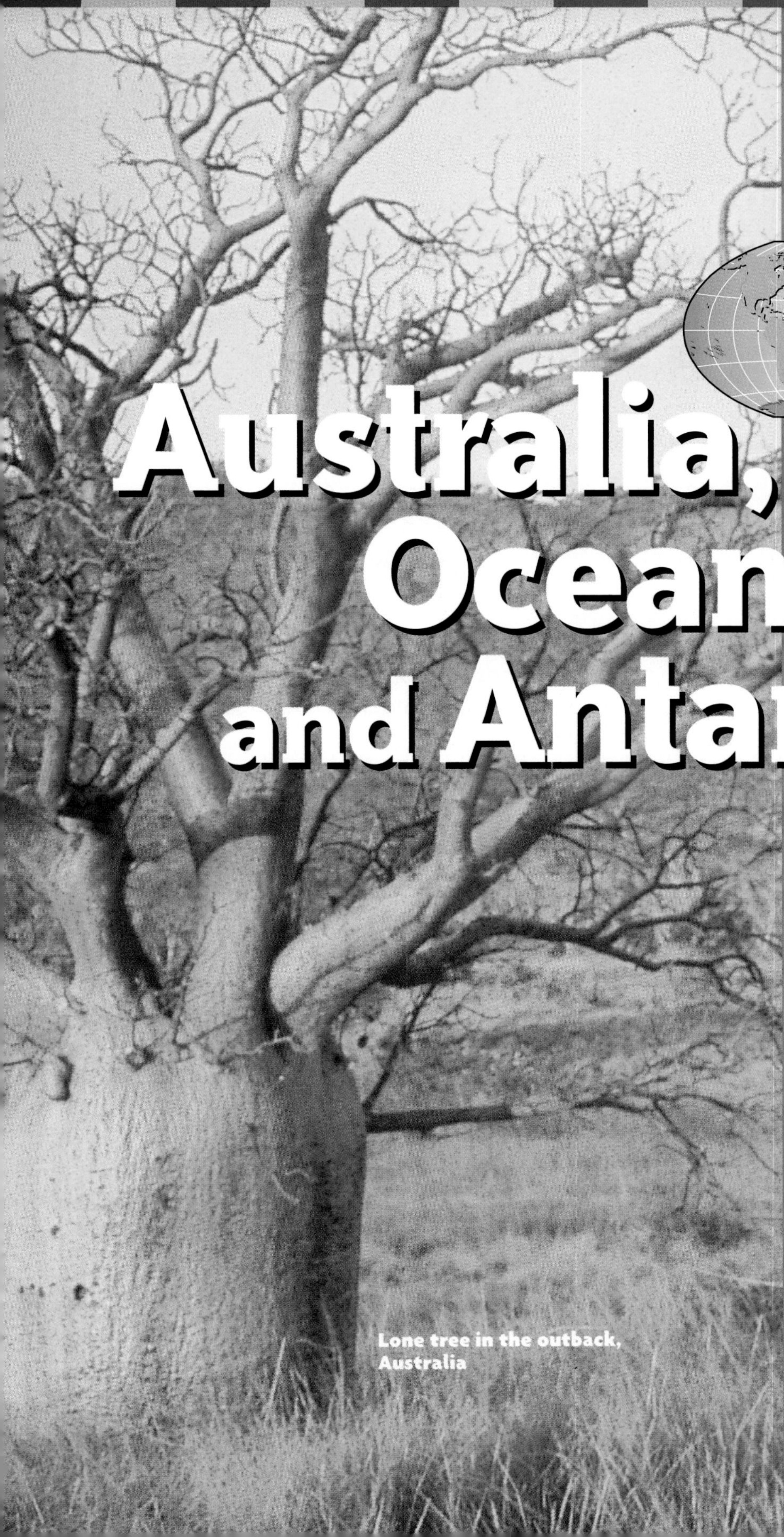

Lone tree in the outback, Australia

NGS ONLINE
www.nationalgeographic.com/education

Focus on:

Australia, Oceania, and Antarctica

LYING ALMOST ENTIRELY in the Southern Hemisphere, this region includes two continents and thousands of islands scattered across the Pacific Ocean. Covering a huge portion of the globe, the region includes landscapes ranging from polar to tropical.

The Land

Both a continent and a single country, Australia is a vast expanse of mostly flat land. The Great Dividing Range runs down the continent's eastern edge. Between this range of mountains and the Pacific Ocean lies a narrow strip of coastal land. West of the Great Dividing Range lies Australia's large—and very dry—interior. Here in the Australian "outback" are seemingly endless miles of scrubland, as well as three huge deserts.

Across the Tasman Sea from Australia lies New Zealand, made up of two main islands— North Island and South Island—and many smaller ones. Mountains and hills dominate the landscape.

North and east of New Zealand is Oceania. Its roughly 25,000 islands lie scattered across the Pacific Ocean on either side of the Equator. Some of these islands are volcanic. Others are huge formations of rock that have risen from the ocean floor. Still others are low-lying coral islands surrounded by reefs.

Antarctica, the frozen continent, covers and surrounds the South Pole. It is almost completely buried under an enormous sheet of ice.

The Climate

Australia is one of the driest continents in the world. Its eastern coast does receive rainfall from the Pacific Ocean. Mountains block this moisture from reaching inland areas, however. Much of Australia's outback has a desert climate.

No place in New Zealand is more than 80 miles (129 km) from the sea. This country has only one climate region: marine west coast. It has mild temperatures and plentiful rainfall throughout the year.

The islands of Oceania have mostly tropical climates, with warm temperatures and distinct wet and dry seasons. Rain forests cover many of the islands.

Antarctica is one of the coldest and windiest places on the earth, as well as one of the driest. It receives so little precipitation that it is considered a desert—the world's largest cold desert.

Sheep grazing near Mount Egmont, New Zealand

◄ Emperor penguins examining the ice, Antarctica

The Economy

Mines dot the Australian landscape. Its ancient rocks and soils are rich in minerals such as uranium, bauxite, iron ore, copper, nickel, and gold. Little of Australia's land is good for growing crops. Instead, vast cattle and sheep ranches—or stations, as the Australians call them—spread across much of the country. Sheep far outnumber people in New Zealand, where pastures are lush and green almost year-round. New Zealand is one of the world's leading producers of lamb and wool.

The people of Oceania depend primarily on fishing and farming. Across much of Oceania, the soil and climate are not favorable for widespread agriculture, and islanders generally raise only enough food for themselves. Yet some larger islands have rich volcanic soil. In such places, cash crops of fruits, sugar, coffee, and coconut products are grown for export.

Antarctica is believed to be rich in mineral resources. To preserve Antarctica for research and exploration, however, many nations have agreed not to mine this mineral wealth.

The People

The first settlers in this region probably came from Asia thousands of years ago. Australia's first inhabitants, the ancestors of today's Aborigines, may have arrived as long as 40,000 years ago. Not until about A.D. 1000, however, did seafaring peoples reach the farthest islands of Oceania.

The British colonized Australia and New Zealand in the 1700s and 1800s. These two countries gained their independence in the early 1900s. Many South Pacific islands were not freed from colonial rule until after World War II. Today Australia and Oceania are a blend of European, traditional Pacific, and Asian cultures.

Despite its vast size, this is the least populous of all the world's regions. It is home to only about 30 million people. More than half of these live in Australia, where they are found mostly in coastal cities such as Sydney and Melbourne. Roughly 4 million people live in New Zealand, which also has large urban populations along its coasts. Oceania is less urbanized. Antarctica has no permanent human inhabitants at all. Groups of scientists live and work on the frozen continent for brief periods to carry out their research.

Exploring the Region

1. **Which two continents lie in this region?**

2. **Why is Antarctica considered a desert?**

3. **Why is so little of Australia's land good for farming?**

4. **Where do most of the region's people live?**

◀ **Girl selling fruit, French Polynesia**

The city of Melbourne, along the southeastern coast of Australia

Australia, Oceania, and Antarctica

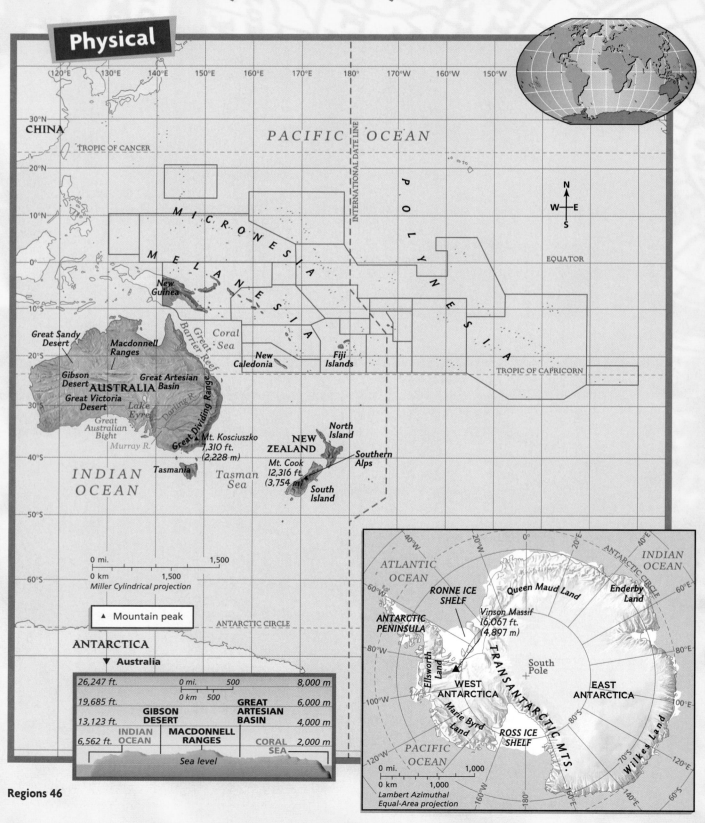

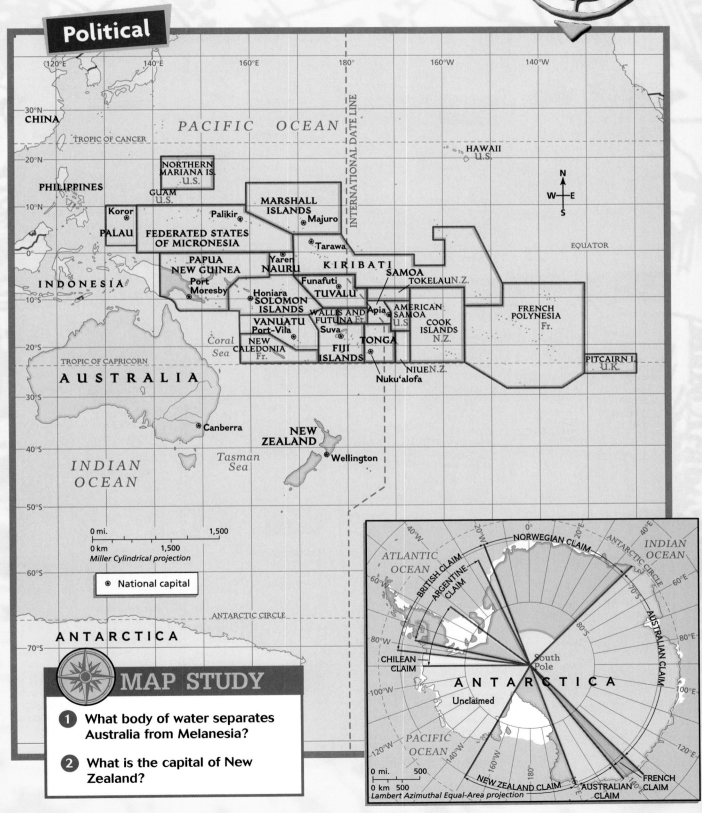

Political

PACIFIC OCEAN

TROPIC OF CANCER

30°N

20°N

CHINA

PHILIPPINES

INDONESIA

AUSTRALIA

TROPIC OF CAPRICORN

INDIAN OCEAN

10°N

0°

10°S

20°S

30°S

40°S

50°S

60°S

70°S

120°E   140°E   160°E   180°   160°W   140°W

HAWAII U.S.

NORTHERN MARIANA IS. U.S.

GUAM U.S.

Koror

PALAU

Palikir

FEDERATED STATES OF MICRONESIA

MARSHALL ISLANDS

Majuro

Tarawa

EQUATOR

KIRIBATI

PAPUA NEW GUINEA

Port Moresby

Yaren NAURU

SAMOA

TOKELAU N.Z.

Honiara

SOLOMON ISLANDS

Funafuti

TUVALU

Apia

AMERICAN SAMOA U.S.

FRENCH POLYNESIA Fr.

WALLIS AND FUTUNA Fr.

VANUATU

Port-Vila

Suva

COOK ISLANDS N.Z.

Coral Sea

NEW CALEDONIA Fr.

FIJI ISLANDS

TONGA

Nuku'alofa

NIUE N.Z.

PITCAIRN I. U.K.

Canberra

NEW ZEALAND

Wellington

Tasman Sea

ANTARCTIC CIRCLE

ANTARCTICA

INTERNATIONAL DATE LINE

0 mi.   1,500
0 km   1,500
Miller Cylindrical projection

⊛ National capital

N
W — E
S

ANTARCTICA inset

ATLANTIC OCEAN

NORWEGIAN CLAIM

ANTARCTIC CIRCLE

INDIAN OCEAN

BRITISH CLAIM

ARGENTINE CLAIM

AUSTRALIAN CLAIM

CHILEAN CLAIM

South Pole

ANTARCTICA

Unclaimed

PACIFIC OCEAN

NEW ZEALAND CLAIM

AUSTRALIAN CLAIM

FRENCH CLAIM

0 mi.   500
0 km   500
Lambert Azimuthal Equal-Area projection

MAP STUDY

1 What body of water separates Australia from Melanesia?

2 What is the capital of New Zealand?

Country Profiles

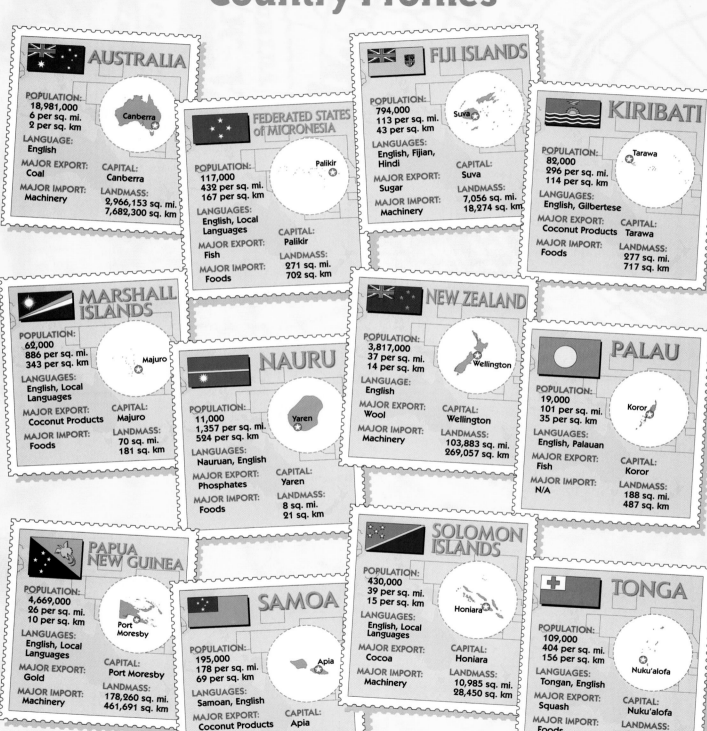

AUSTRALIA

POPULATION:
18,981,000
6 per sq. mi.
2 per sq. km

LANGUAGE:
English

MAJOR EXPORT:
Coal

MAJOR IMPORT:
Machinery

CAPITAL:
Canberra

LANDMASS:
2,966,153 sq. mi.
7,682,300 sq. km

FEDERATED STATES of MICRONESIA

POPULATION:
117,000
432 per sq. mi.
167 per sq. km

LANGUAGES:
English, Local Languages

MAJOR EXPORT:
Fish

MAJOR IMPORT:
Foods

CAPITAL:
Palikir

LANDMASS:
271 sq. mi.
702 sq. km

FIJI ISLANDS

POPULATION:
794,000
113 per sq. mi.
43 per sq. km

LANGUAGES:
English, Fijian, Hindi

MAJOR EXPORT:
Sugar

MAJOR IMPORT:
Machinery

CAPITAL:
Suva

LANDMASS:
7,056 sq. mi.
18,274 sq. km

KIRIBATI

POPULATION:
82,000
296 per sq. mi.
114 per sq. km

LANGUAGES:
English, Gilbertese

MAJOR EXPORT:
Coconut Products

MAJOR IMPORT:
Foods

CAPITAL:
Tarawa

LANDMASS:
277 sq. mi.
717 sq. km

MARSHALL ISLANDS

POPULATION:
62,000
886 per sq. mi.
343 per sq. km

LANGUAGES:
English, Local Languages

MAJOR EXPORT:
Coconut Products

MAJOR IMPORT:
Foods

CAPITAL:
Majuro

LANDMASS:
70 sq. mi.
181 sq. km

NAURU

POPULATION:
11,000
1,357 per sq. mi.
524 per sq. km

LANGUAGES:
Nauruan, English

MAJOR EXPORT:
Phosphates

MAJOR IMPORT:
Foods

CAPITAL:
Yaren

LANDMASS:
8 sq. mi.
21 sq. km

NEW ZEALAND

POPULATION:
3,817,000
37 per sq. mi.
14 per sq. km

LANGUAGE:
English

MAJOR EXPORT:
Wool

MAJOR IMPORT:
Machinery

CAPITAL:
Wellington

LANDMASS:
103,883 sq. mi.
269,057 sq. km

PALAU

POPULATION:
19,000
101 per sq. mi.
35 per sq. km

LANGUAGES:
English, Palauan

MAJOR EXPORT:
Fish

MAJOR IMPORT:
N/A

CAPITAL:
Koror

LANDMASS:
188 sq. mi.
487 sq. km

PAPUA NEW GUINEA

POPULATION:
4,669,000
26 per sq. mi.
10 per sq. km

LANGUAGES:
English, Local Languages

MAJOR EXPORT:
Gold

MAJOR IMPORT:
Machinery

CAPITAL:
Port Moresby

LANDMASS:
178,260 sq. mi.
461,691 sq. km

SAMOA

POPULATION:
195,000
178 per sq. mi.
69 per sq. km

LANGUAGES:
Samoan, English

MAJOR EXPORT:
Coconut Products

MAJOR IMPORT:
Foods

CAPITAL:
Apia

LANDMASS:
1,093 sq. mi.
2,831 sq. km

SOLOMON ISLANDS

POPULATION:
430,000
39 per sq. mi.
15 per sq. km

LANGUAGES:
English, Local Languages

MAJOR EXPORT:
Cocoa

MAJOR IMPORT:
Machinery

CAPITAL:
Honiara

LANDMASS:
10,985 sq. mi.
28,450 sq. km

TONGA

POPULATION:
109,000
404 per sq. mi.
156 per sq. km

LANGUAGES:
Tongan, English

MAJOR EXPORT:
Squash

MAJOR IMPORT:
Foods

CAPITAL:
Nuku'alofa

LANDMASS:
270 sq. mi.
699 sq. km

Countries and flags not drawn to scale

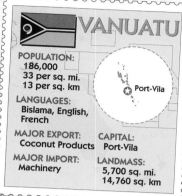

VANUATU

POPULATION:
186,000
33 per sq. mi.
13 per sq. km

Port-Vila

LANGUAGES:
Bislama, English,
French

MAJOR EXPORT: CAPITAL:
Coconut Products   Port-Vila

MAJOR IMPORT: LANDMASS:
Machinery   5,700 sq. mi.
                     14,760 sq. km

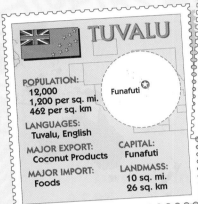

TUVALU

POPULATION:
12,000
1,200 per sq. mi.
462 per sq. km   Funafuti

LANGUAGES:
Tuvalu, English

MAJOR EXPORT: CAPITAL:
Coconut Products   Funafuti

MAJOR IMPORT: LANDMASS:
Foods   10 sq. mi.
                     26 sq. km

Questions From Buzz Bee!

The following questions are taken from National Geographic GeoBees. Use your textbook, the Internet, and other library resources to find the answers.

1. Dried coconut meat is the main export of many South Pacific islands. What is it called?

2. Ross, Amundsen, and Bellingshausen are names of seas that border which country?

3. The world's southernmost city with a metropolitan population greater than 1 million is located north of the Bass Strait. Name this city.

4. What U.S.-administered islands were named for their location about halfway between the continents of North America and Asia?

5. A 1991 international agreement prohibits mining on which continent for at least the next 50 years?

6. If you were swimming off the coast of Australia near Perth, you would be in what body of water?

7. Which continent receives the least amount of rainfall?

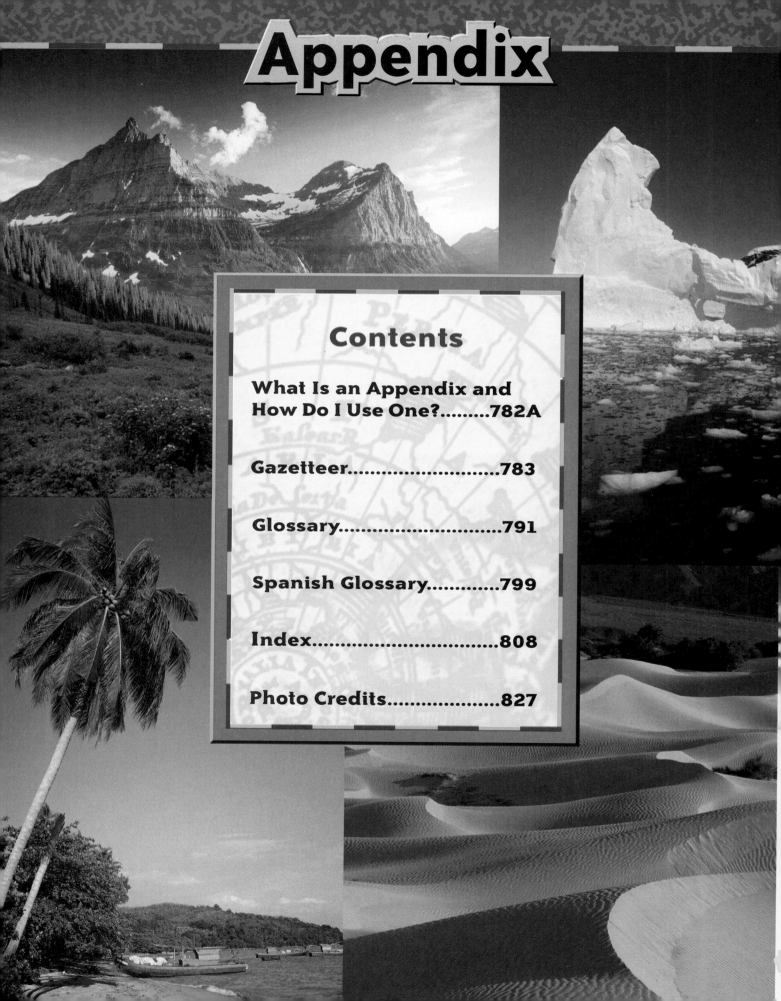

Appendix

Contents

What Is an Appendix and How Do I Use One?

An appendix is the additional material you often find at the end of books. It can help you read the book. The information below and on page 782B will help you learn how to use an appendix.

Gazetteer

A **gazetteer** (GA·zuh·TIHR) is a geographical dictionary. It gives the latitude and longitude measurements for cities and other places of the world. It lists most of the world's largest independent countries, their capitals, and several important geographic features. Each entry also includes a page number telling you where this place can be found in your textbook. Most entries in a gazetteer are spelled out phonetically to help you pronounce the place.

Glossary

A **glossary** is a list of important or difficult terms found in a textbook. The glossary gives a definition of each term as it is used in the book. Since words sometimes have other meanings, you may wish to consult a dictionary to find other uses for the term. The glossary also includes a pronunciation key for each entry and page numbers telling you where in the textbook the term is used.

Spanish Glossary

A **Spanish glossary** does everything that an English glossary does, but it does it in Spanish. A Spanish glossary is especially important to bilingual students, or those Spanish-speaking students who are learning the English language.

Index

An index is an alphabetical listing at the end of a book that includes the subjects of that book and the page numbers where those subjects can be found. The index in this book also lets you know that certain pages contain maps, graphs, photos, or paintings about the subject entry.

Acknowledgments/ Photo Credits

This section lists photo credits and/or literary credits for the book. For example, if there is a photo on page 351, you can look at the photo credits to see where the publisher obtained that photo. Also, some books let you know where they obtained the permission to use excerpts from other books in an acknowledgments section.

Test Yourself!

Do you think you can use an appendix quickly and easily? Try it. Find the answers to the following questions by using the appendix of your textbook.

1. What does *canopy* mean?
2. Where is the word *canopy* used in this book?
3. On what pages can I find information about Cuba?
4. How do you pronounce "Djibouti"?
5. Where exactly is Rabat located?
6. What is the Spanish word for *cassava*?
7. On what page can I find out about the climate of Ecuador?
8. From where did Glencoe get the photo on page 188?
9. Does this book discuss conservation?
10. Colombo is the capital of which country?

Appendix-User Tip

When using an appendix, be sure to notice and use the guide words at the top of each page. There are usually two guide words on the top of each page. These guide words indicate the alphabetically first and last entries on that page.

GAZETTEER

Gazetteer

A Gazetteer (GA•zuh•TIHR) is a geographic index or dictionary. It shows latitude and longitude for cities and certain other places. Latitude and longitude are shown in this way: 48°N 2°E, or 48 degrees north latitude and two degrees east longitude. This Gazetteer lists most of the world's largest independent countries, their capitals, and several important geographic features. The page numbers tell where each entry can be found on a map in this book. As an aid to pronunciation, most entries are spelled phonetically.

Abidjan [AH•bee•JAHN] Capital of Côte d'Ivoire. 5°N 4°W (p. 531)

Abu Dhabi [AH•boo DAH•bee] Capital of the United Arab Emirates. 24°N 54°E (p. 449)

Abuja [ah•BOO•jah] Capital of Nigeria. 8°N 9°E (p. 531)

Accra [ah•KRUH] Capital of Ghana. 6°N 0° longitude (p. 531)

Addis Ababa [AHD•dihs AH•bah•BAH] Capital of Ethiopia. 9°N 39°E (p. 531)

Adriatic [AY•dree•A•tihk] **Sea** Arm of the Mediterranean Sea between the Balkan Peninsula and Italy. 44°N 14°E (p. 280)

Afghanistan [af•GA•nuh•STAN] Central Asian country west of Pakistan. 33°N 63°E (p. 449)

Albania [al•BAY•nee•uh] Country on the Adriatic Sea, south of Yugoslavia. 42°N 20°E (p. 281)

Algeria [al•JIHR•ee•uh] North African country east of Morocco. 29°N 1°E (p. 449)

Algiers [al•JIHRZ] Capital of Algeria. 37°N 3°E (p. 449)

Alps [ALPS] Mountain ranges extending through central Europe. 46°N 9°E (p. 280)

Amazon [A•muh•ZAHN] **River** Largest river in the world by volume and second-largest in length. 2°S 53°W (p. 172)

Amman [a•MAHN] Capital of Jordan. 32°N 36°E (p. 449)

Amsterdam [AHM•stuhr•DAHM] Capital of the Netherlands. 52°N 5°E (p. 281)

Andes [AN•DEEZ] Mountain system extending north and south along the western side of South America. 13°S 75°W (p. 172)

Andorra [an•DAWR•uh] Small country in southern Europe between France and Spain. 43°N 2°E (p. 281)

Angola [ang•GOH•luh] Southern African country north of Namibia. 14°S 16°E (p. 531)

Ankara [AHNG•kuh•ruh] Capital of Turkey. 40°N 33°E (p. 449)

Antananarivo [AHN•tah•NAH•nah•REE•voh] Capital of Madagascar. 19°S 48°E (p. 531)

Arabian [uh•RAY•bee•uhn] **Peninsula** Large peninsula extending into the Arabian Sea. 28°N 40°E (p. 448)

Argentina [AHR•juhn•TEE•nuh] South American country east of Chile on the Atlantic Ocean. 36°S 67°W (p. 173)

Armenia [ahr•MEE•nee•uh] Southeastern European country between the Black and Caspian Seas. 40°N 45°E (p. 449)

Ashgabat [AHSH•gah•BAHT] Capital of Turkmenistan. 38°N 58°E (p. 449)

Asmara [az•MAHR•uh] Capital of Eritrea. 16°N 39°E (p. 531)

Astana Capital of Kazakhstan. 51°N 72°E (p. 449)

Asunción [ah•SOON•see•OHN] Capital of Paraguay. 25°S 58°W (p. 173)

Athens [A•thuhnz] Capital of Greece. 38°N 24°E (p. 281)

Atlas [AT•luhs] **Mountains** Mountain range on the northern edge of the Sahara. 31°N 5°W (p. 448)

Australia [aw•STRAYL•yuh] Country and continent in Southern Hemisphere. 25°S 135°W (p. 741)

Austria [AWS•tree•uh] Western European country east of Switzerland and south of Germany and the Czech Republic. 47°N 12°E (p. 281)

Azerbaijan [A•zuhr•BY•JAHN] European-Asian country on the Caspian Sea. 40°N 47°E (p. 449)

Baghdad [BAG•DAD] Capital of Iraq. 33°N 44°E (p. 449)

Bahamas [buh•HAH•muhz] Country made up of many islands between Cuba and the United States. 23°N 74°W (p. 172)

Bahrain [bah•RAYN] Country located on the Persian Gulf. 26°N 51°E (p. 449)

Baku [bah•KOO] Capital of Azerbaijan. 40°N 50°E (p. 449)

Balkan [BAWL•kuhn] **Peninsula** Peninsula in southeastern Europe. 42°N 20°E (p. 280)

Baltic [BAWL•tihk] **Sea** Sea in northern Europe that is connected to the North Sea. 55°N 17°E (p. 280)

Bamako [BAH•mah•KOH] Capital of Mali. 13°N 8°W (p. 531)

Bangkok [BANG•KAHK] Capital of Thailand. 14°N 100°E (p. 637)

Bangladesh [BAHNG•gluh•DEHSH] South Asian country bordered by India and Myanmar. 24°N 90°E (p. 637)

Bangui [BAHNG•GEE] Capital of the Central African Republic. 4°N 19°E (p. 531)

Banjul [BAHN•JOOL] Capital of Gambia. 13°N 17°W (p. 531)

Barbados [bahr•BAY•duhs] Island country between the Atlantic Ocean and the Caribbean Sea. 14°N 59°W (p. 173)

Beijing [BAY•JIHNG] Capital of China. 40°N 116°E (p. 637)

Beirut [bay•ROOT] Capital of Lebanon. 34°N 36°E (p. 449)

Belarus [BEE•luh•ROOS] Eastern European country west of Russia. 54°N 28°E (p. 281)

Belgium [BEHL•juhm] Western European country south of the Netherlands. 51°N 3°E (p. 281)

Belgrade [BEHL•GRAYD] Capital of Yugoslavia. 45°N 21°E (p. 281)

Belize [buh•LEEZ] Central American country east of Guatemala. 18°N 89°W (p. 173)

Belmopan [BEHL•moh•PAHN] Capital of Belize. 17°N 89°W (p. 173)

Benin [buh•NEEN] West African country west of Nigeria. 8°N 2°E (p. 531)

Berlin [behr•LEEN] Capital of Germany. 53°N 13°E (p. 281)

Bern [BEHRN] Capital of Switzerland. 47°N 7°E (p. 281)

Bhutan [boo•TAHN] South Asian country northeast of India. 27°N 91°E (p. 637)

Bishkek [bihsh•KEHK] Capital of Kyrgyzstan. 43°N 75°E (p. 449)

Bissau [bihs•SOW] Capital of Guinea-Bissau. 12°N 16°W (p. 531)

Black Sea Large sea between Europe and Asia. 43°N 32°E (p. 281)

Bloemfontein [BLOOM•FAHN•TAYN] Judicial capital of South Africa. 26°E 29°S (p. 531)

Bogotá [BOH•goh•TAH] Capital of Colombia. 5°N 74°W (p. 173)

Bolivia [buh•LIHV•ee•uh] Country in the central part of South America, north of Argentina. 17°S 64°W (p. 173)

Bosnia and Herzegovina [BAHZ•nee•uh HEHRT•seh•GAW•vee•nuh] Southeastern European country between Yugoslavia and Croatia. 44°N 18°E (p. 281)

Botswana [bawt•SWAH•nah] Southern African country north of the Republic of South Africa. 22°S 23°E (p. 531)

Brasília [brah•ZEEL•yuh] Capital of Brazil. 16°S 48°W (p. 173)

Bratislava [BRAH•tih•SLAH•vuh] Capital of Slovakia. 48°N 17°E (p. 281)

Brazil [bruh•ZIHL] Largest country in South America. 9°S 53°W (p. 173)

Brazzaville [BRAH•zuh•VEEL] Capital of the Congo. 4°S 15°E (p. 531)

Brunei [bru•NY] Southeast Asian country on northern coast of the island of Borneo. 5°N 114°E (p. 636)

Brussels [BRUH•suhlz] Capital of Belgium. 51°N 4°E (p. 281)

Bucharest [BOO•kuh•REHST] Capital of Romania. 44°N 26°E (p. 281)

Budapest [BOO•duh•PEHST] Capital of Hungary. 48°N 19°E (p. 281)

Buenos Aires [BWAY•nuhs AR•eez] Capital of Argentina. 34°S 58°W (p. 173)

Bujumbura [BOO•juhm•BUR•uh] Capital of Burundi. 3°S 29°E (p. 531)

Bulgaria [BUHL•GAR•ee•uh] Southeastern European country south of Romania. 42°N 24°E (p. 281)

Burkina Faso [bur•KEE•nuh FAH•soh] West African country south of Mali. 12°N 3°E (p. 531)

Burundi [bu•ROON•dee] East African country at the northern end of Lake Tanganyika. 3°S 30°E (p. 531)

Cairo [KY•ROH] Capital of Egypt. 31°N 32°E (p. 449)

Cambodia [kam•BOH•dee•uh] Southeast Asian country south of Thailand and Laos. 12°N 104°E (p. 637)

Cameroon [KA•muh•ROON] Central African country on the northeast shore of the Gulf of Guinea. 6°N 11°E (p. 531)

Canada [KA•nuh•duh] Northernmost country in North America. 50°N 100°W (p. 109)

Canberra [KAN•BEHR•uh] Capital of Australia. 35°S 149°E (p. 741)

Cape Town Legislative capital of the Republic of South Africa. 34°S 18°E (p. 531)

Cape Verde [VUHRD] Island country off the coast of western Africa in the Atlantic Ocean. 15°N 24°W (p. 531)

Caracas [kah•RAH•kahs] Capital of Venezuela. 11°N 67°W (p. 173)

Caribbean [KAR•uh•BEE•uhn] **Sea** Part of the Atlantic Ocean bordered by the West Indies, South America, and Central America. 15°N 76°W (p. 172)

Caspian [KAS•pee•uhn] **Sea** Salt lake between Europe and Asia that is the world's largest inland body of water. 40°N 52°E (p. 448)

Caucasus [KAW•kuh•suhs] **Mountains** Mountain range between the Black and Caspian Seas. 43°N 42°E (p. 448)

Central African Republic Central African country south of Chad. 8°N 21°E (p. 531)

Chad [CHAD] Country west of Sudan in the African Sahel. 18°N 19°E (p. 530)

Chile [CHEE•lay] South American country west of Argentina. 35°S 72°W (p. 173)

China [CHY•nuh] Country in eastern and central Asia, known officially as the People's Republic of China. 37°N 93°E (p. 637)

Chişinău [KEE•shee•NOW] Capital of Moldova. 47°N 29°E (p. 281)

Colombia [kuh•LUHM•bee•uh] South American country west of Venezuela. 4°N 73°W (p. 173)

Colombo [kuh•LUHM•boh] Capital of Sri Lanka. 7°N 80°E (p. 637)

Comoros [KAH•muh•ROHZ] Small island country in Indian Ocean between the island of Madagascar and the southeast African mainland. 13°S 43°E (p. 531)

Conakry [KAH•nuh•kree] Capital of Guinea. 10°N 14°W (p. 531)

Congo [KAHNG•goh] Central African country east of the Democratic Republic of the Congo. 3°S 14°E (p. 531)

Congo, Democratic Republic of the Central African country north of Zambia and Angola. 1°S 22°E (p. 531)

Copenhagen [KOH•puhn•HAY•guhn] Capital of Denmark. 56°N 12°E (p. 281)

Costa Rica [KAWS•tah REE•kah] Central American country south of Nicaragua. 11°N 85°W (p. 173)

Côte d'Ivoire [KOHT dee•VWAHR] West African country south of Mali. 8°N 7°W (p. 531)

Croatia [kroh•AY•shuh] Southeastern European country on the Adriatic Sea. 46°N 16°E (p. 281)

Cuba [KYOO•buh] Island country in the West Indies. 22°N 79°W (p. 172)

Cyprus [SY•pruhs] Island country in the eastern Mediterranean Sea, south of Turkey. 35°N 31°E (p. 281)

Czech [CHEHK] **Republic** Eastern European country north of Austria. 50°N 15°E (p. 281)

Dakar [dah•KAHR] Capital of Senegal. 15°N 17°W (p. 531)

Damascus [duh•MAS•kuhs] Capital of Syria. 34°N 36°E (p. 449)

Dar es Salaam (DAHR EHS sah•LAHM] Capital of Tanzania. 7°S 39°E (p. 531)

Denmark [DEHN•MAHRK] Northern European country between the Baltic and North Seas. 56°N 9°E (p. 281)

Dhaka [DA•kuh] Capital of Bangladesh. 24°N 90°E (p. 637)

Djibouti [jih•BOO•tee] East African country on the Gulf of Aden. 12°N 43°E (p. 531)

Doha [DOH•huh] Capital of Qatar. 25°N 51°E (p. 449)

Dominican [duh•MIH•nih•kuhn] **Republic** Country in the West Indies on the eastern part of Hispaniola. 19°N 71°W (p. 206)

Dublin [DUH•blihn] Capital of Ireland. 53°N 6°W (p. 281)

Dushanbe [doo•SHAM•buh] Capital of Tajikistan. 39°N 69°E (p. 449)

East Timor [TEE•MOHR] Previous province of Indonesia, now under UN administration. 10°S 127°E (p. 637)

Ecuador [EH•kwuh•DAWR] South American country southwest of Colombia. 0° latitude 79°W (p. 173)

Egypt [EE•jihpt] North African country on the Mediterranean Sea. 27°N 27°E (p. 449)

El Salvador [ehl SAL•vuh•DAWR] Central American country southwest of Honduras. 14°N 89°W (p. 173)

Equatorial Guinea [EE•kwuh•TOHR•ee•uhl GIH•nee] Central African country south of Cameroon. 2°N 8°E (p. 531)

Eritrea [EHR•uh•TREE•uh] East African country north of Ethiopia. 17°N 39°E (p. 531)

Estonia [eh•STOH•nee•uh] Eastern European country on the Baltic Sea. 59°N 25°E (p. 281)

Ethiopia [EE•thee•OH•pee•uh] East African country north of Somalia and Kenya. 8°N 38°E (p. 531)

Euphrates [yu•FRAY•TEEZ] **River** River in southwestern Asia that flows through Syria and Iraq and joins the Tigris River. 36°N 40°E (p. 448)

Fiji [FEE•jee] **Islands** Country comprised of an island group in the southwest Pacific Ocean. 19°S 175°E (p. 741)

Finland [FIHN•luhnd] Northern European country east of Sweden. 63°N 26°E (p. 281)

France [FRANS] Western European country south of the United Kingdom. 47°N 1°E (p. 281)

Freetown [FREE•TOWN] Capital of Sierra Leone. 9°N 13°W (p. 531)

French Guiana [gee•A•nuh] French-owned territory in northern South America. 5°N 53°W (p. 173)

Gabon [ga•BOHN] Central African country on the Atlantic Ocean. 0° latitude 12°E (p. 531)

Gaborone [GAH•boh•ROH•NAY] Capital of Botswana. 24°S 26°E (p. 531)

Gambia [GAM•bee•uh] West African country along the Gambia River. 13°N 16°W (p. 531)

Georgetown [JAWRJ•TOWN] Capital of Guyana. 8°N 58°W (p. 173)

Georgia [JAWR•juh] Asian-European country bordering the Black Sea south of Russia. 42°N 43°E (p. 449)

Germany [JUHR•muh•nee] Western European country south of Denmark, officially called the Federal Republic of Germany. 52°N 10°E (p. 281)

Ghana [GAH•nuh] West African country on the Gulf of Guinea. 8°N 2°W (p. 531)

Great Plains The continental slope extending through the United States and Canada. 45°N 104°W (p. 108)

Greece [GREES] Southern European country on the Balkan Peninsula. 39°N 22°E (p. 280)

Greenland [GREEN•luhnd] Island in northwestern Atlantic Ocean and the largest island in the world. 74°N 40°W (p. 109)

Guatemala [GWAH•tay•MAH•lah] Central American country south of Mexico. 16°N 92°W (p. 173)

Guatemala City Capital of Guatemala. 15°N 91°W (p. 173)

GAZETTEER

Guinea [GIH•nee] West African country on the Atlantic coast. 11°N 12°W (p. 531)
Guinea-Bissau [GIH•nee bih•SOW] West African country on the Atlantic coast. 12°N 20°W (p. 531)
Gulf of Mexico Gulf on part of the southern coast of North America. 25°N 94°W (p. 108)
Guyana [gy•AH•nuh] South American country between Venezuela and Suriname. 8°N 59°W (p. 173)

Haiti [HAY•tee] Country in the West Indies on the western part of Hispaniola. 19°N 72°W (p. 206)
Hanoi [ha•NOY] Capital of Vietnam. 21°N 106°E (p. 637)
Harare [hah•RAH•RAY] Capital of Zimbabwe. 18°S 31°E (p. 531)
Havana [huh•VA•nuh] Capital of Cuba. 23°N 82°W (p. 173)
Helsinki [HEHL•SIHNG•kee] Capital of Finland. 60°N 24°E (p. 281)
Himalaya [HI•muh•LAY•uh] Mountain ranges in southern Asia, bordering the Indian subcontinent on the north. 30°N 85°E (p. 636)
Honduras [hahn•DUR•uhs] Central American country on the Caribbean Sea. 15°N 88°W (p. 173)
Hong Kong [HAWNG KAWNG] Port and industrial center in southern China. 22°N 115°E (p. 637)
Hungary [HUHNG•guh•ree] Eastern European country south of Slovakia. 47°N 18°E (p. 281)

Iberian [eye•BIHR•ee•uhn] **Peninsula** Peninsula in southwest Europe, occupied by Spain and Portugal. 41°N 1°W (p. 280)
Iceland [EYES•luhnd] Island country between the North Atlantic and the Arctic Oceans, 65°N 20°W (p. 281)
India [IHN•dee•uh] South Asian country south of China and Nepal. 23°N 78°E (p. 637)
Indonesia [IHN•duh•NEE•zhuh] Southeast Asian island country known as the Republic of Indonesia. 5°S 119°E (p. 637)
Indus [IHN•duhs] **River** River in Asia that begins in Tibet and flows through Pakistan to the Arabian Sea. 27°N 68°E (p. 636)
Iran [ih•RAN] Southwest Asian country that was formerly named Persia. 31°N 54°E (p. 449)
Iraq [ih•RAHK] Southwest Asian country west of Iran. 32°N 43°E (p. 449)
Ireland [EYER•luhnd] Island west of Great Britain occupied by the Republic of Ireland and Northern Ireland. 54°N 8°W (p. 281)
Islamabad [ihs•LAH•muh•BAHD] Capital of Pakistan. 34°N 73°E (p. 637)
Israel [IHZ•ree•uhl] Southwest Asian country south of Lebanon. 33°N 34°E (p. 449)

Italy [IHT•uhl•ee] Southern European country south of Switzerland and east of France. 44°N 11°E (p. 281)

Jakarta [juh•KAHR•tuh] Capital of Indonesia. 6°S 107°E (p. 637)
Jamaica [juh•MAY•kuh] Island country in the West Indies. 18°N 78°W (p. 173)
Japan [juh•PAN] East Asian country consisting of the four large islands of Hokkaido, Honshu, Shikoku, and Kyushu, plus thousands of small islands. 37°N 134°E (p. 636)
Jerusalem [juh•ROO•suh•luhm] Capital of Israel and a holy city for Christians, Jews, and Muslims. 32°N 35°E (p. 449)
Jordan [JAWRD•uhn] Southwest Asian country south of Syria. 30°N 38°E (p. 449)

Kabul [KAH•buhl] Capital of Afghanistan. 35°N 69°E (p. 449)
Kampala [kahm•PAH•lah] Capital of Uganda. 0° latitude 32°E (p. 531)
Kathmandu [KAT•MAN•DOO] Capital of Nepal. 28°N 85°E (p. 637)
Kazakhstan [kuh•ZAHK•STAHN] Large Asian country south of Russia and bordering the Caspian Sea. 48°N 59°E (p. 449)
Kenya [KEHN•yuh] East African country south of Ethiopia. 1°N 37°E (p. 531)
Khartoum [kahr•TOOM] Capital of Sudan. 16°N 33°E (p. 531)
Kiev [KEE•ihf] Capital of Ukraine. 50°N 31°E (p. 281)
Kigali [kee•GAH•lee] Capital of Rwanda. 2°S 30°E (p. 531)
Kingston [KIHNG•stuhn] Capital of Jamaica. 18°N 77°W (p. 173)
Kinshasa [kihn•SHAH•suh] Capital of the Democratic Republic of the Congo. 4°S 15°E (p. 531)
Kuala Lumpur [KWAH•luh LUM•PUR] Capital of Malaysia. 3°N 102°E (p. 637)
Kuwait [ku•WAYT] Country on the Persian Gulf between Saudi Arabia and Iraq. 29°N 48°E (p. 449)
Kyrgyzstan [KIHR•gih•STAN] Central Asian country on China's western border. 41°N 75°E (p. 449)

Laos [LOWS] Southeast Asian country south of China and west of Vietnam. 20°N 102°E (p. 637)
La Paz [lah PAHS] Administrative capital of Bolivia, and the highest capital in the world. 17°S 68°W (p. 173)
Latvia [LAT•vee•uh] Eastern European country west of Russia on the Baltic Sea. 57°N 25°E (p. 281)

Lebanon [LEH•buh•nuhn] Country south of Syria on the Mediterranean Sea. 34°N 34°E (p. 449)

Lesotho [luh•SOH•TOH] Southern African country within the borders of the Republic of South Africa. 30°S 28°E (p. 531)

Liberia [ly•BIHR•ee•uh] West African country south of Guinea. 7°N 10°W (p. 531)

Libreville [LEE•bruh•VIHL] Capital of Gabon. 1°N 9°E (p. 531)

Libya [LIH•bee•uh] North African country west of Egypt on the Mediterranean Sea. 28°N 15°E (p. 449)

Liechtenstein [LIHKT•uhn•SHTYN] Small country in central Europe between Switzerland and Austria. 47°N 10°E (p. 281)

Lilongwe [lih•LAWNG•GWAY] Capital of Malawi. 14°S 34°E (p. 531)

Lima [LEE•mah] Capital of Peru. 12°S 77°W (p. 173)

Lisbon [LIHZ•buhn] Capital of Portugal. 39°N 9°W (p. 281)

Lithuania [LIH•thuh•WAY•nee•uh] Eastern European country northwest of Belarus on the Baltic Sea. 56°N 24°E (p. 281)

Ljubljana [lee•oo•blee•AH•nuh] Capital of Slovenia. 46°N 14°E (p. 281)

Lomé [loh•MAY] Capital of Togo. 6°N 1°E (p. 531)

London [LUHN•duhn] Capital of the United Kingdom, on the Thames River. 52°N 0° longitude (p. 281)

Luanda [lu•AHN•duh] Capital of Angola. 9°S 13°E (p. 531)

Lusaka [loo•SAH•kah] Capital of Zambia. 15°S 28°E (p. 531)

Luxembourg [LUHK•suhm•BUHRG] Small European country between France, Belgium, and Germany. 50°N 7°E (p. 281)

Macau [muh•KOW] Port in southern China. 22°N 113°E (p. 637)

Macedonia [MA•suh•DOH•nee•uh] Southeastern European country north of Greece. 42°N 22°E (p. 281). Macedonia also refers to a geographic region covering northern Greece, the country Macedonia, and part of Bulgaria.

Madagascar [MA•duh•GAS•kuhr] Island in the Indian Ocean off the southeastern coast of Africa. 18°S 43°E (p. 531)

Madrid [muh•DRIHD] Capital of Spain. 41°N 4°W (p. 281)

Malabo [mah•LAH•boh] Capital of Equatorial Guinea. 4°N 9°E (p. 531)

Malawi [mah•LAH•wee] Southern African country south of Tanzania and east of Zambia. 11°S 34°E (p. 531)

Malaysia [muh•LAY•zhuh] Southeast Asian country with land on the Malay Peninsula and on the island of Borneo. 4°N 101°E (p. 636)

Maldives [MAWL•DEEVZ] Island country southwest of India in the Indian Ocean. 5°N 42°E (p. 637)

Mali [MAH•lee] West African country east of Mauritania. 16°N 0° longitude (p. 531)

Managua [mah•NAH•gwah] Capital of Nicaragua. 12°N 86°W (p. 173)

Manila [muh•NIH•luh] Capital of the Philippines. 15°N 121°E (p. 637)

Maputo [mah•POO•toh] Capital of Mozambique. 26°S 33°E (p. 531)

Maseru [MA•zuh•ROO] Capital of Lesotho. 29°S 27°E (p. 531)

Mauritania [MAWR•uh•TAY•nee•uh] West African country north of Senegal. 20°N 14°W (p. 531)

Mauritius [maw•RIH•shuhs] Island country in the Indian Ocean east of Madagascar. 21°S 58°E (p. 531)

Mbabane [uhm•bah•BAH•nay] Capital of Swaziland. 26°S 31°E (p. 531)

Mediterranean [MEH•duh•tuh•RAY•nee•uhn] **Sea** Large inland sea surrounded by Europe, Asia, and Africa. 36°N 13°E (p. 280)

Mekong [MAY•KAWNG] **River** River in southeastern Asia that begins in Tibet and empties into the South China Sea. 18°N 104°E (p. 636)

Mexico [MEHK•sih•KOH] North American country south of the United States. 24°N 104°W (p. 173)

Mexico City Capital of Mexico. 19°N 99°W (p. 173)

Minsk [MIHNSK] Capital of Belarus. 54°N 28°E (p. 281)

Mississippi [MIH•suh•SIH•pee] **River** Large river system in the central United States that flows southward into the Gulf of Mexico. 32°N 92°W (p. 108)

Mogadishu [MOH•guh•DEE•SHOO] Capital of Somalia. 2°N 45°E (p. 531)

Moldova [mawl•DAW•vuh] Small European country between Ukraine and Romania. 48°N 28°E (p. 281)

Monaco [MAH•nuh•KOH] Small country in southern Europe on the French Mediterranean coast. 44°N 8°E (p. 281)

Mongolia [mahn•GOHL•yuh] Country in Asia between Russia and China. 46°N 100°E (p. 637)

Monrovia [muhn•ROH•vee•uh] Capital of Liberia. 6°N 11°W (p. 531)

Montevideo [MAHN•tuh•vuh•DAY•OH] Capital of Uruguay. 35°S 56°W (p. 173)

Morocco [muh•RAH•KOH] North African country on the Mediterranean Sea and the Atlantic Ocean. 32°N 7°W (p. 449)

Moscow [MAHS•KOW] Capital of Russia. 56°N 38°E (p. 401)

Mount Everest [EHV•ruhst] Highest mountain in the world, in the Himalaya between Nepal and Tibet. 28°N 87°E (p. 636)

Mozambique [MOH•zahm•BEEK] Southern African country south of Tanzania. 20°S 34°E (p. 531)

Muscat [MUHS•KAHT] Capital of Oman. 23°N 59°E (p. 449)

Myanmar [MYAHN•MAHR] Southeast Asian country south of China and India, formerly called Burma. 21°N 95°E (p. 637)

Nairobi [ny•ROH•bee] Capital of Kenya. 1°S 37°E (p. 531)

GAZETTEER

Namibia [nuh•MIH•bee•uh] Southern African country south of Angola on the Atlantic Ocean. 20°S 16°E (p. 531)

Nassau [NA•SAW] Capital of the Bahamas. 25°N 77°W (p. 173)

N'Djamena [uhn•jah•MAY•nah] Capital of Chad. 12°N 15°E (p. 531)

Nepal [NAY•PAHL] Mountain country between India and China. 29°N 83°E (p. 637)

Netherlands [NEH•thuhr•lundz] Western European country north of Belgium. 53°N 4°E (p. 281)

New Delhi [NOO DEH•lee] Capital of India. 29°N 77°E (p. 637)

New Zealand [NOO ZEE•luhnd] Major island country southeast of Australia in the South Pacific. 42°S 175°E (p. 741)

Niamey [nee•AHM•ay] Capital of Niger. 14°N 2°E (p. 531)

Nicaragua [NIH•kuh•RAH•gwuh] Central American country south of Honduras. 13°N 86°W (p. 173)

Nicosia [NIH•kuh•SEE•uh] Capital of Cyprus. 35°N 33°E (p. 281)

Niger [NY•juhr] West African country north of Nigeria. 18°N 9°E (p. 531)

Nigeria [ny•JIHR•ee•uh] West African country along the Gulf of Guinea. 9°N 7°E (p. 530)

Nile [NYL] **River** Longest river in the world, flowing north through eastern Africa. 19°N 33°E (p. 448)

North Korea [kuh•REE•uh] East Asian country in the northernmost part of the Korean Peninsula. 40°N 127°E (p. 637)

Norway [NAWR•WAY] Northern European country on the Scandinavian peninsula. 64°N 11°E (p. 280)

Nouakchott [nu•AHK•SHAHT] Capital of Mauritania. 18°N 16°W (p. 531)

Oman [oh•MAHN] Country on the Arabian Sea and the Gulf of Oman. 20°N 58°E (p. 449)

Oslo [AHZ•loh] Capital of Norway. 60°N 11°E (p. 281)

Ottawa [AH•tuh•wuh] Capital of Canada. 45°N 76°W (p. 109)

Ouagadougou [WAH•gah•DOO•goo] Capital of Burkina Faso. 12°N 2°W (p. 531)

Pakistan [PA•kih•STAN] South Asian country northwest of India on the Arabian Sea. 28°N 68°E (p. 637)

Palau [puh•LOW] Island country in the Pacific Ocean. 7°N 135°E (p. 741)

Panama [PA•nuh•MAH] Central American country on the Isthmus of Panama. 9°N 81°W (p. 173)

Panama City Capital of Panama. 9°N 79°W (p. 173)

Papua New Guinea [PA•pyu•wuh NOO GIH•nee] Island country in the Pacific Ocean north of Australia. 7°S 142°E (p. 741)

Paraguay [PAR•uh•GWY] South American country northeast of Argentina. 24°S 57°W (p. 173)

Paramaribo [PAH•rah•MAH•ree•boh] Capital of Suriname. 6°N 55°W (p. 173)

Paris [PAR•uhs] Capital of France. 49°N 2°E (p. 281)

Persian [PUHR•zhuhn] **Gulf** Arm of the Arabian Sea between Iran and Saudi Arabia. 28°N 51°E (p. 448)

Peru [puh•ROO] South American country south of Ecuador and Colombia. 10°S 75°W (p. 173)

Philippines [FIH•luh•PEENZ] Island country in the Pacific Ocean southeast of China. 14°N 125°E (p. 637)

Phnom Penh [puh•NAWM PEHN] Capital of Cambodia. 12°N 106°E (p. 637)

Poland [POH•luhnd] Eastern European country on the Baltic Sea. 52°N 18°E (p. 281)

Port-au-Prince [POHR•toh•PRIHNS] Capital of Haiti. 19°N 72°W (p. 173)

Port Moresby [MOHRZ•bee] Capital of Papua New Guinea. 10°S 147°E (p. 741)

Port-of-Spain [SPAYN] Capital of Trinidad and Tobago. 11°N 62°W (p. 173)

Porto-Novo [POHR•toh•NOH•voh] Capital of Benin. 7°N 3°E (p. 531)

Portugal [POHR•chih•guhl] Country west of Spain on the Iberian Peninsula. 39°N 8°W (p. 280)

Prague [PRAHG] Capital of the Czech Republic. 51°N 15°E (p. 281)

Pretoria [prih•TOHR•ee•uh] Executive capital of South Africa. 26°S 28°E (p. 531)

Puerto Rico [PWEHR•toh REE•koh] Island in the Caribbean Sea; U.S. Commonwealth. 19°N 67°W (p. 173)

Pyongyang [pee•AWNG•YAHNG] Capital of North Korea. 39°N 126°E (p. 637)

Qatar [KAH•tuhr] Country on the southwestern shore of the Persian Gulf. 25°N 53°E (p. 449)

Quito [KEE•toh] Capital of Ecuador. 0° latitude 79°W (p. 173)

Rabat [ruh•BAHT] Capital of Morocco. 34°N 7°W (p. 449)

Reykjavík [RAY•kyah•VEEK] Capital of Iceland. 64°N 22°W (p. 281)

Rhine [RYN] **River** River in western Europe that flows into the North Sea. 51°N 7°E (p. 280)

Riga [REE•guh] Capital of Latvia. 57°N 24°E (p. 281)

Rio Grande [REE•oh GRAND] River that forms part of the boundary between the United States and Mexico. 30°N 103°W (p. 108)

Riyadh [ree•YAHD] Capital of Saudi Arabia. 25°N 47°E (p. 449)

Rocky Mountains Mountain system in western North America. 50°N 114°W (p. 108)

Romania (ru•MAY•nee•uh] Eastern European country east of Hungary. 46°N 23°E (p. 281)

Rome [ROHM] Capital of Italy. 42°N 13°E (p. 281)

Russia [RUH•shuh] Largest country in the world, covering parts of Europe and Asia. 60°N 90°E (p. 401)

Rwanda [ruh•WAHN•duh] East African country south of Uganda. 2°S 30°E (p. 531)

Sahara [suh•HAR•uh] Desert region in northern Africa that is the largest hot desert in the world. 24°N 2°W (p. 448)

Saint Lawrence [LAWR•uhns] **River** River that flows from Lake Ontario to the Atlantic Ocean and forms part of the boundary between the United States and Canada. 48°N 70°W (p. 108)

Sanaa [sahn•AH] Capital of Yemen. 15°N 44°E (p. 449)

San José [SAHNG hoh•SAY] Capital of Costa Rica. 10°N 84°W (p. 173)

San Marino [SAN muh•REE•noh] Small European country located in the Italian peninsula. 44°N 13°E (p. 281)

San Salvador [san SAL•vuh•DAWR] Capital of El Salvador. 14°N 89°W (p. 173)

Santiago [SAN•tee•AH•goh] Capital of Chile. 33°S 71°W (p. 173)

Santo Domingo [SAN•toh duh•MIHNG•goh] Capital of the Dominican Republic. 19°N 70°W (p. 173)

Sao Tome and Principe [SOW•too•MAY PREEN•see•pee] Small island country in the Gulf of Guinea off the coast of central Africa. 1°N 7°E (p. 530)

Sarajevo [SAR•uh•YAY•voh] Capital of Bosnia and Herzegovina. 43°N 18°E (p. 281)

Saudi Arabia [SOW•dee uh•RAY•bee•uh] Country on the Arabian Peninsula. 23°N 46°E (p. 448)

Senegal [SEH•nih•GAWL] West African country on the Atlantic coast. 15°N 14°W (p. 531)

Seoul [SOHL] Capital of South Korea. 38°N 127°E (p. 637)

Seychelles [say•SHEHL] Small island country in the Indian Ocean off eastern Africa. 6°S 56°E (p. 531)

Sierra Leone [see•EHR•uh lee•OHN] West African country south of Guinea. 8°N 12°W (p. 531)

Singapore [SIHNG•uh•POHR] Southeast Asian island country near tip of Malay Peninsula. 2°N 104°E (p. 636)

Skopje [SKAW•PYAY] Capital of the country of Macedonia. 42°N 21°E (p. 281)

Slovakia [sloh•VAH•kee•uh] Eastern European country south of Poland. 49°N 19°E (p. 281)

Slovenia [sloh•VEE•nee•uh] Southeastern European country south of Austria on the Adriatic Sea. 46°N 15°E (p. 281)

Sofia [SOH•fee•uh] Capital of Bulgaria. 43°N 23°E (p. 281)

Solomon [SAH•luh•muhn] **Islands** Island country in the Pacific Ocean northeast of Australia. 7°S 160°E (p. 741)

Somalia [soh•MAH•lee•uh] East African country on the Gulf of Aden and the Indian Ocean. 3°N 45°E (p. 531)

South Africa [A•frih•kuh] Country at the southern tip of Africa, officially the Republic of South Africa. 28°S 25°E (p. 531)

South Korea [kuh•REE•uh] East Asian country on the Korean Peninsula between the Yellow Sea and the Sea of Japan. 36°N 128°E (p. 637)

Spain [SPAYN] Southern European country on the Iberian Peninsula. 40°N 4°W (p. 280)

Sri Lanka [SREE•LAHNG•kuh] Country in the Indian Ocean south of India, formerly called Ceylon. 9°N 83°E (p. 637)

Stockholm [STAHK•HOHLM] Capital of Sweden. 59°N 18°E (p. 281)

Sucre [SOO•kray] Constitutional capital of Bolivia. 19°S 65°W (p. 173)

Sudan [soo•DAN] East African country south of Egypt. 14°N 28°E (p. 531)

Suriname [SUR•uh•NAH•muh] South American country between Guyana and French Guiana. 4°N 56°W (p. 173)

Suva [SOO•vah] Capital of the Fiji Islands. 18°S 177°E (p. 741)

Swaziland [SWAH•zee•LAND] Southern African country west of Mozambique, almost entirely within the Republic of South Africa. 27°S 32°E (p. 531)

Sweden [SWEED•uhn] Northern European country on the eastern side of the Scandinavian peninsula. 60°N 14°E (p. 280)

Switzerland [SWIHT•suhr•luhnd] European country in the Alps south of Germany. 47°N 8°E (p. 280)

Syria [SIHR•ee•uh] Southwest Asian country on the east side of the Mediterranean Sea. 35°N 37°E (p. 449)

Taipei [TY•PAY] Capital of Taiwan. 25°N 122°E (p. 637)

Taiwan [TY•WAHN] Island country off the southeast coast of China, and the seat of the Chinese Nationalist government. 24°N 122°E (p. 637)

Tajikistan [tah•JIH•kih•STAN] Central Asian country east of Turkmenistan. 39°N 70°E (p. 449)

Tallinn [TA•luhn] Capital of Estonia. 59°N 25°E (p. 281)

Tanzania [TAN•zuh•NEE•uh] East African country south of Kenya. 7°S 34°E (p. 531)

Tashkent [tash•KEHNT] Capital of Uzbekistan. 41°N 69°E (p. 449)

T'bilisi [tuh•bih•LEE•see] Capital of the Republic of Georgia. 42°N 45°E (p. 449)

Tegucigalpa [tay•GOO•see•GAHL•pah] Capital of Honduras. 14°N 87°W (p. 173)

Tehran [TAY•uh•RAN] Capital of Iran. 36°N 52°E (p. 449)

Thailand [TY•LAND] Southeast Asian country east of Myanmar. 17°N 101°E (p. 637)

GAZETTEER

GAZETTEER

Thimphu [thihm•POO] Capital of Bhutan. 28°N 90°E (p. 637)

Tigris [TY•gruhs] **River** River in southeastern Turkey and Iraq that merges with the Euphrates River. 35°N 44°E (p. 448)

Tirana [tih•RAH•nuh] Capital of Albania. 42°N 20°E (p. 281)

Togo [TOH•goh] West African country between Benin and Ghana on the Gulf of Guinea. 8°N 1°E (p. 530)

Tokyo [TOH•kee•OH] Capital of Japan. 36°N 140°E (p. 637)

Trinidad and Tobago [TRIH•nuh•DAD tuh•BAY•goh] Island country near Venezuela between the Atlantic Ocean and the Caribbean Sea. 11°N 61°W (p. 173)

Tripoli [TRIH•puh•lee] Capital of Libya. 33°N 13°E (p. 449)

Tunis [TOO•nuhs] Capital of Tunisia. 37°N 10°E (p. 449)

Tunisia [too•NEE•zhuh] North African country on the Mediterranean Sea between Libya and Algeria. 35°N 10°E (p. 449)

Turkey [TUHR•kee] Country in southeastern Europe and western Asia. 39°N 32°E (p. 449)

Turkmenistan [tuhrk•MEH•nuh•STAN] Central Asian country on the Caspian Sea. 41°N 56°E (p. 449)

Uganda [yoo•GAHN•dah] East African country south of Sudan. 2°N 32°E (p. 531)

Ukraine [yoo•KRAYN] Eastern European country west of Russia on the Black Sea. 49°N 30°E (p. 281)

Ulaanbaatar [OO•LAHN•BAH•TAWR] Capital of Mongolia. 48°N 107°E (p. 637)

United Arab Emirates [EH•muh•ruhts] Country made up of seven states on the eastern side of the Arabian Peninsula. 24°N 54°E (p. 449)

United Kingdom Western European island country made up of England, Scotland, Wales, and Northern Ireland. 57°N 2°W (p. 281)

United States of America Country in North America made up of 50 states, mostly between Canada and Mexico. 38°N 110°W (p. 109)

Uruguay [YUR•uh•GWAY] South American country south of Brazil on the Atlantic Ocean. 33°S 56°W (p. 173)

Uzbekistan [UZ•BEH•kih•STAN] Central Asian country south of Kazakhstan. 42°N 60°E (p. 449)

Vanuatu [VAN•WAH•TOO] Country made up of islands in the Pacific Ocean east of Australia. 17°S 170°W (p. 741)

Vatican [VA•tih•kuhn] **City** Headquarters of the Roman Catholic Church, located in the city of Rome in Italy. 42°N 13°E (p. 281)

Venezuela [VEH•nuh•ZWAY•luh] South American country on the Caribbean Sea between Colombia and Guyana. 8°N 65°W (p. 173)

Vienna [vee•EH•nuh] Capital of Austria. 48°N 16°E (p. 281)

Vientiane [vyehn•TYAHN] Capital of Laos. 18°N 103°E (p. 637)

Vietnam [vee•EHT•NAHM] Southeast Asian country east of Laos and Cambodia. 18°N 107°E (p. 637)

Vilnius [VIL•nee•uhs] Capital of Lithuania. 55°N 25°E (p. 281)

Warsaw [WAWR•SAW] Capital of Poland. 52°N 21°E (p. 281)

Washington, D.C. Capital of the United States, in the District of Columbia. 39°N 77°W (p. 109)

Wellington [WEH•lihng•tuhn] Capital of New Zealand. 41°S 175°E (p. 741)

West Indies [IHN•deez] Islands in the Caribbean Sea between North America and South America. 19°N 79°W (p. 172)

Windhoek [VIHNT•HUK] Capital of Namibia. 22°S 17°E (p. 531)

Yamoussoukro [YAH•moo•SOO•kroh] Second capital of Côte d'Ivoire. 7°N 6°W (p. 531)

Yangon [YAHNG•GOHN] Capital of Myanmar. 17°N 96°E (p. 637)

Yangtze [YANG•SEE] **River** Principal river of China that begins in Tibet and flows into the East China Sea near Shanghai, also known as the Chang Jiang [CHAHNG jee•AHNG]. 31°N 117°E (p. 636)

Yaoundé [yown•DAY] Capital of Cameroon. 4°N 12°E (p. 531)

Yellow River River in northern and eastern China, also known as the Huang He [HWAHNG HUH]. 35°N 114°E (p. 636)

Yemen [YEH•muhn] Country south of Saudi Arabia on the Arabian Peninsula. 15°N 46°E (p. 449)

Yerevan [YEHR•uh•VAHN] Capital of Armenia. 40°N 44°E (p. 449)

Yugoslavia [YOO•goh•SLAH•vee•uh] Eastern European country south of Hungary; includes Serbia and Montenegro. 44°N 21°E (p. 281)

Zagreb [ZAH•GREHB] Capital of Croatia. 46°N 16°E (p. 281)

Zambia [ZAM•bee•uh] Southern African country north of Zimbabwe. 14°S 24°E (p. 531)

Zimbabwe [zihm•BAH•bway] Southern African country northeast of Botswana. 18°S 30°E (p. 531)

GLOSSARY*

absolute location exact position of a place on the earth's surface (p. 6)

acid rain rain containing high amounts of chemical pollutants (pp. 97, 127, 154, 368)

adobe sun-dried clay bricks (p. 193)

alluvial plain area that is built up by rich fertile soil left by river floods (p. 497)

altiplano large highland plateau (p. 260)

altitude height above sea level (pp. 184, 242, 583)

anthracite type of hard coal (p. 705)

apartheid system of laws that separated racial and ethnic groups and limited the rights of blacks in South Africa (p. 611)

aquifer underground rock layer so rich in water that water actually flows through it (pp. 51, 465)

archipelago group of islands (pp. 213, 353, 695)

atoll low-lying, ring-shaped island that surrounds a lagoon (pp. 662, 768)

atmosphere layer of air surrounding the earth (p. 30)

autobahn superhighway (p. 307)

autonomy self-government (p. 593)

axis imaginary line that runs through the earth's center between the North and South poles (p. 31); *also* the horizontal (bottom) or vertical (side) line of measurement on a graph (p. 14)

bar graph graph in which vertical or horizontal bars represent quantities (p. 14)

basin low area surrounded by higher land (pp. 227, 564)

bauxite mineral used to make aluminum (pp. 215, 376, 414, 555)

bazaar marketplace (p. 462)

Bedouins nomadic desert peoples of Southwest Asia (p. 489)

bilingual referring to a country that has two official languages (pp. 158, 517)

birthrate number of children born each year for every 1,000 people (p. 85)

Boers name for the Dutch who were the first European settlers in South Africa (p. 610)

bog low swampy land (pp. 296, 365)

boomerang Australian weapon shaped like a bent wing that either strikes a target or curves and sails back to land on the ground near the person who threw it (p. 751)

bush rural areas of Australia (p. 752)

cacao tropical tree whose seeds are used to make chocolate and cocoa (pp. 543, 766)

calligraphy art of beautiful writing (p. 684)

campesino Colombian farmer (p. 259)

canopy umbrella-like covering formed by the tops of trees in a rain forest (pp. 64, 208, 564)

cardinal directions basic directions on the earth: north, south, east, west (p. 11)

casbah older section of Algerian cities (p. 468)

cash crop product grown to be sold for export (pp. 258, 513)

cassava plant with roots that can be ground into flour to make bread or eaten in other ways (pp. 552, 584)

caste social class based on a person's ancestry (p. 648)

caudillo military ruler (p. 242)

channel body of water wider than a strait between two pieces of land (p. 42)

chart graphic way of presenting information clearly (p. 16)

circle graph round or pie-shaped graph showing how a whole is divided (p. 15)

city-state city and its surrounding countryside (p. 333)

civilizations highly developed cultures (p. 81)

civil war fight among different groups within a country (pp. 467, 490, 545)

clan group of people related to one another (pp. 517, 600, 698)

climate usual, predictable pattern of weather in an area over a long period of time (p. 54)

climograph combination bar and line graph giving information about temperature and precipitation (p. 15)

See also the Geographic Dictionary on page 18.

coalition government   government in which two or more political parties work together to run a country (pp. 334, 649)

cold war   period between the late 1940s and late 1980s when the United States and the Soviet Union competed for world influence without actually fighting each other (p. 427)

collection   process in the water cycle during which streams and rivers carry water back to the oceans (p. 50)

colony   overseas territory or settlement tied to a parent country (pp. 131, 156, 190, 215, 327)

commonwealth   partly self-governing territory (p. 219)

communist state   country whose government has strong control over the economy and society as a whole (pp. 216, 308, 367, 426, 675)

compass rose   device drawn on maps to show the directions (p. 11)

compound   group of houses surrounded by walls (p. 544)

condensation   process in which air rises and cools, which makes the water vapor it holds change back into a liquid (p. 50)

conservation   careful use of resources so they are not wasted (p. 96)

constitutional monarchy   government in which a king or queen is the official head of state, but elected officials run the government (pp. 293, 469, 491, 700)

consumer goods   household products, clothing, and other goods people buy to use for themselves (pp. 382, 413, 676)

contiguous   areas that are joined together inside a common boundary (p. 115)

continent   massive land area (p. 35)

continental island   island formed when chunks of land are split off from larger continents or when a piece of land that once linked an island to the mainland is eroded or covered by water (p. 765)

continental shelf   plateau off each coast of a continent that lies under the ocean and stretches for several miles (p. 41)

contour line   line connecting all points at the same elevation on a contour map (p. 12)

cooperative   farm owned and operated by the government (p. 217)

copper belt   large area of copper mines in northern Zambia (p. 618)

copra   dried coconut meat, which is used to make margarine, soap, and other products (p. 776)

coral reef   structure at or near the water's surface formed by the skeletons of small sea animals (pp. 119, 581, 747)

cordillera   group of mountain ranges that run side by side (pp. 147, 255)

core   center of the earth, formed of hot iron mixed with other metals (p. 35)

cottage industry   home- or village-based industry in which family members supply their own equipment to make goods (p. 648)

crevasse   deep crack in the Antarctic ice cap (p. 772)

crop rotation   varying what is planted in a field to avoid using up all the minerals in the soil (p. 97)

crust   uppermost layer of the earth (p. 35)

cultural diffusion   the process of spreading new knowledge and skills to other cultures (p. 81)

culture   way of life of a group of people who share similar beliefs and customs (p. 77)

culture region   area of the world that includes many different countries that all have cultural traits in common (p. 82)

currency   form of money (p. 292)

current   moving streams of water in the world's oceans, which affect the climate of land areas (p. 57)

cyclone   intense storm system with heavy rain and high winds (pp. 623, 654)

czar   name for emperor in Russia's past (p. 424)

death rate   number of people out of every 1,000 who die in a year (p. 84)

deforestation   widespread cutting of forests (pp. 97, 572, 624, 716)

delta   area formed from soil deposited by a river at its mouth (pp. 42, 457, 516, 654, 719)

democracy   form of government in which citizens choose the nation's leaders by voting for them (pp. 79, 131, 431)

desalinization   process used to make seawater drinkable (pp. 95, 493)

desertification   process by which grasslands change to desert (p. 548)

developed country   country in which a great deal of manufacturing is carried out (pp. 94, 609)

developing country   country that is working toward industrialization (p. 94)

devolution   transfer of certain powers from the central government to regional governments (p. 293)

diagram   drawing that shows steps in a process or parts of an object (p. 16)

dialect   local form of a language that differs from the main language in pronunciation or the meaning of words (pp. 78, 327)

Diaspora   collective name for scattered Jewish settlements around the world (p. 485)

dictator   individual who takes control of a government and rules the country as he or she wishes (pp. 79, 568)

dictatorship   government under the control of one all-powerful leader (pp. 465, 729)

dike   high banks of soil built along rivers to control floods (p. 672)

dominion   self-governing nation that accepts the British monarch as head of state (p. 157)

drought   long period of extreme dryness and water shortages (pp. 56, 547, 597)

dry farming   method in which the land is left unplanted every few years so that it can store moisture (p. 324)

dynasty   line of rulers from the same family (pp. 680, 703)

dzong   Buddhist center of prayer and study in Bhutan (p. 659)

earthquake   violent and sudden movement of the earth's crust (p. 36)

economic system   system that sets rules for how people decide what goods and services to produce and how they are exchanged (p. 80)

ecosystem   place where the plants and animals are dependent upon one another and their surroundings for survival (p. 96)

eco-tourist   person who travels to another country to view its natural wonders (pp. 209, 589)

elevation   height above sea level (pp. 12, 40, 336, 517)

elevation profile   cutaway diagram showing changes in elevation of land (p. 17)

El Niño   combination of temperature, wind, and water effects in the Pacific Ocean that causes heavy rains in some areas and drought in others (p. 57)

embargo   order that restricts or prohibits trade with another country (pp. 218, 498)

emigrate   to move to another country (pp. 88, 349)

empire   group of lands under one ruler (pp. 261, 687)

enclave   small territory entirely surrounded by a larger territory (pp. 511, 612)

endangered species   plant or animal under the threat of completely dying out (p. 594)

environment   natural surroundings (p. 25)

equinox   day when day and night are of equal length in both hemispheres (p. 32)

erg   huge area of shifting sand dunes in the Sahara (p. 466)

erosion   process of moving weathered material on the earth's surface (pp. 37, 97)

escarpment   steep cliff between higher and lower land (pp. 228, 582, 608)

estancia   ranch (p. 236)

ethnic cleansing   forcing people from a different ethnic group to leave their homes (p. 383)

ethnic group   people who share a common culture, language, or history (pp. 78, 133, 427)

evaporation   process in which the sun's heat turns liquid water into water vapor (p. 49)

exclave   small part of a country that is separated from the main part (p. 614)

exile   inability to live in one's own country because of political beliefs (p. 683)

export   to trade goods to other countries (p. 92)

famine   lack of food (pp. 85, 707)

fault   crack in the earth's crust (pp. 37, 510, 382, 673)

favela   slum area (p. 231)

federal republic   government divided between national and state powers (pp. 131, 191, 309, 431)

fellahin   farmers in Egypt who live in villages and work on small plots of land that they rent from landowners (p. 462)

fjord   steep-sided valley cut into mountains by the action of glaciers (pp. 345, 755)

foothill   low hill at the base of a mountain range (p. 261)

fossil fuel   coal, oil, or natural gas (p. 127)

free enterprise system   economic system in which people start and run businesses with limited government intervention (pp. 123, 430, 583)

free port   place where goods can be loaded, stored, and shipped again without needing to pay any import taxes (p. 723)

free trade taking down trade barriers so that goods flow freely among countries (pp. 93, 128)

free trade zone area where people can buy goods from other countries without paying extra taxes (p. 218)

gaucho cowhand (p. 236)

geographic information systems (GIS) special software that helps geographers gather and use information about a place (pp. 10, 26)

geography the study of the earth in all its variety (p. 23)

geothermal energy electricity produced by natural underground sources of steam (pp. 356, 756)

geyser spring of water heated by molten rock inside the earth so that, from time to time, it shoots hot water into the air (pp. 356, 754)

glacier giant slow-moving sheets of ice (pp. 38, 50, 144)

Global Positioning System (GPS) group of satellites that travels around the earth, which can be used to locate exact places on the earth (p. 25)

globe spherical model of Earth (p. 4)

great circle route ship or airplane route following a great circle; the shortest distance between two points on the earth (p. 6)

greenhouse effect buildup of certain gases in the atmosphere that, like a greenhouse, hold more of the sun's warmth (p. 60)

grid system network of imaginary lines on the earth's surface, formed by the crisscrossing patterns of the lines of latitude and longitude (p. 6)

groundwater water that fills tiny cracks and holes in the rock layers below the earth's surface (p. 51)

habitat type of environment in which a particular animal species lives (p. 589)

hacienda large ranch (p. 190)

hajj religious journey to Makkah that Muslims are expected to make at least once during their lifetime if they are able to do so (p. 494)

harmattan dry, dusty wind that blows south from the Sahara (p. 541)

heavy industry manufactured goods such as machinery, mining equipment, and steel (pp. 350, 413)

hemisphere one-half of the globe; the Equator divides the earth into Northern and Southern Hemispheres; the Prime Meridian divides it into Eastern and Western Hemispheres (p. 5)

hieroglyphics form of writing that uses signs and symbols (pp. 189, 461)

high island Pacific island formed by volcanic activity (p. 768)

high-technology industry industry that produces computers and other kinds of electronic equipment (p. 685)

high veld flat, grass-covered plains on the interior plateau of South Africa (p. 608)

Holocaust systematic murder of more than 6 million European Jews by Adolf Hitler and his followers during World War II (pp. 308, 485)

human rights basic freedoms and rights that all people should enjoy (p. 683)

humid continental climate weather pattern characterized by long, cold, and snowy winters and short, hot summers (p. 66)

humid subtropical climate weather pattern characterized by hot, humid, rainy summers and short, mild winters (p. 67)

hurricane violent tropical storm with high winds and heavy rains (pp. 56, 185, 207)

hydroelectric power electricity generated by flowing water (pp. 242, 414, 565, 756)

iceberg chunk of a glacier that has broken away and floats free in the ocean (p. 773)

ice shelf layer of ice above water in Antarctica (p. 773)

immigrant person who moves to a new country to make a permanent home (pp. 132, 751)

import to buy goods from another country (p. 93)

industrialize to change an economy to rely more on manufacturing and less on farming (pp. 195, 425, 571)

inflation overall increase in the price of goods across the entire economy (p. 232)

infrastructure transportation and communication networks on which an economy depends (p. 309)

intensive cultivation growing crops on every available piece of land (p. 698)

intermediate direction　　any direction between the cardinal directions, such as southeast or northwest (p. 11)

invest　　to put money into a business (p. 676)

irrigation　　farming practice followed in dry areas to collect water and bring it to crops (p. 97)

Islamic republic　　government run by Muslim religious leaders (p. 499)

island　　body of land smaller than a continent and surrounded by water (p. 40)

isthmus　　narrow piece of land that connects two larger pieces of land (pp. 40, 205)

jute　　plant fiber used for making rope, burlap bags, and carpet backing (p. 647)

kibbutz　　settlement in Israel where the people share property and produce goods (p. 483)

krill　　tiny, shrimplike animal that lives in the waters off Antarctica and is the food for many other creatures (p. 774)

lagoon　　shallow pool of water surrounded by reefs, sandbars, or atolls (p. 662)

land bridge　　narrow strip of land that joins two larger landmasses (p. 181)

landfill　　area where trash companies dump the waste they collect (p. 127)

landform　　individual features of the land (p. 24)

landlocked　　country with no land bordering a sea or an ocean (pp. 247, 266, 375, 511)

La Niña　　pattern of unusual weather in the Pacific Ocean that has the opposite effects of El Niño (p. 57)

latitude　　location north or south of the Equator, measured by imaginary lines (parallels) numbered in degrees north or south (pp. 5, 184)

leap year　　year that has an extra day; occurs every fourth year (p. 31)

life expectancy　　the number of years that an average person is expected to live (p. 431)

light industry　　making of such goods as clothing, shoes, furniture, and household products (p. 413)

line graph　　graph in which one or more lines represent changing quantities over time (p. 14)

literacy rate　　percentage of people who can read and write (p. 208)

llanos　　grassy plains (pp. 242, 256)

local wind　　pattern of wind caused by landforms in a particular area (p. 58)

loch　　narrow bay that reaches far inland (p. 290)

loess　　fertile, yellow-gray soil deposited by wind and water (p. 672)

longitude　　location east or west of the Prime Meridian, measured by imaginary lines (meridians) numbered in degrees east or west (p. 6)

low island　　Pacific island formed of coral and having little vegetation (p. 768)

magma　　hot melted rock that sometimes flows to the earth's surface in a volcanic eruption (p. 35)

mainland　　the major part of a country (p. 336)

mangrove　　tropical tree with roots that extend both above and beneath the water (p. 541)

mantle　　rock layer about 1,800 miles (2,897 km) thick between the core and the crust (p. 35)

manuka　　small shrub of New Zealand (p. 754)

map key　　code that explains the lines, symbols, and colors used on a map (p. 10)

maquiladora　　factory that assembles parts made in other countries (p. 185)

marine west coast climate　　weather pattern characterized by rainy and mild winters and cool summers (p. 65)

marsupial　　mammal that carries its young in a pouch (p. 749)

Mediterranean climate　　weather pattern characterized by mild, rainy winters and hot, dry summers (p. 66)

megalopolis　　pattern of heavy urban settlement over a large area (pp. 116, 700)

mestizo　　person with mixed Spanish and Native American or African background (pp. 190, 258)

migrant worker　　person who travels from place to place when extra help is needed to plant or harvest crops (p. 196)

migrate　　to move from one place to another (p. 480)

GLOSSARY

GLOSSARY

monarchy form of government in which a king or queen inherits the right to rule a country (p. 79)

monotheism belief in one God (p. 486)

monsoon seasonal wind that blows over a continent for months at a time (pp. 56, 646, 705, 715)

moor treeless, windy highland area with damp ground (pp. 290, 354)

moshav settlement in Israel where people share property but also own some private property (p. 484)

mosque place of worship for followers of Islam (pp. 384, 462, 479)

multilingual able to speak several languages (p. 316)

multinational firm that does business in several countries (p. 316)

mural wall painting (p. 189)

nationalism desire of a territory or colony to become an independent nation (p. 351)

national park area set aside to protect wilderness and wildlife and for recreation (p. 135)

nature preserve protected areas for plants and animals (pp. 378, 581)

natural resource product of the earth that people use to meet their needs (p. 90)

navigable describes a body of water wide and deep enough to allow the passage of ships (pp. 126, 260, 301, 324)

neutrality refusing to take sides in disagreements and wars between countries (p. 310)

newsprint type of paper used for printing newspapers (p. 153)

nomads people who move from place to place with herds of animals (pp. 377, 516, 687)

nonrenewable resource natural resource such as metals or minerals that cannot be replaced (p. 91)

nuclear energy power made by creating a controlled atomic reaction (p. 431)

oasis green area in a desert fed by underground water (pp. 459, 493, 518)

oil shale rock that contains oil (p. 372)

orbit path that a body in the solar system travels around the sun (p. 29)

outback inland regions of Australia (p. 748)

overgraze to allow animals to strip areas so bare that plants cannot grow back (p. 547)

ozone type of oxygen that forms a layer in the atmosphere and protects all living things on the earth from certain harmful rays of the sun (p. 775)

parliamentary democracy government in which voters elect representatives to a law-making body, which chooses a prime minister to head the government (pp. 157, 210, 292)

parliamentary republic see *parliamentary democracy* (p. 327)

peat wet ground with decaying plants that can be dried and used for fuel (pp. 296, 372)

peninsula piece of land with water on three sides (pp. 40, 146, 182, 408)

permafrost permanently frozen lower layers of soil in the tundra and subarctic regions (pp. 68, 410)

pesticides powerful chemicals that kill crop-destroying insects (pp. 96, 648)

phosphate mineral salt used in fertilizers (pp. 461, 555, 769)

pictograph graph in which small symbols represent quantities (p. 15)

pidgin language language formed by combining elements of several different languages (p. 767)

plain low-lying stretch of flat or gently rolling land (p. 40)

plantain kind of banana (p. 592)

plantation large farm that grows a single crop for sale (pp. 186, 207)

plateau flat land with higher elevation than a plain (pp. 40, 324)

plate huge slab of rock that makes up the earth's crust (pp. 599, 726)

plate tectonics theory that the earth's crust is not an unbroken shell but consists of plates, or huge slabs of rock, that move (p. 35)

plaza public square (p. 193)

poaching illegal hunting of protected animals (p. 581)

polder area of land reclaimed from the sea (p. 315)

pope head of the Roman Catholic Church (pp. 334, 369)

population density   average number of people living in a square mile or square kilometer (p. 86)

potash   type of mineral salt that is often used in fertilizers (p. 388)

prairie   rolling, inland grassy area with very fertile soil (p. 146)

precipitation   water that falls back to the earth as rain, snow, sleet, or hail (p. 50)

prime minister   official who heads the government in a parliamentary democracy (p. 157)

privatize   to transfer the ownership of factories from the government to individual citizens (p. 380)

projection   in mapmaking, a way of drawing the round Earth on a flat surface (p. 7)

province   regional political division similar to states (p. 143)

quota   number limit on how many items of a particular product can be imported from a particular country (p. 93)

rain forest   dense forest that receives high amounts of rain each year (p. 61)

rain shadow   dry area on the inland side of coastal mountains (p. 59)

recycling   reusing materials instead of throwing them out (p. 128)

refugee   person who flees to another country to escape persecution or disaster (pp. 88, 383, 568, 595)

region   area that shares common characteristics (p. 25)

relief   differences in height in a landscape; how flat or rugged the surface is (p. 12)

Renaissance   period of great achievement in art and learning that began in Italy in the 1300s and spread throughout Europe (p. 333)

renewable resource   natural resource that cannot be used up or can fairly quickly be replaced naturally or grown again (p. 91)

republic   strong national government headed by elected leaders (pp. 210, 233, 258, 303, 369, 461)

reunification   bringing together the two parts of Germany under one government (p. 309)

revolution   one complete orbit around the sun (p. 31)

rural   area in the countryside (pp. 134, 433)

saga   long story (p. 357)

samurai   powerful land-owning warriors in Japan (p. 699)

sauna   wooden room heated by water sizzling on hot stones (p. 351)

savanna   broad grassland in the tropics with few trees (pp. 64, 541, 563)

scale   relationship between distance on a map and actual distance on the earth (p. 11)

scale bar   on a map, a divided line showing the map scale, usually in miles or kilometers (p. 11)

secede   to withdraw from a national government (pp. 131, 154)

secular   nonreligious (pp. 467, 480)

selva   tropical rain forests in Brazil (p. 228)

serf   farm laborer who could be bought and sold along with the land (p. 425)

service industry   business that provides services to people instead of producing goods (pp. 124, 150, 195)

shah   title given to kings who ruled Iran (p. 499)

shogun   military leaders in Japan (p. 699)

silt   small particles of rich soil (p. 458)

sirocco   hot, dry winds that blow across Italy from North Africa (p. 331)

sisal   plant fiber used to make rope and twine (p. 588)

skerry   rocky island (p. 349)

slash-and-burn farming   method of clearing land for planting by cutting and burning forest (p. 624)

smog   thick haze of fog and chemicals (p. 198)

socialism   economic system in which many businesses are owned and run by the government (p. 717)

solar system   Earth, eight other planets, and thousands of smaller bodies that all revolve around the sun (p. 29)

sodium nitrate   chemical used in fertilizer and explosives (p. 269)

sorghum   tall grass with seeds that are used as grain and to make syrup (p. 619)

spa   resort that has hot mineral springs that people bathe in to regain their health (p. 378)

GLOSSARY

GLOSSARY

station   cattle or sheep ranch in Australia (p. 748)

steppe   partly dry grassland often found on the edges of a desert (pp. 69, 386, 406, 515, 570, 687)

strait   narrow body of water between two pieces of land (pp. 42, 721)

subarctic climate   weather pattern characterized by severely cold and bitter winters and short, cool summers (p. 67)

subcontinent   large landmass that is part of another continent but distinct from it (p. 645)

subsistence farm   small plot where a farmer grows only enough food to feed his own family (pp. 186, 207, 261, 542)

suburb   smaller community that surrounds a city (pp. 134, 340, 432)

summer solstice   day with the most hours of sunlight and the fewest hours of darkness (p. 32)

taiga   huge forests of evergreen trees that grow in subarctic regions (p. 410)

tannin   substance used in processing leather (p. 236)

tariff   tax added to the value of goods that are imported (p. 93)

terraced field   strips of land cut out of a hillside like stair steps so the land can hold water and be used for farming (pp. 678, 724)

timberline   elevation along mountains above which no trees grow (p. 70)

tornado   funnel-shaped windstorm that sometimes forms during a severe thunderstorm (p. 56)

townships   crowded neighborhoods outside cities in South Africa where most nonwhites live (p. 611)

trench   valley in the ocean floor (p. 41)

tributary   small river that flows into a larger river (p. 653)

tropics   low-latitude region between the Tropic of Cancer and the Tropic of Capricorn (p. 55)

trust territory   area temporarily placed under control of another nation (p. 769)

tsetse fly   insect whose bite can kill cattle or humans with a deadly disease called sleeping sickness (p. 571)

tsunami   huge sea wave caused by an earthquake on the ocean floor (pp. 36, 695)

tundra   vast rolling treeless plain in high latitude climates in which only the top few inches of ground thaw in summer (pp. 68, 146, 410)

tungsten   metal used in electrical equipment (p. 677)

typhoon   name for hurricane in Asia (pp. 56, 673, 769)

urban   area in the city (pp. 134, 432)

urbanization   movement to cities (p. 87)

vaquero   cowhand (p. 185)

wadi   dry riverbed filled by rainwater from rare downpours (p. 493)

water cycle   process in which water moves from the oceans to the air to the ground and finally back to the oceans (p. 49)

watershed   region drained by a river (p. 594)

water vapor   water in the form of gas (p. 49)

weather   unpredictable changes in the air that take place over a short period of time (p. 53)

weathering   process that breaks surface rocks into boulders, gravel, sand, and soil (p. 37)

welfare state   country that uses tax money to support people who are sick, needy, jobless, or retired (pp. 246, 349)

winter solstice   day with the fewest hours of sunlight and the most hours of darkness (p. 32)

yurt   large circle-shaped tent made of animal skins that can be packed up and moved from place to place (p. 688)

absolute location/ubicación absoluta
posición exacta de un lugar en la superficie de la Tierra (pág. 6)

acid rain/lluvia ácida   lluvia que contiene grandes cantidades de contaminantes químicos (págs. 97, 127, 154, 368)

adobe/adobe   ladrillos secados al Sol (pág. 193)

alluvial plain/llanura aluvial   área creada por el suelo fértil que se acumula después de las inundaciones causadas por los ríos (pág. 497)

altiplano/altiplano   meseta grande y muy elevada; también se llama altiplanicie (pág. 260)

altitude/altitud   altura sobre el nivel del mar (págs. 184, 242, 583)

anthracite/antracita   tipo de carbón mineral muy duro (pág. 705)

apartheid/apartheid   sistema de leyes que separaba los grupos raciales y étnicos y limitaba los derechos de la población negra (pág. 611)

aquifer/manto acuífero   capa de rocas subterránea en que el agua es tan abundante que corre entre las rocas (págs. 51, 465)

archipelago/archipiélago   grupo de islas (págs. 213, 353, 695)

atoll/atolón   isla de muy poca elevación que se forma alrededor de una laguna en la forma de un anillo (págs. 662, 768)

atmosphere/atmósfera   capa de aire que rodea la Tierra (pág. 30)

autobahn/autobahn   autopista muy rápida (pág. 307)

autonomy/autonomía   gobernarse por sí mismo (pág. 593)

axis/eje terrestre   línea imaginaria que atraviesa el centro de la Tierra entre el Polo Norte y el Polo Sur (pág. 31); también la línea vertical (del lado) u horizontal (de abajo) de una gráfica que se usa para medir (pág. 14)

bar graph/gráfica de barras   gráfica en que franjas verticales u horizontales representan cantidades (pág. 14)

basin/cuenca   área baja rodeada de tierras más elevadas (págs. 227, 564)

Véase también el Diccionario Geográfico en la página 18.

bauxite/bauxita   mineral que se usa para hacer aluminio (págs. 215, 376, 414, 555)

bazaar/bazar   mercado (pág. 462)

Bedouins/beduinos   gente nomádica del desierto del sudoeste de Asia (pág. 489)

bilingual/bilingüe   se refiere a un país que tiene dos idiomas oficiales (págs. 158, 517)

birthrate/índice de natalidad   número de niños que nace cada año por cada mil personas (pág. 85)

Boers/bóers   los holandeses que fueron los primeros colonos en Sudáfrica (pág. 610)

bog/ciénaga   tierra baja y pantanosa (págs. 296, 365)

boomerang/bumerán   arma australiana con forma de ala que se lanza para que golpee un objetivo o de la vuelta y caiga a los pies de la persona que la lanzó (pág. 751)

bush/campo   áreas rurales de Australia (pág. 752)

cacao/cacao   árbol tropical cuyas semillas se usan para hacer chocolate y cocoa (págs. 543, 766)

calligraphy/caligrafía   el arte de escribir con letra muy bella (pág. 684)

campesino/campesino   agricultor (pág. 259)

canopy/bóveda   techo formado por las copas de los árboles en los bosques húmedos (págs. 64, 208, 564)

cardinal directions/puntos cardinales   cuatro direcciones básicas en la Tierra: norte, sur, este, oeste (pág. 11)

casbah/casbah   la sección antigua de las ciudades de Argelia; también se llama *alcazaba* (pág. 468)

cash crop/cultivo comercial   producto que se cultiva para exportación (págs. 258, 513)

cassava/yuca   planta con raíces que se pueden convertir en harina para hacer pan o que se pueden cocinar de otras formas (págs. 552, 584)

caste/casta   clase social basada en la ascendencia de una persona (pág. 648)

caudillo/caudillo   gobernante militar (pág. 242)

channel/canal   una masa de agua entre dos tierras que tiene más anchura que un estrecho (pág. 42)

chart/cuadro   manera gráfica de presentar información con claridad (pág. 16)

circle graph/gráfica de círculo gráfica redonda que muestra como un todo es dividido (pág. 15)

city-state/ciudad estado ciudad junto con las tierras que la rodean (pág. 333)

civilizations/civilizaciones culturas altamente desarrolladas (pág. 81)

civil war/guerra civil pelea entre distintos grupos dentro de un país (págs. 467, 490, 545)

clan/clan grupo de personas que están emparentadas (págs. 517, 600, 698)

climate/clima el patrón que sigue el estado del tiempo en un área durante muchos años (pág. 54)

climograph/gráfica de clima gráfica que combina barras y líneas para dar información sobre la temperatura y la precipitación (pág. 15)

coalition government/gobierno por coalición gobierno en que dos o más partidos trabajan juntos para dirigir un país (págs. 334, 649)

cold war/guerra fría período entre los fines de los 1940 y los fines de los 1980 en que los Estados Unidos y la Unión Soviética compitieron por tener influencia mundial sin pelear uno contra el otro (pág. 427)

collection/drenaje proceso durante el ciclo hidrológico en que los ríos llevan el agua de regreso a los océanos (pág. 50)

colony/colonia territorio o poblado con lazos a un país extranjero (págs. 131, 156, 190, 215, 327)

commonwealth/estado libre asociado territorio que en parte se gobierna por sí solo (pág. 219)

communist state/estado comunista país cuyo gobierno mantiene mucho control sobre la economía y la sociedad en su totalidad (págs. 216, 308, 367, 426, 675)

compass rose/rosa náutica dibujo en los mapas que muestra las direcciones; también se llama rosa de los vientos (pág. 11)

compound/complejo residencial grupo de viviendas rodeada por una muralla (pág. 544)

condensation/condensación proceso en que el aire sube y se enfría, lo cual hace que el vapor de agua que contiene se convierta de nuevo en líquido (pág. 50)

conservation/conservación uso juicioso de los recursos para no malgastarlos (pág. 96)

constitutional monarchy/monarquía constitucional gobierno en que un rey o reina es el jefe de estado oficial pero los gobernantes son elegidos (págs. 293, 469, 491, 700)

consumer goods/bienes de consumo productos para la casa, ropa y otras cosas que la gente compra para su uso personal (págs. 382, 413, 676)

contiguous/contiguas áreas adyacentes dentro de la misma frontera (pág. 115)

continent/continente masa de tierra inmensa (pág. 35)

continental island/isla continental isla formada cuando un pedazo de tierra se separa de un continente o cuando la erosión hace que el agua cubra la parte del continente que antes unía la isla a éste (pág. 765)

continental shelf/plataforma continental meseta formada por parte de un continente que se extiende por varias millas debajo del mar (pág. 41)

contour line/curva de nivel línea que conecta todos los puntos a la misma elevación en un mapa de relieve (pág. 12)

cooperative/cooperativa granja que es propiedad y es operada por el gobierno (pág. 217)

copper belt/cinturón de cobre área extensa de minas de cobre en el norte de Zambia (pág. 618)

copra/copra pulpa seca del coco que se usa para hacer margarina, jabón y otros productos (pág. 776)

coral reef/arrecife coralino estructura formada al nivel del mar o cerca de éste por los esqueletos de pequeños animales marinos (págs. 119, 581, 747)

cordillera/cordillera grupo de cadenas paralelas de montañas (págs. 147, 255)

core/núcleo centro de la Tierra, que está formado de hierro caliente y otros metales (pág. 35)

cottage industry/industria familiar industria basada en una casa o aldea en que los miembros de la familia usan sus propias herramientas para hacer productos (pág. 648)

crevasse/grieta rajadura profunda en el casquete de hielo de la Antártida (pág. 772)

crop rotation/rotación de cultivos variar lo que se siembra en un terreno para no agotar todos los minerales que tiene el suelo (pág. 97)

crust/corteza capa de afuera de la Tierra (pág. 35)

cultural diffusion/difusión cultural el proceso de esparcir nuevos conocimientos y habilidades a otras culturas (pág. 81)

culture/cultura modo de vida de un grupo de personas que comparten creencias y costumbres similares (pág. 77)

culture region/región cultural área del mundo que incluye muchos países que tienen los mismos rasgos culturales (pág. 82)

currency/moneda tipo de dinero (pág. 292)

current/corriente movimiento de las aguas del mar que afecta el clima de las masas de tierra (pág. 57)

cyclone/ciclón   tormenta violenta con vientos muy fuertes y mucha lluvia (págs. 623, 654)

czar/zar   título de los antiguos emperadores rusos (pág. 424)

death rate/índice de mortalidad   número de personas de cada mil que mueren en un año (pág. 84)

deforestation/deforestación   la extensa destrucción de los bosques (págs. 97, 572, 624, 716)

delta/delta   área formada por el suelo que deposita un río en su desembocadura (págs. 42, 457, 516, 654, 719)

democracy/democracia   tipo de gobierno en que los ciudadanos seleccionan los líderes de la nación por medio del voto (págs. 79, 131, 431)

desalinization/desalinización   proceso de hacer el agua de mar potable (págs. 95, 493)

desertification/desertización   proceso por el cual los pastos se convierten en desiertos (pág. 548)

developed country/país desarrollado   país donde hay mucha manufactura de productos (págs. 94, 609)

developing country/país en vías de desarrollo   país que está industrializándose (pág. 94)

devolution/devolución   transferencia de ciertos poderes del gobierno central a los gobiernos regionales (pág. 293)

diagram/diagrama   dibujo que muestra los pasos en un proceso o las partes de un objeto (pág. 16)

dialect/dialecto   forma local de un idioma que se diferencia del idioma normal por su pronunciación o por el sentido de algunas palabras (págs. 78, 327)

Diaspora/diáspora   nombre dado a los poblados judíos alrededor del mundo (pág. 485)

dictator/dictador   individuo que toma control de un país y lo gobierna como quiere (págs. 79, 568)

dictatorship/dictadura   gobierno bajo el control de un líder que tiene todo el poder (págs. 465, 729)

dike/dique   muros de tierra muy altos construidos a lo largo de los ríos para controlar las inundaciones (pág. 672)

dominion/dominio   naciones que se gobiernan por sí solas que aceptan al monarca británico como jefe de estado (pág. 157)

drought/sequía   largos períodos de sequedad y de escasez de agua (págs. 56, 547, 597)

dry farming/agricultura en seco   método de cultivar en que la tierra se deja sin sembrar cada varios años para que almacene humedad (pág. 324)

dynasty/dinastía   serie de gobernantes de la misma familia (págs. 680, 703)

dzong/dzong   centro budista en Bután para rezar y estudiar (pág. 659)

earthquake/terremoto   movimiento violento e inesperado de la corteza de la Tierra (pág. 36)

economic system/sistema económico   sistema que establece reglas que determinan cómo las personas deciden cuáles bienes y servicios van a producir y cómo los van a intercambiar (pág. 80)

ecosystem/ecosistema   lugar en el cual las plantas y animales dependen unos de otros y de sus alrededores para sobrevivir (pág. 96)

eco-tourist/ecoturista   persona que viaja a otro país para ver sus bellezas naturales (págs. 209, 589)

elevation/elevación   altura por encima del nivel del mar (págs. 12, 40, 336, 517)

elevation profile/perfil de elevaciones   diagrama que muestra los cambios en la elevación de la tierra como si se hubiera hecho un corte vertical del área (pág. 17)

El Niño/El Niño   combinación de la temperatura, los vientos y los efectos del agua en el océano Pacífico que causa lluvias fuertes en algunas áreas y sequía en otras (pág. 57)

embargo/embargo   orden que limita o prohibe el comercio con otro país (págs. 218, 498)

emigrate/emigrar   mudarse a otro país (págs. 88, 349)

empire/imperio   grupo de países bajo un gobernante (págs. 261, 687)

enclave/enclave   territorio pequeño totalmente rodeado por un territorio más grande (págs. 511, 612)

endangered species/especie en vías de extinción   planta o animal que está en peligro de desaparecer completamente (pág. 594)

environment/medio ambiente   alrededores naturales (pág. 25)

equinox/equinoccio   día en que el día y la noche tienen la misa duración en los dos hemisferios (pág. 32)

erg/ergio   inmensas áreas en el Sahara en que se mueven las dunas de arena (pág. 466)

erosion/erosión   proceso de mover los materiales desgastados en la superficie de la Tierra (págs. 37, 97)

escarpment/escarpa   acantilado empinado entre una área baja y una alta (págs. 228, 582, 608)

SPANISH GLOSSARY

estancia/estancia　rancho (pág. 236)

ethnic cleansing/limpieza étnica　forzar a personas de un grupo étnico distinto a abandonar el lugar donde viven (pág. 383)

ethnic group/grupo étnico　personas que comparten una cultura, idioma o historia en común (págs. 78, 133, 427)

evaporation/evaporación　proceso mediante el cual el calor del sol convierte el agua líquida en vapor de agua (pág. 49)

exclave/territorio externo　parte pequeña de un país que está separada de la parte principal (pág. 614)

exile/exilio　tener que vivir fuera de su país nativo por causa de sus creencias políticas (pág. 683)

export/exportar　comerciar y mandar bienes a otros países (pág. 92)

famine/hambruna　falta de alimentos (págs. 85, 707)

fault/falla　fractura en la corteza de la Tierra (págs. 37, 510, 582, 673)

favela/favela　barrio pobre y deteriorado (pág. 231)

federal republic/república federal　nación en que el poder está dividido entre el gobierno nacional y el de los estados (págs. 131, 191, 309, 431)

fellahin/felás　granjeros en Egipto que viven en aldeas y cultivan pequeños terrenos que arriendan de un hacendado (pág. 462)

fjord/fiordo　valle creado por el movimiento de glaciares en las montañas que deja laderas sumamente empinadas (págs. 345, 755)

foothill/estribaciones　colinas bajas al pie de una cadena de montañas (pág. 261)

fossil fuel/combustibles fósiles　carbón, petróleo o gas natural (pág. 127)

free enterprise system/sistema de libre empresa　sistema económico en que la gente empieza y administra negocios con poca intervención del gobierno (págs. 123, 430, 583)

free port/puerto libre　lugar donde las mercancías se pueden cargar, almacenar y embarcar de nuevo sin tener que pagar derechos de importación (pág. 723)

free trade/libre comercio　eliminar las barreras al comercio para que se puedan mover productos libremente entre países (págs. 93, 128)

free trade zone/zona de cambio libre　área donde la gente puede comprar bienes de otros países sin pagar impuestos adicionales (pág. 218)

gaucho/gaucho　vaquero (pág. 236)

geographic information systems (GIS)/ sistemas de información geográfica (SIG)　programas de computadoras especiales que ayudan a los geógrafos a obtener y usar la información geográfica sobre un lugar (págs. 10, 26)

geography/geografía　el estudio de la Tierra y de toda su variedad (pág. 23)

geothermal energy/energía geotérmica　electricidad producida por fuentes de vapor subterráneas naturales (págs. 356, 756)

geyser/géiser　manantial de agua calentado por rocas fundidas dentro de la Tierra que, de vez en cuando, arroja agua caliente al aire (págs. 356, 754)

glacier/glaciar　capa de hielo inmensa que se mueve muy lentamente (págs. 38, 50, 144)

Global Positioning System (GPS)/Sistema global de posición (GPS)　grupo de satélites que le dan la vuelta a la Tierra y se usan para localizar lugares exactos en la Tierra (pág. 25)

globe/globo terráqueo　modelo esférico de la Tierra (pág. 4)

great circle route/línea de rumbo　ruta que sigue un círculo máximo; usada por aviones y barcos porque es la distancia más corta entre dos puntos en la Tierra (pág. 6)

greenhouse effect/efecto invernadero　la acumulación de ciertos gases en la atmósfera que mantienen más del calor del Sol, como hace un invernadero (pág. 60)

grid system/sistema de coordenadas geográficas　red de líneas imaginarias en la superficie de la Tierra, formada por las líneas de latitud y longitud que se cruzan (pág. 6)

groundwater/agua subterránea　agua que llena las rajaduras y hoyos en las capas de roca debajo de la superficie de la Tierra (pág. 51)

habitat/hábitat　tipo de ambiente en que vive una especie animal en particular (pág. 589)

hacienda/hacienda　un rancho grande (pág. 190)

hajj/*hajj*　viaje religioso a La Meca que todo musulmán debe hacer por lo menos una vez en la vida si puede (pág. 494)

harmattan/harmattan　viento seco y lleno de polvo que sopla hacia el sur desde el Sahara (pág. 541)

heavy industry/industria pesada　manufactura de productos como maquinaria, equipo de minería y acero (págs. 350, 413)

hemisphere/hemisferio　una mitad del globo terráqueo; el ecuador divide la Tierra en los hemisferios norte y sur; el primer meridiano la divide en hemisferios este y oeste (pág. 5)

hieroglyphics/jeroglíficos　forma de escribir que usa signos y símbolos (págs. 189, 461)

high island/isla oceánica　isla del Pacífico formada por actividad volcánica (págs. 768)

high-technology industry/industria de alta tecnología　industria que produce computadoras y otras clases de equipo electrónico (pág. 685)

high veld/veld　pastos llanos de la meseta interior de Sudáfrica (pág. 608)

Holocaust/Holocausto　la matanza sistemática de más de 6 millones de judíos europeos por Adolfo Hitler y sus seguidores durante la Segunda Guerra Mundial (págs. 308, 485)

human rights/derechos humanos　libertades y derechos básicos que todas las personas deben disfrutar (pág. 683)

humid continental climate/clima húmedo continental　patrón del estado del tiempo con inviernos largos, fríos y con mucha nieve y veranos cortos y calurosos (pág. 66)

humid subtropical climate/clima húmedo subtropical　patrón del estado del tiempo con veranos calurosos, húmedos y lluviosos e inviernos cortos y templados (pág. 67)

hurricane/huracán　tormenta tropical violenta con vientos y lluvias fuertes (págs. 56, 185, 207)

hydroelectric power/energía hidroeléctrica　electricidad generada por una corriente de agua (págs. 242, 414, 565, 756)

iceberg/iceberg　pedazo de un glaciar que se ha desprendido y flota libremente en los océanos (pág. 773)

ice shelf/plataforma de hielo　capa de hielo sobre el mar en la Antártida (pág. 773)

immigrant/inmigrante　persona que se muda permanentemente a un país nuevo (págs. 132, 751)

import/importar　comprar productos de otro país (pág. 93)

industrialize/industrializar　cambiar una economía de manera que dependa más de la manufactura que de la agricultura (págs. 195, 425, 571)

inflation/inflación　aumento en el precio de productos en toda la economía (pág. 232)

infrastructure/infraestructura　redes de transporte y comunicación de las cuales depende una economía (pág. 309)

intensive cultivation/cultivo intensivo　labrar toda la tierra posible (pág. 698)

intermediate direction/puntos intermedios　cualquier dirección entre los puntos cardinales, como sudeste o noroeste (pág. 11)

invest/invertir　poner dinero en un negocio (pág. 676)

irrigation/irrigación　práctica agrícola en áreas secas de colectar agua y llevarla a los cultivos (pág. 97)

Islamic republic/república islámica　gobierno dirigido por líderes musulmanes (pág. 499)

island/isla　masa de tierra más pequeña que un continente, rodeada de agua (pág. 40)

isthmus/istmo　lengua de tierra que conecta a dos masas de tierra más grandes (págs. 40, 205)

jute/yute　fibras de una planta que se usan para hacer soga, sacos y el revés de alfombras (pág. 647)

kibbutz/kibutz　poblado en Israel donde las personas comparten la propiedad y producen bienes (pág. 483)

krill/krill　animales diminutos parecidos a los camarones que viven en las aguas alrededor de la Antártida y sirven de alimento para muchos otros animales (pág. 774)

lagoon/laguna　masa de agua poco profunda rodeada por arrecifes, bancos de arena o un atolón (pág. 662)

land bridge/puente de tierra　franja de tierra que une a dos masas de tierra mayores (pág. 181)

landfill/vertedero de basura　lugar donde las compañías que recogen la basura botan los residuos que colectan (pág. 127)

landform/accidente geográfico　característica particular de la tierra (pág. 24)

landlocked/rodeado de tierra　país que no tiene tierras bordeadas por un mar u océano (págs. 247, 266, 375, 511)

La Niña/La Niña　patrón infrecuente en el estado del tiempo del océano Pacífico que tiene los efectos contrarios a los de El Niño (pág. 57)

SPANISH GLOSSARY

SPANISH GLOSSARY

latitude/latitud   posición al norte o al sur del ecuador, medida por medio de líneas imaginarias (paralelos) numeradas con grados norte o sur (págs. 5, 184)

leap year/año bisiesto   año que tiene un día adicional; cada cuarto año (pág. 31)

life expectancy/expectativas de vida   el número de años que se espera que viva la persona promedio (pág. 431)

light industry/industria ligera   fabricación de productos como muebles, ropa, zapatos y artículos para el hogar (pág. 413)

line graph/gráfica lineal   gráfica en que una o varias líneas representan cambios de cantidad a través del tiempo (pág. 14)

literacy rate/índice de alfabetización   porcentaje de personas que saben leer y escribir (pág. 208)

llanos/llanos   planicie cubierta de hierba (págs. 242, 256)

local wind/vientos locales   patrones en los vientos causados por los accidentes geográficos de un área en particular (pág. 58)

loch/rías   bahías estrechas que llegan hasta muy dentro de la tierra (pág. 290)

loess/loes   suelo amarillento y fértil depositado por el viento y el agua (pág. 672)

longitude/longitud   posición al este o el oeste del primer meridiano, medida por medio de líneas imaginarias (meridianos) numeradas con grados este u oeste (pág. 6)

low island/isla coralina   isla del Pacífico formada por coral que tiene poca vegetación (pág. 768)

magma/magma   roca caliente y fundida que a veces fluye hasta la superficie de la Tierra en erupciones volcánicas (pág. 35)

mainland/territorio continental   la parte principal de un país (pág. 336)

mangrove/mangle   árbol tropical con raíces que se extienden por encima y por debajo del agua (pág. 541)

mantle/manto   capa de rocas de 1,800 millas (2,897 km.) de grueso entre el núcleo y la corteza (pág. 35)

manuka/manuka   pequeño arbusto de Nueva Zelanda (pág. 754)

map key/leyenda   explicación de las líneas, símbolos y colores usados en un mapa; también se llama clave del mapa (pág. 10)

maquiladora/maquiladora   fábrica donde se ensamblan piezas hechas en otros países (pág. 185)

marine west coast climate/clima húmedo marítimo   patrón del estado del tiempo con inviernos lluviosos y templados y veranos frescos (pág. 65)

marsupial/marsupial   mamífero que lleva a sus crías en una bolsa (pág. 749)

Mediterranean climate/clima húmedo mediterráneo   patrón del estado del tiempo con inviernos lluviosos y templados y veranos calurosos y secos (pág. 66)

megalopolis/megalópolis   área extensa de mucha urbanización (págs. 116, 700)

mestizo/mestizo   persona cuya ascendencia incluye indios americanos o africanos y españoles (págs. 190, 258)

migrant worker/trabajador itinerante   persona que viaja a distintos lugares donde hacen falta trabajadores para sembrar y cosechar cultivos (pág. 196)

migrate/migrar   mudarse de un lugar a otro (pág. 480)

monarchy/monarquía   tipo de gobierno en que un rey o reina hereda el derecho de gobernar un país (pág. 79)

monotheism/monoteísmo   creencia en un solo Dios (pág. 486)

monsoon/monzón   vientos que soplan en un continente por varios meses seguidos en ciertas estaciones del año (págs. 56, 646, 705, 715)

moor/páramo   área elevada y sin árboles pero con mucho viento y tierra húmeda (págs. 290, 354)

moshav/*moshav*   poblados en Israel en que la gente comparte alguna propiedad pero también tiene propiedad privada (pág. 484)

mosque/mezquita   edificio de devoción islámico (págs. 384, 462, 479)

multilingual/multilingüe   que puede hablar varios idiomas (pág. 316)

multinational/multinacional   compañía que hace negocios en varios países (pág. 316)

mural/mural   pintura hecha sobre una pared (pág. 189)

nationalism/nacionalismo   deseo de un territorio o colonia de hacerse independiente (pág. 351)

national park/parque nacional   área reservada para proteger la flora y la fauna y para la recreación (pág. 135)

nature preserve/santuario natural   área protegida para las plantas y animales (págs. 378, 581)

natural resource/recurso natural   producto de la Tierra que la gente usa para satisfacer sus necesidades (pág. 90)

navigable/navegable   describe una masa de agua ancha y profunda suficiente para que los barcos puedan viajar por ella (págs. 126, 260, 301, 324)

neutrality/neutralidad   negarse a ponerse a favor de uno de los adversarios en un desacuerdo o una guerra entre países (pág. 310)

newsprint/papel de periódico   tipo de papel en que se imprimen los periódicos (pág. 153)

nomads/nómadas   gente que se muda de un lugar a otro con sus manadas o rebaños de animales (págs. 377, 516, 687)

nonrenewable resource/recurso no renovable   recurso natural, como metales o minerales, que no puede reemplazarse (pág. 91)

nuclear energy/energía nuclear   energía producida por medio de una reacción atómica controlada (pág. 431)

oasis/oasis   área verde en medio de un desierto a donde llegan aguas subterráneas (págs. 459, 493, 518)

oil shale/esquistos grasos   rocas que contienen aceite (pág. 372)

orbit/órbita   trayectoria que los cuerpos en el sistema solar siguen alrededor del Sol (pág. 29)

outback/tierra adentro   el interior de Australia (pág. 748)

overgraze/pastar excesivamente   cuando el ganado despoja los pastos hasta tal punto que las plantas no pueden crecer de nuevo (pág. 547)

ozone/ozono   tipo de oxígeno que forma una capa en la atmósfera que protege a todas las cosas vivas de ciertos rayos del Sol que son peligrosos (pág. 775)

parliamentary democracy/democracia parlamentaria   gobierno en que los votantes eligen a representantes a un cuerpo que hace las leyes y que selecciona a un primer ministro para que sea el jefe del gobierno (págs. 157, 210, 292)

parliamentary republic/república parlamentaria   *véase* parliamentary democracy/democracia parlamentaria (pág. 327)

peat/turba   suelo mojado con plantas en descomposición que se puede secar y usar para combustible (págs. 296, 372)

peninsula/península   masa de tierra con agua alrededor de tres lados (págs. 40, 146, 182, 408)

permafrost/permafrost   capa de suelo congelada en la tundra y las regiones subárticas; también se llama permagel (págs. 68, 410)

pesticides/pesticidas   sustancias químicas poderosas que matan a los insectos que destruyen los cultivos (págs. 96, 648)

phosphate/fosfato   sal mineral que se usa en los abonos (págs. 461, 555, 769)

pictograph/pictograma   gráfica en que pequeños símbolos representan cantidades (pág. 15)

pidgin language/idioma rudimentario   lenguaje formado al combinar elementos de varios idiomas distintos (pág. 767)

plain/llanura   extensión de tierra plana u ondulante a elevaciones bajas (pág. 40)

plantain/plátano de cocinar   tipo de banano (pág. 592)

plantation/plantación   granja grande en que se siembra un solo cultivo para venderse (págs. 186, 207)

plateau/meseta   planicie a elevaciones más altas que las llanuras (págs. 40, 324)

plate/placa   plancha de roca inmensa que forma parte de la corteza de la tierra (págs. 599, 726)

plate tectonics/tectónica de placas   teoría que dice que la corteza de la Tierra no es una envoltura enteriza, sino que está formada por placas, o planchas de roca inmensas, que se mueven (pág. 35)

plaza/plaza   sitio donde se reúne el público (pág. 193)

poaching/caza furtiva   cacería ilegal de animales protegidos (pág. 581)

polder/pólder   área de tierra ganada del mar (pág. 315)

pope/papa   líder de la Iglesia Católica Apostólica Romana (págs. 334, 369)

population density/densidad de población   promedio de personas que viven en una milla cuadrada o kilómetro cuadrado (pág. 86)

potash/potasa   tipo de sal mineral que a menudo se usa en los abonos (pág. 388)

prairie/pradera   área de pastos ondulantes en el interior con suelo muy fértil (pág. 146)

precipitation/precipitación   agua que regresa a la Tierra en la forma de lluvia, nieve, aguanieve o granizo (pág. 50)

prime minister/primer ministro   líder del gobierno en una democracia parlamentaria (pág. 157)

privatize/privatizar   transferir la propiedad de fábricas de las manos del gobierno a las de individuos (pág. 380)

SPANISH GLOSSARY

projection/proyección　una de las maneras de dibujar la Tierra redonda en una superficie plana para hacer un mapa (pág. 7)

province/provincia　división política regional, parecida a un estado (pág. 143)

quota/cuota　límite en la cantidad de un producto que se puede importar de un país en particular (pág. 93)

rain forest/bosque húmedo　bosque denso que recibe grandes cantidades de lluvia todos los años (pág. 61)

rain shadow/sombra pluviométrica　área seca en el lado interior de montañas costeras (pág. 59)

recycling/reciclaje　usar materiales de nuevo en vez de botarlos (pág. 128)

refugee/refugiado　persona que huye de un país a otro para evitar la persecución o un desastre (págs. 88, 383, 568, 595)

region/región　área destacada por ciertas características (pág. 25)

relief/relieve　las diferencias en altitud de una zona; lo plana o accidentada que es una superficie (pág. 12)

Renaissance/Renacimiento　período de grandes logros en el arte y el estudio de la antigüedad que comenzó en Italia en los años 1300 y continuó por toda Europa (pág. 333)

renewable resource/recurso renovable　recurso natural que no se puede gastar, que la naturaleza puede reemplazar o que se puede cultivar de nuevo (pág. 91)

republic/república　gobierno nacional fuerte encabezado por líderes elegidos (págs. 210, 233, 258, 303, 369, 461)

reunification/reunificación　juntar de nuevo las dos partes de Alemania bajo un mismo gobierno (pág. 309)

revolution/revolución　una órbita completa alrededor del Sol (pág. 31)

rural/rural　área en el campo (págs. 134, 433)

saga/saga　historia larga (pág. 357)

samurai/samurai　propietarios y guerreros poderosos del Japón (pág. 699)

sauna/sauna　cuarto de madera calentado por agua que hierve sobre piedras calientes (pág. 351)

savanna/sabana　pastos extensos en los trópicos con pocos árboles (págs. 64, 541, 563)

scale/escala　relación entre las distancias en un mapa y las distancias verdaderas en la Tierra (pág. 11)

scale bar/barra de medir la escala　en un mapa, línea con divisiones que muestra la escala del mapa, generalmente en millas o kilómetros (pág. 11)

secede/secesión　separarse de un gobierno nacional (págs. 131, 154)

secular/secular　no religioso (págs. 467, 480)

selva/selva　bosque húmedo tropical, como el de Brasil (pág. 228)

serf/siervo　labrador que podía ser comprado y vendido con la tierra (pág. 425)

service industry/industria de servicio　negocio que proporciona servicios a la gente en vez de producir productos (págs. 124, 150, 195)

shah/sha　título de los reyes que gobernaban Irán (pág. 499)

shogun/shogun　líder militar en Japón (pág. 699)

silt/cieno　pequeñas partículas de suelo fértil (pág. 458)

sirocco/siroco　vientos calurosos y secos que soplan a través de Italia desde el norte de África (pág. 331)

sisal/sisal　fibra de una planta que se usa para hacer soga y cordel (pág. 588)

skerry/*skerry guard*　isla rocosa (pág. 349)

slash-and-burn farming/agricultura por tala y quema　método de limpiar la tierra para el cultivo en que se cortan y se queman los bosques (pág. 624)

smog/smog　neblina espesa compuesta de niebla y sustancias químicas (pág. 198)

socialism/socialismo　sistema económico en que muchos negocios son propiedad y están dirigidos por el gobierno (pág. 717)

solar system/sistema solar　la Tierra, ocho planetas adicionales y miles de astros más pequeños que giran alrededor del Sol (pág. 29)

sodium nitrate/nitrato de sodio　sustancia química usada en abonos y explosivos (pág. 269)

sorghum/sorgo　cereal de tallo alto cuyas semillas sirven de alimento y del cual se hace un jarabe para endulzar (pág. 619)

spa/termas　balneario con manantiales de agua mineral caliente en que la gente se baña para recobrar su salud (pág. 378)

station/estación　rancho donde se crían ganado vacuno u ovejas en Australia (pág. 748)

steppe/estepa　pastos parcialmente secos que a menudo se encuentran en los bordes de un desierto (págs. 69, 386, 406, 515, 570, 687)

strait/estrecho　masa de agua delgada entre dos masas de tierra (págs. 42, 721)

subarctic climate/clima subártico patrón del estado del tiempo con inviernos extremadamente fríos y veranos cortos y frescos (pág. 67)

subcontinent/subcontinente masa de tierra grande que forma parte de un continente pero se puede diferenciar de él (pág. 645)

subsistence farm/granja de subsistencia terreno pequeño en el cual un granjero cultiva sólo lo suficiente para alimentar a su propia familia (págs. 186, 207, 261, 542)

suburb/suburbio comunidad pequeña en los alrededores de una ciudad (págs. 134, 340, 432)

summer solstice/solsticio de verano día con más horas de sol y menos horas de oscuridad (pág. 32)

taiga/taiga bosques enormes de árboles de hoja perenne en regiones subárticas (pág. 410)

tannin/tanino sustancia usada en el procesamiento del cuero (pág. 236)

tariff/arancel impuesto sobre el valor de bienes importados (pág. 93)

terraced field/terrazas franjas, parecidas a escalones, que se cortan en la ladera de una colina para que el suelo aguante el agua y se pueda usar para la agricultura (págs. 678, 724)

timberline/límite de los árboles elevación por encima de la cual no crecen árboles en las montañas (pág. 70)

tornado/tornado tormenta en forma de un torbellino que a veces se forma durante una tormenta eléctrica fuerte (pág. 56)

townships/municipios barrios abarrotados de gente en las afueras de las ciudades de Sudáfrica donde viven la mayoría de las personas que no son blancas (pág. 611)

trench/fosa marina valle en el fondo del mar (pág. 41)

tributary/afluente río pequeño que desagua en un río más grande (pág. 653)

tropics/trópicos región entre el Trópico de Cáncer y el Trópico de Capricornio (pág. 55)

trust territory/territorio en fideicomiso área que está bajo el control temporario de otra nación (pág. 769)

tsetse fly/mosca tsetsé insecto cuya picada puede matar al ganado o a los seres humanos por medio de la enfermedad del sueño (pág. 571)

tsunami/tsunami ola inmensa causada por un terremoto en el fondo del mar (págs. 36, 695)

tundra/tundra inmensas planicies ondulantes y sin árboles en latitudes altas con climas en que sólo varias pulgadas del suelo de la superficie se deshielan (págs. 68, 146, 410)

tungsten/tungsteno metal usado en equipos eléctricos (pág. 677)

typhoon/tifón nombre para un huracán en Asia (págs. 56, 673, 769)

urban/urbano parte de una ciudad (págs. 134, 432)

urbanization/urbanización movimiento hacia las ciudades (pág. 87)

vaquero/vaquero pastor de ganado vacuno (pág. 185)

wadi/uadi lecho de un río seco que llenan los aguaceros poco frecuentes (pág. 493)

water cycle/ciclo hidrológico proceso mediante el cual el agua se mueve de los océanos al aire, del aire a la tierra y de la tierra a los océanos una vez más (pág. 49)

watershed/cuenca fluvial región drenada por un río (pág. 594)

water vapor/vapor de agua agua en forma de gas (pág. 49)

weather/estado del tiempo cambios en la atmósfera que son difíciles de pronosticar y tienen lugar durante un período de tiempo corto (pág. 53)

weathering/desgaste proceso que rompe la superficie rocosa en peñas, grava, arena y suelo (pág. 37)

welfare state/estado de bienestar social estado que usa el dinero recaudado por los impuestos para mantener a personas que están enfermas, pobres, sin trabajo o retiradas (págs. 246, 349)

winter solstice/solsticio de invierno día con menos horas de sol y más horas de oscuridad (pág. 32)

yurt/*yurt* tienda de campaña grande y circular hecha de pieles de animales que se puede desmantelar y llevar de un lugar a otro (pág. 688)

SPANISH GLOSSARY

INDEX

INDEX

INDEX

INDEX

INDEX

INDEX

INDEX

INDEX

INDEX

Xhosa people, 610
Xi River, 671, 672

Yalu River, 706
Yangon, Myanmar, 716
Yangtze River, 23, 24, 632, *c639,* 671, 672, 674
Yaoundé, Cameroon, 571
Yarapa River, 253
Yeats, William Butler, 299
Yellow River, 632, 671, 672, 712
Yellow Sea, 74
Yellowstone National Park, 126, 202
Yeltsin, Boris, 432
Yemen, *m449, m478,* 496; climate, *m483;* economy, *p446;* landforms, *m448, m479;* population, *m497;* profile, *c454*

Yenisey River, 396, 409
Yerevan, Armenia, 510
Yokohama, Japan, 700
Yoruba people, 543
Yucatán Peninsula, 170, 182, 186, *p188,* 188–89
Yugoslavia, *m281, m366,* 383–84; climate, *m372;* communism in, 383; economy, *m376;* landforms, *m280, m367;* language, *m282;* population, *m387;* profile, *c287*
Yugoslav republics, 381; climate, *m372;* economy, *m376;* landforms, *m367;* population, *m387.* *See also* Former Yugoslav Republics
Yukon River, 119
Yukon Territory, 144, 154
yurt, 517, *p517*

Zagreb, Croatia, 383–84
Zagros Mountains, 444, 499
Zaire, 568

Zambezi River, 526, *p527,* 618, 619, 623
Zambia, *m531, c538, m608,* 618–19; agriculture, 619; climate, *m615;* economy, 618–19, *m620;* history, 619; landforms, *m530, m609,* 618; language, 619; natural resources, 618–19; people, 619; population, *m625, m625;* profile, *c538*
Zanzibar, 588, 589–90
Zapata, Emiliano, 190–91
Zealand, 354
Zeppelin, Led, 294
Zhang Qian, 692
Zimbabwe, *m531, c538, m608,* 618, 619–20; AIDS in, 620; agriculture, 619–20; climate, *m615;* economy, 619–20, *m620;* history, 620; landforms, *m530, m609,* 619; natural resources, 619–20; people, 620; population, *m625;* profile, *c538*
Zionists, 485
Zulu people, 610
Zurich, Switzerland, 310

INDEX

Cover (tl)PhotoDisc, (tr)Carol Beckwith & Angela Fisher/Robert Estall Photo Agency, (c)PhotoDisc, (bl)PhotoDisc, (br)CORBIS; **iv-v** Ken Stimpson/Panoramic Images, Chicago, All Rights Reserved; **ix** Owen Franken/CORBIS; **vi-vii** PhotoDisc; **xvi; 1** PhotoDisc; **2** Craig Aurness/CORBIS, (bl)Todd Gipstein/CORBIS, (br)Craig Lovell/CORBIS; **3** (tr)Robert Caputo/AURORA, (bl)Peter Turnley/CORBIS, (br)Roger Ressmeyer/CORBIS; **20** (l)Robert Landau/CORBIS, (r)Arne Dedert/AFP; **20-21**Stone/S. Purdy Matthews; **22** Norman Kent/oi2.com; **23** NASA/National Geographic Image Collection; **24** (l)Richard T. Nowitz/National Geographic Image Collection, (r)Yann Arthus-Bertrand/CORBIS; **26** Galen Rowell/CORBIS; **29** Maria Stenzel/National Geographic Image Collection; **30** Timothy G. Laman/National Geographic Image Collection; **34** Natalie Fobes/National Geographic Image Collection; **37** Michael K. Nichols/National Geographic Image Collection; **39** David Doubilet/National Geographic Image Collection; **40** (l)Kenneth Garrett/National Geographic Image Collection, (t)Michael K. Nichols/National Geographic Image Collection; **46** AP/Wide World Photos; **47** Matt Meadows; **48** Annie Griffiths Belt/National Geographic Image Collection; **49** George Grall/National Geographic Image Collection; **52** Southampton Oceanography Centre; **53** J. Blair/National Geographic Image Collection; **54** Medford Taylor/National Geographic Image Collection; **60** (l)Bruce Dale/National Geographic Image Collection, (r)Jodi Cobb/National Geographic Image Collection; **63** Phil Schermeister/National Geographic Image Collection; **64** (l)Michael K. Nichols/National Geographic Image Collection, (r)Beverly Joubert/National Geographic Image Collection; **67** (t)Annie Griffiths Belt/National Geographic Image Collection, (l)James P. Blair/National Geographic Image Collection, (bc)Jodi Cobb/National Geographic Image Collection, (br)Raymond K. Gehman/National Geographic Image Collection; **68** (t)Natalie Fobes/National Geographic Image Collection, (c)George F. Mobley/National Geographic Image Collection, (l)Maria Stenzel/National Geographic Image Collection; **69** (l)James L. Stanfield/National Geographic Image Collection, (r)Phil Schemeister/National Geographic Image Collection; **70** Pat Jerrold, Papilio/CORBIS; **71** Bill Curtsinger/National Geographic Image Collection; **74** (l)Calvin Larsen/Photo Researchers, (r)Stone/Joseph Sohm; **74-75** Planet Earth Pictures/FPG International/PictureQuest; **75** (tl)Francois Gohier/Photo Researchers, (tr, b)©1995 Jan Sonnenmair/AURORA; **76** Steve McCurry/National Geographic Image Collection; **77** Kenneth Garrett/National Geographic Image Collection; **79** (l)Black Star/National Geographic Image Collection, (r)Steven L. Raymer/National Geographic Image Collection; **80** (l)Steve McCurry/National Geographic Image Collection, (r)Steve McCurry/National Geographic Image Collection; **84** Gerd Ludwig/National Geographic Image Collection; **87** AFP/CORBIS; **89** Bettmann/CORBIS; **90** Jim Sugar Photography/CORBIS; **91** (t)Ric Ergenbright/CORBIS, (b)Steve McCurry/National Geographic Image Collection; **93** (l)H.Edward Kim/National Geographic Image Collection, (r)George F. Mobley/National Geographic Image Collection; **95** Bryan & Cherry Alexander; **96** James P. Blair/National Geographic Image Collection; **97** Jodi Cobb/National Geographic Image Collection; **102** (l)Steve McCurry, (r)Michael Lewis; **102-103** David R. Stoecklein; **105** Susie Post, (l)Norbert Rosing; **106** Richard Nowitz/PhotoTake.NYC/Picture Quest; **107** Eugene Fisher & Barbara Brundege; **114** Owen Franken/CORBIS; **115** David Hiser/National Geographic Image Collection; **118** (l)Steven L. Raymer/National Geographic Image Collection, (r)Vincent Musl/National Geographic Image Collection; **121** Michael S. Yamashita/National Geographic Image Collection; **122** Chris Johns/National Geographic Image Collection; **123** Karen Kasmauski/Matrix; **125** Gregory Scott Doramus/Omnigraphix; **126** Annie Griffiths Belt/CORBIS; **127** Joel Satore/National Geographic Image Collection; **130** The Stock Market; **131** Ira Block/National Geographic Image Collection; **134** Mike Habermann, courtesy Music of the World; **136** CORBIS; **140** Ray Pfortner, Peter Arnold Inc.; **140-141** Michael Mathers, Peter Arnold Inc.; **141** (l)Stone/David Young-Wolff, (r)David Schmidt; **142** David A. Harvey/National Geographic Image Collection; **143** Raymond K. Gehman/National Geographic Image Collection; **146** Richard T. Nowitz/National Geographic Image Collection; **150** Anna Susan/National Geographic Image Collection; **153** David A. Harvey/National Geographic Image Collection; **155** Marie-Louise Brimberg/National Geographic Image Collection; **156** Michael Evan Sewell/Visual Pursuit; **157** Marie-Louise Brimberg/National Geographic Image Collection; **159** Paul A. Souders/CORBIS; **160** (l)Daniel J. Wiener/National Geographic Image Collection, (r)Bruce Dale/National Geographic Image Collection; **164** Giraudon/Art Resource, New York; **165** Michael S. Yamashita; **166** (l)Stone/David Levy, (r)Stone/Oliver Benn; **166-167** Kenneth Garrett/NGS; **169** Stone/Norbert Wu, Stone/William J. Hebert; **170** Sisse Brimberg; **171** Stone/Chad Ehlers; **180** Randy Faris/CORBIS; **181** Bettmann/CORBIS; **188** Tomasz Tomaszewski/National Geographic Image Collection; **190** Cotton Coulson/National Geographic Image Collection; **192** James L. Amos/National Geographic Image Collection; **193** David A. Harvey/National Geographic Image Collection; **194** Nik Wheeler/CORBIS; **195** (l)Joel Satore/National Geographic Image Collection, (r)Tomas Tomaszewski/National Geographic Image Collection; **198** Albert Moldvay/National Geographic Image Collection; **202** Nik Wheeler/CORBIS; **203** Matt Meadows; **204** SuperStock; **205** Art Wolfe; **208** (t)Jan Butchofsky-Houser/CORBIS, (b)Vincent Musl/National Geographic Image Collection; **210** Michael S. Yamashita/CORBIS; **213** Jonathan Blair/National Geographic Image Collection; **215** ©Robert A. Tyrrell; **217** (l)Tony Arruza/CORBIS, (r)Michael K. Nichols/National Geographic Image Collection; **221** George Mobley/National Geographic Image Collection; **224** (l)Artville, (r)Giraudon/Art Resource, New York; **225** Loren McIntyre; **226** PhotoDisc; **227** Alex Webb/Magnum; **232** (l)Jim Zuckerman/CORBIS, (r)Yann Arthus-Bertrand/CORBIS; **234** Jeremy Horner/CORBIS; **235** Robert van der Hilst/CORBIS; **237** Winfield I. Parks, Jr./National Geographic Image Collection; **239** Pablo Corral V/CORBIS; **240** Kit Houghton Photography/CORBIS; **241** Michael K. Nichols/National Geographic Image Collection; **242** Pablo Corral V/CORBIS; **243** Robert Caputo/Aurora & Quanta Productions; **245** Louis O. Mazzatenta/National Geographic Image Collection; **247** (l)Jack Fields/CORBIS, (r)Louis O. Mazzatenta/National Geographic Image Collection; **252** (t)Michael & Patricia Fogden, (b)William Albert Allard; **252-253** Stuart Franklin; **253** (l)Michael Doolittle, (r)Marc Van Roosmalen Bat Conservation International; **254** Owen Franken/CORBIS; **258** Richard S. Durrance/National Geographic Image Collection; **260** Frank & Helen Schreider/National Geographic Image Collection; **263** Tiziana and Gianni Baldizzone/CORBIS; **264** Johan Reinhard/National Geographic Image Collection; **265** Art Wolfe; **266** Maria Stenzel/National Geographic Image Collection; **268** (l)Richard T. Nowitz/National Geographic Image Collection, (r)James L. Stanfield/National Geographic Image Collection; **271** William A. Allard/National Geographic Image Collection; **274** (l)IFA-Bilderteam-Travel/Bruce Coleman Inc., (r)Stone/Robert Everts; **274-275** SuperStock; **277** Stone/D.C. Lowe; (l)Stone/Chris Haigh; **278** Stone/Bert Blokhuis; **279** Stone/Ron Sanford; **288** James L. Stanfield/National Geographic Image Collection; **289** London Aerial Photo Library/CORBIS; **293** Emory Kristof/National Geographic Image Collection; **295** Adam Woolfitt/CORBIS; **296** Thad Samuels Abell II/National Geographic Image Collection; **298** (l)Thad Samuels Abell II/National Geographic Image Collection, (r)Tim Thompson/CORBIS; **299** Michael St. Maur Sheil/CORBIS; **300** James L. Stanfield/National Geographic Image Collection; **301** (l)James P. Blair/National Geographic Image Collection, (r)Felix Zaska/CORBIS; **303** Ric Ergenbright/CORBIS; **306** Owen Franken/CORBIS; **308** Ric Ergenbright/CORBIS; **310** Sisse Brimberg/National Geographic Image Collection; **311** Bob Krist/CORBIS; **313** Michael John Kielty/CORBIS; **315** SuperStock; **320** AP/Wide World Photos; **321** Reuters/CORBIS; **322** Mark L. Stephenson/CORBIS; **323** ©David Cumming, Eye Ubiquitous/CORBIS; **326** (l)Medford Taylor/National Geographic Image Collection, (r)Charles O'Rear/CORBIS; **327** Adam Woolfitt/CORBIS; **330** National Geographic Image Collection; **332** (t)Richard T. Nowitz/National Geographic Image Collection, (b)Vittoriano Rastelli/CORBIS; **333** Louis O. Mazzatenta/National Geographic Image Collection; **334** Thad Samuels Abell II/National Geographic Image Collection; **335** (l)Bettmann/CORBIS, (r)Gianni Dagli Orti/CORBIS; **336** Ira Block/National Geographic Image Collection; **339** Panoramic Images; **341** James L. Stanfield/National Geographic Image Collection; **344 345** Tomasz Tomaszewski/National Geographic Image Collection; **348** (l)©Buddy May/CORBIS, (r)Richard S. Durrance/National Geographic Image Collection; **349** Bryan & Cherry Alexander; **352** Jonathan Blair/CORBIS; **353** ©Ian Yates; Eye Ubiquitous/CORBIS; **355** SuperStock; **357** Sisse Brimberg/National Geographic Image Collection; **358** SuperStock; **359** Richard T. Nowitz/CORBIS; **362** (l)Robert Winslow, (r)Stone/Johan Elzenga; **362-363** Stone/Oliver Strewe; **363** (l)Martin Bond/Science Photo Library/Photo Researchers, (r)Mike Lewis, Northamptonshire Grammar School, United Kingdom; **364** Bob Krist/CORBIS; **365** Tomasz Tomaszewski/National Geographic Image Collection; **368 369 370** James L. Stanfield/National Geographic Image Collection; **371** Priit J. Vesilind/National Geographic Image Collection; **373** Steven L. Raymer/National Geographic Image Collection; **375** James Stanfield/National Geographic Image Collection;

TIME
REPORTS

A New Kind of War

Battling Terrorism in the Land of the Free

TIME REPORTS

FOCUS ON WORLD ISSUES

1995
Terror strikes Oklahoma City.

1996
Terrorists in Saudi Arabia blow up housing for U.S. troops.

LISA RUDY HOKE—BLACK STAR

A Day for Heroes

September 11, 2001, was a day John Jonas will never forget. A New York City firefighter, Jonas was the captain of Ladder Company 6. That morning two **hijacked** commercial airplanes were deliberately flown into the World Trade Center's twin towers.

Jonas and five other men in his company rushed to the scene. An hour later they were walking down the fire stairs of the south tower. An older woman they were rescuing told them she couldn't go on. They told her they wouldn't leave her. And just then the building collapsed around them in clouds of dust. "I'm thinking," Jonas said later, "'I can't believe this is how it ends for me.'"

But life didn't end either for Jonas or for the five firefighters with him. Nor did it end for Josephine Harris, the woman they saved. Above and below them and in the nearby north tower, more than 2,600 people died. But somehow the part of the stairway they were on didn't collapse. "It was a freak of timing," said Jonas. Another minute,

JAMES NACHTWEY—VII FOR TIME

▲ New York City firefighters battle the World Trade Center disaster on September 11, 2001.

either way, and the crumpling building would have crushed them like the others.

More than 300 rescue workers lost their lives trying to save others that day. All of them were heroes.

No Ordinary Crime

On September 11, hijackers also flew an airplane into the Pentagon—the headquarters of the U.S. military. Another 125 people died. Other hijackers seized a fourth airplane, but the passengers heroically fought back. Instead of hitting a building, the plane crashed in a field in Pennsylvania leaving no survivors.

Clearly, the horrible crashes were not ordinary crimes. They were acts of **terrorism**. Terrorism is the illegal use of violence against people or property to make a point. The point may involve a particular belief, such as religion or politics. ■

EXPLORING THE ISSUES

1. **Interpreting Points of View** Do you agree with the article's view of heroes? Why or why not?

2. **Making Inferences** How is the violence of terrorism different from the violence of war?

TR2

1998
Terrorists destroy the U.S. Embassy in Nairobi, Kenya.

2000
Terrorists in Yemen batter a U.S. warship, the USS Cole.

2001
Suicide pilots level the World Trade Center's twin towers.

Al-Qaeda

The terrorists who hijacked the airplanes belonged to a group called al-Qaeda (al•KAY•dah). The group was founded by Osama bin Laden, a wealthy Saudi Arabian.

Al-Qaeda was created to fight the Russian invasion of Afghanistan. After the Russians left Afghanistan, al-Qaeda members changed their goals. They wanted to force all non-Muslims out of the Middle East. They hated the U.S. troops based in Saudi Arabia and the Jewish people living in Israel.

Al-Qaeda's members also believed Muslims were being changed too much by modern ideas. They hated freedom of religion and wanted strict religious leaders to control Muslim countries.

Al-Qaeda's beliefs were not shared by all Muslims. The attacks on the United States horrified people around the world, including millions of Muslims who live in the Middle East, the United States, and elsewhere.

Closing in on America

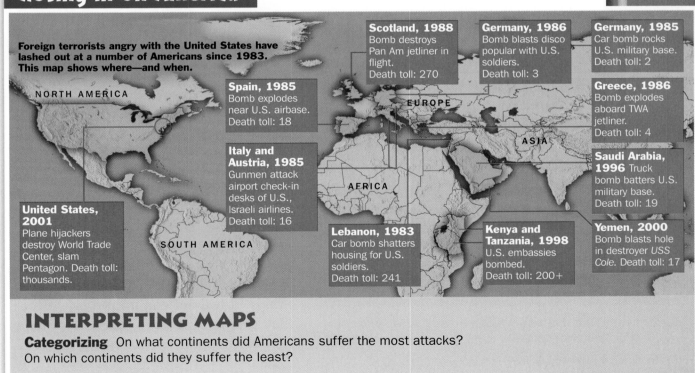

Foreign terrorists angry with the United States have lashed out at a number of Americans since 1983. This map shows where—and when.

Scotland, 1988
Bomb destroys Pan Am jetliner in flight.
Death toll: 270

Germany, 1986
Bomb blasts disco popular with U.S. soldiers.
Death toll: 3

Germany, 1985
Car bomb rocks U.S. military base.
Death toll: 2

Spain, 1985
Bomb explodes near U.S. airbase.
Death toll: 18

Greece, 1986
Bomb explodes aboard TWA jetliner.
Death toll: 4

Italy and Austria, 1985
Gunmen attack airport check-in desks of U.S., Israeli airlines.
Death toll: 16

Saudi Arabia, 1996 Truck bomb batters U.S. military base.
Death toll: 19

United States, 2001
Plane hijackers destroy World Trade Center, slam Pentagon. Death toll: thousands.

Lebanon, 1983
Car bomb shatters housing for U.S. soldiers.
Death toll: 241

Kenya and Tanzania, 1998
U.S. embassies bombed.
Death toll: 200+

Yemen, 2000
Bomb blasts hole in destroyer USS Cole. Death toll: 17

INTERPRETING MAPS

Categorizing On what continents did Americans suffer the most attacks? On which continents did they suffer the least?

Behind the Hatred

What makes the United States the target of so much deadly anger? One answer is its support for Israel. Israel was founded in 1948. Soon afterwards, an Arab-Israeli war forced about 750,000 Palestinian Arabs from their homes.

Today many of those Palestinians live in refugee camps. So do their children and their grandchildren. Those 4 million Palestinians want a nation of their own. Israel has offered to exchange land for a promise of peace. But so far the Palestinians have rejected that offer.

U.S. Troops in Saudi Arabia

Another source of anger is the presence of U.S. troops in Saudi Arabia. The Saudi government asked the United States to station troops there. But the holy cities of Makkah (Mecca) and Madinah (Medina) are in Saudi Arabia. To many Muslims, U.S. troops on Saudi soil are an insult to Islam.

To terrorist Osama bin Laden and his followers, the solution to these problems was violence. In 1996, he urged Muslims to kill U.S. troops in Saudi Arabia. In 1998, he called for attacks on American **civilians**. Civilians are people

▲ Fanatics hail terrorist Osama bin Laden in 2001.

not in the armed forces or diplomatic services. By the end of 2001, several thousand people had been killed.

The United States responded to September 11 with a determination and resolve bin Laden surely didn't expect. "Our war on terror begins with al-Qaeda," President George W. Bush said. "It will not end until every terrorist group of global reach has been found, stopped, and defeated." ■

EXPLORING THE ISSUES

1. **Drawing Conclusions** What are some reasons many Americans support Israel's presence in the Middle East?

2. **Making Inferences** Why are acts of terror against civilians often effective?

War on All Fronts

After the terrorist attacks on September 11, 2001, President George W. Bush spoke to the nation and announced a war on terrorism. He warned the world, "We will make no distinction between the terrorists who committed these acts and those who harbor them."

Al-Qaeda had followers all over the world, but it was based in Afghanistan. The Taliban, a strict religious party that controlled Afghanistan's government, protected Osama bin Laden and al-Qaeda. They refused to help the U.S. So in October, the President ordered the U.S. military to attack Afghanistan.

Aid for Children

The U.S. was not at war with the Afghan people, but with the Taliban and al-Qaeda. During the attack, U.S. planes dropped food and medicine to the men, women, and children in the civilian population.

Nations around the world backed the United States and began arresting al-Qaeda members hiding in their countries. Some sent troops to help the Americans fight in Afghanistan.

A few weeks after the attack began, the Taliban government collapsed. With the aid of the U.S. and its allies, the Afghan people created a new government. Meanwhile, American and allied troops began hunting for al-Qaeda forces in Afghanistan. The U.S. also sent troops to the Philippines, Yemen, and the nation of Georgia to train local troops to fight terrorists.

MARK RICHARDS

▲ Safety checks help prevent terrorism. But the cost—less freedom—worries many Americans.

Liberty and Security

At home, federal agencies stepped up their efforts to find terrorists. President Bush ordered banks to hold money belonging to groups linked to terrorists. Congress passed a new antiterrorist law making it easier to tap phones, intercept e-mail, and search homes.

Some people worried that the new antiterrorist law would chip away at our **liberties**—such as freedom of speech and the right to privacy. For this reason, Congress set a five-year time limit on parts of the new law. ▨

EXPLORING THE ISSUE

1. Analyzing Information Shortly after September 11, 2001, President Bush said, "No one should be singled out for unfair treatment or unkind words because of their ethnic background or religious faith." What do you think he meant by that statement?

2. Problem Solving What liberties, if any, might you be willing to give up in order to ensure national security?

Stopping Terrorism: What Can One Person Do?

The rescue workers who responded to the attacks on the World Trade Center and the Pentagon were true heroes. In the months that followed, Americans honored them for their courage and sacrifice.

The response of Americans to tragedy showed the world the nation's hidden strengths—its people. Wherever they lived, Americans reacted. They gave blood. They held candlelight **vigils** to honor the victims. They flew flags to show their unity. They cut deeply into their budgets, contributing more than $200 million in the first week to help victims' families.

They all made it clear, as a girl from Ohio told TIME For Kids, that no terrorist can weaken the nation's spirit. "They bent steel," said Danielle, 12, of the World Trade Center murderers, "but they can't break the U.S."

Be a Local Hero

Wherever you live, you can help keep that spirit alive. And you can do it even years after the disasters of September 2001 took place.

Learn all you can about terrorism. Learn what it is, why it exists, and how people at all levels of government are fighting it.

Then join that fight any way you can. With posters and letters, report successful efforts to combat this evil. Raise money for groups that help out the victims of terrorism everywhere.

▲ Terrorist attacks in September 2001 trigger a burst of patriotism everywhere in America.

Finally, refuse to give in to fear. Terrorists use fear as a weapon. If you can keep fear from changing your life, you will have taken a big bite out of terrorism.

The novelist Stephen King—who often writes about human fears—agreed. "If everybody continues working," he said, "they [the terrorists] don't win." ■

EXPLORING THE ISSUE

1. **Problem Solving** What might people do to stop the fear of terrorism from keeping them from doing what they want to do?

2. **Summarizing the Main Idea** Write a new title for this piece. Share it with your classmates. Explain why you think your title fits the story.

REVIEW AND ASSESS

UNDERSTANDING THE ISSUE

1. Defining Key Terms In your own words, define the following terms: *terrorism*, *hijacker*, *the Pentagon*, *al-Qaeda*, *Taliban*, *liberty*, *security*, *principles*, *ideals*, and *vigil*.

2. Writing to Inform In a 300-word article, describe a terrorist act you heard or read about. Describe how you reacted when you heard about it.

3. Writing to Persuade What do you think Americans should know about terrorism? Put your answer in a 250-word letter to the editor of your local newspaper. Support your answer with facts. Use at least five of the terms listed above.

INTERNET RESEARCH ACTIVITY

4. Use Internet resources to find information on what individuals and organizations are doing today to help victims of terrorism. Use what you learn to write a report on current efforts, and share it with the class.

5. With your teacher's help, use Internet resources to learn more about how the tragedies of September 11 resulted in an increase of visible patriotism in the United States. Focus your research on finding personal stories of how the attacks increased individual Americans' beliefs in and loyalty to the United States. Prepare a brief report on your findings.

BEYOND THE CLASSROOM

6. Study the map on page 513. Research one of the terrorist attacks noted there. What does the attack tell you about the goals, thoughts, and methods terrorists have? Describe the attack and answer those questions in a brief oral report.

TERRY BARNER/SILVER IMAGE

▲ Muslims mourn victims of the World Trade Center attack.

7. Visit your school or local library. Research a country, such as Israel, Northern Ireland, or Bosnia where the people have suffered from terrorist attacks. Find out what programs have been started by groups or individuals to bring an end to the violence. Present your findings to the whole class.

TIME/CNN Poll
Fighting Terrorism: How Far Would You Go?

What are Americans willing to do to fight terrorism? These pie graphs show what a TIME/CNN Poll found out.

To prevent terrorist attacks, would you favor or oppose the government doing each of the following?

	Favor		Oppose
1. Allow police to wire-tap phone conversations of suspected terrorists without a court's okay:	68%		29%
2. Let courts jail, for as long as they want, people suspected of links to terrorist groups:	59%		38%
3. Let police intercept e-mail messages sent by anyone in the United States and scan them for suspicious words or phrases:	55%		42%
4. Require everyone in the United States to carry an identification card issued by the Federal Government:	57%		41%
5. Let police stop people on the street and search them:	29%		69%

Source: TIME magazine, October 8, 2001; Gray slices indicate respondents who were not sure.

BUILDING GRAPH READING SKILLS

1. Analyzing the Data The U.S. Constitution bars the government from making "unreasonable searches and seizures" of citizens. Which of the graphs show how Americans think about this right? What's your thinking on this issue?

2. Making Inferences The U.S. Constitution bars the government from taking away a person's "life, liberty, or property" without a fair trial. What do most people who took part in this poll seem to think about this right? Why do you think they hold that view?

FOR UPDATES ON WORLD ISSUES GO TO
www.timeclassroom.com/glencoe